AQUATIC SURFACE CHEMISTRY

AQUATIC SURFACE CHEMISTRY

Chemical Processes at the Particle–Water Interface

Edited by

WERNER STUMM

Swiss Federal Institute of Technology, Zurich, Switzerland

A WILEY-INTERSCIENCE PUBLICATION

JOHN WILEY & SONS

New York · Chichester · Brisbane · Toronto · Singapore

A NOTE TO THE READER:
This book has been electronically reproduced from digital information stored at John Wiley & Sons, Inc. We are pleased that the use of this new technology will enable us to keep works of enduring scholarly value in print as long as there is a reasonable demand for them. The content of this book is identical to previous printings.

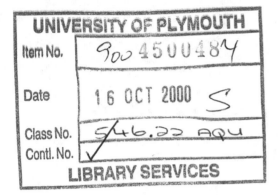

Library of Congress Cataloging in Publication Data:

Aquatic surface chemistry.

(Environmental science and technology)
"A Wiley-Interscience publication."
Includes bibliographies and index.
1. Surface chemistry. 2. Water chemistry.
3. Biogeochemical cycles 4. Soils
I. Stumm, Werner, 1924– . II. Series.

GB855.A65 1987 551.48'0154 86–28078
ISBN 0-471-82995-1

Printed in the United States of America

10 9 8 7 6

CONTRIBUTORS

R. SCOTT ALTMANN, Research Fellow, Department of Inorganic, Analytical, and Applied Chemistry, University of Geneva, Geneva, Switzerland

ALEX E. BLUM, Research Fellow, Yale University, Department of Geology and Geophysics, New Haven, Connecticut

GERARD H. BOLT, Professor, Department of Soil Sciences and Plant Nutrition, Agricultural University, Wageningen, The Netherlands

JACQUES BUFFLE, Senior Lecturer, Department of Inorganic, Analytical, and Applied Chemistry, University of Geneva, Geneva, Switzerland

GERHARD FURRER, Research Fellow, Institute for Water Resources and Water Pollution Control (EAWAG), Zurich, Switzerland; Swiss Federal Institute of Technology (ETH), Zurich, Switzerland

WILLIAM S. FYFE, Professor, Department of Geology, The University of Western Ontario, London, Ontario, Canada

PHILIP M. GSCHWEND, Professor, Department of Civil Engineering, Massachusetts Institute of Technology, Cambridge, Massachusetts

ANTONIO C. LASAGA, Professor, Department of Geology and Geophysics, Kline Geology Laboratory, Yale University, New Haven, Connecticut

JACOB A. MARINSKY, Professor, Department of Chemistry, State University of New York at Buffalo, Buffalo, New York

FRANÇOIS M. M. MOREL, Professor, Department of Civil Engineering, Massachusetts Institute of Technology, Cambridge, Massachusetts

JAMES J. MORGAN, Professor, Department of Environmental Engineering Science, California Institute of Technology, Pasadena, California

HERBERT MOTSCHI, Dr., Ciba-Geigy AG, Basel, Switzerland

CHARLES R. O'MELIA, Professor, Department of Geography and Environmental Engineering, The Johns Hopkins University, Baltimore, Maryland

ROGER PARSONS, Professor, Department of Chemistry, University of Southampton, Southampton Hampshire, Great Britain

JEAN-CLAUDE PETIT, Senior Research Fellow, DRDD/SESD, CEN-FAR, BP6, Fontenay-aux-Roses, France

PAUL W. SCHINDLER, Professor, Department of Inorganic Chemistry, Institute for Inorganic, Analytical, and Physical Chemistry, University of Bern, Bern, Switzerland

WALTER SCHNEIDER, Professor, Laboratory for Inorganic Chemistry, Swiss Federal Institute of Technology (ETH), Zurich, Switzerland

JACQUES SCHOTT, Director of Research at CNRS, Laboratory of Mineralogy and Crystallography, University Paul Sabatier, Toulouse Cedex, France

BERNHARD SCHWYN, Research Fellow, Department of Biochemistry, University of California, Berkeley, California

LAURA SIGG, Head of Analytical Section, Institute for Water Resources and Water Pollution Control (EAWAG), Zurich, Switzerland; Senior Lecturer, Swiss Federal Institute of Technology (ETH), Zurich, Switzerland

ALAN T. STONE, Professor, Department of Geography and Environmental Engineering, The Johns Hopkins University, Baltimore, Maryland

WERNER STUMM, Director, Institute for Water Resources and Water Pollution Control (EAWAG), Zurich, Switzerland; Professor, Swiss Federal Institute of Technology (ETH), Zurich, Switzerland

DAVID R. TURNER, Senior Research Scientist, The Marine Biological Association of the United Kingdom, The Laboratory, Citadell Hill, Plymouth, Great Britain

WILLEM H. VAN RIEMSDIJK, Research Fellow, Department of Soil Sciences and Plant Nutrition, Agricultural University, Wageningen, The Netherlands

JOHN C. WESTALL, Professor, Department of Chemistry, Oregon State University, Corvallis, Oregon

MICHAEL WHITFIELD, Senior Research Scientist, The Marine Biological Association of the United Kingdom, The Laboratory, Citadell Hill, Plymouth, Great Britain

N. LEE WOLFE, Research Fellow, Environmental Research Laboratory, U.S. Environmental Protection Agency, Athens, Georgia

RICHARD G. ZEPP, Professor, Environmental Research Laboratory, U.S. Environmental Protection Agency, Athens, Georgia

SERIES PREFACE
Environmental Science and Technology

The Environmental Science and Technology Series of Monographs, Textbooks, and Advances is devoted to the study of the quality of the environment and to the technology of its conservation. Environmental science therefore relates to the chemical, physical, and biological changes in the environment through contamination or modification, to the physical nature and biological behavior of air, water, soil, food, and waste as they are affected by man's agricultural, industrial, and social activities, and to the application of science and technology to the control and improvement of environmental quality.

The deterioration of environmental quality, which began when man first collected into villages and utilized fire, has existed as a serious problem under the ever-increasing impacts of exponentially increasing population and of industrializing society. Environmental contamination of air, water, soil, and food has become a threat to the continued existence of many plant and animal communities of the ecosystem and may ultimately threaten the very survival of the human race.

It seems clear that if we are to preserve for future generations some semblance of the biological order of the world of the past and hope to improve on the deteriorating standards of urban public health, environmental science and technology must quickly come to play a dominant role in designing our social and industrial structure for tomorrow. Scientifically rigorous criteria of environmental quality must be developed. Based in part on these criteria, realistic standards must be established and our technological progress must be tailored to meet them. It is obvious that civilization will continue to require increasing amounts of fuel, transportation, industrial chemicals, fertilizers, pesticides, and countless other products; and that it will continue to produce waste products of all descriptions. What is urgently needed is a total systems approach to modern civilization through which the pooled talents of scientists and engineers, in cooperation with social scientists and the medical profession, can be focused on the development of order and equilibrium in the presently disparate segments of the human environment. Most of the skills and tools that are needed are already in existence. We surely have a right to hope a technology that has created such manifold environment problems is also capable of solving them. It is our hope that this Series in Environmental Sciences and Technology will not only serve to make this challenge more explicit to the established professionals, but that it also will help to stimulate the student toward the career opportunities in this vital area.

Robert L. Metcalf
Werner Stumm

PREFACE

The aim of this book is to give an account of current research and applications on chemical processes occurring at the interfaces of water with natural occurring solids. The processes discussed and the concepts presented are applicable to all natural waters (oceans and fresh waters as well as soil water systems and sediment water systems) and to the surfaces of natural solids such as minerals, soils, sediments, biota, and humus. An appreciation of the physical chemistry of these interfaces and the reactions controlled by them is a prerequisite for understanding many of the important processes in natural systems. The geochemical fate of most trace elements is controlled by the reaction of solutes with solid surfaces; simple chemical models for the residence time of reactive elements in oceans and lakes are based on the partitioning of species between soluble and sedimenting aquatic particles.

The authors—electrochemists, surface and colloid chemists, geochemists, oceanographers, aquatic chemists, soil chemists, and environmental engineers—have attempted to write their chapters in such a way as to assist the readers (students, geochemists, water and soil scientists, environmental engineers) in understanding general principles as well as to guide research in aquatic surface chemistry. The interactions of solutes with the solid surfaces are looked at from a mechanistic and dynamic point of view rather than a descriptive one. Emphasis is on explanation and intellectual stimulation rather than on extensive documentation.

In this volume we progress from theoretical models and laboratory studies to applications in natural water, soil, and geochemical systems, emphasizing those processes that regulate the distribution and concentration of elements and compounds. The interaction of a solute with a surface (adsorption) requires a characterization in terms of the physical and chemical properties of the solvent, the solute, and the sorbent. For the past few decades the electric double layer model has been the conceptual framework to interpret surface chemical phenomena. But the fundamental chemical interaction of solutes with natural surfaces is through the formation of coordinative bonds. Thus, specific chemical factors need to be considered in addition to the theory of the electric double layer in order to explain many phenomena in natural systems and to derive rate laws on geochemical processes such as the dissolution and weathering of rocks and the formation of minerals. The complementary modes of the different mechanisms that contribute to the overall energy of adsorption is examined through a survey of recent chemical models of adsorption, paying special attention to the coordination chemistry of the oxide–water interface and discussing aspects of molecular structure in surface complexes on the basis of spectroscopic investigation.

The dynamics of particles, especially the role of particle–particle interactions (coagulation), is critically assessed. The effect of particles on the catalysis of redox reactions and heterogeneous organic reactions and on photochemically induced processes involving particles are reviewed. The behavior of the elements in oceans and lakes are closely linked to the extent to which they are involved in particle–water interactions either via biological utilization or by adsorption–desorption reactions at active surfaces. The elements associated with the different modes are found in coherent groups of the periodic table and their cycling in seawater can be related to the inorganic parameters that have been used to rationalize their speciation in natural waters and their utilization by organisms. This is also evident in lakes where field data on the composition of the settling particles document that biological particles are particularly important for the transport of certain heavy metals to the sediments.

Understanding how geochemical cycles in natural water systems are coupled by particles and organisms may aid our understanding of global ecosystems and on how the interacting systems may become disturbed by civilization. After all, the inorganic and physical processes and the biological processes that are analyzed here at the microlevel are of influence on the major geochemical cycles.

Most of the authors met in January 1986 in Switzerland for a workshop. Background papers formed the basis for the discussions. This volume is not a "proceedings of a conference," but instead is the offspring of the workshop and its stimulating discourses and interactions.

Acknowledgments

I am most grateful to many colleagues who have reviewed individual chapters and have given useful advice on the organization of the book. I am especially indebted to Diana Hornung, who has carried a large burden in helping me edit this book. Credit for the creation of this volume is, of course, primarily due to its authors.

WERNER STUMM

Zurich, Switzerland
January 1987

CONTENTS

PART ONE. THE SOLID–SOLUTION INTERFACE

PART TWO. THE FORMATION AND DISSOLUTION OF SOLID PHASES

AQUATIC SURFACE CHEMISTRY

PART ONE

THE SOLID–SOLUTION INTERFACE

1

ADSORPTION MECHANISMS IN AQUATIC SURFACE CHEMISTRY

John C. Westall

Department of Chemistry, Oregon State University, Corvallis, Oregon

Abstract

Adsorption reactions influence three important processes within the scope of aquatic chemistry: (1) the distribution of elements between the aqueous phase and particulate matter, which affects their transport through the hydrogeosphere; (2) the electrostatic properties of suspended particles, which affect their aggregation and transport; (3) reactivity at surfaces, including dissolution, precipitation, and surface-catalyzed reactions of solutes. A full understanding of adsorption requires that the interaction of a solute with a surface be characterized in terms of the fundamental physical and chemical properties of the solvent, solute, and sorbent. It should be borne in mind that it is not only the interaction between the solute and the surface that is important, but also all possible pairs of interactions between surface, solvent, and solutes:

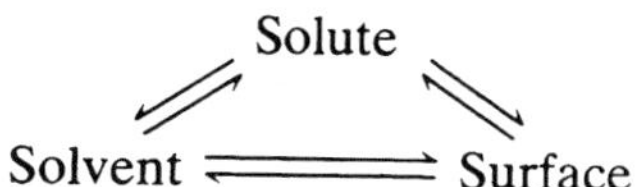

1. INTRODUCTION

In recognition of the complexity of natural surfaces, we expect that a number of different mechanisms contribute to the overall energy of adsorption and that it is often not easy to distinguish among mechanisms. Many questions arise in this regard. For example: Is adsorption to a natural material described better as adsorption to a two-dimensional surface or as absorption into three-dimensional phase? To what extent is adsorption of organic acids and bases driven by the hydrophobic interaction and to what extent by specific chemical interactions between acid and base functional groups and the surface? How appropriate is it to describe the surface chemistry of oxides in terms of one or two types of functional groups?

The goal of characterizing adsorption reactions in terms of fundamental properties of solute, solvent, and sorbent can be approached by alternating between observations of effects of adsorption in natural systems [as described, for example, by Sigg (1987) and Whitfield and Turner (1987)] and studies of mechanisms in relatively simple, well-controlled laboratory systems, for which only a few adsorption mechanisms are important. In this chapter we concentrate on the interpretation of adsorption mechanisms in reasonably well-controlled laboratory systems.

The discussion presented in this chapter will cover the following topics: (1) fundamental electrical and chemical intermolecular interactions; (2) adsorption mechanisms usually associated with organic compounds and organic surfaces; and (3) adsorption mechanisms associated with inorganic surfaces. There is an overlap between the organic and inorganic mechanisms; a mathematical approach to describing these reactions is presented, which allows many kinds of adsorption mechanisms to be formulated within the same framework.

2. ADSORPTION MECHANISMS

The term "mechanism" is used here in an equilibrium sense, as the sum of electrochemical energies responsible for the equilibrium state of the system, rather than in the kinetic sense, as the series of steps leading to the equilibrium state of the system. Mechanisms are considered first at the generic level of intermolecular interactions and then at the specific level of particular solute–solvent–surface systems. A summary of the intermolecular interactions that might reasonably be considered to be of importance in aquatic surface chemistry is given in Table 1.1. Examples of each of these interactions are abundant in the field of liquid chromatography, in which properties of solvents and surfaces are manipulated to obtain different adsorption behavior for different solutes (Snyder and Kirkland, 1979).

The chemical interactions of primary importance are those of the solute with the surface and those of the solute with the solvent. The interaction of the solvent with the surface is of less importance in understanding the aquatic surface chemistry under normal circumstances, when water is the only solvent.

Although the fundamental chemical interaction of solutes with the surface is through the formation of coordinate bonds, several types of reactions in which these bonds are formed can be distinguished. Examples of these reactions are given in Table 1.2: surface hydrolysis, surface complexation, and

Table 1.1. Intermolecular Interactions of Importance in Aquatic Surface Chemistry

A. Chemical Reactions with Surfaces
- Surface hydrolysis
- Surface complexation
- Surface ligand exchange
- Hydrogen bond formation

B. Electrical Interactions at Surfaces
- Electrostatic interactions
- Polarization interactions

C. Interactions with Solvent
- Hydrophobic expulsion

ligand exchange reactions. Reactions of surface hydroxyl groups are formulated by analogy with reactions already known to occur in homogeneous solution; they are discussed in detail by Schindler and Stumm (1987). These sorts of chemical interactions are all short-range; adsorption of ionic species by chemical interactions often leads to a development of an electric charge and long-range effects, but the chemical interactions themselves require intimate contact between adsorbent and adsorbate.

Hydrogen bonding is generally weaker than the specific chemical interactions cited above, but it is often important in the adsorption of many polar organic molecules to natural organic and oxidic surfaces.

Chemical interactions were described above in terms of the surface and

Table 1.2. Reactions of Surface Hydroxyl Groups, XOH

Hydrolysis (Acid–Base)

$$XOH_2^+ \rightleftharpoons XOH + H^+$$

$$XOH \rightleftharpoons XO^- + H^+$$

Complexation

$$XOH + M^{2+} \rightleftharpoons XOM^+ + H^+$$

$$\begin{matrix} XOH \\ | \\ XOH \end{matrix} + M^{2+} \rightleftharpoons \begin{matrix} XO \\ | \\ XO \end{matrix}\!\!>\!M + 2H^+$$

Ligand Exchange

$$XOH + ROH \rightleftharpoons XOR + H_2O$$

$$\begin{matrix} XOH \\ | \\ XOH \end{matrix} + R\!<\!\begin{matrix} OH \\ OH \end{matrix} \rightleftharpoons \begin{matrix} XO \\ | \\ XO \end{matrix}\!\!>\!R + 2H_2O$$

the solute; similar interactions occur between solvent and solutes and between different solutes. The distribution of species in solution that results from these reactions is generally referred to as *chemical speciation.* In practice, solution speciation is usually determined independently of adsorption reactions, and adsorption reactions are interpreted in terms of particular species in solution.

One notable exception to the dominance of the usual solute–solvent and solute–surface interactions in adsorption is the case of the hydrophobic effect. This effect can be viewed as the result of the attraction not of the solute to the surface or the solvent, but of the solvent for itself, which restricts the entry of the hydrophobic solute into the aqueous phase (Tanford, 1980).

Other solvent–solvent or solvent–surface interactions become particularly important in dealing with adsorption from mixed solvents. These other interactions will not be dealt with at a mechanistic level here, but they should not be overlooked.

Electrical energies are of two types: electrostatic and polarization (Murrell and Boucher, 1982). Electrostatic energies arise from interactions between permanent charges (ions) and permanently separated charges within a molecule (dipoles). Polarization energies arise from induced dipoles when molecules come under each other's influence.

The strongest electrostatic energies are the coulombic energies between charged species; weaker ones arise between dipoles. For most natural surfaces, the electrostatic state of the interface is determined primarily by the pH of the solution; the effects of electrostatic interactions are attenuated by solutions of high ionic strength. Most natural surfaces are negatively charged at environmental pH values; thus the electrostatic interaction is generally favorable for adsorption of cations and unfavorable for adsorption of anions. Since electrostatic adsorption is long-range, the location of counterions is critical. If the adsorbing phase is three-dimensional, such as a gel layer at an oxide surface or a porous organic coating, and if counterions can be contained within this adsorbing phase, the electric field will not extend significantly beyond the particle, and electrokinetic effects will not be observed. If adsorption occurs at a planar (two-dimensional) surface, electrokinetic effects will be observed and will be seen to vary with ionic strength.

Polarization energies are generally much weaker than other forces in aqueous systems but are quite significant in nonaqueous systems. These interactions may operate in concert with the hydrophobic effect.

This array of intermolecular interactions leads to the six adsorption mechanisms shown in Table 1.3. Certainly no single one of these mechanisms is entirely responsible for adsorption in a natural system, but consideration of each of these models provides a convenient framework for understanding the sensitivity of adsorption to environmental parameters and a basis for mathematical models of the adsorption process.

The mechanisms are considered in both three-dimensional and two-dimensional versions. For adsorption of nonpolar compounds, for which the long-range interactions are of little importance, the presence of a third di-

Table 1.3. Representative Adsorption Mechanisms in Aquatic Surface Chemistry

I. Hydrophobic transfer of a nonpolar organic compound to a two-dimensional surface or a three-dimensional nonaqueous phase

Two-dimensional surface	Three-dimensional phase
R	R R
R	R R
R	R R

II. Transfer of an ionic organic compound by ligand exchange or hydrophobic interaction to a two-dimensional surface with inorganic counterion in the aqueous double layer, or coexistence of organic ion and counterion in a three-dimensional Donnan phase.

Two-dimensional surface	Three-dimensional phase
R^- Na^+	R^- Na^+
R^- Na^+	Na^+ R^-
R^- Na^+	R^- Na^+

III. Reaction of surface hydroxyl groups to yield a charged two-dimensional surface with counterions in the aqueous double layer, or charged hydroxyl groups of a porous surface layer with counterions in a three-dimensional Donnan phase.

Two-dimensional surface	Three-dimensional phase
XO^- Na^+	XO^- Na^+
XO^- Na^+	Na^+ XO^-
XO^- Na^+	XO^- Na^+

mension in the adsorbing phase is of little significance; for adsorption of ionic compounds, the two- or three-dimensional structure at the interface has a dramatic effect not only on the adsorption of the compound but also on the electrokinetic behavior of the particle.

3. ORGANIC COMPOUNDS

3.1. Hydrophobic Organic Compounds

The mechanism that is most easily understood from first principles and the one that can be generalized most easily is hydrophobic adsorption (Tanford, 1980). In very simple terms, this mechanism is based on the fact that most natural organic sorbents, regardless of source, have been observed to be more or less comparable in their ability to accept nonpolar compounds. Hydrophobic adsorption is driven by the incompatibility of the nonpolar compounds with water, not by the attraction of the compounds to organic matter. Thus

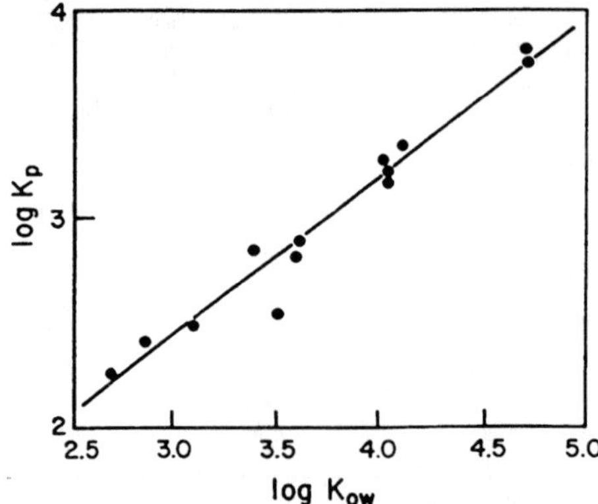

Figure 1.1. Partition constant as a function of a property of the compound being adsorbed. The partition constants (Reaction 1.1) are for the mono-, di-, tri-, and tetramethylbenzenes and chlorobenzenes, between water and an aquifer material of 0.15% organic carbon. The values of log K have been adjusted to be correct for a sorbent of 100% organic carbon.

it is possible to characterize the adsorption of a wide range of organic compounds on a wide range of sorbents based on a single property of the compound (e.g., its octanol–water partition constant, K_{ow}) and a property of the sorbent (the fraction of the sorbent that is organic carbon, f_{oc}) (Karickhoff, 1984; Karickhoff et al., 1979; Schwarzenbach and Westall, 1981).

The validity of the first relationship is illustrated in Figure 1.1, which shows the partition constants of substituted benzenes on a river sediment as a function of the K_{ow} of the compounds. The validity of the relationship between K_p and f_{oc} is illustrated in Figure 1.2. The natural materials with which the experiments were carried out range from an aquifer material with less than 0.1% organic carbon to an activated sludge with 33% organic carbon.

Then the distribution of a nonpolar organic compound, R, between an aqueous phase and a nonaqueous phase (designated by the overbar) can be represented by the reaction

$$\mathrm{R} \rightleftharpoons \overline{\mathrm{R}} \qquad K_p \tag{1.1}$$

and the distribution constant K_p can be calculated from

$$\log K_p = a \log K_{ow} + \log f_{oc} + b \tag{1.2}$$

Although the values determined for the coefficients a and b show a slight dependence on the set of compounds used in the regression (Lyman, 1982), the relationship given by Eq. 1.2 appears to be one of the most useful and widely applicable in the investigation of adsorption on natural surfaces. The relationships are also reasonably valid for some compounds other than the

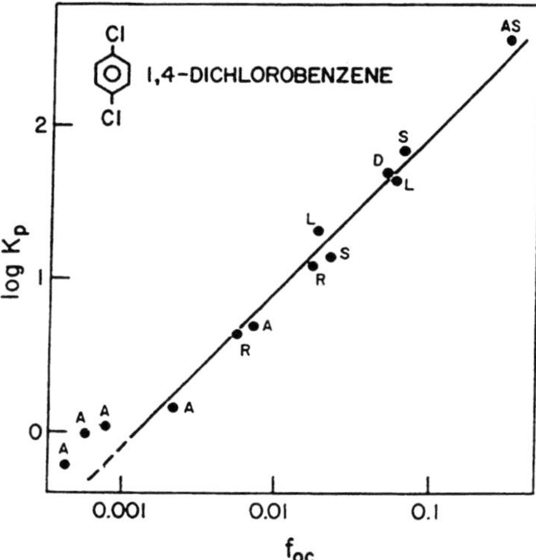

Figure 1.2. Partition constant as a function of a property of the sorbent. The partition constants (Reaction 1.1) are given for *p*-dichlorobenzene between water and a set of sorbents from different sources and different contents of organic carbon. The letters refer to the source of the sorbents: A, aquifer material; R, river sediment; L, lake sediment; S, soil; AS, activated sludge.

nonpolar compounds for which the relationship was originally developed and for which the theoretical basis is valid.

The application of the model to a field study is discussed by Schwarzenbach et al. (1983).

3.2. Hydrophobic Ionizable Organic Compounds

Hydrophobic ionizable organic compounds (HIOC) exist in the aqueous phase as a neutral species and an ionized species. Examples of these compounds include phenols and anilines. The adsorption of the neutral species can be characterized by the same paradigm as that used for the nonpolar compounds. The adsorption of the ionic form requires additional considerations.

The behavior of HIOCs on natural sorbents has been described by Schellenberg et al. (1984). The mechanisms can be understood by considering the distribution in the octanol–water model system (Westall et al., 1985). The distribution of the neutral form of a phenol, AH, between octanol and water is represented by the reaction

$$AH \rightleftharpoons \overline{AH} \qquad K_{ow} \tag{1.3}$$

The neutral form will ionize in the aqueous phase according to the reaction

$$AH \rightleftharpoons A^- + H^+ \qquad K_a \tag{1.4}$$

and the anion A^-, along with a counterion, can also be transferred to the nonaqueous phase. The transfers of ions of the organic compound and of a salt MX are represented by the following:

$$A^- \rightleftharpoons \overline{A^-} \qquad K_{\text{ion}} \tag{1.5}$$

$$M^+ \rightleftharpoons \overline{M^+} \qquad K_M \tag{1.6}$$

$$X^- \rightleftharpoons \overline{X^-} \qquad K_X \tag{1.7}$$

The sum of the charges of all ions in the nonaqueous phase must preserve electroneutrality in the bulk of the phase:

$$\sum_i z_i \bar{c}_i = 0 \tag{1.8}$$

If the affinity of the nonaqueous phase is different for different ions, as is usually the case, a Donnan potential, ψ_D (Bard, 1966), will develop between the two phases. The mass law equations for each of the reactions 1.5–1.7, with the term reflecting the Donnan potential, are written

$$c_i \exp\left(\frac{-z_i F \psi_D}{RT}\right) K_i = \bar{c}_i \tag{1.9}$$

The exponential term represents the energy necessary to move a charged species from the reference phase into the Donnan phase at potential ψ_D. The conditional partition constant for species i is denoted by K_i.

Equations 1.8 and 1.9 for reactions 1.3–1.7 define the Donnan equilibrium for partition between these two three-dimensional phases. The equations are solved for the Donnan potential ψ_D such that the concentrations satisfy Eq. 1.8. Marinsky (1987) discusses the use of a Donnan model for association of ionic species with humic substances.

The concentration distribution ratio for the HIOCs,

$$D = \frac{[\overline{A^-}] + [\overline{AH}]}{[A^-] + [AH]} \tag{1.10}$$

is more complicated than the simple partition constant used for nonpolar compounds. As shown in Eq. 1.10, it contains the concentrations of both the ionic and neutral species.

Experimental data for the distribution of the chlorinated phenols in the octanol–water system as a function of pH are shown in Figure 1.3. Three regions of partition behavior are apparent. At very low pH, the neutral species AH is predominant in both the aqueous and nonaqueous phases, and the apparent log D is equal to log K_{ow}. At pH values just above the pK_a, the

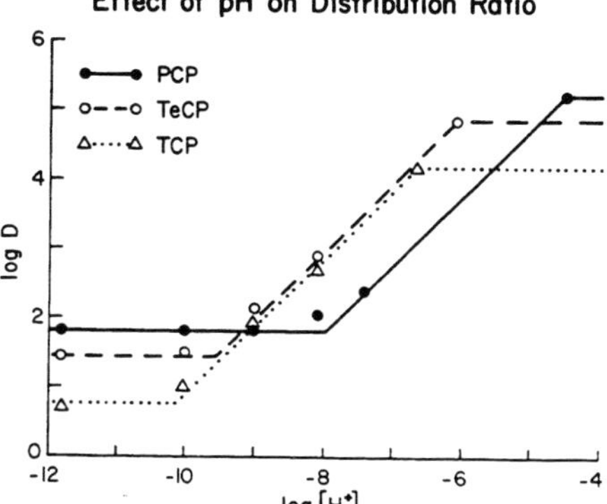

Figure 1.3. The effect of pH on the concentration distribution ratio of chlorophenols (PCP, pentachlorophenol; TeCP, 2,3,4,5-tetrachlorophenol; TCP, 2,4,5-trichlorophenol). The distribution ratio as a function of pH reflects the K_{ow} and K_a of each compound.

main species in the aqueous phase becomes the ionized form A^-. Then the apparent distribution ratio varies with pH, and the distribution can be characterized as equilibrium between the ionic species A^- in the aqueous phase and the neutral species $\overline{AH}$ in the nonaqueous phase. At still higher values of pH, when the pH is greater than the pK_a in the nonaqueous phase, the species A^- becomes predominant in both phases, and the apparent distribution ratio is again independent of pH. Whether the ionic form exists in the nonaqueous phase as a free ion or as an ion pair is not apparent from these data, since only one ionic strength is considered; however, other data show that both forms are important.

It is significant that the anionic form is the predominant species in the nonaqueous phase at pH values as low as pH 7. This result has been observed for adsorption to natural materials as well.

The preceding discussion has dealt with the partition of ionizable organic compounds in model systems in which partition was driven exclusively by the hydrophobic effect. In the following section, other model systems are considered in which adsorption is driven by a combination of the hydrophobic effect and a specific chemical interaction (ligand exchange) of the HIOC with the surface.

Ulrich et al. (1986) have compared the adsorption of a series of alkylcarboxylic acids (butanoic–dodecanoic) at two model surfaces: mercury and δ-Al_2O_3. The mercury surface is nonpolar and has little tendency for specific interactions with the carboxylate group. Adsorption of the neutral alkylcarboxylic acids to the mercury surface was found to follow the hydrophobic effect, and adsorption constants of the series of compounds followed their octanol–water partition coefficients. Adsorption decreased with increasing pH and ionization of the acid, due to the unfavorable energy of attracting counterions to the surface. This effect can be regarded as the two-dimensional analog of the Donnan effect discussed above for the three-dimensional partitioning into octanol.

The alumina surface is polar and can react with the carboxylate groups by ligand exchange. The adsorption of the lower molar mass carboxylic acids was ascribed primarily to ligand exchange; a maximum in the curve of distribution ratio versus pH was found to coincide with the value predicted from the mass action law and the acidity constants of the acids and the surface; this pH was approximately the pK_a of the acids. The adsorption of the higher molar mass carboxylic acids appeared to be dominated by the hydrophobic effect.

3.3. Hydrophobic Ionic Organic Compounds

Hydrophobic ionic molecules contain a permanent charge in addition to a hydrophobic group. Ionic surfactants are common members of this class of compounds. In contrast to the carboxylic acids discussed above, the charge on these compounds is independent of pH. However, pH-dependent adsorption of these molecules on oxide surfaces is still expected, since the charge on oxide surfaces depends on pH.

The primary question with respect to adsorption is whether the energy of adsorption is related primarily to ionic interactions, in which case ionic strength and pH are key experimental variables, or hydrophobic interactions, in which case the nature of the hydrophobic group on the solute molecule and the fraction of organic carbon of the sorbent are key variables. The sulfonate functional group that is considered in this example is much less likely to exhibit strong ligand-exchange interactions with the surface than the carboxylate functional group considered in the last example. Thus the sulfonate group is less representative of natural organic matter and more likely to be subject to predominantly electrostatic adsorption than other ligands that form coordinate bonds with the surface.

In Figure 1.4*a* the adsorption of homologs of 4-(n'-dodecyl)benzenesulfonate (often referred to as linear alkylbenzenesulfonate, LAS) on alumina, of which the pH of zero protonic charge (pH_{zpc}) is approximately pH 8.8, is shown (Dick et al., 1971). It is seen that branching of the dodecyl groups reduces the adsorption energy of this negatively charged molecule in a systematic way. Thus one might assume that the adsorption of this group is primarily hydrophobic in nature. However, inspection of the adsorption of one homolog at a series of different pH values (Fig. 1.4*b*) shows that adsorption decreases systematically with increase in pH, and above the pH_{zpc} the LAS is hardly adsorbed at all. This complementary behavior reflects the fact that adsorption of these molecules on the oxide surface is influenced by hydrophobic interactions, but ultimately electrostatic interactions are stronger and control adsorption.

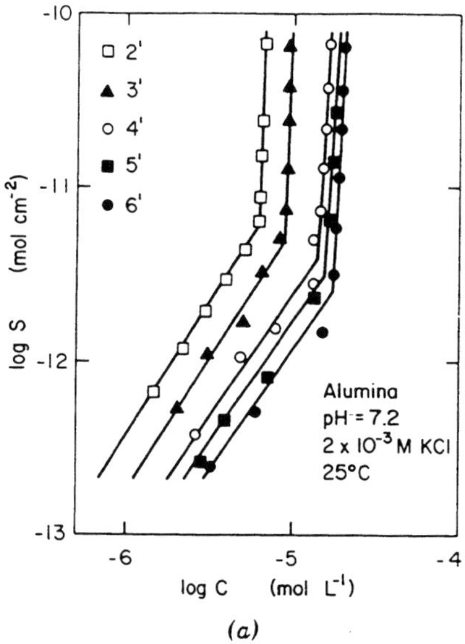

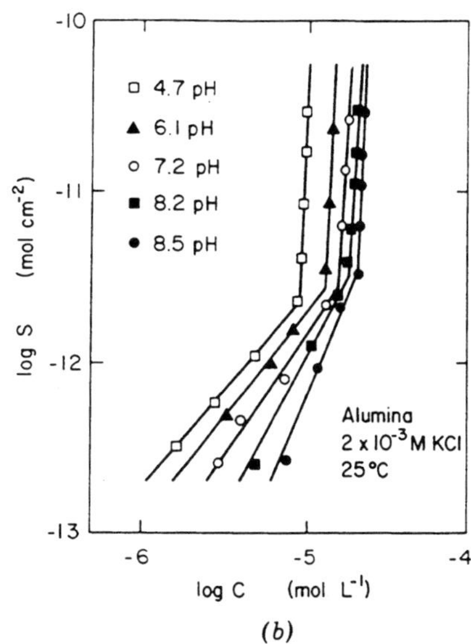

Figure 1.4. Effect of structure of hydrophobic group and pH on adsorption of anionic surfactants on alumina. (*a*) The logarithm of the surface concentrations as a function of the logarithm of solution concentration is shown for a series of 4-(n'-dodecyl)benzenesulfonate. The numbers in the legend indicate the position on the dodecyl chain at which the benzene ring is attached. Increased branching of the dodecyl group leads to weaker adsorption. (*b*) The logarithm of the surface concentrations as a function of the logarithm of solution concentration is shown for 4-(2'-dodecyl)benzenesulfonate as a function of pH. An increase in pH leads to weaker adsorption, consistent with the onset of electrostatic repulsion at the surface, of which the pH_{zpc} (in the absence of surfactant) is 8.9. After Dick et al. (1971).

It is not surprising that specific chemical interactions and electrostatic effects dominate adsorption at a pure oxide surface. However, for a natural surface with a higher content of organic carbon, one might expect a much greater contribution from hydrophobic effects.

4. INORGANIC COMPOUNDS

Adsorption driven by hydrophobic interactions is relatively easy to characterize since interactions are rather nonspecific. In contrast, adsorption at hydrous oxide surfaces is generally quite specific and correspondingly more difficult to characterize. Although the only adsorption reaction treated in this discussion is surface hydrolysis, a similar approach is used for the more complicated adsorption reactions of metals and ligands.

Virtually all surface hydrolysis experiments are carried out in the presence of a "background electrolyte," many of which appear to exhibit weak specific chemical interactions with the surface. While the description of these interactions fits directly into the conceptual framework developed here, they will not be considered explicitly. Interpretation of the experimental evidence for the adsorption of background electrolyte ions is a subject in itself, in which the concepts presented here can be used.

The complete reaction for the adsorption of a hydrogen (or hydronium) ion on an oxide surface is shown in Figure 1.5. Usually this reaction is written only for the transfer of hydrogen ion from the bulk of the solution to the surface and the reaction of hydrogen ion with the surface; however, the transfer of the countercharge from the bulk of solution to the solution part of the double layer is an integral part of the reaction. Failure to consider the energy of this part of the reaction leads to an incomplete understanding of the surface chemistry.

It is noteworthy that of the four steps of the reaction that are listed, only the overall energy of the reaction can be measured. In order to extract the intrinsic chemical energy of reaction (step 1) from the observed energy, it is necessary to calculate from some electrostatic model the energy of transfer of the hydrogen ion to the surface (step 2) and the energy to move the countercharge into the solution part of the double layer (steps 3 and 4). The energy associated with step 2 is related to the surface potential, and the energy of steps 3 and 4 depends on the model used for the electric double layer.

This ambitious undertaking of measuring one value and calculating three others to arrive at a fourth does result in some covariance in the values obtained for the chemical and electrostatic parameters of the model, as discussed by Westall and Hohl (1980).

This example of an adsorption reaction can be formulated more generally in terms of the basic elements of an electrochemical equilibrium model of the surface solution interface: structure, chemical interactions, and electrostatic interactions. The structure is of primary importance for the electrical model, since the long-range electrostatic effects depend on the location of charge

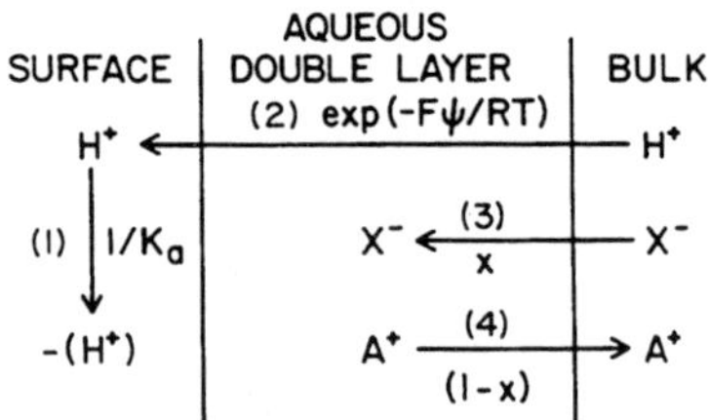

Figure 1.5. The complete reaction for adsorption of a proton. Step 1 is the chemical energy of adsorption at the surface; steps 2–4 involve electrostatic energies.

through the interface. The chemical energies of interaction are usually simply assigned a value, without any explicit description of structure. However, it is important that the structure be qualitatively consistent with the chemical model.

It has been difficult to separate the observed energy of reaction (as represented by the logarithm of the apparent mass action quotient, log Q) into chemical and electrostatic components. The approach that is usually used is to consider surface chemical groups to be of a few discrete energies, to which is added a continuously variable electrostatic energy. Then one expects the apparent acidity quotient of the surface to be distributed about a discrete value. However, the question still remains as to how much of the width observed in the distribution of values is due to electrostatic effects and how much is due to the heterogeneity of the functional groups on the surface.

Therefore, since the separation between electrostatic and chemical energies is more or less indeterminate, the prospect of interpreting experimental data simply in terms of the observed distribution of log Q becomes worthy of consideration. A similar approach has been taken for reactions of humic acids (Buffle and Altmann, 1987). However, a close examination of this approach reveals that the transformation of experimental data to a distribution of adsorption energies provides little insight that is not already available from the titration curve (log a_H versus T_H) or the curve of the buffer intensity versus log a_H, which is found from the inverse of the derivative of the titration curve.

For elucidating the nature of adsorption energies, the use of data from additional kinds of experiments proves to be more useful than mathematical manipulation of adsorption data themselves. Following a discussion of chemical and electrostatic models for the surface–solution interface, data for both surface charge and surface potential will be used with the models for interpreting adsorption energies.

4.1. Chemical Models

The customary way to represent reactions at oxide surfaces has been through reactions of the type (Schindler and Stumm, 1987)

$$XOH_2^+ \rightleftharpoons XOH + H^+ \qquad K_{a1} \tag{1.11}$$

$$XOH \rightleftharpoons XO^- + H^+ \qquad K_{a2} \tag{1.12}$$

This representation is appealing since it is reasonably simple and allows a convenient representation of the condition of zero protonic charge at the surface, $n_{XOH_2^+} = n_{XO^-}$.

A chemically more explicit representation of the same model with diprotic acid groups is through the surface species

$$\left[=X\begin{matrix}OH_2\\ OH_2\end{matrix}\right]^{+} \underset{K_{a1}}{\overset{-H^+}{\rightleftharpoons}} \left[=X\begin{matrix}OH_2\\ OH\end{matrix}\right]^{0} \underset{K_{a2}}{\overset{-H^+}{\rightleftharpoons}} \left[=X\begin{matrix}OH\\ OH\end{matrix}\right]^{-} \qquad (1.13)$$

The surface is thus represented as an ensemble of N_s of these groups per unit surface area, which accommodate N_s exchangeable protons when the surface is electrically neutral. Since the surface group can exist in a positive, neutral, or negative state, this model is referred to as a three-state model. These three states arise from the fact that every surface oxygen atom can exist in either a protonated or nonprotonated state, and the energy of dissociation of a particular proton is influenced by the state of one nearest neighbor.

It has been shown that it is possible to use a wide range of combinations of acidity constants to fit the reaction model to experimental data (Westall and Hohl, 1980). In particular, if K_{a1} is made very small and K_{a2} is made very large, the neutral species becomes insignificant. Then reaction 1.13 becomes

$$\left[=X\begin{matrix}OH_2\\ OH_2\end{matrix}\right]^{+} \underset{K_{a1}K_{a2}}{\overset{-2H^+}{\rightleftharpoons}} \left[=X\begin{matrix}OH\\ OH\end{matrix}\right]^{-} \qquad (1.14)$$

or, if it is written for the loss of one proton, with a Z surface group equivalent to one-half of an X surface group,

$$ZOH^{1/2+} \underset{(K_{a1}K_{a2})^{1/2}}{\overset{-H^+}{\rightleftharpoons}} ZO^{1/2-} \qquad (1.15)$$

Reaction 1.15 can be represented symbolically as

$$ZOH^{1/2+} \longrightarrow ZO^{1/2-} + H^+ \qquad K_a = (K_{a1}K_{a2})^{1/2} \qquad (1.16)$$

As derived above, the model represented by reaction 1.15 seems to be based on the physically unreasonable condition that the neutral species is more acidic than the positively charged species. However, an alternative derivation of this model, which is completely plausible physically, is possible.

Instead of basing the model on the metal centers to which two oxygen atoms are coordinated, one can consider simply the chemistry of the oxygen atoms themselves. The surface can be considered as an ensemble of $2N_s$

surface oxygen atoms, which, at electroneutrality, are covered by N_s protons. If each group is considered to have zero nearest-neighbor interactions, the model that results is the one given by reaction 1.16. Thus this model is referred to as a two-state model, since each surface group can exist in either a protonated or a deprotonated state. Van Riemsdijk et al. (1986) and Bolt and Van Riemsdijk (1987) discuss other aspects of the two-state model, referred to by them as the one-pK model.

Thus the significant difference in the derivation of the models given by reactions 1.11 and 1.12 and the model of reaction 1.16 is simply the number of nearest-neighbor interactions: the three-state model allows one, and the two-state model allows zero. Furthermore, since the electrostatics are considered separately and explicitly, a model with zero nearest-neighbor interactions is a quite reasonable approximation or at least not significantly more unreasonable than the assumption of only one nearest-neighbor interaction.

The two-state and three-state models described by reactions 1.11–1.16 are defined mathematically by the species, mass action, and material balance equations in Table 1.4A–C. The surface density of exchangeable protons and surface density of charge, which are to be discussed later, are given in Table 1.4D and E. To solve the chemical model in Table 1.4, an additional equation is required to relate σ^H and ψ; this equation is derived from an electrostatic model.

Table 1.4. Equations Defining the Three-State and Two-State Models

Three-State Model	Two-State Model
A. Species	
XOH_2^+, XOH, XO^-	$ZOH^{+1/2}$, $ZO^{-1/2}$
B. Mass Action[a]	
$n_{X^+}K_{a1} = n_X \; a_H(w) \exp(-F\psi/RT)$	$n_{Z^+}K_a = n_{Z^-} a_H(w) \exp(-F\psi/RT)$
$n_X K_{a2} = n_{X^-} \; a_H(w) \exp(-F\psi/RT)$	
C. Material Balance	
$N_s = n_{X^+} + n_X + n_{X^-}$	$2N_s = n_{Z^+} + n_{Z^-}$
D. Surface Density of Exchangeable Protons	
$N_H = 2n_{X^+} + n_X$	$N_H = n_{Z^+}$
E. Surface Density of Protonic Charge	
$\sigma^H = e(n_{X^+} - n_{X^-})$	$\sigma^H = e(n_{Z^+} - n_{Z^-})/2$
$= e(N_H - N_s)$	$= e(N_H - N_s)$

[a]Species are represented by the following subscripts: $XOH_2^+ = X^+$; $XOH = X$; $XO^- = X^-$; $ZOH^{1/2+} = Z^+$; $ZO^{1/2-} = Z^-$.

4.2. Electrostatic Models

The structure of the solid–solution interface is the key to the electrostatic model for the interface. Customarily, models for the oxide–electrolyte interface have been taken from variations of the Gouy–Chapman–Stern (GCS) model. While the GCS theory provides an excellent mathematical basis for the formulation of interfacial models, it is incorrect to attach too much physical significance to the value of parameters.

The interpretation of results of adsorption at smooth homogeneous interfaces, such as that of mercury and NaF, is not completely explained even by the GCS theory. Parsons (1987) discusses surface chemistry at single crystal and polycrystalline solid metal surfaces, showing that different crystal faces of the same pure metal exhibit dramatically different surface chemical properties. For the solid–solution interfaces usually considered in aquatic surface chemistry, such as those of oxides and minerals, the surfaces are not only polycrystalline and amorphous but are also composed of different atoms in different states of hydration.

The form of the GCS model is too simple to afford a strict, literal interpretation of all these effects. However, as will be shown, the GCS model is of the correct form to allow interpretation of data in terms of operational parameters. The relationships among different models for the interface will be seen by formulating them as different limiting cases of the GCS model.

The framework for the discussion of the oxide–electrolyte interface is the triple-layer model, as the extended GCS model has been described (Fig. 1.6). The charging of an oxide surface, rather than that of a conducting metal surface, is considered. According to this model the surface is divided into several planes or layers of mean charge and potential.

The potential-determining ions that are most intimately associated with the surface are assigned to the surface plane, where they experience the

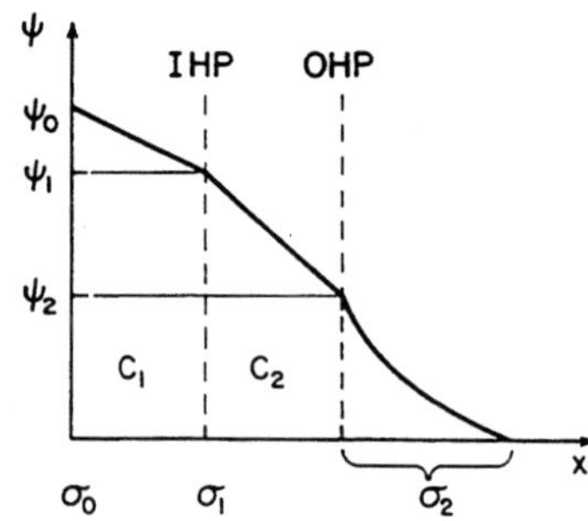

Figure 1.6. The triple-layer model for the structure at the solid–solution interface. The model is represented by the electrostatic potential profile through the interface. All the idealized models represented in Figure 1.7 are mathematically identical to limiting cases of the triple-layer model with particular values of the capacitances C_1 and C_2.

potential ψ_0 and contribute to the charge σ_0. The primary potential-determining ion at the oxide surface is generally considered to be H^+. At the inner Helmholtz plane (IHP) are assigned ions which are less intimately associated with the surface than the primary potential-determining ions but still experience chemical interaction with the surface. These ions contribute to the charge σ_1 and experience the potential ψ_1. The ions that might be appropriate for assignment to the IHP include strongly adsorbed metals and ligands as well as weakly adsorbed ions from the background electrolyte.

The outer Helmholtz plane (OHP) is the innermost boundary of the diffuse layer, in which ions are influenced by electrostatic forces only. The distribution of ions, charge, and potential in this diffuse layer is described by the Poisson–Boltzmann equation, and the overall relationship between charge in the diffuse layer, σ_2, and potential at the OHP, ψ_2, is given by the Gouy–Chapman equation:

$$\sigma_2 = -(8\epsilon\epsilon_0 RTc_\infty)^{1/2} \sinh\left(\frac{F\psi_2}{2RT}\right) \tag{1.17}$$

where ϵ is the dielectric constant of water, ϵ_0 is the permittivity of free space, and c_∞ is the concentration of a monovalent electrolyte in the bulk of solution.

Between the surface and the IHP is a capacitance C_1, and between the IHP and OHP a capacitance C_2. The charge in each of the surface planes is related to the potential drop between the planes by the equation for a parallel-plate capacitor.

$$\sigma_0 = C_1(\psi_0 - \psi_1) \tag{1.18}$$

$$\sigma_1 = C_2(\psi_1 - \psi_2) \tag{1.19}$$

The final electrostatic constraint on the model is electroneutrality throughout the interface.

$$\sigma_0 + \sigma_1 + \sigma_2 = 0 \tag{1.20}$$

Equations 1.17–1.20 completely define the general electrostatic model for the interface. Solution of these four equations in six unknowns (ψ_0, ψ_1, ψ_2, σ_0, σ_1, σ_2) requires two other equations, which are values of σ_0 and σ_1 derived from chemical considerations, such as those presented in Table 1.5.

In practice this multilayer model of the oxide–electrolyte interface contains too many details to be compatible with the resolution of the experimental data. However, the mathematical form of the equations does allow the relationship among the physically imaginable states of the interface to be formulated.

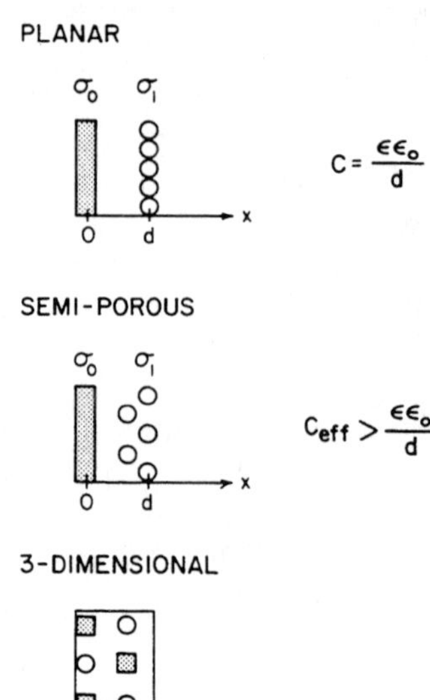

Figure 1.7. Three idealized cases for structure at the solid–solution interface. In the planar case, ions are lined up in such a way that the charge–potential relationship resembles that of a parallel-plate capacitor. In the semiporous case, there is some penetration of counterions toward the oxide surface, yielding higher apparent charges and effective capacitances than would be possible for a parallel-plate capacitor. In the three-dimensional gel, or fully porous case, the mobile counterions are interspersed with the charge that is fixed to the solid lattice. This structure results in a Donnan equilibrium.

Three simple models for the interfacial structure are depicted in Figure 1.7. The simplest structure is the surface charge in one plane and the countercharge in a single parallel plane. The electrostatic model of this structure is that of a parallel-plate capacitor, with capacitance

$$C = \frac{\epsilon\epsilon_0}{d} \tag{1.21}$$

where d is the separation between the planes. The triple-layer model reduces to this simple constant-capacitance model if $C_1 \ll C_2$ and $C_1 \ll C_d$, where C_d is the effective capacitance of the diffuse layer found by differentiating Eq. 1.17:

$$C_d = \frac{\partial\sigma_2}{\partial\psi_2} \tag{1.22}$$

The physical condition that leads to this limiting case is a very high concentration of supporting electrolyte.

If counterions penetrate the oxide lattice, or if the surface roughness is such that the effective separation of charge is reduced as shown in Figure 1.7, the effective capacitance is increased. The capacitances found for this model might be significantly larger than those anticipated by comparison with other solid–electrolyte interfaces, but these large values are ascribed to the small separation of charge.

As the surface roughness increases, the separation of charge between the solid and solution decreases. Ultimately the two-dimensional surface becomes a three-dimensional surface phase, in which both the potential-determining ion and the counterions are subject to the same potential. In this case the surface becomes a Donnan phase.

The triple-layer model becomes mathematically equivalent to a Donnan model as C_1 becomes much greater than C_2 and C_d; then there is almost no attenuation of the electric field between the surface and the IHP, but the potential is reduced to almost zero between the IHP and the OHP. Thus potential-determining H^+ ions as well as electrolyte ions at the IHP experience the potential $\psi_0 = \psi_1$, and electroneutrality conditions become $\sigma_0 + \sigma_1 = 0$.

Strictly speaking the Donnan model allows no external electric field and therefore no electrokinetic potential. However a quasi-Donnan model could be imagined in which C_1 was large enough to result in almost complete, but not total, neutralization of charge at the surface, allowing a small electrokinetic potential.

4.3. Formulation of Electrochemical Equilibrium Problems by Matrix Algebra

As has been discussed, there is a continuum of ways in which to conceive of surface electrochemical equilibrium models. Thus it is quite important that particular models be defined precisely and concisely. The formulation of such problems by matrix algebra is well suited not only for the definition of the problems but also for their solution.

The application of matrix algebra to the solution of surface chemical equilibrium problems has been discussed in detail by Westall (1980). This method is summarized in Table 1.5. The key to the procedure is the definition of the species (every chemical entity in the system), the components (a minimal set of reactants out of which all the species can be formed), and a stoichiometry matrix (denoted the A matrix), which relates the components to the species. The rows of the A matrix are the coefficients in the mass action equations, and the columns are the coefficients in the mass balance equations. For conventional chemical components, either the total concentration or the free concentration is found analytically, while for electrostatic components the total concentration is found from the electrostatic equations, that is, from some form of Eqs. 1.17–1.20.

In Table 1.6 are formulated the A, T, and K matrices for four surface

Table 1.5. Electrochemical Equilibrium Equations: Definitions[a]

Type of Equations	Scalar	Matrix
Mass action	$\log C_i = \log K_i + \sum_j a_{ij} \log X_i$	$\mathbf{C}^* = \mathbf{K}^* + \mathbf{A}\mathbf{X}^*$
Mass balance	$Y_j = \sum_i a_{ij} C_i - T_j$	$\mathbf{Y} = {}^t\mathbf{A}\mathbf{C} - \mathbf{T}$
Iteration	$z_{jk} = \sum_i \left(\frac{a_{ij} a_{ik} C_i}{X_k}\right) = \frac{\partial Y_j}{\partial X_k}$	$\mathbf{Z}\,\Delta\mathbf{X} = \mathbf{Y}$

Scalar Symbols	Description	Matrix or Vector Symbols	Description
a_{ij}	Stoichiometric coefficient of component j in species i	**A**	Matrix of a_{ij}
C_i	Free concentration of species i	**C**	Vector of C_i
		C*	Vector of $\log C_i$
K_i	Stability constant of species i	**K***	Vector of $\log K_i$
T_j	Total analytical concentration of component j	**T**	Vector of T_j
X_j	Free concentration of component j	**X**	Vector of X_j
		X*	Vector of $\log X_j$
Y_j	Residual in material balance equation for component j	**Y**	Vector of Y_j
z_{jk}	Partial derivative $(\partial Y_j/\partial X_k)$	**Z**	Jacobian of Y with respect to X

[a]After Westall (1980).

electrochemical equilibrium models: the three-state chemical model with constant capacitance and Gouy–Chapman electrostatic models, a two-state chemical model with Stern electrostatic model, and a Donnan model. These matrices completely define the electrochemical equilibria.

The only difference between the two three-state models (Table 1.6A) is in the formulation of the total charge by the electrostatic equations. The constant-capacitance model has been used extensively in studies of oxide surface chemistry; the parameters must be operationally defined for each electrolyte concentration. The Gouy–Chapman model has been used less frequently, but it allows parameters to be valid over a range of electrolyte concentrations, and it has recently been employed for a reinterpretation of a wide variety of adsorption data (Dzombak, 1986).

The two-state model (Table 1.6B) is formulated in the same way as the three-state models but without the neutral surface species. The neutral surface component does not appear as a species and thus has no effect on the speciation. The Stern double-layer model is used here with the two-state chemical model; since there are two unknown electrostatic potentials with this model, there are two electrostatic components. Hence the electrostatic model is more complicated for the Stern model than it is for the simple constant-capacitance or diffuse-layer models.

The two-state model with the Stern electrostatic model has been used to represent data over a wide range of electrolyte concentrations (Fig. 1.8). It offers the advantage of extreme simplicity, since the only adjustable parameters are the pK_a, which is equal to the pH_{zpc}, and the inner-layer capacitance C_1. As will be shown, the two-state model is also consistent with measured oxide surface potential.

The Donnan model (Table 1.6C) has been proposed for oxide surfaces and used extensively for interpretation of the interaction of metals with polymeric gels and natural humic matter (Marinsky, 1987). In the Donnan model there

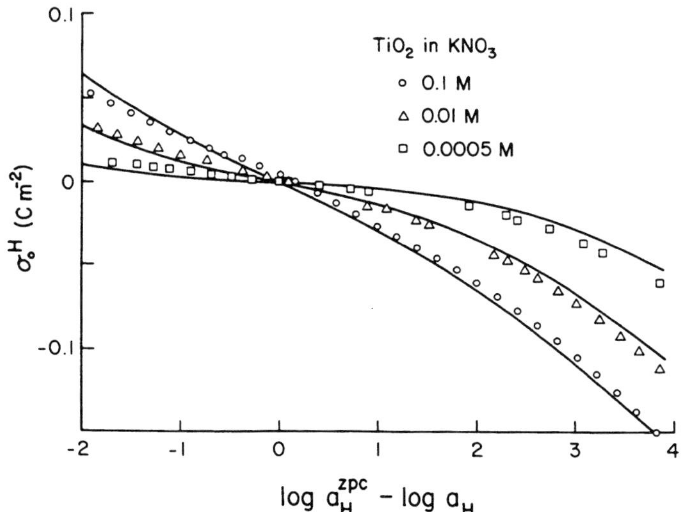

Figure 1.8. Titration of a suspension of TiO_2 in various concentrations of KNO_3. The protonic surface charge $[\sigma_0^H = (T_H - [H^+] + [OH^-]) \cdot (F/s\ a)]$ is shown. The lines are calculated from the two-state chemical model with the Stern electrostatic model with inner-layer capacitance $C = 1.23$ F m^{-2}. The TiO_2 is Degussa P-25 (primarily anatase.) (Westall and Hindagolla, 1986)

Table 1.6. Algebraic Formulation of *A*, *K*, and *T* Matrices of Surface Electrochemical Equilibrium Models[a]

A. Three-State Chemical Model with Constant Capacitance and Gouy–Chapman Electrostatic Models

	Components			
Species	XOH	H^+	$X(\psi_0)$	K
H^+	0	1	0	1
OH	0	−1	0	K_w
XOH	1	0	0	1
XOH_2^+	1	1	1	$1/K_{a1}$
XO^-	1	−1	−1	K_{a2}
	T_{XOH}	T_H	T_σ	

$T_{XOH} = N_s\, sa/N_A$

$T_H = C_A - C_B$

$T_\sigma = C_1\psi_0\, sa/F$ (Constant capacitance model)

or

$T_\sigma = -\sigma_2\, sa/F$ (Gouy–Chapman model)

B. Two-State Chemical Model with Stern Electrostatic Model

	Components				
Species	$ZOH^0_{1/2}$	H^+	$X(\psi_0)$	$X(\psi_1)$	K
H^+	0	1	0	0	1
OH^-	0	−1	0	0	K_w
$ZOH^{+1/2}$	1	$\frac{1}{2}$	$\frac{1}{2}$	0	$K_a^{-1/2}$
$ZO^{-1/2}$	1	$-\frac{1}{2}$	$-\frac{1}{2}$	0	$K_a^{1/2}$
	T_{ZOH}	T_H	T_{σ_0}	T_{σ_1}	

$T_{ZOH} = 2N_s\, sa/N_A$

$T_H = C_A - C_B$

$$T_{\sigma_0} = \frac{sa}{F} C_1(\psi_0 - \psi_1)$$

$$T_{\sigma_1} = -\frac{sa}{F}[C_1(\psi_0 - \psi_1) + \sigma_2]$$

Table 1.6. (*Continued*)

C. Donnan Model

Species	Components of Model: HA	H^+	Na^+	Cl^-	$X(\psi_D)$	V_D/V_w	K
H^+	0	1	0	0	0	0	1
OH^-	0	−1	0	0	0	0	K_w
Cl^-	0	0	0	1	0	0	1
Na^+	0	0	1	0	0	0	1
HA	1	0	0	0	0	0	1
A^-	1	−1	0	0	0	0	K_a
$\overline{H^+}$	0	1	0	0	1	1	K_H
$\overline{OH^-}$	0	−1	0	0	−1	1	K_{OH}
$\overline{Cl^-}$	0	0	0	1	−1	1	K_{Cl}
$\overline{Na^+}$	0	0	1	0	1	1	K_{Na}
$\overline{HA}$	1	0	0	0	0	1	K_{HA}
$\overline{A^-}$	1	−1	0	0	−1	1	$K_a K_A$
	$\overline{T}_{HA}$	T_H	T_{Na}	T_{Cl}	$\overline{T}_\sigma = 0$	T_1	

[a]The following symbols are used:

a	concentration of solid
C_A	concentration of strong acid
C_B	concentration of strong base
N_A	Avogadro's number
s	specific surface area
$X(\psi)$	the electrostatic component, $\exp(-F\psi/RT)$
V_D/V_w	the ratio of the volumes of the Donnan and aqueous phases
σ_2	the charge in the diffuse layer, given by Eq. 1.17.

is only one unknown potential. Since all of the surface charge is expressed explicitly in the Donnan model, the balance of the charge outside of the Donnan phase is zero, and T_σ is set equal to zero. A dummy component is introduced to correct for the relative volumes of the aqueous and Donnan phases.

4.4. Direct Measurement of Change in Interfacial Potential Difference

Where mathematical manipulation of adsorption data has failed to resolve electric and chemical energies at the surface, additional data can. Two other kinds of data that can be obtained for oxide surfaces are electrophoretic mobility and the change in oxide surface potential with solution composition.

The interpretation of electrophoretic mobility is impeded by uncertainty in the relationships of the electrophoretic mobility to the ζ-potential (the effective potential at the mean "slipping plane") and of the ζ-potential to the structure at the interface. In contrast, the change in oxide *surface* potential (actually the interfacial potential difference between the oxide surface and solution) with change in solution composition lends itself to an unambiguous interpretation.

The change in surface potential with solution composition can be measured directly with electrochemical cells of the type

Reference electrode | Solution | Oxide | Semiconductor | Metal

or

Reference electrode | Solution | Semiconducting oxide | Metal

The capacitance of the semiconductor is potential-dependent and of the same order of magnitude as the capacitance of the oxide. Thus the differential capacitance can be used as a null detector for changes in surface potential as a function of solution composition (Westall and Hindagolla, 1986).

Results of such experiments by Watanabe et al. (1974) for TiO_2 are shown in Figure 1.9. It is seen that the change in surface potential with log a_H is almost the Nernstian value of $(RT \ln 10)/F$ = 59 mV/decade at 298 K. This fact has significant implications for the values of surface acidity constants and surface speciation, as will be shown.

Electrochemical equilibrium between the surface and the solution requires that electrochemical potentials of species in both phases be equal [see Stumm

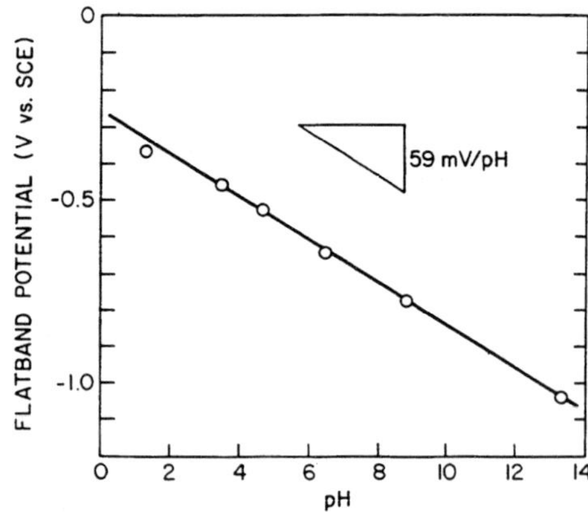

Figure 1.9. The change in interfacial potential difference with pH for TiO_2. Determined from TiO_2 semiconducting electrodes in 0.5 *M* KCl by shift in flatband potential. After Watanabe et al., (1974).

and Morgan (1981) for details]:

$$\overline{\mu}_{H^+}(w) = \overline{\mu}_{H^+}(s) \tag{1.23}$$

where the electrochemical potential for a species i in phase α is

$$\overline{\mu}_i(\alpha) = \mu_i^0(\alpha) + RT \ln a_i(\alpha) + z_i F\psi(\alpha) \tag{1.24}$$

Expansion of Eq. 1.23 in terms of Eq. 1.24 for both phases and rearrangement yields the interfacial potential difference in terms of hydrogen ion activities:

$$\psi = \psi(s) - \psi(w) = \mu_{H^+}^0(w) - \mu_{H^+}^0(s) + \frac{RT}{F} \ln \frac{a_{H^+}(w)}{a_{H^+}(s)} \tag{1.25}$$

Then for the Nernstian interfacial potential difference to be observed with hydrogen ion activity in water, a necessary and sufficient condition is that the surface activity of hydrogen ion must be virtually constant throughout the range of surface charge over which the observations are made. Such is the case for analogous systems such as AgI, at which surface activities of Ag^+ and I^- are constant due to the buffering of the solid phase.

Then the key question is, how much does the surface hydrogen ion activity change as the surface charge develops? This question can be reexpressed in terms of the surface buffer intensity, β', defined by

$$\beta' = \frac{dN_H}{d \log a_H(s)} \tag{1.26}$$

where N_H is the total density (m^{-2}) of exchangeable protons on the surface as defined in Table 1.4. In the following section, values of β' calculated from the models in Table 1.4 will be compared to those found experimentally from Figures 1.8 and 1.9 to ascertain what values of the adjustable parameters N_s and K_{a1}, K_{a2} are appropriate.

The surface buffer characteristics of the three-state and the two-state model, calculated from Eq. 1.26 and Table 1.4B–D, are shown in Figure 1.10. The key variable for the buffer intensity of the three-state model is the surface association constant, K_A, for the reaction

$$\frac{1}{2}\,XOH_2^+ + \frac{1}{2}\,XO^- \rightleftharpoons XOH \qquad K_A = \left(\frac{K_{a1}}{K_{a2}}\right)^{1/2} \tag{1.27}$$

If K_A is very large, the neutral XOH group dominates the surface at the pH_{zpc}, the concentrations of both of the charged species are very low, and the buffer intensity is very low. If K_A is very small, the species XOH_2^+ and XO^- are dominant at the pH_{zpc}, and the surface buffer intensity is very large. At the

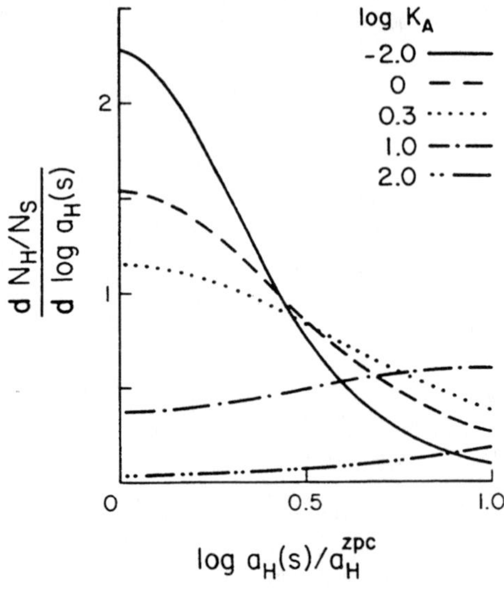

Figure 1.10. Surface buffer intensity as a function of hydrogen ion activity relative to the pH_{zpc}. The curves are shown for the three-state model with various values of the surface association constant $K_A = (K_{a1}/K_{a2})^{1/2}$. The buffer intensity of the two-state model is *identical* to that of the three-state model with $K_A = 2$. Since there is often uncertainty about the value of the surface site density N_s, the buffer intensity has been normalized to unit surface site density. The values of β' were calculated from Eq. 1.26, where N_H is as defined in Table 1.4.

value $K_A = 2$, the buffer characteristics of the three-state model are *identical* to those of the two-state model. (The surface speciation under this condition is not identical, although the buffer intensity is.)

Since the total capacity of the surface to donate or accept protons is independent of the value of K_A, a surface that is a very good buffer at the pH_{zpc} must be a very poor buffer far from the pH_{pzc}, and vice versa. This effect can be seen in Figure 1.10.

The qualitative conclusion that comes from this discussion is that systems with three-state models with small values of K_A and the two-state model provide better buffering of surface hydrogen ion activity and thus would lead to a more nearly Nernstian response. A rigorous examination of this hypothesis requires a full discussion of the chemical models in combination with the electrostatic models (Westall and Hindagolla, 1986) and will not be given here. However, an estimate of the magnitude of the surface buffer intensity that is necessary for the surface charge data (Fig. 1.8) to be consistent with the surface potential data (Fig. 1.9) can be developed.

The value of the surface buffer intensity β' can be found experimentally from the conventional buffer intensity β,

$$\beta = \frac{dN_H}{d \log a_H(w)} = \frac{1}{e} \frac{d\sigma^H}{d \log a_H(w)} \tag{1.28}$$

and the surface potential response, $d\psi/[d \log a_H(w)]$, through a function η:

$$\eta = \frac{d \log a_H(s)}{d \log a_H(w)} = 1 - \frac{F}{RT \ln 10} \frac{d\psi}{d \log a_H(w)} \tag{1.29}$$

which is derived from Eq. 1.25. Both β and η can be found experimentally from Figures 1.8 and 1.9, respectively.

In this analysis, data obtained from surfaces prepared under completely different circumstances—pyrogenic particles and electronic devices—are combined. The question arises, how similar are the surfaces? Ideally the similarity of the surfaces would be verified by independent physical measurements. However, in the absence of such measurements, the validity of the approach used here is supported by the relative insensitivity of the response of both types of surfaces to their preparation.

The equation relating β' to β and η is

$$\beta' = \frac{\beta}{\eta} \tag{1.30}$$

From the slope of the curve of surface charge versus log a_H (Fig. 1.8) for 0.1 *M* KNO_3 at the pH_{zpc} (0.03 C m^{-2}), the value of $\beta = 0.19 \times 10^{18}$ m^{-2} is found from Eq. 1.28. The value of $\eta = 0.05$ is found from Figure 1.9. With N_s at a maximum value of 12×10^{18} m^{-2}, the value of $\beta'/N_s = 0.3$. Then, according to Figure 1.10, the value of log K_A must be no greater than approximately 1 for the surface charge data to agree with the surface potential data. This analysis is very sensitive to the value of $d\psi/[d \log a_H(w)]$, which was set to a lower bound for this analysis; higher values of $d\psi/[d \log a_H(w)]$ would lead to still smaller values of K_A.

This illustration bears out what is found from the more rigorous approach, that a two-state model or a three-state model with reasonably small values of K_A is required to correspond simultaneously with surface charge and surface potential data.

5. SUMMARY

The mathematical form of the Gouy–Chapman–Stern theory is appropriate for most surface electrochemical equilibrium models involving the adsorption of both organic and inorganic compounds. The matrix algebraic approach to solving these electrochemical equilibrium problems has been advanced, not only because of the ease of solving the problem, but also because of the conciseness and exactness of the expression of the problem.

Several examples of the use of model systems to investigate mechanisms have been presented. For these model systems to be of maximum use for application to more complicated natural systems, they must be fully characterized and understood, and the mathematical models should be appropriate. In this regard, drawing conclusions about the physical nature of the surface through the use of multilayer double layer models is inappropriate.

Models for oxide surfaces have been reviewed with respect to separation of electrical and chemical components of adsorption energy. Combination of measurements of surface charge with direct measurements of surface potential from semiconductor oxide electrodes favor the two-state model or the three-state model with relatively small values of K_A. The use of continuous distribution to characterize oxide surface chemistry is of limited value.

Acknowledgments

The author thanks the United States Department of Energy (through Battelle Pacific Northwest Laboratories Contract B-N8267-A-H) for partial support of this work and Gerhard Furrer for review of the manuscript.

SYMBOLS

a	Adjustable parameter
a	Concentration of solid (g m^{-3}) (as used in Table 1.6 A–C)
$a_i(\alpha)$	Activity of species i in phase α (referenced to mol m^{-3})
b	Adjustable parameter
c_i	Concentration of species i (mol m^{-3})
c_∞	Concentration of a monovalent electrolyte in the bulk of solution (mol m^{-3})
C	Specific capacitance (F m^{-3})
d	Separation between plates of parallel capacitor (m)
D	Concentration distribution ratio
e	Elementary charge (C)
f_{oc}	Fraction of organic carbon in sorbent
F	Faraday constant (C mol^{-1})
IHP	Inner Helmholtz plane
K	Equilibrium constant
K_{ow}	Octanol–water partition constant
K_p	Partition constant (cm^3 g^{-1})
n_i	Surface concentration of species i (m^{-2})
N_A	Avogadro's number (mol^{-1})
N_H	Surface concentration of exchangeable protons (m^{-2})
N_s	Surface site density (m^{-2})

OHP	Outer Helmholtz plane
pH_{zpc}	pH at which there is zero protonic charge on the surface
Q	Activity or concentration quotient (microscopic equilibrium constant)
R	Gas constant (J mol^{-1} K^{-1})
s	Specific surface area (m^2 g^{-1})
T	Absolute temperature (K)
T_H	Total (analytical) concentration of H^+ (strong acid minus strong base) (mol m^{-3})
z_i	Charge on species i (unitless)
β	Buffer intensity (m^3 mol^{-1})
ϵ	Dielectric constant (unitless)
ϵ_0	Permittivity of free space (F m^{-1})
η	Function of surface potential response (unitless)
$\mu_i^0(\alpha)$	Standard chemical potential of species i in phase α (J mol^{-1})
$\overline{\mu}_i(\alpha)$	Electrochemical potential of species i in phase α (J mol^{-1})
σ	Surface charge (C m^{-2})
ψ	Surface potential, relative to bulk of solution (V)
ψ_D	Donnan potential (V)

REFERENCES

Bard, A. J. (1966), *Chemical Equilibrium,* Harper & Row, New York.

Bolt, G. H., and Van Riemsdijk, W. H. (1987), "Surface Chemical Processes in Soil," Chapter 6 in this book.

Buffle, J. and Altmann, R. S. (1987), "Interpretation of Metal Complexation by Heterogeneous Complexants," Chapter 13 in this book.

Dick, S. G., Fuerstenau, D. W., and Healy, T. W. (1971), Adsorption of alkylbenzene sulfonate surfactants at the alumina–water interface, *J. Colloid Interface Sci.* **37,** 595–602.

Dzombak, D. (1986), "Toward a Uniform Model for the Sorption of Inorganic Ions on Hydrous Oxides," Ph.D. Thesis, Massachusetts Institute of Technology, Cambridge, MA.

Karickhoff, S. W., Brown, D. S., and Scott, T. A. (1979), Sorption of hydrophobic pollutants on natural sediments, *Water Res.* **13,** 241–248.

Karickhoff, S. W. (1984), Organic pollutant sorption in aquatic systems, *J. Hydraul. Eng.* **110,** 707–735.

Lyman, W. J. (1982), "Adsorption Coefficient for Soils and Sediments," in W. J. Lyman, W. F. Reehl, and D. H. Rosenblatt Eds., *Handbook of Chemical Property Estimation Methods,* McGraw-Hill, New York.

Marinsky, J. A. (1987), "A Two-Phase Model for the Interpretation of Proton and Metal Ion Interaction with Charged Polyelectrolyte Gels and Their Linear Analogs," Chapter 3 in this book.

Murrell, J. N., and Boucher, R. N. (1982), *Properties of Liquids and Solutions,* Wiley, Chichester, Chapter 2.

Parsons, R. (1987), "The Electric Double Layer at the Solid–Solution Interface," Chapter 2 in this book.

Schellenberg, K., Leuenberger, C., and Schwarzenbach, R. P. (1984), Sorption of chlorinated phenols by natural sediments and aquifer materials, *Environ. Sci. Technol.* **18,** 652–657.

Schindler, P. W., and Stumm, W. (1987), "The Surface Chemistry of Oxides, Hydroxides, and Oxide Minerals," Chapter 4 in this book.

Schwarzenbach, R. P., and Westall, J. (1981), Transport of nonpolar organic compounds from surface water to groundwater. Laboratory sorption studies, *Environ. Sci. Technol.* **15,** 1360–1367.

Schwarzenbach, R. P., Giger, W., Hoehn, E., and Schneider, J. K. (1983), Behavior of organic compounds during infiltration of river water to groundwater. Field studies, *Environ. Sci. Technol.* **17,** 472–479.

Sigg, L. (1987), "Surface Chemical Aspects of the Distribution and Fate of Metal Ions in Lakes," Chapter 12 in this book.

Snyder, L. R., and Kirkland, J. J. (1979), *Introduction to Modern Liquid Chromatography,* Wiley, New York.

Stumm, W., and Morgan, J. (1981), *Aquatic Chemistry,* Wiley, New York, Chapter 1.

Tanford, C. (1980), *The Hydrophobic Effect,* Wiley, New York.

Ulrich, H.-J., Cosovic, B., and Stumm, W. (1986), Comparison between two model surfaces: The mercury electrode and δ-Al_2O_3 colloids, *Environ. Sci. Technol.,* submitted.

Van Riemsdijk, W., Bolt, G. H., Koopal, L. K., and Blaakmeer, J. (1986), Electrolyte adsorption on heterogeneous surfaces, *J. Colloid Interface Sci.,* **109,** 219–228.

Watanabe, T., Fujishima, A., and Honda, K. (1974), Potential variation at the semiconductor–electrolyte interface through a change in pH of the solution, *Chem. Lett.* 897–900.

Westall, J. (1980), "Chemical Equilibrium Including Adsorption on Charged Surfaces," in M. C. Kavanaugh and J. O. Leckie, Eds., *Particulates in Water* (*Adv. Chem. Ser. No. 189*), American Chemical Society, Washington, D.C.

Westall, J., and Hindagolla, S. (1986), unpublished data.

Westall, J., and Hohl, H. (1980), A comparison of electrostatic models for the oxide/solution interface, *Adv. Colloid Interface Sci.* **12,** 265–294.

Westall, J., Leuenberger, C., and Schwarzenbach, R. P. (1985), Influence of pH and ionic strength on the aqueous–nonaqueous distribution of chlorinated phenols, *Environ. Sci. Technol.* **19,** 193–198.

Whitfield, M., and Turner, D. R. (1987), "The Role of Particles in Regulating the Composition of Seawater," Chapter 17 in this book.

2

THE ELECTRIC DOUBLE LAYER AT THE SOLID–SOLUTION INTERFACE

Roger Parsons

Department of Chemistry, University of Southampton, Hampshire, Great Britain

Abstract

The influence of the surface crystal structure on the structure of the electrical double layer at the metal–electrolyte interface is discussed, and its effect on the absorption of ions and molecules is outlined. The subject is introduced by indicating why the electronic work function is dependent on surface crystal structure and then how this quantity is related to the potential of zero charge (PZC). From this discussion it follows that a polycrystalline metal at a given potential will have a local charge that varies over the surface, and consequently its adsorption behavior will be a complex average of the behavior of contributions from elements of the surface having different exposed crystal structures. Since variations in adsorption behavior with surface crystal structure are often quite wide, fundamental analysis starting with a study of polycrystalline electrodes is virtually impossible. The study of well-defined surfaces of single crystals is essential if real progress in understanding is to be achieved.

1. INTRODUCTION

Although this volume is concerned principally with the interface between nonmetallic solids and aqueous solutions, the fundamental study of such interfaces has not yet proceeded far beyond the use of a model assuming uniform behavior of the solid surface. This is, in effect, the adaptation of the classical double-layer model for mercury surfaces with allowance for particular types of specific adsorption, that is, acid–base or complexation reactions. Such an approach tends to neglect one of the basic properties of a solid surface—its geometrical structure—and the way it may influence behavior of the surface in contact with an electrolyte. Some progress has now been made in this direction using metallic surfaces; this will be outlined in the hope that such information may help in the consideration of surfaces more relevant to aquatic surface chemistry.

2. ELECTRONIC WORK FUNCTION AND POTENTIAL OF ZERO CHARGE

It is useful to begin with the simplest situation in which the geometrical structure of the metal surface influences its properties in a way that may be considered physical rather than chemical. This may be approached by considering the emission of electrons from a metal. This occurs when the energy of the electrons in the highest occupied level in the metal increases by an amount Φ, known as the *electronic work function*. The simplest model of a metal, the free electron model, leads to an energy diagram of the form shown in Figure 2.1, where electron levels inside the energy well, representing the potential energy inside the metal, are filled up to a level called the *Fermi level*, which lies Φ below the vacuum level. (At finite temperatures there is a somewhat smeared-out change from occupied to empty levels around the Fermi level, but this does not affect the discussion here.) Electrons at the Fermi level can gain excess energy Φ by absorption of photons or of thermal energy and escape—by photoemission or thermal emission, respectively. The energy required to escape may also be modified by charging the metal electrostatically; electrons escape more readily from a negatively charged metal and less readily from a positively charged metal.

The energy at the Fermi level determines the equilibrium distribution of electrons; it is, in fact, equal to the *electrochemical potential of electrons* $\tilde{\mu}_e$, so that if metals M and N are brought into contact, the condition for equilibrium of the mobile electrons is

$$\tilde{\mu}_e^{\mathrm{M}} = \tilde{\mu}_e^{\mathrm{N}} \tag{2.1}$$

Figure 2.1. Schematic energy diagram for a free electron metal.

where the superscripts indicate the metal. Electrochemical potentials are most conveniently referred to the state in vacuum so that $\tilde{\mu}_e$ will depend on the charged state of the metal. This is expressed as (see, e.g., Parsons, 1954)

$$\tilde{\mu}_e = \alpha_e - e\psi \tag{2.2}$$

where α_e (the *real potential*) is the value of $\tilde{\mu}_e$ when the metal is uncharged, and ψ (the *outer potential*) is the electrostatic potential of the metal due to its charge. It follows that

$$\alpha_e = -\Phi \tag{2.3}$$

because Φ also refers to an uncharged metal.

When two pieces of metal α and β are brought into contact (Fig. 2.2), electrons flow from the metal with the lower value of Φ (here chosen as M) to that with the higher value of Φ until the negative charge built up on the latter and the positive charge on the former bring the two Fermi levels to the same position. From Eqs. 2.1 and 2.2,

$$\alpha^{M} - e\psi^{M} = \alpha^{N} - e\psi^{N} \tag{2.4}$$

and the resulting potential difference between the two metals is

$$\begin{aligned} \Delta\psi &= \psi^{M} - \psi^{N} = (\alpha_e^{N} - \alpha_e^{M})L/e \\ &= (\Phi^{M} - \Phi^{N})L/e \end{aligned} \tag{2.5}$$

This is known as the *contact potential* or Volta potential difference.

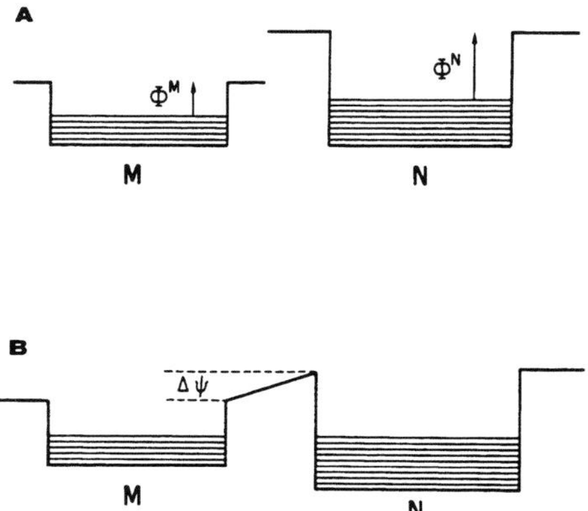

Figure 2.2. Equilibrium condition between two free electron metals. (*A*) Before contact; (*B*) after contact.

The model depicted in these diagrams is much oversimplified. To proceed further, it must be considered in more detail in one respect. The energy α_e or $-\Phi$ can be regarded as arising from two causes. One is the interaction of the electron with the *bulk* of the metal and is called the *chemical potential* μ_e. The other results from the nonuniform distribution of charge at the metal–vacuum surface (even in the absence of net charge); this is the energy $-e\chi$ arising from the *surface potential difference* χ. Thus

$$\alpha_e^e = \mu_e^e - e\chi \tag{2.6}$$

The surface potential χ of a metal arises because the mobility of electrons is much higher than that of the positive ion cores making up the metal lattice. Thus the electrons tend to "overshoot" the lattice, and a surface double layer is formed that is equivalent to a sheet of negative charge outside and a sheet of positive charge inside the nominal surface (Fig. 2.3). It is this surface potential that accounts for the different values of the work function for the different *crystal faces* of a given metal, as shown, for example, in the field electron microscope picture of a clean metal surface. The different geometric structure of the surface leads to different amounts of electron overshoot. In general, the more closely packed the surface, the higher the work function.

The surface potential is also strongly affected by foreign atoms or molecules adsorbed on the metal surface. Hence work function measurements are widely used to study adsorption. From this, it might be expected that putting a metal into an electrolyte would substantially alter the surface potential. Although this quantity is not directly measurable, indications are that there is no great change. This can be seen from the following analysis. The cell shown in Figure 2.4 consists of two metal electrodes dipping into a dilute electrolyte solution and arranged so that there is no separation of free charge across each metal–electrolyte junction; that is, each electrode is at its potential of zero charge. A simple thermodynamic analysis shows that the potential difference across this cell is given by

$$e(E_{\sigma=0}^{\mathrm{N}} - E_{\sigma=0}^{\mathrm{M}}) = -(\mu_e^{\mathrm{N}} - eg_{\mathrm{dipol}}^{\mathrm{N/S}}) + (\mu_e^{\mathrm{M}} - eg_{\mathrm{dipol}}^{\mathrm{M/S}}) \tag{2.7}$$

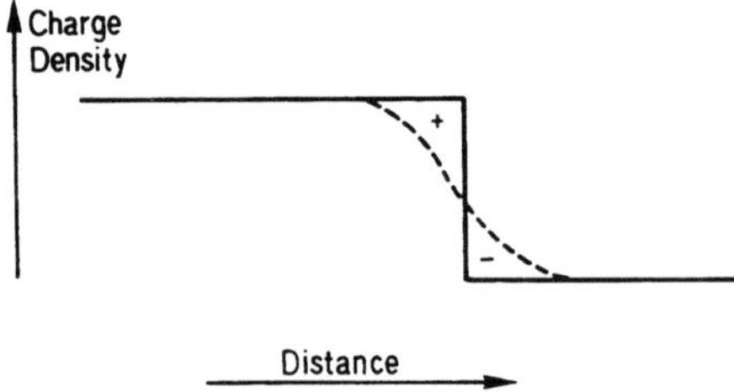

Figure 2.3. Distribution of charge at a metal surface. Solid line: Profile of charge due to positive ions in the metal lattice. Dashed line: Profile of charge due to free electrons.

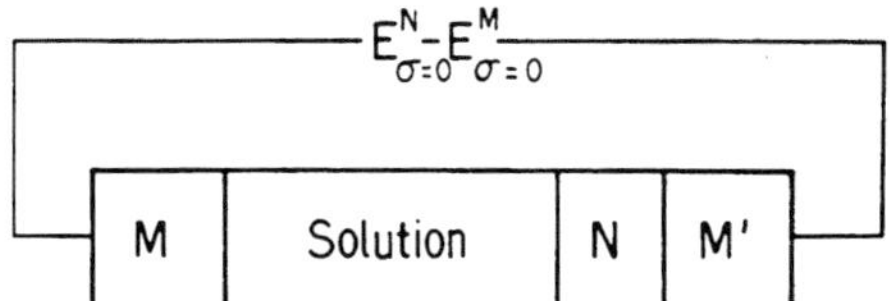

Figure 2.4. Schematic diagram of electrochemical cell with two metal electrodes each at its potential of zero charge.

where $g^{M/S}_{dipol}$ is the *total dipolar potential difference* across the metal–electrolyte junction. If this dipolar potential difference were identical to the metal dipolar potential χ^M, then Eq. 2.7 would become simply

$$e(E^N_{\sigma=0} = E^M_{\sigma=0}) = \Phi^N - \Phi^M \tag{2.8}$$

and there would be a direct proportionality between the *potential of zero charge* (PZC) and the electronic work function. Such a relation was first proposed by Frumkin and Gorodetzkaya (1928) and analyzed thermodynamically by Parsons (1954). A detailed assessment by Trasatti (1971) showed some evidence in favor of Eq. 2.8. This observation that $\chi^M = g^{M/S}_{dipol}$ means *either* that the electron overshoot potential is unaffected by the presence of solvent and that there is no appreciable orientation of solvent dipoles perpendicular to the interface *or* that any change in the electron overshoot potential is compensated by solvent orientation. The former seems more probable in view of evidence from models that dipoles in a monolayer near an interface tend to orient parallel to an imaging plane (Parsons and Reeves, 1981) and also that adsorbed zwitterionic molecules are found experimentally to cause negligible interfacial potential differences (Baugh and Parsons, 1973). Supporting evidence for this view comes from the observation that water is very weakly adsorbed (physisorbed) on a number of clean metal surfaces (see, for example, Sass et al., 1981).

3. SINGLE-CRYSTAL ELECTRODES

The correlation between electronic work function and PZC of Eq. 2.8 combined with the fact that the work function depends on the geometrical structure of the metal surface leads immediately to the conclusion that the PZC also depends on the structure of the metal surface (Parsons, 1964). This is now amply confirmed by experimental evidence (see the review by Hamelin et al., 1983).

The PZC of a solid metal may be obtained experimentally from measurements of the electrical capacity of the metal–electrolyte interface, for example, using an ac bridge. The differential capacitance, $C = \partial\sigma/\partial E$, where σ is the *free charge per unit area* on the double layer and E is the *applied potential*

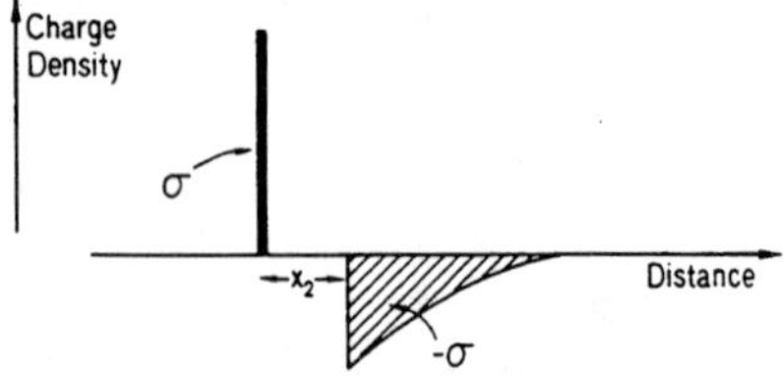

Figure 2.5. Schematic diagram of the electrical double layer at a metal–electrolyte interface.

difference, is strongly dependent on E. It is usually interpreted in terms of a model that is shown in its simplest form in Figure 2.5. The charge σ on the metal due to an excess or deficiency of electrons is balanced by an equal and opposite charge $-\sigma$ in the solution, made up of the ions present in the solution which can approach the metal only up to a distance of closest approach when their centers are x_2 from the metal surface. Analysis of this model leads to an equivalent circuit consisting of two capacitors in series: C^i is independent of the solution composition but may be quite strongly dependent on σ; and C^d, according to the analysis of Gouy (1910) and Chapman (1913), is given by

$$C^d = \frac{\epsilon}{L_D}\left[1 + \left(\frac{\sigma e L_D}{2\epsilon kT}\right)^2\right]^{1/2} \tag{2.9}$$

where L_D is the Debye length

$$L_D = \left(\frac{\epsilon kT}{2e^2 n}\right)^{1/2} \tag{2.10}$$

ϵ is the permittivity of the solvent, k is Boltzmann's constant, T is the thermodynamic temperature, e is the electronic charge, n is the number of cations or anions in unit volume of a uniunivalent electrolyte. In dilute solutions where L_D is large and so ϵ/L_D is small, the minimum value of C^d can be smaller than C^i. This minimum then appears on the curve and can be used to locate the value of E where $\sigma = 0$, that is, the PZC.

In Figure 2.6, capacity–potential curves are shown for the three low-index planes of silver in contact with a dilute aqueous solution of NaF. These three curves each show a pronounced minimum that can be closely identified with the condition $\sigma = 0$ (Valette and Hamelin, 1973). It is evident that, as would be expected from Eq. 2.8, the closest packed surface (111) (see Fig. 2.7) has the highest PZC, while the least close-packed (110) has the lowest PZC, and that these differ by approximately 300 mV.

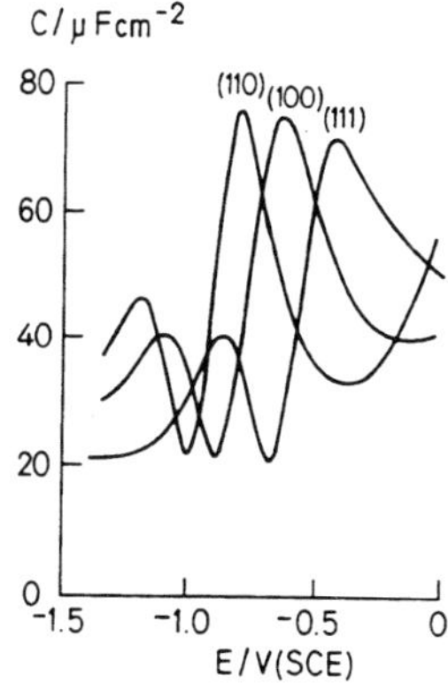

Figure 2.6. Capacity–potential curves for three low-index planes of silver in contact with aqueous 0.01 *M* NaF at 25°C. (Redrawn from Valette and Hamelin, 1973.)

4. THE PROBLEM OF POLYCRYSTALLINE ELECTRODES

The surface of a polycrystalline electrode may be regarded as made up of patches of surfaces of different orientations. Valette and Hamelin (1973) suggested that a reasonable approximation was to consider only the three low-index planes. It is evident from Figure 2.6 that such a surface at a uniform potential will consist of patches carrying different charges, even charges of different sign. For example, at −0.8 V (SCE*) the (111) patches will be negatively charged while the (100) and (110) patches will be positively charged. Thus the structure of the diffuse layer in the solution presents a difficulty. Valette and Hamelin suggested that the diffuse layer opposite each patch should be treated independently and neglected any problems at the boundaries of the patches. Thus they took the equivalent circuit of a polycrystalline electrode to be that shown in Figure 2.8*A* and estimated the proportions of the three low-index planes from a metallographic survey of an etched polycrystal. This analysis gave moderate agreement with experiment. A very similar analysis was given by Grigoriev (1976), but Bagotskaya et al. (1980) suggested that if the patches were small the diffuse layer charge could be smeared out and the equivalent circuit would then become that shown in Figure 2.8*B*. Vorotyntsev (1981) has discussed these two models and given the conditions for their applicability. Essentially the conditions are that Figure 2.8*A* is usually valid when the characteristic size of the patches (e.g., their diameter) is larger than the Debye length L_D, whereas Figure 2.8*B* is valid if it is smaller than L_D.

While the synthesis of results for a polycrystal seems feasible using these

*SCE = Saturated calomel electrode.

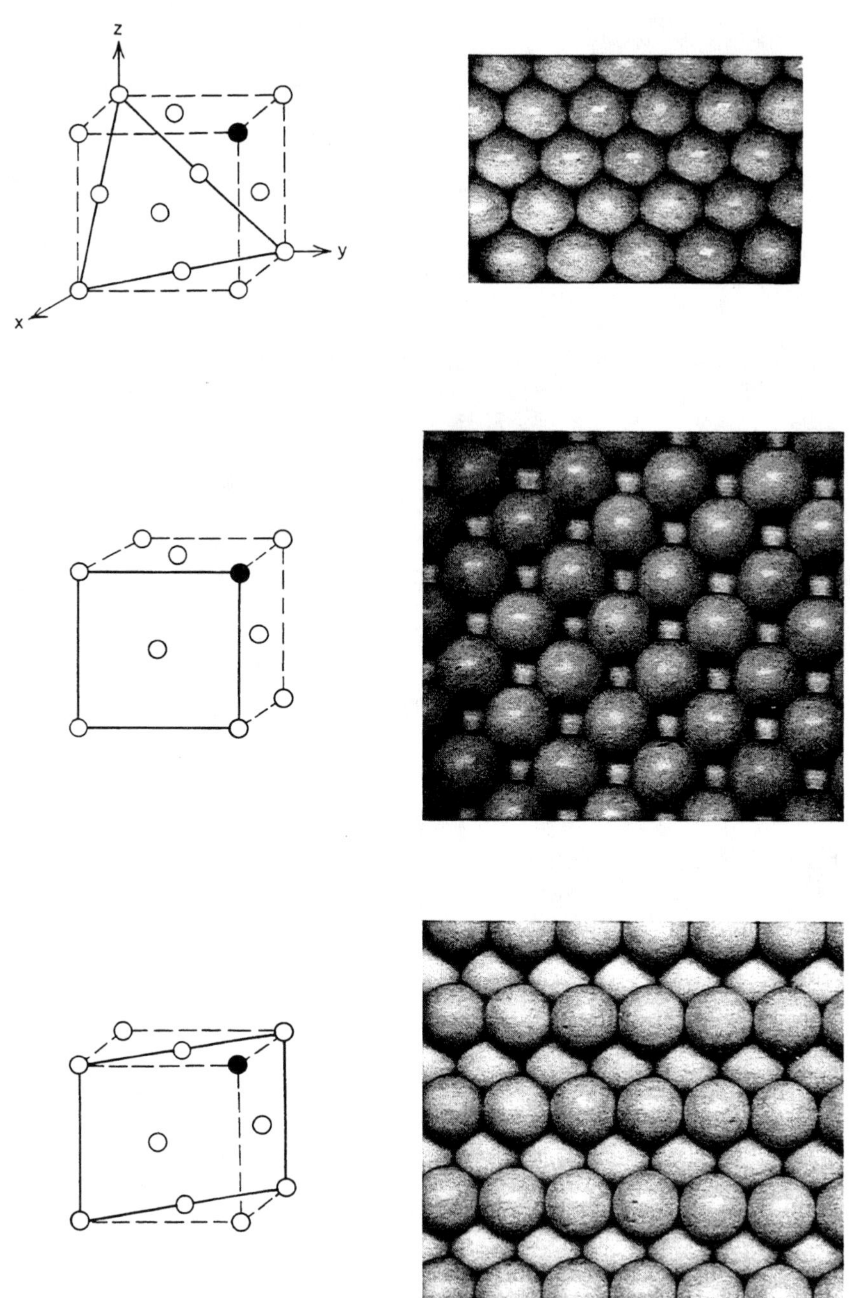

Figure 2.7. Arrangement of atoms in the surface plane of three low-index planes of a face-centered cubic structure.

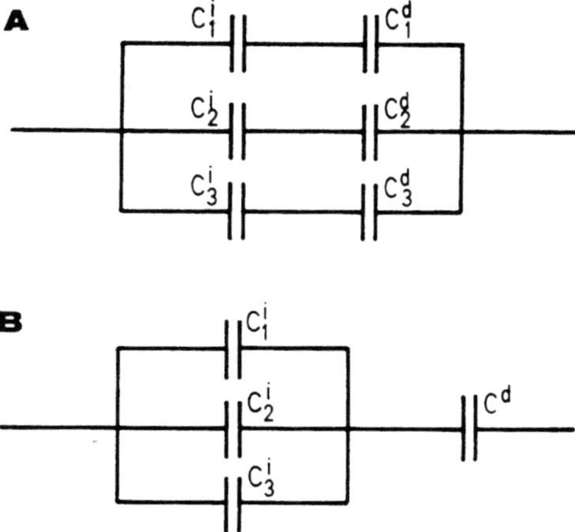

Figure 2.8. (*A*) Equivalent circuit for a double layer at a polycrystalline electrode having an independent patch structure. (*B*) Equivalent circuit for a double layer at a polycrystalline electrode having a patch structure but a common diffuse layer.

models, the process of attempting to understand the behavior of a polycrystalline electrode directly seems likely to be difficult, if not impossible, even in the simplest case when there is no specific adsorption.

5. SPECIFIC IONIC ADSORPTION

In the systems described above, the ions from the solution are present in the diffuse layer charge as a result of the electrostatic attraction of the charge on the metal electrode. This effect is nonspecific, depending only on the charge of the ion and not on its nature. Deviations from this purely electrostatic behavior are found in many systems and are attributed to specific interactions between the ions and the metal. These specific forces are usually of much shorter range than the electrostatic forces, and so it is often assumed that the specific effects are confined to the monolayer of species next to the metal surface. Consequently it might be expected that adsorption of this type is sensitive to the structure of the metal surface, and indeed this is found to be so. An example is shown in Figure 2.9, where the capacity curves for the low-index surfaces in contact with aqueous solutions containing chloride ion are shown. It is immediately evident that there are quite large differences in behavior. These arise from differences in the strengths of adsorption of Cl^- on the different Ag faces, in their potential dependence, and in their PZCs. The interplay of these factors is quite complex and can be unraveled only for single crystal faces. Such an analysis was attempted for the (110) face of silver by Valette et al. (1978).

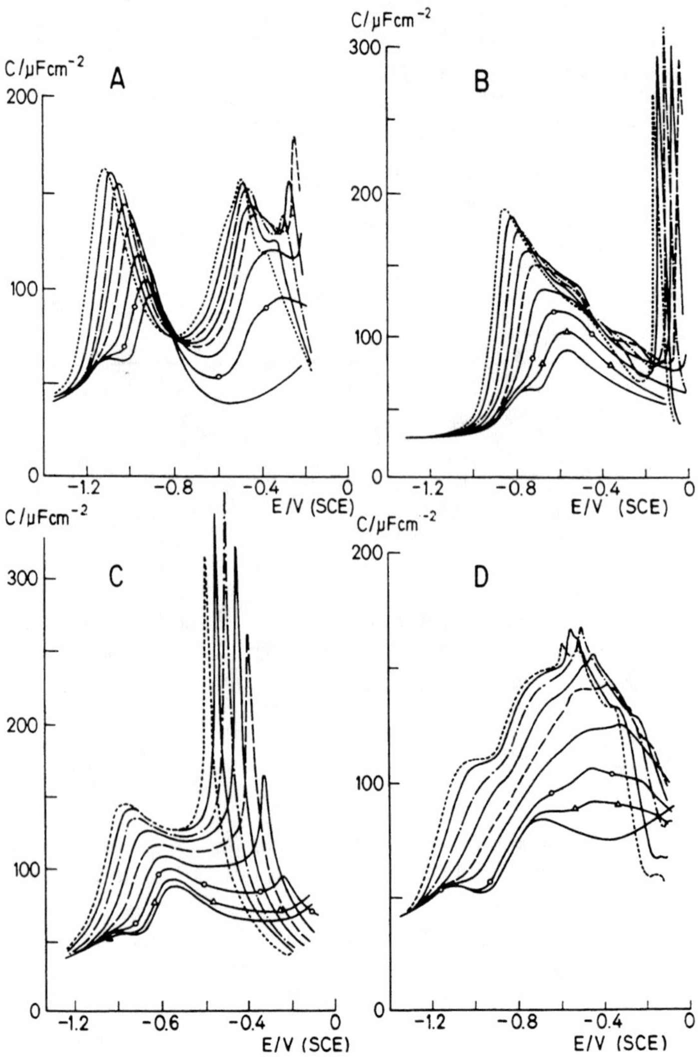

Figure 2.9. Capacity–potential curves for a silver electrode in contact with aqueous solutions of x M NaCl + 0.1 M NaF at 25°C. (A) (110) plane. (B) (111) plane. (C) (100) plane. (D) Polycrystalline electrode. (From Fleischmann et al., 1981.)

6. ADSORPTION OF NEUTRAL MOLECULES

Similar problems arise with the adsorption of neutral molecules, even with a substance as chemically inert as diethyl ether. The capacity curves of the low-index planes of a gold electrode in contact with aqueous NaF containing ether are shown in Figure 2.10. It is evident that each plane exhibits quite distinctive

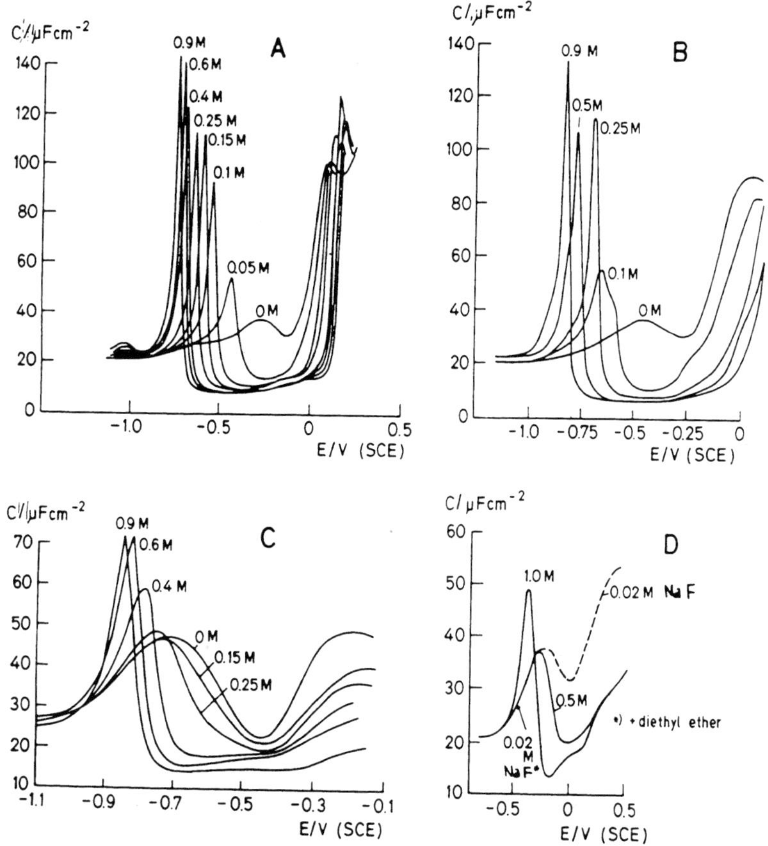

Figure 2.10. Capacity–potential curves for a gold electrode in aqueous 10 mM NaF at 25°C. (A) (111) plane. (B) (100) plane. (C) (110) plane. (D) Polycrystalline electrode. (From Lipkowski et al., 1983, and Nguyen Van Hong, 1981.)

features that do not appear in the corresponding curves for the polycrystalline electrode also shown in the figure. Again any attempt to understand these curves directly seems doomed to failure even when, as in this example, the system is simplified by choosing an electrolyte that is not specifically adsorbed.

The complications become much greater when the electrolyte is specifically adsorbed and a strongly adsorbed neutral species is also present as in the system studied by Fleischmann et al. (1981). Here the adsorption of pyridine from chloride solution on silver single crystals can be seen to have features characteristic of each crystal face, but calculation of the adsorption on the various faces would require a large number of experiments. On the polycrystalline electrode, most of these features vanish in the complex interplay of contributions from the different parts of the surface.

7. CHEMISORPTION AND SURFACE MOBILITY

The demarcation between physical and chemical adsorption is difficult to define, and systems like the pyridine adsorption mentioned above probably involve chemical bond formation with the metal electrode. The adsorption of hydrogen and oxygen on noble metals like platinum, however, seems to be clearly chemical in nature. The remarks made above about the need for single-crystal data in order to understand the adsorption features are equally forcefully illustrated by recent results of Motoo and Furuya (1984) for iridium and the now well-known results for platinum. The latter provide also a clear demonstration of the way surface structure may be changed rather easily by the process of adsorption and/or the change of electrode potential.

A platinum (111) surface annealed at temperatures above 800°C yields a

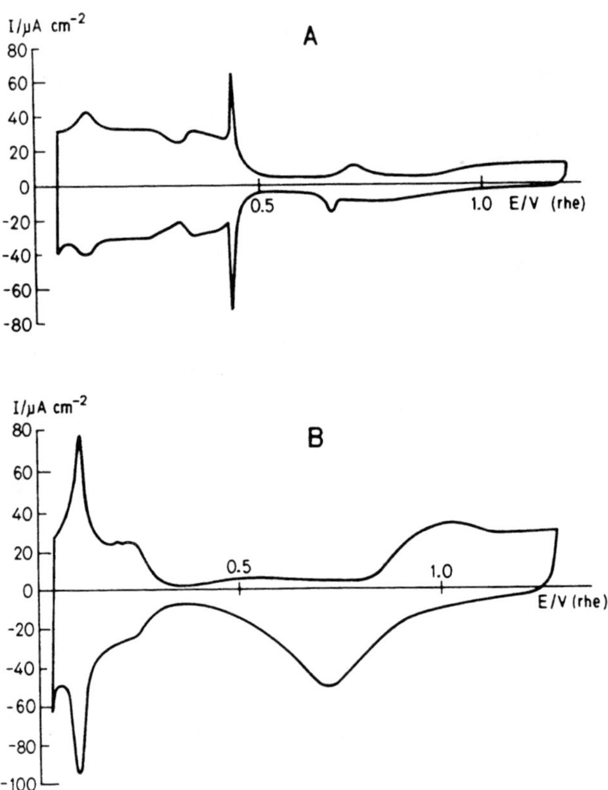

Figure 2.11. Linear sweep voltammograms of a platinum electrode with the (111) face exposed to aqueous 0.5 *M* H_2SO_4 at 25°C. (*A*) First cycle after annealing at ~900°C quenching, and contacting with the solution. (*B*) First cycle after the electrode in (*A*) was subjected to a more positive potential during which a charge equivalent to a monolayer of oxygen was passed and then recovered. (From Clavilier et al., 1980).

linear sweep voltammogram* of the form shown in Figure 2.11*A* when it is brought immediately into contact with an aqueous solution of sulfuric acid. A single excursion of the potential into the region of oxygen adsorption, such that a charge equivalent to the adsorption of a monolayer of oxygen is passed and then recovered on the negative sweep, results in a striking modification of the form of the voltammogram (Clavilier et al., 1980) as illustrated in Figure 2.11*B*. The charge transferred at potentials more negative than 0.4 V (RHE†) is usually attributed to the adsorption or desorption of atomic hydrogen according to the reaction

$$H^{+}_{(soln)} + e^{-} \rightleftharpoons H_{ads} \qquad (2.11)$$

The change in the profile in this region consists essentially in the replacement of strongly bound hydrogen (at potentials between 0.3 and 0.5 V (RHE) (in Fig. 2.11*A*) by an equivalent amount of weakly bound hydrogen [the peak centered at 0.13 V (RHE) in Fig. 2.11*B*]. Low-energy electron diffraction (LEED) experiments by Wagner and Ross (1983) have indicated that the electrode surface annealed at high temperature is characterized by long-range order that is completely destroyed by the electrochemical treatment described above. The resulting surface, though still of course nominally (111), has many steps, and when it is compared with the (110) surface, it appears that the sharp peak at 0.13 V (RHE) in Figure 2.11*B* is due to hydrogen adsorption on the (110) sites on these steps.

These and other recent experiments suggest that even quite high-melting metal electrodes are surprisingly mobile under electrochemical conditions. It also appears that this mobility is often dependent on the potential and the adsorption of not only species like hydrogen and oxygen, but also of anions from the solution. Thus the geometric structure of the electrode surface may depend not only on the orientation of the crystal but also on the treatment that surface has received.

8. CONCLUSIONS

The complications outlined here for solid metal electrodes seem likely to have their parallel at other solid surfaces. For example, it is to be expected that a crystalline oxide will have different PZC values at its various types of exposed faces. To add this degree of complexity to the other difficulties of dealing

*The current flowing through the electrode is recorded as a function of the potential when the latter is changed linearly with time. Usually positive and negative sweeps are used; if used repetitively, it becomes a cyclic voltammogram. The current density $j = d\sigma/dt$ is the rate of charge flow, or $j = (d\sigma/\partial E)(\partial E/dt) = Cv$, where v is the sweep rate dE/dt. Hence the capacity is j/v. Note, however, that when chemisorption occurs, the charge σ is stored at the interface, not just as free charge but in the form of neutralized ions.

†Reversible hydrogen electrode.

with practical solid–aqueous solution interfaces might lead to immediate despair. However, two remarks may qualify this to some extent. First, the progress in understanding surface science has been accelerating, particularly with the new techniques devised in the last two decades. Second, if the surface of a solid has a very high mobility, many of the differences in behavior of single crystal faces may be smeared out, and it may be possible to use a simpler treatment more like that used for mercury electrodes.

Acknowledgments

This chapter could not have been written without the experimental work of my friends and former colleagues at the Laboratoire d'Electrochimie Interfaciale du C.N.R.S. at Meudon. I am also grateful for many discussions with them.

REFERENCES

Bagotskaya, I. A., Damaskin, B. B., and Levi, M. D. (1980), The influence of crystallographic inhomogeneity of a polycrystalline electrode surface on the behavior of the electric double layer, *J. Electroanal. Chem.* **115,** 189–209.

Baugh, L. M., and Parsons, R. (1973), The adsorption of amino acids at the mercury–water interphase. Part 1: Glycine in acid neutral and base solutions, *J. Electroanal. Chem.* **41,** 311–328.

Chapman, D. L. (1913), A contribution to the theory of electrocapillarity, *Phil. Mag.* **25** [6], 475–481.

Clavilier, J., Faure, R., Guinet, G., and Durand, R. (1980), Preparation of monocrystalline Pt microelectrodes and electrochemical study of the plane surfaces cut in the direction of the {111} and {110} planes, *J. Electroanal. Chem.* **107,** 205–216.

Fleischmann, M., Robinson, J., and Waser, R. (1981), An electrochemical study of the adsorption of pyridine and chloride ions on smooth and roughened silver surfaces, *J. Electroanal. Chem.* **117,** 257–266.

Frumkin, A. N., and Gorodetskaya, A. V. (1928), Kapillarelektrische Erscheinung an Amalgamen, I Thalliumamalgame, *Z. physik. Chem.* **136,** 451–472.

Gouy, M. (1910), Sur la constitution de la charge electrigue à la surface d'un électrolyte, *J. Phys.* **9,** 457–468.

Grigoriev, N. B. (1976), Relation between the characteristics of the electrical double layer of a polycrystalline electrode and separate faces of a single crystal in electrolyte solutions, *Dokl. Akad. Nauk SSR* **229,** 647–650.

Hamelin, A., Vitanov, T., Sevastyanov, E., and Popov, A. (1983), The electrochemical double layer on *sp* metal single crystals: The current status of data, *J. Electroanal. Chem.* **145,** 225–264.

Lipkowski, J., Nguyen Van Huong, C., Hinnen, C., Parsons, R., and Chevalet, J. (1983), Adsorption of diethyl ether on single crystal gold electrodes. Calculation of adsorption parameters, *J. Electroanal. Chem.* **143,** 375–396.

Motoo, S., and Furuya, N. (1984), Hydrogen and oxygen adsorption on Ir (111), (100), and (110) planes, *J. Electroanal. Chem.* **167,** 308–315.

Nguyen Van Huong, C., Hinnen, C., Dalbera, J. P., and Parsons, R. (1981), Adsorption of ethyl ether on polycrystal and the (110) single crystal face of gold by admittance measurements and modulated reflection spectroscopy, *J. Electroanal. Chem.* **125,** 177–192.

Parsons, R. (1954), "Equilibrium Properties of Electrified Interphases," in J. O'M Bockris and B. E. Conway, Eds., *Modern Aspects of Electrochemistry,* Butterworths, London, pp. 103–178.

Parsons, R. (1964), The kinetics of electrode reactions and the electrode material, *Surface Sci.* **2,** 418–435.

Parsons, R., and Reeves, R. (1981), Molecular models for the structure of solvent in an interphase, *J. Electroanal. Chem.* **123,** 141–149.

Sass, J. K., Kretzschmar, K., and Holloway, S. (1981), Water adsorption on metal surfaces: An electrochemical viewpoint, *Vacuum* **31,** 483–486.

Trasatti, S. (1971), Work function, electronegativity and electrochemical behavior of metals. II. Potentials of zero charge and "electrochemical" work functions, *J. Electroanal. Chem.* **33,** 351–378.

Valette, G., and Hamelin, A. (1973), Structure et propriétés de la couche double electrochimique á l'interphase argent/solution aqueuse de fluorure de sodium, *J. Electroanal. Chem.* **45,** 301–319.

Valette, G., Hamelin, A., and Parsons, R. (1978), Specific adsorption on silver single crystals in aqueous solutions, *Z. physik. Chem.* **113,** 71–89.

Vorotyntsev, M. A. (1981), A connection between the capacitance characteristics of polycrystalline and single crystalline electrodes; an analysis of the applicability of equivalent circuits, *J. Electroanal. Chem.* **123,** 379–387.

Wagner, F. T., and Ross, P. N., Jr. (1983), LEED analysis of electrode surfaces, structural effects of potentiodynamic cycling on Pt single crystals, *J. Electroanal. Chem.* **150,** 141–164.

3

A TWO-PHASE MODEL FOR THE INTERPRETATION OF PROTON AND METAL ION INTERACTION WITH CHARGED POLYELECTROLYTE GELS AND THEIR LINEAR ANALOGS

Jacob A. Marinsky

Department of Chemistry, State University of New York at Buffalo, Buffalo, New York

Abstract

The physicochemical properties of weakly acidic (basic) polymeric gels and their linear polyelectrolyte analogs, when dispersed in aqueous media, have been shown to be characterized by two kinds of behavior. In the first kind the polymeric macromolecule is permeable to simple neutral salt (hydrophilic), and a Donnan potential term has to be taken into account to understand fully the experimental results. In the second kind the macromolecule is impermeable to salt (hydrophobic), and the co-ion of the salt is not involved in the potential term that has to be taken into account to facilitate interpretation of these experimental results. The same two-phase model is seen to apply to the gel and its linear polyelectrolyte analog equally well in both categories (hydrophilic and hydrophobic) in all the examples presented.

1. INTRODUCTION

High sensitivity of the physicochemical properties of charged polymeric molecules to ionic strength is always encountered, no matter whether the polyelectrolyte molecule is suspended in the aqueous medium as a physically discernible separate phase (gel) or is completely and uniformly dispersed in solution (linear polyelectrolyte) (Marinsky, 1985). Examples of this phenomenon are presented in Figures 3.1 and 3.2, where the potentiometric properties of the weakly acidic polyelectrolyte gel (Sephadex CM-50-125) and the linear polyelectrolyte, polyacrylic acid, are examined at several different ionic strength levels (Marinsky and Slota, 1980; Nagasawa, 1965). In both figures the apparent pK $\{pK^{app}_{(HA)_v} = pH - \log[\alpha/(1 - \alpha)]\}$ plotted versus α, the degree of dissociation, is displaced upward approximately one pK unit for each tenfold decrease in ionic strength. This behavior is a direct consequence of the polyelectrolyte molecule's existence as a separate phase even when phase separation is not physically discernible.

Such control of the measurable averaged properties of the charged macromolecule by the separate phase it defines is analyzed in Section 2. In the course of the presentation, each step in the development of the unified model

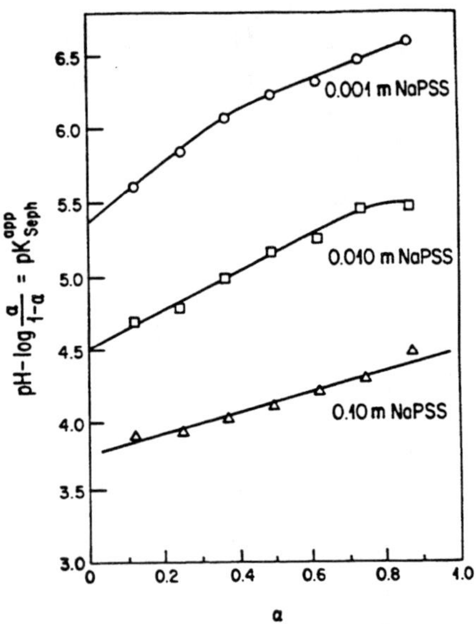

Figure 3.1. Potentiometric properties of the flexible Sephadex CM50-120 gel measured at three different sodium polystyrenesulfonate (NaPSS) concentration levels (Donnan potential term neglected).

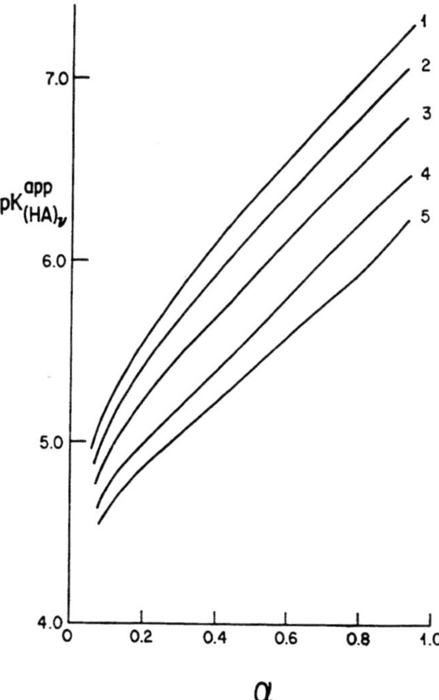

Figure 3.2. Potentiometric titration curves of poly(acrylic acid) at 15 ± 2°C, from Figure 4 of Nagasawa et al., 1965. (NaCl concentrations: 1, 0.00500 *N*; 2, 0.00100 *N*; 3, 0.0200 *N*; 4, 0.0500, *N*; 5, 0.0100 *N*.)

that evolves is fully documented experimentally. In Section 3, examples of the useful application of the unified model to the interpretation of proton and metal ion interactions with gels and linear polyelectrolytes are presented with emphasis on the protolysis properties of weakly acidic polymers.

2. DEVELOPMENT OF THE UNIFIED MODEL

The step-by-step development of the unified model presented here first examines a weakly acidic gel–simple salt system. Two scenarios are considered. In the first, the charged matrix of the gel is presumed to define a membrane that is permeable to the diffusible components of the system: simple salt MX; simple acid HX; and water. In the second, the charged matrix is assigned hydrophobic properties and resists diffusion of the components.

2.1. The Permeable Gel; A Donnan-Based Interpretation of the Sensitivity of the Potentiometric Properties of Weakly Acidic Hydrophilic Gels to the Presence of Neutral Salt*

At equilibrium during each step of the potentiometric titration of a weakly acidic (or weakly basic) gel $(HA)_v$, in the presence of a simple electrolyte MX, the chemical potential μ, of each diffusible component is the same in both phases. That is,

$$\mu_{HX} = \bar{\mu}_{HX}, \qquad \mu_{MX} = \bar{\mu}_{MX}, \qquad \text{and} \qquad \mu_{H_2O} = \bar{\mu}_{H_2O} \tag{3.1}$$

where the bar over the μ identifies the gel phase.

In the gel phase,

$$\bar{\mu}_{HX} = \mu^{\circ}_{HX} + RT \ln \bar{a}_{HX} + \pi V_{HX} \tag{3.2a}$$

$$\bar{\mu}_{MX} = \mu^{\circ}_{MX} + RT \ln \bar{a}_{MX} + \pi V_{MX} \tag{3.2b}$$

where μ° represents the chemical potential of each component in the standard state, π is the osmotic pressure of the water in the gel phase, and V is the partial molar volume of the diffusible components. The osmotic pressure is related to the activity of water, a_w, in the two phases by

$$\pi = \frac{RT}{V_w} \ln \left(\frac{\bar{a}_w}{a_w}\right) \tag{3.3}$$

The distribution, at equilibrium, of HX and MX during each step of the potentiometric titration is defined by the reaction

$$\overline{HX} + MX \rightleftharpoons \overline{MX} + HX \tag{3.4}$$

where the bar over HX or MX identifies the gel phase, as used in Eq. 3.1 already. Recalling that the activity of each component in the solution phase is defined by

$$\mu = \mu^{\circ} + RT \ln a \tag{3.5}$$

and that by choosing the standard state to be the same in both phases the sum of all μ° terms is zero, we obtain

$$RT \ln \left(\frac{a_{HX}\bar{a}_{MX}}{\bar{a}_{HX}a_{MX}}\right) = \frac{\pi}{RT}(V_{HX} - V_{MX}) \tag{3.6}$$

*This section calls upon the following works: Alegret el at., 1984; Marinsky, 1985; Marinsky, Gupta, and Schindler, 1982; Marinsky, Lim, and Chung, 1983; Marinsky and Merle, 1984; Marinsky and Slota, 1980; Marinsky, Wolf, and Bunzl, 1980.

as the expression of this equilibrium. The activity of the common co-ion, a_X, cancels, and Eq. 3.6 can be simplified to

$$\ln\left(\frac{a_H \bar{a}_M}{\bar{a}_H a_M}\right) = \frac{\pi}{RT}(V_H - V_M) \tag{3.7a}$$

or

$$pM - pH = p\overline{M} - p\overline{H} + \frac{\pi}{RT}(0.4343)(V_H - V_M) \tag{3.7b}$$

Even in quite rigid gel systems, π does not exceed a value of 200 atm and the $(\pi/RT)(0.4343)(V_H - V_M)$ term is small enough to neglect. For example, the difference between the partial molar volume of Na^+ and H^+ is only 0.0012 L (Mukerjee, 1961) so that when π = 200 atm the value of the term is less than -0.005; Eq. 3.7b may be simplified without penalty to

$$pM - pH = p\overline{M} - p\overline{H} \tag{3.7c}$$

At equilibrium the $p\overline{H}$ (of the gel phase) is then given by

$$p\overline{H} = p\overline{K}^{int}_{(HA)_\nu} + \log\frac{\alpha}{1-\alpha} + \log \bar{y}_{A^-} \tag{3.8}$$

where

$$p\overline{H} = -\log \overline{C}_H - \log \bar{y}_H \tag{3.9}$$

$\overline{K}^{int}_{(HA)_\nu}$ is the intrinsic dissociation constant of the repeating acidic group in the macromolecule, α is the degree of dissociation (neutralization), $\overline{C}_H$ denotes the concentration of H^+ in the gel, and $\bar{y}_{A^-}$ and $\bar{y}_H$ are activity coefficients of the designated species in the gel phase. Correction for deviation from ideality of the associated species HA is considered negligible, and activity coefficients are assigned only to the charged species H^+ and A^-. With this representation of $p\overline{H}$, Eq. 3.7c becomes

$$pM - pH = p\overline{M} - p\overline{K}^{int}_{(HA)_\nu} - \log\frac{\alpha}{1-\alpha} - \log \bar{y}_{A^-} \tag{3.10}$$

By definition,

$$p\overline{M} = -\log \overline{C}_M - \log \bar{y}_M \tag{3.11}$$

$$pM - pH = -\log \overline{C}_M - \log \bar{y}_M - p\overline{K}^{int}_{(HA)_\nu} - \log\frac{\alpha}{1-\alpha} - \log \bar{y}_{A^-} \tag{3.12}$$

For each molecular unit (HA) that dissociates, one molecule of positive ions (M^+ and H^+) must enter the gel phase to preserve electroneutrality. Since $\overline{C}_M > \overline{C}_H$,

$$\overline{C}_M \approx \frac{\alpha\nu}{V_p} \tag{3.13}$$

where ν denotes the number of ionizable groups in $(HA)_\nu$, and V_p is the effective volume of the gel domain. Substituting this definition of $\overline{C}_M$ into Eq. 3.12 yields

$$\text{pH} - \text{pM} - \log\frac{\alpha^2}{1-\alpha} + \log\frac{V_p}{\nu} = \text{p}\overline{K}^{\text{int}}_{(HA)_\nu} + \log \bar{y}_{A^-} + \log \bar{y}_M \tag{3.14}$$

Consider now with Eq. 3.14, potentiometric data obtained during titration with standard NaOH in the presence of the simple uniunivalent salt NaX:

$$\text{pH} - \text{pNa} - \log\frac{\alpha^2}{1-\alpha} - \log\frac{\nu}{V_p} = \text{p}\overline{K}^{\text{int}}_{(HA)_\nu} + \log \bar{y}_{A^-} + \log \bar{y}_{Na} \tag{3.14a}$$

There is experimental evidence in the ion-exchange literature to support the estimate that $\bar{y}_H = \bar{y}_{Na}$ in these gel systems. At a relatively low crosslinking percentage (2% divinylbenzene by weight), $\bar{y}_H$ is about equal to $\bar{y}_{Na}$, as evidenced by the ion-exchange distribution of Na^+ and H^+ between polystyrene sulfonate–based resin and a simple dilute electrolyte mixture of Na^+ and H^+ (Na^+, H^+, X^-). The selectivity coefficient measured over the complete composition range of the resin deviates very little from unity to demonstrate this as an experimental fact ($K^{Na}_H = 1.02 \pm 0.02$ at $\overline{X}_{Na} = 0$, $K^{Na}_H = 1.07 \pm 0.02$ at $\overline{X}_{Na} = 0.5$, and $K^{Na}_H = 1.12 \pm 0.03$ at $\overline{X}_{Na} = 1$) (Myers and Boyd, 1956).

Since $\bar{y}_{Na} \approx \bar{y}_H$, the right-hand side of Eq. 3.14 can be rewritten as

$$\text{p}\overline{K}^{\text{int}}_{(HA)_\nu} + \log \bar{y}_{A^-} + \log \bar{y}_H$$

From Eqs. 3.8 and 3.9, we have

$$\text{p}\overline{K}^{\text{int}}_{(HA)_\nu} + \log \bar{y}_H + \log \bar{y}_{A^-} = \text{p}\overline{C}_H - \log\frac{\alpha}{1-\alpha} \tag{3.15}$$

and

$$\text{p}\overline{K}^{\text{int}}_{(HA)_\nu} + \log \bar{y}_H + \log \bar{y}_{A^-} = \text{p}\overline{K}^{\text{app}}_{(HA)_\nu} \tag{3.16}$$

Equation 3.14a can thus be rewritten as

$$\mathrm{pH} - \mathrm{pNa} - \log \frac{\alpha^2}{1 - \alpha} - \log \frac{\nu}{V_p} = \mathrm{p}\overline{K}^{\mathrm{app}}_{(\mathrm{HA})_\nu} \tag{3.17}$$

Recall that the intrinsic dissociation constant, $K^{\mathrm{int}}_{\mathrm{HA}}$, of a weakly dissociable acid expressed as $\mathrm{p}K^{\mathrm{int}}_{(\mathrm{HA})}$ is related to the standard free energy change of the dissociation process, $\Delta G°$, as follows:

$$\mathrm{p}K^{\mathrm{int}}_{(\mathrm{HA})} = -\log K^{\mathrm{int}}_{(\mathrm{HA})} = \frac{0.434\Delta G°}{RT} \tag{3.18}$$

The dissociation of a weakly acidic polyelectrolyte gel, however, requires an additional amount of energy (ΔG_{add}), first to remove H^+ held by the strong electrostatic force developed by the charged surface of the gel matrix and second to transfer H^+ ion out of the gel phase into the solution phase. The second energy term has been accounted for by the Donnan potential term, $\mathrm{p}\overline{\mathrm{Na}} - \mathrm{pNa}$, that is incorporated in Eq. 3.14a as shown

$$\begin{aligned} -\log \alpha - \log \frac{\nu}{V_p} - \log \overline{y}_{\mathrm{Na}} - \mathrm{pNa} &= -\log \frac{\alpha \cdot \nu \overline{y}_{\mathrm{Na}}}{V_p} - \mathrm{pNa} \\ &= -\log (\overline{\mathrm{Na}})\overline{y}_{\mathrm{Na}} - \mathrm{pNa} \\ &= \mathrm{p}\overline{\mathrm{Na}} - \mathrm{pNa} \end{aligned}$$

The first, less amenable to direct analysis, can be described in the following manner:

$$\frac{0.4343\Delta G_{\mathrm{el}}}{RT} = \mathrm{p}\overline{\mathrm{H}} - \log \frac{\alpha}{1 - \alpha} - \mathrm{p}K^{\mathrm{int}}_{(\mathrm{HA})_\nu} \tag{3.19}$$

ΔG_{el}, which can be identified with $e\psi_{(a)}$, the electrostatic potential at the polyion site of protonation, may also be described by

$$\Delta G_{\mathrm{el}} = \frac{\partial G_{\mathrm{el}}}{\Delta \psi} \tag{3.20}$$

where G_{el} is the electrostatic free energy of a polyion with ν ionized groups. We see from this that the deviation term, $\log \overline{y}_{\mathrm{H}} + \log \overline{y}_{\mathrm{A}^-}$, of Eq. 3.16 is due to electrostatic interaction between the simple ions in the gel phase and the charge on the polyion surface and may be identified with $\mathrm{p}\Delta K_{(\mathrm{HA})_\nu}$, where

$$\mathrm{p}\Delta K_{(\mathrm{HA})_\nu} = -0.4343 \frac{e\psi_{(a)}}{kT} \tag{3.21}$$

with e the unit electrical charge of the mobile ion, $\psi_{(a)}$ the electric potential at the surface of the polyion matrix as noted above, k the Boltzmann constant, and T the absolute temperature. As α approaches a value of zero, $\psi_{(a)}$, and consequently $\mathrm{p}\Delta K_{(\mathrm{HA})_\nu}$ or the sum $\log \bar{y}_\mathrm{H} + \log \bar{y}_{\mathrm{A}^-}$ must also approach zero as their limiting value.

2.2. The Impermeable Gel; A Charged Surface–Controlled Distribution of Ions for Interpretation of the Sensitivity of the Potentiometric Properties of Weakly Acidic Hydrophobic Gels to the Presence of Neutral Salt

In weakly acidic (basic) gels impermeable to salt, the distribution profile of M, the potential-controlling ion, and H ion during the gel's neutralization with base is presumed to be well described by Poisson–Boltzmann statistics (Marinsky, 1985) as shown: In these equations M, H, e, Ψ, k, and T are once again defined as before.

$$\vec{M} = M \exp\left(\frac{-e\vec{\Psi}}{kT}\right) \qquad \vec{H} = H \exp\left(-\frac{e\vec{\Psi}}{kT}\right) \tag{3.22}$$

The arrows above M, H, and Ψ are used to identify these quantities with the surface domain of the gel. The ratio of $\vec{M}$ to $\vec{H}$ is related then to their measurable ratio in solution, the exponential term being cancelable, since the H ion is also exposed to the potential $\vec{\Psi}$ defined by the M ion.

$$\frac{\vec{M}}{\vec{H}} = \frac{M}{H} \tag{3.23}$$

It is immediately apparent that the logarithmic form of Eq. 3.23 is identical to Eq. 3.7c. Indeed, at each equilibrium reached during the course of a potentiometric titration of the salt-impermeable gel, the $\mathrm{p}K^{\mathrm{app}}_{(\mathrm{HA})_\nu}$ is once again related to $\mathrm{p}K^{\mathrm{int}}_{(\mathrm{HA})_\nu}$ by the term ($\mathrm{p}\vec{\mathrm{M}} - \mathrm{pM} + \log y_{\mathrm{A}^-}$). In this instance, of course, the $\mathrm{p}\vec{\mathrm{M}} - \mathrm{pM}$ term is not accessible to direct computation as it is with the salt-permeable system because of the inaccessibility of the polymer domain volume to measurement. However, the duplication of sensitivity of the $\mathrm{p}K^{\mathrm{app}}_{(\mathrm{HA})_\nu}$ versus α plots to ionic strength is predictable with Eq. 3.23, with the predicted behavior expected to prevail over a greater salt concentration range in the hydrophobic gel.

2.3. Model-Based Predictions with Respect to the Sensitivity of the Potentiometric Properties of Weakly Acidic Gels to Salt Concentration Levels

In both hydrophilic and hydrophobic gels, the gel phase controls the measured potentiometric response during the neutralization process. If the gel is rigid, the quantity $\overline{M}$ (or $\vec{M}$) is uniquely defined by α. This in turn defines the quantity $\overline{H}$ (or $\vec{H}$), since the H^+ ion activity must be a unique function of the free energy of dissociation of the repeating acid group in the fixed $\overline{M}(\vec{M})$ environment. At a particular MX concentration level, the ratio of [M] to [H] has to duplicate its ratio in the gel. Thus a tenfold change in the concentration of MX has to be reflected in a tenfold change in [H] and a change of one pK unit in the value of $pK^{app}_{(HA)_{\nu,\alpha}}$ at that α value. This predicted response of H_α and $pK^{app}_{(HA)_{\nu,\alpha}}$ to ionic strength is valid as long as the value of $\overline{M}$ (or $\vec{M}$) is at least 10 times as great as its value in solution at the highest ionic strength employed.

One can deduce from the above that there is a greater potential for divergence from the model-based predictions with the hydrophilic gel. In such gel–salt systems, some diffusion of MX always occurs until Donnan equilibrium is reached. At higher concentration levels of MX, the uniqueness of $\overline{M}$ *as a function of* α, $\overline{M}_\alpha$, the basis of the model, is impaired: Diffusion of significant quantities of MX into the gel phase eventually changes $\overline{M}$ by too much for the model to be strictly applicable. Since there is no diffusion of salt into the hydrophobic gel, the applicable concentration range of the model is more extended.

The $pK^{app}_{(HA)_\nu}$ versus α curves obtained at different MX concentration levels are predicted by the model to converge at low α values. Such behavior is anticipated because the reduced value of $\overline{M}$ (or $\vec{M}$) as α approaches zero must eventually approach the concentration of M defined by the ionic strength employed. At this point, $p\overline{K}^{app}_{(HA)_\nu} = pK^{app}_{(HA)_\nu}$. The sharpness of such convergence, that is, the rate of change in the slope of the curve, must be exaggerated at the lowest ionic strength levels as α approaches zero.

The model also predicts that extrapolation of all curves to intercept the ordinate at $\alpha = 0$ must eventually yield the intrinsic pK of the repeating weakly acidic group of the gel as long as the data are compiled at sufficiently low α values to define the region of rapid change in the slope of the curves. In this region where curves converge, deviation from ideality which is attributable to charge interaction between counterions and the oppositely charged surface of the gel matrix is rapidly diminishing. Since nonideality, as defined by the model, no longer persists at $\alpha = 0$, the intrinsic pK is resolved from such extrapolation. In actuality, however, the extrapolated pK will have to be corrected for an ionic strength effect in the salt-permeable case.

Finally, the model predicts that if the gel is salt-permeable and inflexible, the value of pH − log[α/(1 − α)]($pK^{app}_{(HA)_{\nu,\alpha}}$) should be a unique function of (pH + pX); if it is impermeable and rigid, the $pK^{app}_{(HA)_{\nu,\alpha}}$ term is predicted to be a unique function of pH alone. That this is indeed the case may be seen from the following. With the rigid, salt-permeable matrix we know from Eq. 3.6 that

$$\mathrm{pH} + \mathrm{pX} - \mathrm{pM} - \mathrm{pX} = \mathrm{p}\overline{\mathrm{H}} + \mathrm{p}\overline{\mathrm{X}} - \mathrm{p}\overline{\mathrm{M}} - \mathrm{p}\overline{\mathrm{X}} \tag{3.6a}$$

ignoring the πV term. By rearranging this equation, we obtain

$$\mathrm{pH} + \mathrm{pX} = \mathrm{p}\overline{\mathrm{H}} - \mathrm{p}\overline{\mathrm{M}} + \mathrm{pM} + \mathrm{pX} \tag{3.6b}$$

If we now add the −log[α/(1 − α)] term to both sides of Eq. 3.6b, we are left with the expression

$$\begin{aligned}\mathrm{pH} - \log\frac{\alpha}{1-\alpha} + \mathrm{pX} &= \mathrm{p}\overline{\mathrm{H}} - \mathrm{p}\overline{\mathrm{M}} + \mathrm{pM} - \log\frac{\alpha}{1-\alpha} + \mathrm{pX} \\ &= \mathrm{p}K^{app}_{(HA)_\nu} + \mathrm{pX}\end{aligned} \tag{3.24}$$

With the rigid, permeable matrix we have already noted that $\mathrm{p}\overline{\mathrm{H}} - \mathrm{p}\overline{\mathrm{M}}$ has to be a unique function of α. This means that pH is directly proportional to pM, which in turn is essentially equal to pX. From this we learn that $pK^{app}_{(HA)_\nu}$ must be a unique function of pH + pX.

In the rigid, impermeable gel, on the other hand, we learn from Eq. 3.23 that

$$\mathrm{pH} = \mathrm{p}\vec{\mathrm{H}} - \mathrm{p}\vec{\mathrm{M}} + \mathrm{pM} \tag{3.23a}$$

so that

$$\mathrm{pH} - \log\frac{\alpha}{1-\alpha} = \mathrm{p}\vec{\mathrm{H}} - \mathrm{p}\vec{\mathrm{M}} + \mathrm{pM} - \log\frac{\alpha}{1-\alpha} = \mathrm{p}K^{app}_{(HA)_\nu} \tag{3.24b}$$

and $pK^{app}_{(HA)_\nu}$ has to be a unique function of pH alone.

2.4. Experimental Verification of Model Predictions

2.4.1. Salt-Permeable Gels. Potentiometric data obtained with Amberlite IRC-50, a crosslinked polymethacrylic acid gel (Marinsky and Slota, 1980) distributed by the Rohm and Haas Company, have been analyzed with eq. 3.14a to facilitate our first test of the Donnan model. The sum of the terms on the left-hand side of this equation are plotted versus α for this purpose in Figure 3.3. The intercept value of the ordinate at α = 0 is

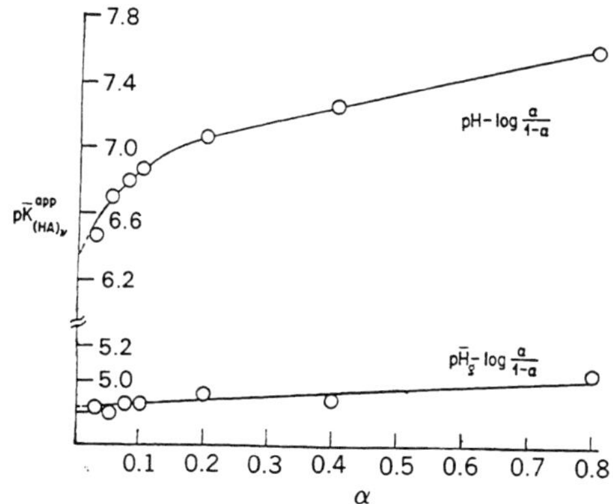

Figure 3.3. Analysis of a potentiometric study of the protonation of Amberlite IRC-50.

4.83 ± 0.05. As predicted by the model, this pK value is in excellent accord with the pK value of 4.8 reported in the literature (Sillén and Martell, 1964) for isobutyric acid, the weakly acidic molecule closely identifiable with the repeating functional unit of polymethacrylic acid. The value ordinarily assigned to p$K^{app}_{(HA)_\nu}$ by presuming pH − log[a/(1 − α)] to be a measure of this parameter has been included for comparison to demonstrate how sizable the effect of neglecting the Donnan potential term in three-dimensional gel systems can be.

A more detailed display of the effect of ionic strength on the measurable potentiometric properties of weakly acidic gel–salt systems is provided in Figures 3.1 and 3.4. In Figure 3.1, the p$K^{app}_{(HA)_\nu}$ computed for the highly flexible Sephadex CM-50 gel from the pH measured at three different concentration levels of sodium polystyrene sulfonate (0.10, 0.010, and 0.0010 molal, monomer basis) is plotted versus α, the degree of dissociation (Marinsky and Merle, 1984). As the concentration level of polyelectrolyte salt increases, the apparent pK decreases at each particular α value to yield three separate curves. This result is in qualitative accord with predictions of the Donnan model. Similarly, plots of pH − log[α/(1 − α)] versus α obtained with the inflexible CM-25 Sephadex gel in 0.010 and 0.0010 m $NaClO_4$ (Alegret et al., 1984) and presented in Figure 3.4 are displaced upwards as the ionic strength of the system is decreased, with the vertical displacement of the curves somewhat greater than observed in Figure 3.1. The lowest set of points in Figure 3.4 are obtained when correction for the Donnan potential term is included (Eq. 3.14a) in the computation of p$K^{app}_{(HA)_\nu}$ using both sets of data. The p$K^{int}_{(HA)_\nu}$ value of 3.3 ± 0.07 resolved with the essentially straight line of zero slope that best describes these points is once again in good agreement with the intrinsic pK of 3.3 reported for methoxycarboxylic acid (Sillén and Martell, 1964), the weakly acidic molecule identifiable with the functional unit repeated throughout the Sephadex molecule. The fact that the p$K^{app}_{(HA)_\nu}$ value is

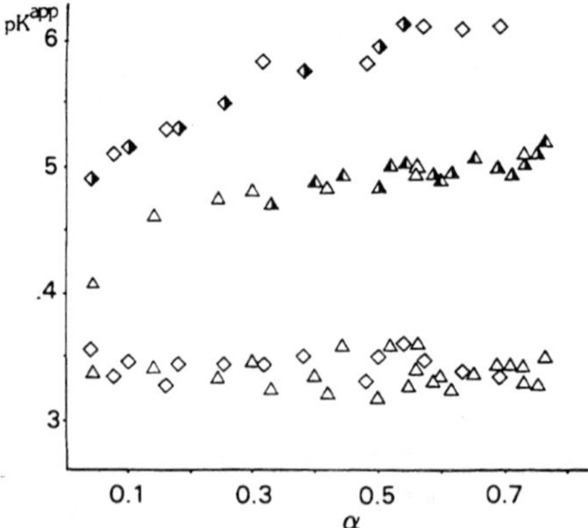

Figure 3.4. Potentiometric properties of the rigid Sephadex CM-25 gel measured in 0.0010 (◇, ◈) and 0.010 M (△, ▲) sodium perchlorate; upper two sets of data points, Donnan potential term neglected; lowest set of data points, Donnan potential term included by using Eq. 3.14a for computation of $pK^{app}_{(HA)_\nu}$, using all 0.0010 M (◇, ◈) and 0.010 M (△, ▲) sodium perchlorate–based data.

insensitive to the degree of dissociation after correcting for the Donnan effect suggests that effective screening of the charged gel matrix is provided by its rigidly defined geometry. A similarly effective screening of charge is noticeable in the $pK^{app}_{(HA)_\nu}$ versus α plot presented in Figure 3.3 for the polymethacrylic acid gel.

When the data obtained with the flexible Sephadex gel (Fig. 3.1) are reexamined with Eq. 3.14, a $pK^{int}_{(HA)_\nu}$ value of 3.3 is resolved once again from

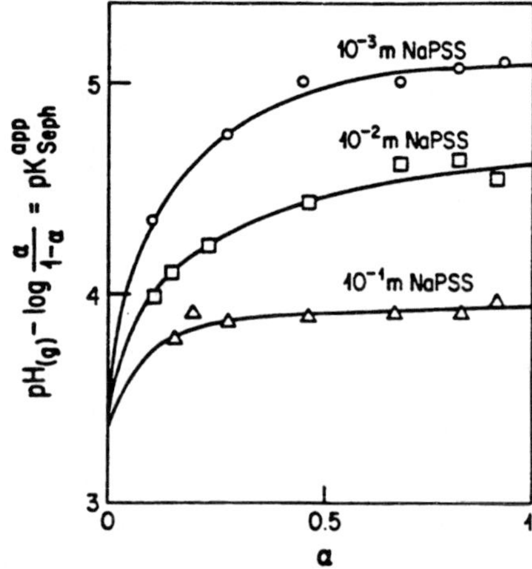

Figure 3.5. Potentiometric data of Figure 3.1 analyzed with inclusion of Donnan potential term.

plots of $pK^{app}_{(HA)_v}$ versus α at the three different concentration levels of sodium polystyrene sulfonate (see Fig. 3.5). In this instance, however, the plots of pK versus α yield three separate curves which diverge increasingly with increasing α from the common intercept value of 3.3 at α = 0 (Marinsky and Merle, 1984). The sizable but different degrees of expansion of the gel with α (Fig. 3.6) at the different ionic strengths employed are responsible for the less effective screening of the macromolecule surface charge that this result implies. Indeed, the protonation behavior of the flexible gel mimics that of the linear polyelectrolyte analog pictured in Figure 3.7, where the extrapolated curves once again converge to a value of 3.3 (Gekko and Naguchi, 1975). The very sharp drop in $pK^{app}_{(HA)_v}$ as α approaches zero that has to be drawn to force the convergence of the curves to a common point is justified by the model's anticipation of this property of the potentiometric data.

The Donnan model also predicts that $pK^{app}_{(HA)_v}$ $\{= \text{pH} - \log[\alpha/(1 - \alpha)]\}$ should be a unique function of (pH + pX) when the gel is inflexible. That this prediction is also a valid deduction of the model is shown in Figure 3.8. Here the $pK^{app}_{(HA)_v}$ data obtained with the rigid Sephadex gel (CM-25) and plotted versus α to yield the upper two sets of data points presented in Figure 3.4 are replotted against (pH + $pClO_4$) to yield the single line predicted (Marinsky, 1985). The uniqueness of such presentation of the data is lost with the more flexible gel (Fig. 3.4), for which a similar plot yields curves that

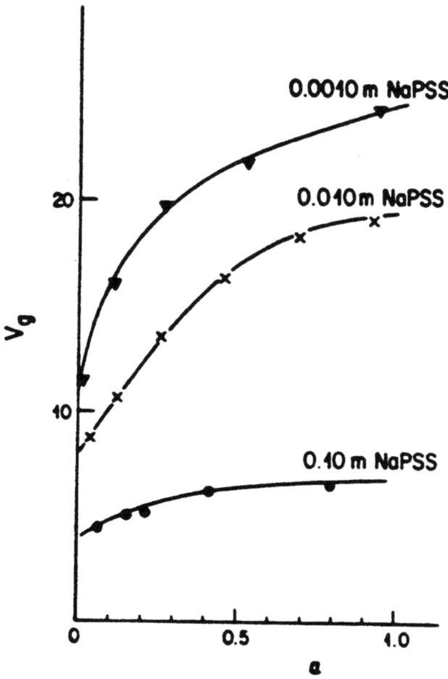

Figure 3.6. Variation of gel volume (Sephadex CM-50-120) as a function of degree of ionization (α) at different concentrations of sodium polystyrrenesulfonate.

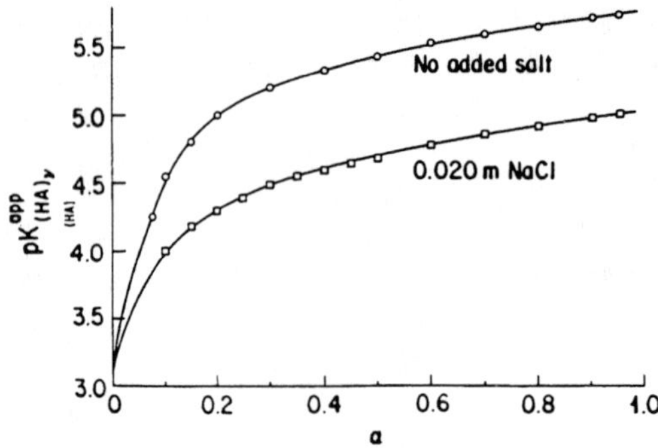

Figure 3.7. Dependence of potentiometric properties of carboxymethyldextran, the linear polyelectrolyte analog of Sephadex, on salt-medium concentration.

eventually diverge and separate to varying degrees from the curve obtained with the rigid gel. This result was also predictable, since the unique response of $(\overline{pH} - \overline{pM})$ to α in a rigid gel is lost in the flexible gel.

2.4.2. Salt-Impermeable Gels. The important research of Paterson and Rahman (1984) has shown that predictions deduced from the statistical approach used earlier to describe the counterion distribution profile affected by the variable charged surface of salt-impermeable gels are also fully realized. In their investigation of hydrous oxide gels, they have found, for example, that the anion capacity of β-FeOOH is a single-valued function of (pH + pX) in the presence of NaCl but becomes a single-valued function of pH when $NaClO_4$ is used to vary the ionic strength of the system. This uniqueness of Cl^- ion capacity dependence on pH alone in the presesnce of $NaClO_4$ is accountable with the model if ClO_4^- ion is not able to enter the gel channel

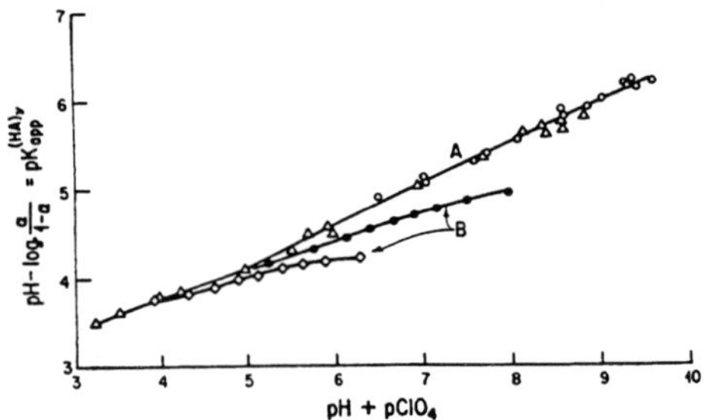

Figure 3.8. (A) Uniqueness of the $pK^{app}_{(HA)_\nu}$ function plotted versus ($pH + pClO_4^-$) by using potentiometric data obtained with the rigid Sephadex gel (CM-25) over a $NaClO_4$ concentration level ranging from 0.0010 to 0.10 M (△ and ○). (B) Removal of the $pK^{app}_{(HA)_\nu}$ function uniqueness in a flexible Sephadex gel (CM-50); (●) 0.010 M $NaClO_4$; (◇) 0.10 M $NaClO_4$.

because of its size. That this is indeed the case has been experimentally verified by the fact that only that fraction of the Cl^- ion capacity that is associated with the gel surface (~15%) is exchangeable with ClO_4^- ion.

2.4.3. Experimental Verification of Applicability of Two Models to Linear Polyelectrolyte–Salt Mixtures. In the introduction to this chapter, it was claimed that in simple salt–polyelectrolyte mixtures the domain of the polyelectrolyte molecule defines a separate phase that can be either permeable or impermeable to the diffusible components of the system just like the gel analog. That this is indeed the case is demonstrated in the several examples that are presented next in Figures 3.9 to 3.12, where $pK^{app}_{(HA)_\nu}$ is plotted versus (pH + pX) to facilitate this objective. The potentiometric data obtained at a number of different salt concentration levels with several hydrophilic polyelectrolytes—poly(acrylic) acid (PAA); poly(methacrylic) acid (PMA); D,L-poly(glutamic) acid (D,L-PGA); and L-poly(glutamic) acid (L-PGA)—by Nagasawa, Murose, and Kondo (1965) and by Olander and Holtzer (1968) are used for this purpose in these figures. Anion activity coefficients employed to evaluate pX were obtained from the single-ion activity coefficient values based upon computations due to Kielland (1937) up to an ionic strength of 0.10, the upper range of their projected validity. Above this ionic strength the mean molal activity coefficient published for the uniunivalent salt was used. The resolution of a single unique curve that is obtained with this approach for poly(acrylic) and D,L-poly(glutamic) acid in the course of their neutralization with standard base (Figs. 3.9 and 3.10) shows conclusively that the Donnan model is as applicable to the linear polyelectrolyte as it is to the salt-permeable gel. The divergence in $pK^{app}_{(HA)_\nu}$ from the otherwise uniquely resolved curve (Fig. 3.11) that is obtained with the poly(methacrylic acid)

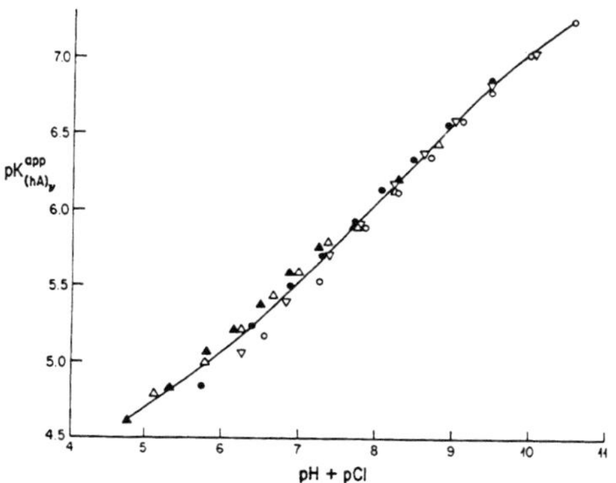

Figure 3.9. Apparent pK values obtained for poly(acrylic acid) at five different NaCl concentrations plotted versus (pH + pCl). The $pK^{app}_{(HA)_\nu}$ values are interpolated from the potentiometric titration curves presented in Figure 4 of Nagasawa et al., 1965; NaCl concentrations: (○) 0.0050 N, (▽) 0.010 N, (●) 0.020 N, (△) 0.050 N, and (▲) 0.100 N.

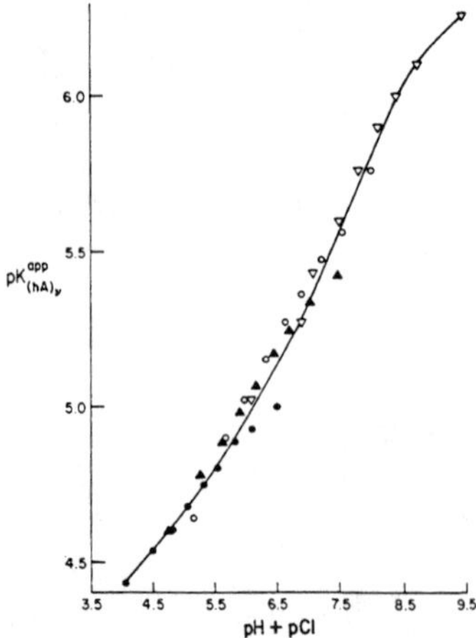

Figure 3.10. Apparent pK values obtained for poly(D,L-glutamic acid) at four different sodium chloride concentrations plotted vs. (pH + pCl). The $pK^{app}_{(HA)_\nu}$ values are interpolated from the potentiometric titration curves presented in Figure 1 of Olander and Holtzer (1968). NaCl concentrations: (▽) 0.010 N, (○) 0.050 N, (▲) 0.10 N, and (●) 0.40 N.

arises from the discontinuity introduced by conformational change in the molecule from a compact to a random configuration. The resultant curve is uniquely defined before and after the conformational change in the molecule. With the poly-L-(glutamic acid) molecule there is, using data due to Olander and Holtzer (Fig. 3.12), a unique description of $pK^{app}_{(HA)_\nu}$ as a function of (pH + pX) in the helix region. Divergence of the curves from each other is noticeable after passing the discontinuity in the curves due to the transition from the α-helix to the single-stranded configuration. Once the transition is complete, reconvergence does not occur in the PGA as it does with the poly(methacrylic acid) molecule.

We infer from these results that the PAA, PMA, D,L-PGA, and L-PGA in the α-helix configuration are rigid, inflexible hydrophilic macromolecules; the L-PGA molecule, on transition to the single-stranded configuration from its helical conformation, apparently loses some of its rigidity, presuming equilibrium has been reached in the potentiometric titration procedure described by the experimenters.

Potentiometric data obtained at different salt concentration levels with the hydrophobic polyelectrolytes poly(ethyleneimine) and poly(vinylamine) in earlier research by Bloy von Teslong and Staverman (1974) are examined in Figures 3.13 and 3.14 to show that the potentiometric properties of these kinds of polymers are uniquely a function of pH. The apparent pK data used in these figures were interpolated from the earlier plots of $pK^{app}_{(HA)_\nu}$ versus α compiled for these polyelectrolytes by the above researchers. The resolution of a single unique curve that results from this approach serves a twofold purpose. It substantiates the hydrophobicity and rigidity assignment to these

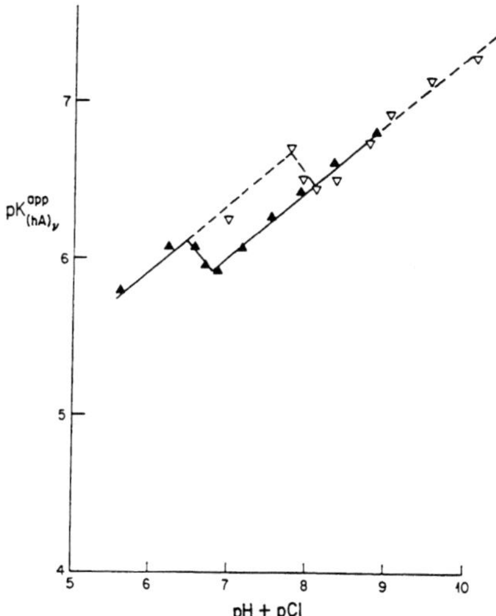

Figure 3.11. Apparent pK values obtained for the isotactic poly(methacrylic acid) in 0.100 and 0.100 and 0.100 N NaCl plotted versus (pH + pCl). The pK^{app} values are interpolated from the potentiometric titration curves presented in Figure 5 of Nagasawa et al., 1965. NaCl concentrations: (▽) 0.0100 N and (▲) 0.100 N.

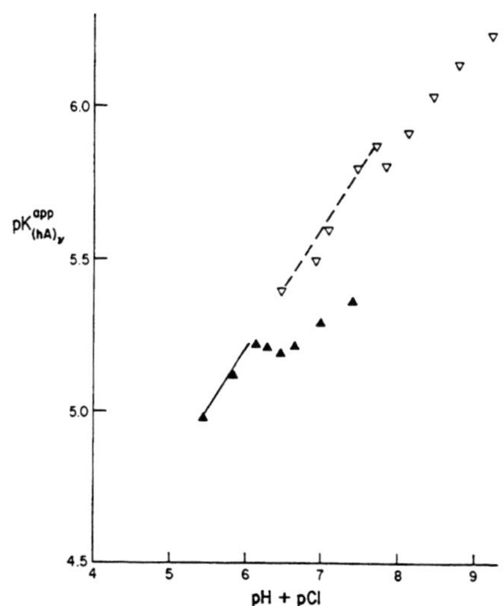

Figure 3.12. Apparent pK values obtained for poly(L-glutamic acid) in 0.010 and 0.10 N NaCl plotted versus (ph + pCl). The p$K^{app}_{(HA)_\nu}$ values are interpolated from the potentiometric titration curves presented in Figures 2 and 4 of Olander and Holtzer (1968). NaCl concentrations: (▽) 0.010 N and (▲) 0.103 N.

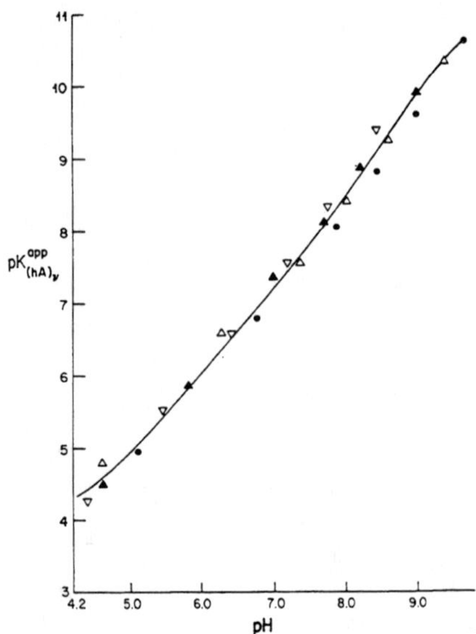

Figure 3.13. Apparent pK values obtained for poly(ethlyeneimine) in 0.001, 0.010, 0.10, and 1.0 N NaCl plotted versus the experimental pH. The $pK^{app}_{(HA)_\nu}$ values are interpolated from the potentiometric titration curves presented in Figure 1 of Bloy von Treslong and Staverman (1974). NaCl concentrations: (▽) 0.0010 N, (▲) 0.010 N, (△) 0.10 N, and (●) 1.0 N.

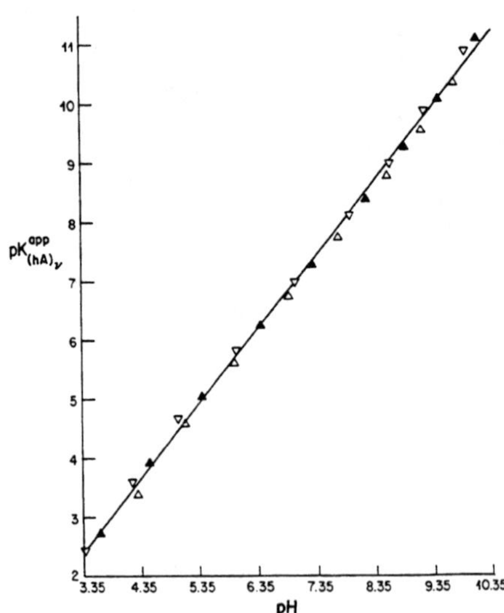

Figure 3.14. Apparent pK values obtained for poly(vinylamine) in 0.050, 0.10, and 1.0 N NaCl plotted versus the experimental pH. The $pK^{app}_{(HA)_\nu}$ values are interpolated from the potentiometric titration curves presented in Figure 3 of Bloy von Treslong and Staverman (1974). NaCl concentrations: (▽) 0.050 N, (△) 0.10 N, and (▲) 1.0 N.

polymers while providing convincing evidence for the claim that charged polymeric molecules in solution define a separate phase in the system to mimic the physicochemical properties of their gel–solution phase analogs.

3. APPLICATION OF THE UNIFIED MODEL TO THE INTERPRETATION OF PROTON AND METAL-ION INTERACTION WITH VARIOUS CHARGED POLYMERIC SYSTEMS

3.1. Hydroxylated Oxides

Even though the sensitivity of the protolytic behavior of hydroxylated oxide to salt concentration levels (Schindler et al., 1976; Atkinson et al., 1967) is the same as that of weakly acidic (basic) gels, investigators of these systems have failed to take note of this parallel in their attempts to interpret this behavior. Instead, they have resorted to various modifications (Schindler and Gamsjäger, 1972; Hohl and Stumm, 1976; Huang and Stumm, 1973; Bowden et al., 1977) of the classical double-layer model (Stern, 1924) for this purpose. Most effective rationalization of the salt effect by this kind of approach has led to the introduction of a triple-layer model by Yates and coworkers (Yates et al., 1974). In this model the surface (S) complexation of counterions and co-ions such as SO–Na and SOH_2–Cl, in NaCl, is included in the estimate of surface charge. Davis, James, and Leckie (1978; see also Davis and Leckie, 1978) modified this approach in their extension of the surface complexation concept.

In a review article by Westall and Hohl (1980) five electrostatic models that had been proposed to describe the electrical double layer of oxide suspensions were examined. By introducing a mathematical procedure that obtained optimal fit of the experimental data through adjustment of parameters associated with each model, they showed that each model tested could represent the data equally well. As noted by Davis, James, and Leckie (1978), on comparing their parameters with those derived from the triple-layer model of Yates et al. (1974), the values of corresponding parameters in *all* five models were not comparable. This result led Westall and Hohl to conclude that the models, while of correct mathematical form to represent the data, do not necessarily provide an accurate physical description of the oxide surface–electrolyte solution interface.

One can conclude from the above that the parameter-crowded double- and triple-layer charged surface models do not lead to an intellectually satisfying analysis of the phenomena encountered with hydroxylated oxides in the presence of excess neutral salt. Indeed, the introduction of surface complexation models, for example, seems completely unnecessary. One need only return to the fact that the sensitivity of the protolytic behavior of hydroxylated oxides to excess neutral salt is the same as with weakly acidic gels to realize that the unified model developed here must be applicable to these systems as well. There is absolutely no need to hypothesize surface complexation reactions between such unlikely candidates as Na^+ and SiO_3^- ions.

In order to determine whether our prognosis that the sensitivity of the protolytic properties of silica to excess salt is indeed a consequence of either a rigid, salt-permeable or a rigid, salt-impermeable interface, potentiometric data obtained in our laboratory with silica–salt systems are examined below with the unified approach.

In the batch equilibration experiments carried out, weighed quantities of silica (60–200 mesh, Fisher Scientific Company, S-162, Lot 782494, Grade 950 for gas chromatography) were suspended in aqueous media (0.010, 0.030, 0.080, 0.30, 1.0, 2.0, and 5.0 *M* in NaCl) to which well-defined quantities of standard NaOH had been added. The estimate of α for use in the computation of $pK^{app}_{Si(OH)_4}$ was reached by first measuring the quantity of base consumed in each equilibration through titration of carefully measured aliquot portions of the supernatant solution with standard HCl. After accounting for dissolution of the silica in the course of the equilibration, the computation of the initial acidic capacity of the residue was based on the estimate that the silica was completely hydroxylated; that is, a capacity of 10 mmoles of hydrogen per gram of silica was the basis for estimates of α and $1 - \alpha$ in the computation of the apparent p*K* values employed in Figure 3.15, where $pK^{app}_{Si(OH)_4}$ is plotted versus pH + pCl. The curve, as drawn, passes within ±0.05 p*K* unit of all but four of the points so obtained. We have chosen to disregard the lowest point because the experimental apparent p*K* value should only approach and

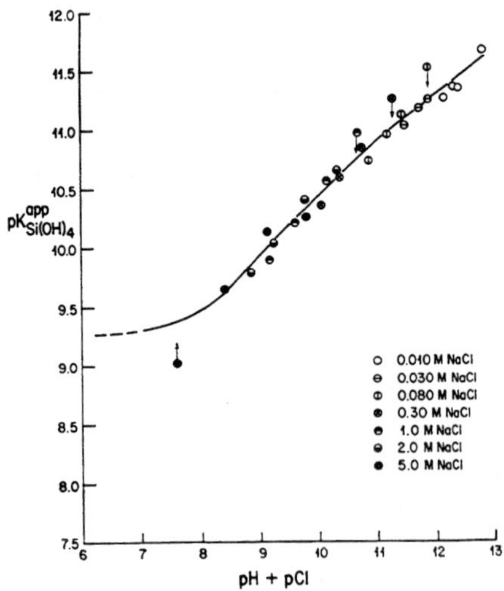

Figure 3.15. Plot of pH + pCl versus apparent p*K* values obtained for hydroxylated silica in NaCl at different concentration levels.

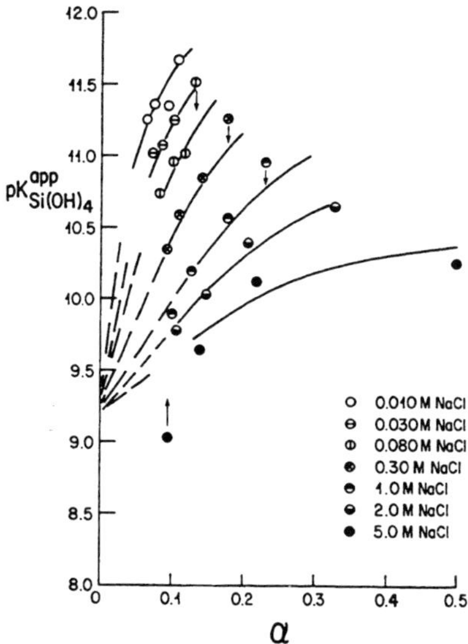

Figure 3.16. Apparent pK versus α plots obtained from potentiometric examination of hydroxylated silica at different NaCl concentration levels.

must never fall below the intrinsic pK value of the repeating functional unit of the gel phase. In any case, the great majority of points fall on or are very close to the single line drawn, and the correlation observed is quite sufficient to support the conclusions that (1) the unified model is applicable and (2) its use shows that the rigid silica gel employed in these studies is permeable to simple salt.

A separate analysis of the potentiometric data is presented in the Figure 3.16, where the apparent pK values are now plotted versus α. The data, so examined, yield separate curves that diverge from a common point at $\alpha = 0$, the rate of divergence increasing with decreasing ionic strength. According to our unified model, the extrapolated value of pK at $\alpha = 0$ should correspond to the intrinsic pK of the silicic acid unit repeated in the hydroxylated oxide gel. The value of 9.2 that this extrapolation procedure resolves is in rather good agreement with the pK value of 9.3–9.4 published for silicic acid (Busey and Mesmer, 1977).

A careful examination of Figures 3.15 and 3.16 shows that the several points that do not fall on the single curve drawn in Figure 3.15 also yield anomalous results in the individual apparent pK versus α plots resolved in Figure 3.16. The common points of discrepancy, marked by the arrows drawn in the figures, suggest that the measurement may have been faulty in these instances.

A final test for internal consistency of our characterization of the hydroxylated oxide as a rigid, salt-permeable gel has been provided by computation of the apparent pK at the site of the dissociation reaction with Eq. 3.14a. A gel volume of 0.325 cm^3 per dry gram of silica, measured in the experimental program, was used to determine the concentration of Na^+ ion in the gel phase at each point of the neutralization carried out at the different salt concentration levels. The results of these computations are listed in Table 3.1. The four rows of this table that are marked by an asterisk correspond to the anomalous points marked with an arrow in Figures 3.15 and 3.16. If we disregard these points, the average p$\overline{K}$ resolved is 9.30, in good agreement with the extrapolated pK in Figure 3.16.

Because of the high counterion concentration level at the reaction site, p$\overline{K}$ is expected to be unaffected by change in α as in the case of the rigid

Table 3.1. Results of Computations with Eq. 3.14a of the p$\overline{K}$ Characterizing the Protolysis of $Si(OH)_4$ at the Reaction Site

Experiment	Salt Conc. (M)	p$\overline{K}$
A-0	0.010	9.56
A-1	0.010	9.02
A-2	0.010	9.14
A-3	0.010	9.07
B-1	0.030	9.18
B-2	0.030	9.22
B-3	0.030	9.17
C-0*	0.080	9.73
C-1	0.080	9.34
C-2	0.080	9.26
C-3	0.080	9.17
D-0*	0.30	9.66
D-1	0.30	9.44
D-2	0.30	9.38
D-3	0.30	9.23
E-0*	1.0	9.63
E-1	1.0	9.46
E-2	1.0	9.33
E-3	1.0	9.22
G-0	2.0	9.11
G-1	2.0	9.44
G-2	2.0	9.33
G-3	2.0	9.25
F-0	5.0	9.03
F-1	5.0	9.53
F-2	5.0	9.25
F-3*	5.0	8.76
	Average:	9.30

*Anomalous points of Figures 3.15 and 3.16.

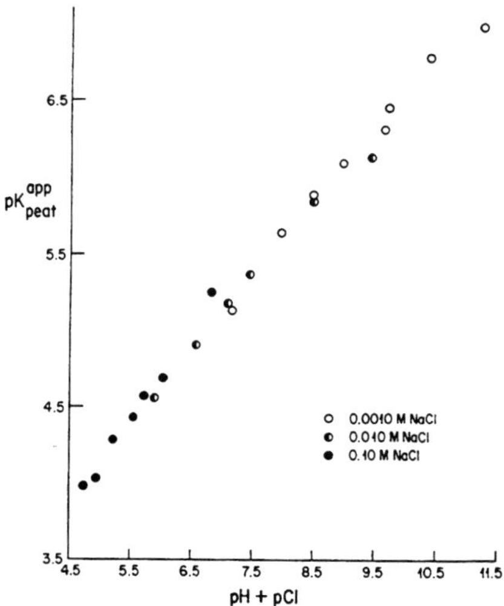

Figure 3.17. Apparent pK values obtained for peat in 0.0010, 0.010, and 0.10 M NaCl plotted versus (pH + pCl).

Sephadex gel (see Fig. 3.4); the fairly extended range of scatter in the computed value of p$\overline{K}$ is undoubtedly a consequence of the accumulation of uncertainties that arise in the assignment of activity coefficient values essential to treatment of the extra parameters that are contained in Eq. 3.14a.

3.2. Naturally Occurring Organic Acid Gels; Peat and Humic Acid

Potentiometric data obtained in earlier research by Marinsky and co-workers (1980) during the neutralization with standard NaOH of a sphagnum peat dispersed in aqueous media at three different NaCl concentration levels (0.0010, 0.010, and 0.10 M) have been reexamined to determine whether this rigid gel is permeable or impermeable to NaCl. When apparent pK is plotted versus (pH + pCl), all the data points fall on a single line, whereas when it is plotted against pH alone, three distinctly separate curves are obtained. The plot of apparent pK versus (pH + pCl) is presented in Figure 3.17 to show the uniqueness of the Donnan-based plot. Apparently the concentration level of Na^+ counterion associated with the gel phase is sufficiently large over the complete α range studied to produce a Donnan term ($pNa^+ - p\overline{Na}^+$) large enough to maintain the uniqueness of the $\overline{Na}/\overline{H}^+$ ratio with respect to α that is required by the unified model to yield a single curve like the one shown in Figure 3.17.

This was not the case with the humic acid gel, which is also rigid and hydrophilic (Marinsky, 1982). In this system the uniqueness of the apparent pK versus (pH + pNO_3) plot presented in Figure 3.18 is reasonably main-

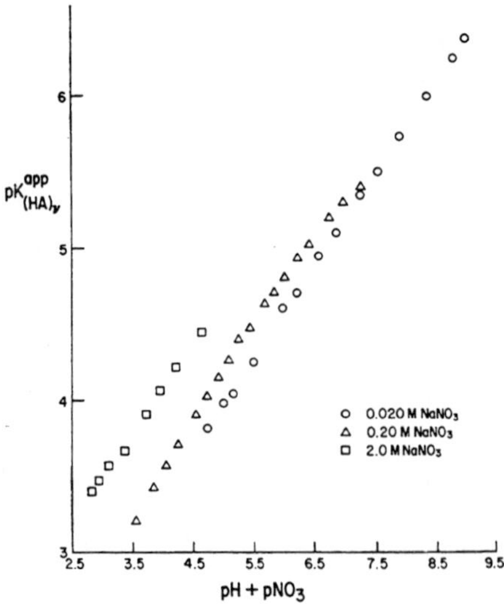

Figure 3.18. Apparent pK values obtained for humic acid in 0.020, 0.20, and 2.0 M $NaNO_3$.

tained for experiments conducted in 0.020 and 0.20 M $NaNO_3$ but is lost for the experiments carried out in 2.0 M $NaNO_3$.

This result is not unexpected, however. In our earlier examination of these potentiometric data, apparent pK was plotted versus α for the three systems (Marinsky, 1982). In the 0.020 and 0.20 M $NaNO_3$, curves separated by 0.85–1.0 pK unit over their full length were resolved. The curve obtained with the 2.0 M $NaNO_3$ system was separated by only 0.10–0.15 pK unit from the 0.20 M curve. It was suggested that the concentration level of Na^+ ion in the rigid gel phase, before additional Na^+ entered by diffusion of the salt into the gel, had to approach a concentration level of about 0.1–0.2 M to yield these results.

In Figure 3.18, the small displacement of the points obtained in 0.20 M $NaNO_3$, which is more exaggerated in the lower α range and the larger displacement of those obtained in 2.0 M $NaNO_3$ compared to those obtained in 0.020 M $NaNO_3$ are consistent with this earlier estimate. The magnitude of the displacement correlates very well indeed with rough estimates of the change in the α-controlled pattern of the $\overline{Na}^+$ to $\overline{H}^+$ ratio as the salt concentration level is increased from 0.020 to 2.0 M.

3.3. Some Properties of Bentonite in the Presence of Excess Salt

The substitution of lower-valence metal atoms in the silicon oxide framework of bentonite endows it with cationic exchange capacity of the order of 0.7 meq per dry gram. In addition, the silicon oxide component of the gel surface is hydroxylated. As a consequence, the ratio of cation (e.g., Na^+) to H^+ ion, whose concentration is reproducibly controlled by the free energy of dissociation of the $Si(OH)_4$, must in turn control the Na^+/H^+ ion ratio in the solution phase equilibrated with the bentonite. Thus if the salt concentration level of a solution in equilibrium with bentonite is changed, the level of H^+ ion in the solution will be observed to change as well.

The extent of that change is not predictable without the availability of gel volume data. Unlike the earlier gel examples, bentonite does not maintain a fixed volume as salt concentration levels in contact with it are changed. Bentonite is smectic. It has a layered structure that separates to varying degrees to accommodate more or less solvent when equilibrated with aqueous media at different salt concentration levels. It is this response to changes in the water activity of the solution in contact with the bentonite that complicates anticipation of pH response to different salt concentration levels.

In our laboratory we have measured the volume of bentonite as a function of the salt concentration level of solutions in contact with it. These data are presented in columns 1 and 2 of Table 3.2. The concentration of Na^+ ion in the Donnan equilibrated gel phase has been computed using these volume data and is listed in the third column. The equilibrium pH measured in the aqueous phase is given in the fourth column. Finally, in the last column of the table we have computed with the deduced Donnan potential term ($pNa^+ - p\overline{Na}^+$) the $p\overline{H}$ level in the bentonite phase. It is essentially constant at a value of 8.77 ± 0.06.

The fact that the pH value of the gel is predicted to be independent of the ionic strength of the solution equilibrated with it is to be expected, of course, if our attempts to account for deviations from ideality are reasonably adequate. Nonideality contributed by the concentration level of salt is the only factor that is expected to lead to variability in this parameter, since the degree of dissociation of the hydroxylated silicon oxide must be determined by the free energy of dissociation of the $Si(OH)_4$, which in turn has been defined by the method of preparation of the bentonite for study.

We believe that while it provides another test of the validity of the unified model this example also leads to additional insight with respect to the model's wide and useful applicability.

Table 3.2. The $p\overline{H}$ of a Bentonite Gel Phase after Equilibration with NaCl at Different Concentration Levels

NaCl Concentration (*M*)	Gel Volume per Dry Gram of Bentonite (mL)	$\overline{Na}^+$ Concentration	pH (Measured)	$p\overline{H}$
0.010	7.3 ± 0.5	0.097	9.82	8.83
0.10	1.7 ± 0.2	0.435	9.35	8.71
1.0	1.4 ± 0.2	1.28	8.93[a]	8.84
5.0	0.64 ± 0.1	5.53	8.69[a]	8.73

[a]The pH reported in 1.0 and 5.0 *M* NaCl has been corrected by 0.36 and 0.13 unit for error introduced into the measurement at these high Na^+-ion concentration levels.

3.4. The Potentiometric Properties of Fulvic Acid in Aqueous Medium

The potentiometric properties of four different fulvic acid sources [Armadale Horizons Bh (Gamble, 1970), Suwannee River (Thurman and Malcolm, 1983), a Swedish aquatic fulvic acid (Paxeus and Wedborg, 1984), and a Laurentide fulvic acid (kindly provided by D. S. Gamble, Department of Agriculture, Ottawa, Canada)] examined in our laboratory at different salt (Na(K)NO_3) and fulvic concentration levels, yield the characteristic pattern shown in Figure 3.19. Their permeability and rigidity has been examined in plots of apparent pK versus (pH + pNO_3) and pH alone (Figs. 3.20 and 3.21) using the potentiometric data compiled at the different salt concentration levels employed (Ephraim and Marinsky, 1986). The curves, separated in Figure 3.20, essentially merge into a single curve in Figure 3.21 to demonstrate that the polymeric fulvic acid molecule defines an essentially rigid, salt-impermeable phase in aqueous medium.

According to the unified model, the ratio of $\vec{N}a^+$ ($\vec{K}^+$) to $\vec{H}^+$ at the surface of the salt-impermeable molecule, uniquely defined by α, the degree of dissociation, should be mimicked in the aqueous solution as long as the activity of Na^+ (K^+) in the solution phase is lower than its activity at the molecular surface.

In addition, the separation of apparent pK versus α plots similar to those presented in Figure 3.19 should be one pK unit for each tenfold change in the ionic strength as long as the concentration differential defined by exp $[-e\Psi(a)/kT]$ is at least a factor of 10. However, the separations of 0.45 and

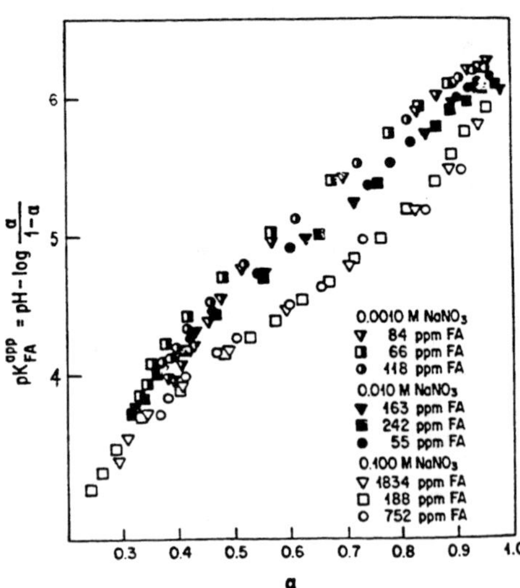

Figure 3.19. The variation of pK_{FA}^{app} in Armadale Horizons Bh fulvic acid with degree of dissociation at three different ionic strength levels. Ionic medium, $NaNO_3$.

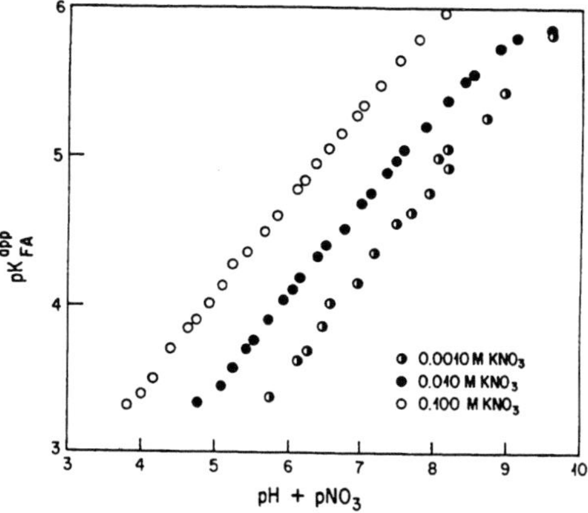

Figure 3.20. The variation of pK^{app}_{FA} with the pH + pNO_3 term in Armadale Horizons Bh fulvic acids. Ionic strength ranges from 0.0010 to 0.10.

0.2 pK unit between curves in Figure 3.19 are far less than the predicted value.

This discrepancy between experiment and theory can be explained as follows. The fulvic acid molecule is small; for example, the Armadale Horizons Bh fulvic acid has been determined to have an average molecular weight of about 1000. Because of this, end effects become important in the linear molecule, with the functional sites nearest the two ends of the molecule tending to.behave like a normal monomer molecule. One can, on this basis, envisage the extent of departure from predicted behavior to depend on the relative contribution of the fulvic acid (FA) molecule extremities and its rigid impermeable charged surface to the potentiometric properties.

In order to examine this explanation for the smaller expected curve sep-

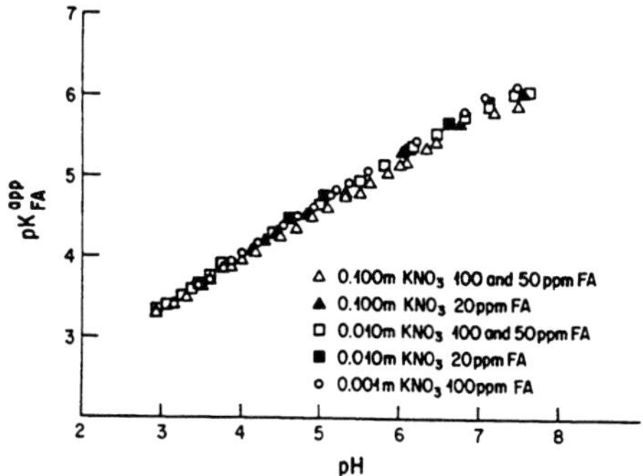

Figure 3.21. The variation of pK^{app}_{FA} with pH in Armadale Horizon Bh fulvic acid. Ionic strength ranges from 0.0010 to 0.10.

aration that was encountered with changing ionic strength levels, additional potentiometric studies were carried out at elevated salt concentration levels (1.0 and 5.0 *M*) with the Armadale Horizons Bh FA to determine first the salt concentration level at which discrepancy between the activity of Na^+ (K^+) at the molecule's surface and in the solution proper no longer persists. This condition was expected to be signaled by the insensitivity of the apparent pK versus α plot to any further increase in the salt concentration level.

These data are plotted together with the earlier pK_{FA}^{app} versus α plots compiled at the lower I values (Fig. 3.19) in Figure 3.22. The curves obtained at I = 1.0 and 5.0 *M* are nearly identical, showing that the anticipated removal of difference between the activity of the Na^+ (K^+) ion at the surface of the FA molecule and in the solution proper occurs at a salt concentration level of 1.0 *M*. It is interesting to note as well that the separation between the I = 0.1 and I = 1.0 curves is somewhat larger (0.55 pK unit) than between the I = 0.010 and I = 0.10 curves (0.45 pK unit), which in turn is larger than the separation (0.20 pK unit) between the I = 0.010 and I = 0.0010 curves.

With the ionic strength beyond which $pK_{FA_{I,\alpha}}^{app}$ is no longer a function of the neutral salt concentration level identified as described, the reasonableness of the proposed rationalization of the discrepancy between expectation and observation is accessible to examination. With our revised model, the variation of $pK_{FA_{I,\alpha}}^{app}$ with α and I, where I < 1.0, is directly related to the fractional (x)

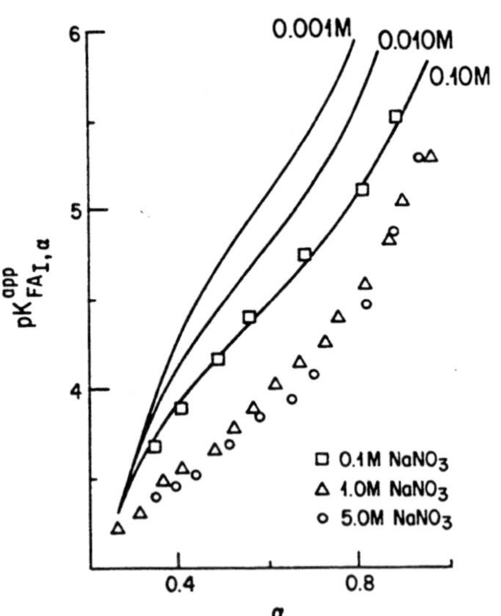

Figure 3.22. The insensitivity of $pK_{FA_{I,\alpha}}^{app}$ to ionic strength at high neutral salt concentration levels; Armadale Horizons Bh fulvic acid.

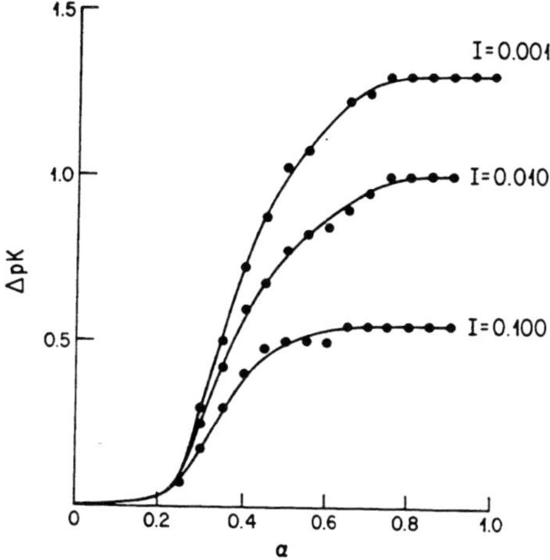

Figure 3.23. Variation of electrostatic deviation term (ΔpK) at three different ionic strength levels; Armadale Horizons Bh fulvic acid.

and $1 - x$ contribution of end and charged surface effects in the molecule as shown in Eq. 3.25.

$$K^{\text{app}}_{\text{FA},I,\alpha} = [x + (1 - x)(a_I/\vec{a}_{I_c}\alpha)](K^{\text{app}}_{\text{FA},I_c,\alpha}) \tag{3.25}$$

The experimental potentiometric plot obtained at and beyond I_c, the critical ionic strength (1.0 M in this case), provides $K^{\text{app}}_{\text{FA},I_c,\alpha}$ for use in Eq. 3.25 whose range of utility is restricted to the experimental situation where $a_I/\vec{a}_{I_c,\alpha}$ is less than unity and a represents the activity of the potential-determining ion, Na^+ (K^+). One can expect this condition to be met only after an α value large enough to produce a noticeable surface charge effect is reached. Before this limiting α value is reached, $a_I/\vec{a}_{I_c,\alpha} = 1$ and

$$K^{\text{app}}_{\text{FA},I,\alpha} = K^{\text{app}}_{\text{FA},I_c,\alpha} \tag{3.25a}$$

That this is indeed the case may be seen in Figure 3.23, where the vertical displacement of the $pK^{\text{app}}_{\text{FA},I,\alpha}$ versus α curves (I = 0.10, 0.010, and 0.0010 M) from the $I = 1.0(5.0)M$ curve presented in Figure 3.22 is plotted versus α. The curves start to diverge at $I \approx 0.2$ to define the point at which the charged molecular surface is sufficiently developed to lead to $a_I/\vec{a}_{I_c,\alpha}$ values smaller than unity.

By using values of $K^{\text{app}}_{\text{FA},I,\alpha}$ and $K^{\text{app}}_{\text{FA},I_c,\alpha}$ interpolated from Figure 3.23 and the $a_I/\vec{a}_{I_c,\alpha}$ term developed with this revised model, Eq. 3.25 has been used to estimate the fractional contribution of end effects to the reduced displacement of the $pK^{\text{app}}_{\text{FA},I,\alpha}$ curves with ionic strength in the high-α region (0.5–1.0);

for the three different salt concentration levels (0.0010, 0.010, and 0.10*M*), the fractional contribution of end effects is about 0.06, 0.09, and 0.17, respectively.

The slow increase with ionic strength in the estimated fractional contribution of end effects from this interpretation of the displacement of the $pK^{app}_{FA,I,\alpha}$ curves is believed to indicate that head-to-tail interaction between FA molecules may be promoted by dilution of the neutral salt medium.

3.5. Metal-Ion Binding to Gels and Their Linear Analogs

Correction can be made, as described earlier in the protonation studies, for the effect of salt (NaX) concentration levels on the measurable quantities of a particular metal ion, M^{2+}, whose interaction with the repeating ligand of a gel or its linear polyelectrolyte analog may be under investigation. For the computation of $\overline{C}_{H^+}$ in the protolysis studies, a_{H^+} was multiplied by $\overline{C}_{Na^+}/a_{Na^+}$ (see Eq. 3.7). The activity coefficient ratios of these ions in the polymer domain are approximately equal to each other and can be assumed to cancel. To duplicate this approach in the metal ion complexation studies, the activity of the metal ion in solution, $a_{M^{z+}}$, must be multiplied by the $\overline{C}_{Na}/a_{Na}$ factor taken to the z power and by the activity coefficient ratio $(\overline{\gamma}_{Na^+})^z/\overline{\gamma}_{M^{z+}}$, which no longer cancels. Because this activity coefficient ratio is difficult to assess, the direct approach to estimation of $\overline{C}_{M^{z+}}$ in the polymer domain at the site of the complexation reaction is more susceptible to uncertainty than the estimate of $\overline{C}_{H^+}$ in the protolysis studies.

For this reason, such direct assessment of $\overline{C}_{M^{z+}}$ has been avoided in most of the metal ion–polymer ligand complexation studies that we have conducted. Instead, the competitive reaction of H^+ and M^{z+} for the repeating ligand of the various polymer gels (and their linear polyelectrolyte analogs) that we have investigated has been examined. By this approach (Marinsky, 1976) the ratio of $(\overline{a}_H)^z$ to $\overline{a}_{M^{z+}}$ in the equilibrium quotient for the competing equilibria

$$z\overline{HA} + \overline{M}^{z+} \rightleftharpoons \overline{MA}^{(z-1)+} + z(\overline{H^+}) + \overline{A}^{z-1} \tag{3.26}$$

can be equated with the activity ratio in the solution phase because the Donnan potential correction terms cancel as shown:

$$\frac{(\overline{a}_{H^+})^z}{\overline{a}_{M^{z+}}} = \frac{(a_{H^+})^z}{a_{M^{z+}}} \frac{(\overline{a}_{Na^+})^z}{(\overline{a}_{Na^+})^z} \frac{(a_{Na^+})^z}{(a_{Na^+})^z} = \frac{(a_{H^+})^z}{a_{M^{z+}}} \tag{3.27}$$

and

$$K = \frac{\overline{MA}^{(z-1)+}}{M^{z+}} \frac{(H^+)^z\,(\overline{A})^{z-1}}{(\overline{HA})^z} = \frac{\beta_{MA^{(z-1)+}}}{(\beta_{HA})^z} \tag{3.28}$$

In the event that M^{z+} is bound to one and two ligands concurrently, the stability constant of each complexed species can be resolved through a plot of K versus $A^{(z-1)}/V_g$, where V_g is the volume of the gel (polymer) domain. The slope and ordinate axis intercept values of the resultant straight line lead to resolution of the parameters $\beta_{MA_2^{(z-2)+}}/(\beta_{HA})^z$ and $\beta_{MA^{(z-1)+}}/(\beta_{HA})^z$, respectively. Since the intrinsic stability constant of the repeating weakly acidic unit is known, the formation constants of $MA^{(z-1)+}$ and $MA_2^{(z-2)+}$ are immediately accessible with this approach.

Examples of the extension of this approach to the study of metal ion complexation encountered in gels and their polyelectrolyte analogs are provided by our investigations of the interaction of divalent metal ions with polymethacrylic acid (Anspach and Marinsky, 1975; Marinsky and Anspach, 1975); with Ca(II)-polymethacrylic acid gel (Marinsky, 1982), with polyacylic acid (Travers and Marinsky, 1974), polyglutamic acid (Imai and Marinsky, 1980; Marinsky et al., 1973), alginic acid (Lim, 1981), fulvic acid (Ephraim and Marinsky, 1986), humic acid (Marinsky et al., 1980), and peat (Marinsky et al., 1982). The results of these investigations have led to the observation that the probability of two repeating monomer units of a polyelectrolyte gel or its linear analog binding to one metal ion is quite low. Indeed, on the basis of results obtained with the Cu(II)-polymethacrylic acid gel, the accessibility of ligand was estimated to be about 3% of its value in the monomer analog system, Cu(II)-isobutyric acid (Marinsky, 1982).

4. CONCLUSION

The extreme sensitivity of the protonation and metal ion complexation properties of weakly acidic (basic) gels and their linear polyelectrolyte analogs to excess salt has been found to be attributable to a common source. Both kinds of macromolecules, whether they appear to be completely in solution or are present as a separate gel phase, behave as if they were present as a separate phase. Their particular response to experimental conditions depends on their permeability or lack of permeability to simple electrolyte and on their degree of rigidity. With the insight gained from their treatment as a separate phase, quantitative interpretation of their interaction with protons and metal ions becomes possible.

REFERENCES

Alegret, S., Marinsky, J. A., and Escaleas, M. J. (1984), *Talanta,* **31,** 693.

Anspach, W. M., and Marinsky, J. A. (1975), *J. Phys. Chem.* **79,** 433.

Atkinson, R. J., Posner, A. M., and Quirk, J. P. (1967), *J. Phys. Chem.* **71,** 550.

Bloy von Treslong, C. J., and Staverman, A. J. (1974), *Rec. Trav. Chim.* **93,** 171.

Bowden, J. W., Posner, A. M., and Quirk, J. P. (1977), *Aust. J. Soil Res.* **51,** 121.

Busey, R. H., and Mesmer, R. E. (1977), *Inorg. Chem.* **16,** 2444.

Davis, J. A., James, R. O., and Leckie, J. O. (1978), *J. Colloid Interface Sci.* **63,** 480.

Davis, J. A., and Leckie, J. O. (1978), *J. Colloid Interface Sci.* **63,** 480.

Ephraim, J., Alegret, S., Mathuthu, A., Bicking, M., Malcolm, R. A., and Marinsky, J. A. (1986), *J. Environ. Sci. Technol.*, **20,** 354.

Ephraim, J., and Marinsky, J. A. (1986), The influence of polyelectrolyte properties and functional heterogeneity on the copper binding equilibrium in an Armadale Horizons Bh fulvic acid sample, *J. Environ. Sci. Technol.*, **20,** 367.

Gamble, D. S. (1970), *Can. J. Chem.* **48,** 2662.

Gekko, K., and Naguchi, H. (1975), *Biopolymers,* **14,** 2555.

Hohl, H., and Stumm, W. (1976), *J. Colloid Interface Sci.* **55,** 281.

Huang, C. P., and Stumm, W. (1973), *J. Colloid Interface Sci.* **43,** 409.

Imai, N., and Marinsky, J. A. (1980), *Macromolecules* **13,** 275.

Kielland, J. (1937), *J. Amer. Chem. Soc.* **59,** 1675.

Lim, F. G. (1981), "Studies of Hydrogen and Metal Ion Equilibria in Polysaccharide Systems—Alginic Acid and Chondroiton Sulfate," Ph.D. thesis, State University of New York at Buffalo.

Marinsky, J. A. (1976), *Coordination Chem. Rev.* **19,** 125.

Marinsky, J. A. (1982), *J. Phys. Chem.* **86,** 3318.

Marinsky, J. A. (1985), *J. Phys. Chem.* **89,** 5294.

Marinsky, J. A., and Anspach, W. M. (1975), *J. Phys. Chem.* **79,** 439.

Marinsky, J. A., Gupta, S., and Schindler, P. (1982), *J. Colloid Interface Sci.* **89,** 401, 1102.

Marinsky, J. A., Imai, N., and Lim, M. C. (1973), *Israel J. Chem.* **11,** 601.

Marinsky, J. A., Lim, F. A., and Chung, K. S. (1983), *J. Phys. Chem.* **87,** 3139.

Marinsky, J. A., and Merle, Y. (1984), *Talanta* **31,** 199.

Marinsky, J. A., and Slota, P. (1980), "An Electrochemical Method for the Determination of the Effective Volume of Charge Polymers in Solution," A. Eisenberg, Ed., *Ions in Polymers,* American Chemical Society, Washington, DC, (Adv. Chem. Ser. **187**), p. 811.

Marinsky, J. A., Wolf, A., and Bunzl, K. (1980), *Talanta* **27,** 461.

Mukerjee, P. (1961), *J. Phys. Chem.* **65,** 740.

Myers, G. E., and Boyd, G. E. (1956), *J. Phys. Chem.* **60,** 521.

Nagasawa, M., Murase, T., and Kondo, K. (1965), *J. Phys. Chem.* **69,** 4005.

Olander, D. S., and Holtzer, A. (1968), *J. Am. Chem. Soc.* **90,** 4549.

Paterson, R., and Rahman, H. (1984), *J. Colloid Interface Sci.* **97,** 423.

Paxeus, N., and Wedborg, M. (1984), "Acid Base Properties of Aquatic Fulvic Acid and its low Molecular Weight Fraction," International Humic Substance Society Second International Conference, The University of Birmingham, England, July 22–28, 1984.

Schindler, P. W., and Gamsjäger, H. (1972), *Kolloid Z. Z. Polymere* **250,** 759.

Schindler, P. W., Wälti, E., and Fürst, (1976), *Chimica* **30**(2), 107.

Sillén, G., and Martell, A. (1964), *Stability Constants of Metal Ion Complexes,* The Chemical Society, Burlington House, London.

Stern, O. (1924), *Z. Elektrochem.* **30,** 508.

Thurman, D. M., and Malcolm, R. L. (1983), "Structural Study of Humic Substances, New Approaches and Methods," in R. F. Christman, and E. T. Gjessing, Eds., *Aquatic and Terrestrial Humic Materials,* Ann Arbor Science, Ann Arbor, MI, pp. 1–23.

Travers, L., and Marinsky, J. A. (1974), *J. Polymer Sci, Polymer Phys. Ed.* **47,** 285.

Westall, J., and Hohl, H. (1980), *Adv. Colloid Interface Sci.* **12,** 265.

Yates, D. E., Levine, S., and Healy, T. W. (1974), *Trans. Faraday Soc.* **70,** 1807.

4

THE SURFACE CHEMISTRY OF OXIDES, HYDROXIDES, AND OXIDE MINERALS

Paul W. Schindler

Department of Inorganic Chemistry, Institute for Inorganic, Analytical, and Physical Chemistry, University of Bern, Switzerland

and

Werner Stumm

Institute for Water Resources and Water Pollution Control (EAWAG), Zurich, Switzerland; Swiss Federal Institute of Technology (ETH), Zurich, Switzerland

Abstract

The initial and reversible steps of the interaction of (hydr)oxide surfaces with electrolytes, that is, adsorption–desorption equilibria with H^+ (OH^-), metal ions, and ligands (anions), are discussed in terms of the coordination chemistry of the oxide–water interface. This interface is characterized by different types of surface hydroxyl groups, and metallic centers (Lewis acids). Competitive complex formation equilibria (metal ion versus H^+, or anion versus OH^-) explain the strong dependence of metal ion and anion binding, respectively, on pH. Some experimentally determined representative intrinsic surface equilibrium constants reflecting the acid–base characteristics of surface hydroxyl groups and the stability of metal complexes, of anion complexes, and of ternary surface complexes, respectively, are tabulated. Such constants can also be estimated from equilibrium constants for corresponding reactions in solution. Additional support for the existence of inner-sphere surface complexes comes *from molecular methods (spectroscopy) and investigations on the kinetics of adsorption of ligands and of metal ions.*

(Hydr)oxides are abundant components of the earth's crust. The rates of processes occurring at these surfaces, such as precipitation (heterogeneous nucleation) and the dissolution of mineral phases—of importance in the weathering of rocks, the formation of soils and sediments, the corrosion of metals, and their inhibition—are critically dependent on the coordinative interactions taking place on these surfaces.

1. INTRODUCTION

Surface chemistry of oxides, hydroxides, and oxide minerals in aquatic environments includes the reactions of hydrous oxide surfaces with electrolyte solutions. (Hydrous oxides include oxides, hydroxides, and oxide hydroxides.) It thus specifically includes interactions of hydrous oxide surfaces with H^+ (resp. OH^-) ions, with dissolved metal ions, ligands, and metal–ligand complexes. It should be noted that the ligands present in natural aquatic systems range from simple monodentate species such as Cl^- to high molecular weight polyelectrolytes.

In a first step these interactions consist of adsorption of dissolved species from the bulk of the solution to the (hydrated) oxide–water interface. This step is fast and essentially reversible. It is followed by a series of slow and at least partially irreversible processes. The nature of these consecutive steps is dependent on the prevailing surface concentrations of adsorbed species (Fig. 4.1).

Adsorption of charged species results in changes in both surface charge and surface potential. High surface concentrations may even produce charge reversals. The surface potential of suspended particles is of primary importance for colloid stability (O'Melia, 1987). Changes in particle charge may then give rise to coagulation (or peptization) phenomena and thus to alteration in particle size distribution and sedimentation characteristic of particulate matter in aquatic systems. On a global scale such changes in sedimentation rates are observed in estuaries that act as an efficient trap for the particulate load carried in by rivers. In both freshwater and seawater systems, settling particles are important carriers for adsorbed metal ions and anions; they thus contribute to the regulation of the chemical composition of natural aquatic systems (Sigg, 1987; Whitfield and Turner, 1987).

Adsorption of H^+ ions may induce dissolution of the adsorbing oxides and oxide minerals. Such weathering reactions are of primary importance for the acquisition of solutes. In the early period of aquatic science (Stumm, 1967), the extent of weathering (and thus the composition of natural waters) was assumed to be controlled by thermodynamics, that is, by pertinent solubility equilibria. Chemical thermodynamics is a reliable guide in well-mixed systems

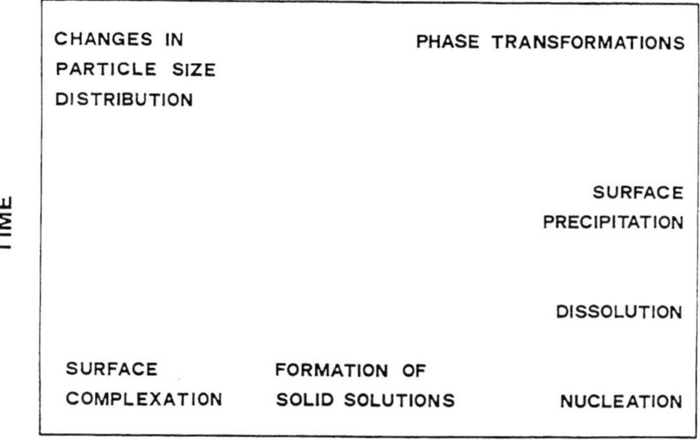

Figure 4.1. Topics in surface chemistry. The region of low surface coverage and limited (1 day) observation time is the domain of surface complexation. Enhanced surface coverage leads to nucleation and precipitation (or dissolution) processes. Extended observation time is required to follow changes in colloid stability.

where the residence time exceeds the time required to establish equilibrium. For silicate weathering, the contact time of the corroding water is not sufficient to approach equilibrium. With the notable exception of $CaCO_3$, the extent of weathering of oxides and oxide minerals is kinetically controlled. Although the mechanisms of dissolution of oxides and silicates are not yet fully understood, there is general agreement that surface processes are the rate-controlling steps (Stumm and Furrer, 1987; Schott and Petit, 1987). Moreover, it has been found that comparatively simple rate laws are obtained if the observed rate is related to the concentrations of adsorbed species and formed surface complexes (Pulfer et al., 1984; Stumm and Furrer, 1986).

Adsorption of metal ions exceeding a critical surface concentration leads to formation of hydroxide clusters of the adsorbed metal on the adsorbing surface (Bleam and McBride, 1985), a process that has also been termed *surface precipitation* (Farley et al., 1985). Surface precipitation may result in complete coating of the initial surface and thus in a fundamental change in surface properties.

Simultaneous adsorption of both cations and anions is apparently a prerequisite for heteronucleation (Hohl et al., 1985).

This chapter discusses the initial and reversible steps of hydrous oxide–electrolyte interaction, that is, adsorption–desorption equilibria. Because there exist recently published compilations (Hingston, 1981; Kinniburgh and Jackson, 1981), we do not intend to present a comprehensive review. Emphasis will be placed on a presentation of the main concepts of the coordination chemistry of the oxide–water interface.

2. COORDINATION CHEMISTRY OF THE OXIDE–WATER INTERFACE

2.1. Fundamentals

In humid environments, oxides and oxide minerals are covered with surface hydroxyl groups, S—OH. The presence of two lone electron pairs and a dissociable hydrogen ion indicates that these groups are ampholytes. Adsorption of H^+ and OH^- ions is thus based on protonation and deprotonation of surface hydroxyls:

$$\text{S—OH} + \text{H}^+ \rightleftharpoons \text{S—OH}_2^+ \tag{4.1}$$

$$\text{S—OH}\ (+\text{OH}^-) \rightleftharpoons \text{S—O}^- + \text{H}^+(+\text{H}_2\text{O}) \tag{4.2}$$

Deprotonated surface hydroxyls exhibit Lewis base behavior. Adsorption of metal ions is therefore understood as competitive complex formation involving one or two surface hydroxyls.

$$\text{S—OH} + \text{M}^{z+} \rightleftharpoons \text{S—OM}^{(z-1)+} + \text{H}^+ \tag{4.3}$$

$$2\text{S—OH} + \text{M}^{z+} \rightleftharpoons (\text{S—O})_2\text{M}^{(z-2)+} + 2\text{H}^+ \tag{4.4}$$

Since the coordination sphere of the adsorbed metal ion is only partially occupied by the surface ligands, further ligands may be acquired:

$$\text{S—OH} + \text{M}^{z+} + l\text{L} \rightleftharpoons \text{S—OML}_l^{(z-1)+} + \text{H}^+ \tag{4.5}$$

The species formed by reaction 4.5 is denoted as a *type A ternary surface complex* with the metal between the surface and the ligand. Although reaction 4.5 contributes to anion adsorption, the main mechanism for anion adsorption is ligand exchange, again involving one or two surface hydroxyls:

$$\text{S—OH} + \text{L} \rightleftharpoons \text{S—L}^+ + \text{OH}^- \tag{4.6}$$

$$2\text{S—OH} + \text{L} \rightleftharpoons \text{S}_2\text{L}^{2+} + 2\text{OH}^- \tag{4.7}$$

For the case where L is a polydentate ligand, *type B ternary surface complexes* may be formed:

$$\text{S—OH} + \text{L} + \text{M}^{z+} \rightleftharpoons \text{S—L—M}^{(z+1)+} + \text{OH}^- \tag{4.8}$$

For some years the above-postulated reactions were mainly the offspring of classical equilibrium analysis. There is now increasing support from spectroscopy (von Zelewsky and Bemtgen, 1982; Motschi, 1987) elucidating the structure of both binary and ternary surface complexes. Studies on the kinetics of adsorption of metal ions on γ-Al_2O_3 have shown that the adsorption rate constants correlate significantly with the rate constants for the release of water molecules from the hydrated metal ions (Hachiya et al., 1984). This observation strongly supports the concept of formation of inner-sphere complexes as indicated by Eqs. 4.3 and 4.4.

2.2. The Structure of the (Hydr)oxide–Water Interface

The origin of the surface hydroxyl groups is schematically depicted in Figure 4.2*a*–*c*. The surface of a dry oxide is characterized by the presence of low-coordinated metal ions (Fig. 4.2*a*) giving rise to Lewis acidity. (The presence

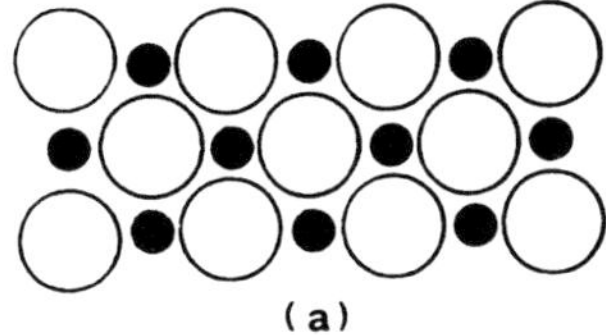

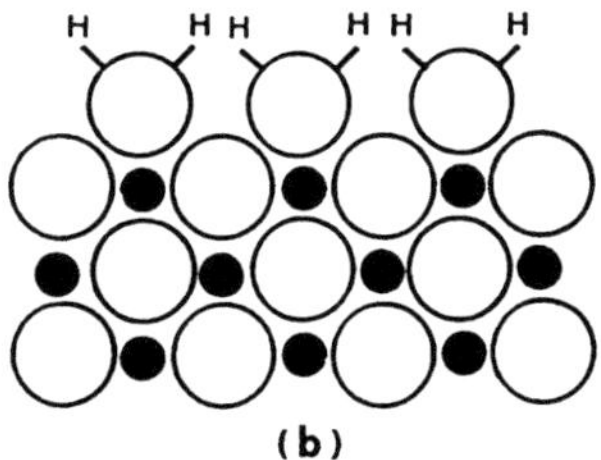

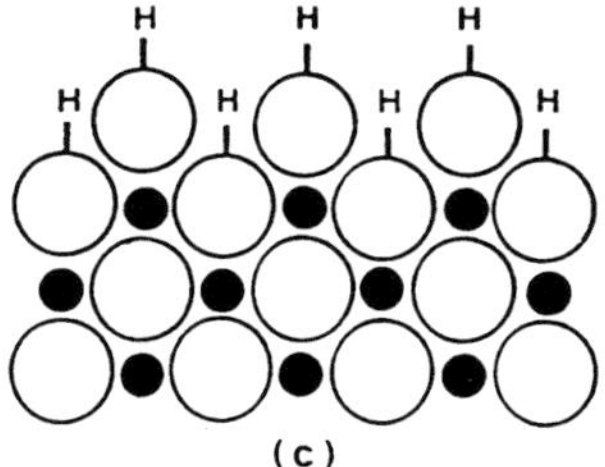

Figure 4.2. Cross section of the surface layer of a metal oxide. (●) metal ions; (○) oxide ions. (a) Surface ions show low coordination and exhibit Lewis acidity. (b) In the presence of water, the surface metal ions may coordinate H_2O molecules. (c) Dissociative chemisorption leads to a hydroxylated surface. (Reproduced by permission from Schindler, 1981.)

Table 4.1. Some Coordinative Environments of Metal Ions S^{z+} in Hydrated Surfaces of (Hydr)oxides

Coordination Unit[a]	Stoichiometry	Pertinent Surface Group
	(a) S^{3+}, Coordination Number 6	
$SO_{4/4}(OH)_{2/2}$	$S_2O_3 \cdot H_2O$	I
$SO_{2/4}(OH)_{4/2}$	$S_2O_3 \cdot 2H_2O$	I
$S(OH)_{6/2}$	$S_2O_3 \cdot 3H_2O$	I
$SO_{2/4}(OH)_{2/2}(OH)(OH_2)$	$S_2O_3 \cdot 4H_2O$	IV
$S(OH)_{4/2}(OH)(OH_2)$	$S(OH)_3 \cdot H_2O$	IV
	(b) S^{4+}, Coordination Number 4	
$SO_{3/2}OH$	$SO_2 \cdot \frac{1}{2}H_2O$	II
$SO_{2/2}(OH)_2$	$SO_2 \cdot H_2O$	III
$SO_{1/2}(OH)_3$	$SO_2 \cdot \frac{3}{2}H_2O$	IV

[a]The suffixes indicate structurally the nearest neighboring atoms, that is, in $S(OH)_{6/2}$ every metal ion S^{3+} has 6 neighboring OH group and every OH group has 2 neighboring S^{3+} metal ions.

of Lewis acid sites at surfaces of dry oxides is well established.) Addition of water leads in a first step to coordination of H_2O molecules to these coordinatively undersaturated metallic centers (Fig. 4.2*b*). Dissociative chemisorption (Fig. 4.2*c*) under formation of hydroxyl groups seems, however, to be energetically favored. Geometrical considerations and chemical observations (in reactions with D_2O, CH_2N_2, CH_3MgI, etc.) indicate an average surface density of 5 (2–10) hydroxyls per square nanometer. The crucial question is whether the formed surface hydroxyls are chemically equivalent. For some surfaces, infrared spectroscopy reveals the presence of different types of S—OH groups. Two bands observed at silica were attributed to free and hydrogen-bonded groups, respectively (Kiselev, 1971). Five different groups were detected at partially dehydroxylated γ-Al_2O_3, the small differences in valence vibration being attributed to differences in the numbers of adjacent oxygen ions (Peri, 1965). Inspection of feasible coordinative environments of some metal ions in hydrated surfaces (Table 4.1) suggests various possibilities such as

I	II	III	IV	V
S–O(H)–S	S—OH	S(OH)(OH)	S(OH₂)(OH)	S(OH)(OH)(OH)

There is little doubt that differences in structure would be reflected in differences in acid–base characteristics as well as in ligand properties. Recent quantum-chemical calculations (on the effect of the coordinative environments upon the Brønsted acidity of surface hydroxyl groups at alumina) indicate that bridging hydroxyls **I** are stronger acids than terminal groups **II** (Kawakami and Yoshida, 1985). It is therefore very likely that a given oxide sample carries different types of surface hydroxyls. The above-mentioned IR data suggest that the different groups can be assigned to a limited number of classes (the groups in a given class being identical). The possibility of hydroxyl groups covering a continuous spectrum of chemical properties cannot be positively ruled out.

There remains the question of the relevance of surface heterogeneity for the chemistry of natural aquatic systems. Weathering reactions usually involve a vast variety of solids. The problem of identifying those minerals that are primarily responsible for the acquired solutes leaves the question of exploring the detailed surface structure of a given solid sample as a topic of minor importance. The same applies to processes involving suspended particulate matter where our main concern is still to gain insight into the chemical composition of the adsorbing surface. The actual significance of surface heterogeneity is seen in connection with the problem of formulating stability constants of surface complexes.

2.3. Evaluating Equilibrium Constants

The concept of surface complexation permits us to handle adsorption equilibria in the same way as equilibria in solutions. Hence uptake and release of H^+ ions in solutions of constant ionic strength (Eqs. 4.1 and 4.2) can be described by the acidity constants

$$K_{a1}^{s} = \frac{\{SOH_2^+\}}{\{SOH\}[H^+]} \quad (dm^3\ mole^{-1}) \tag{4.1a}$$

$$K_{a2}^{s} = \frac{\{SO^-\}[H^+]}{\{SOH\}} \quad (mole\ dm^{-3}) \tag{4.2a}$$

where { } denotes the concentrations of surface species in moles per kilogram of adsorbing solid. Similarly, adsorption equilibria involving metal ions (Eqs. 4.3 and 4.4) are conveniently characterized by

$$^*K_1^s = \frac{\{SOM^{(z-1)+}\}[H^+]}{\{SOH\}[M^{z+}]} \tag{4.3a}$$

$$^*\beta_2^s = \frac{\{(SO)_2M^{(z-2)+}\}[H^+]^2}{\{SOH\}^2[M^{z+}]} \quad (kg\ dm^{-3}) \tag{4.4a}$$

In some cases it is desirable to express the concentrations of the surface species in the same units as the concentrations of dissolved species. The conversion is easily accomplished with the equation

$$[\text{SOH}] = \frac{A}{V}\{\text{SOH}\} \quad (\text{mole dm}^{-3}) \tag{4.9}$$

where A = quantity of oxide in kilograms and V = volume of aqueous phase in cubic decimeters.

The quotients K^s_{a1}, K^s_{a2}, $^*K^s_1$, $^*\beta^s_2$ introduced above and similar expressions related to ligand exchange and formation of ternary surface complexes are experimentally accessible quantities. They have the rank of conditional stability constants whose values (at constant temperature, pressure, and ionic strength) are dependent on the prevailing surface coverage. A frequently encountered empirical relation is

$$K^s(\text{X}) = K^s_{(\text{int})}(\text{X}) \exp(-\alpha\{\text{X}\}) \tag{4.10}$$

$$\log K^s(\text{X}) = \log K^s_{(\text{int})}(\text{X}) - \frac{\alpha}{\ln(10)}\{\text{X}\} \tag{4.10a}$$

$K^s(\text{X})$ is the *conditional stability constant of the surface species X*. $K^s_{(\text{int})}$ is the *intrinsic constant*, and α is an empirical constant whose value is dependent on the system under consideration. Some examples for systems where Eqs. 4.10 and 4.10a have been found to apply are collected in Table 4.2.

Equation 4.10a demonstrates the possibility of evaluating intrinsic constants by simple linear extrapolation of the experimentally available conditional constants. The procedure is illustrated by Figure 4.3.

A further implication of Eqs. 4.10 and 4.10a is the existence of a region of low surface coverage (Fig. 4.4) where conditional and intrinsic constants become experimentally indistinguishable. Although frequently observed, Eq. 4.10 is not generally valid. There are systems (Fig. 4.5) where plots of log $K^s(\text{X})$ versus $\{\text{X}\}$ deviate markedly from linearity. The main conclusions, that is, the possibility of evaluating intrinsic constants by extrapolating conditional constants and the existence of regions of low surface coverage where conditional and intrinsic constants become equivalent, remain valid.

The above-mentioned experimental observations indeed call for interpretation. There are obviously three factors that could contribute to the observed decrease of surface stability constants with increasing surface coverage:

1. The existence of a surface potential produced by the adsorbed charged species.
2. Lateral interactions between adsorbed species resulting in an increase of their activity coefficients.
3. Surface heterogeneity (see Section 2.2).

Table 4.2. Dependence of Surface Stability Constants on the Concentrations of Formed Surface Species

Solid	Reaction	Surface Stability Constant	Reference
$Fe(OH)_3$(am)	$>Fe{-}OH + H^+ \rightleftharpoons >Fe{-}OH_2^+$	$\log K_{a1}^s = 6.6 - 1.12\{FeOH_2^+\}^a$	Farley et al. (1985)
	$>Fe{-}OH - H^+ \rightleftharpoons >Fe{-}O^-$	$\log K_{a2}^s = -9.1 - 1.12\{FeO^-\}^a$	Farley et al. (1985)
TiO_2 (rutile)	$\gg Ti{-}OH + H^+ \rightleftharpoons \gg Ti{-}OH_2^+$	$\log K_{a1}^s = 4.13 - 176\{TiOH_2^+\}^b$	Gisler (1980)
	$\gg Ti{-}OH - H^+ \rightleftharpoons \gg Ti{-}O^-$	$\log K_{a2}^s = -7.39 - 69\{TiO^-\}^b$	Gisler (1980)
Fe_3O_4	$>Fe{-}OH + Co^{2+} - H^+ \rightleftharpoons >FeOCo^+$	$\log {}^*K_1^s = -2.44 - 384\{FeOCo^+\}^c$	Tamura et al. (1983)
	$2>Fe{-}OH + Co^{2+} - 2H^+ \rightleftharpoons (>FeO)_2Co$	$\log {}^*\beta_2^s = -6.71 - 1898\{(FeO)_2Co\}^c$	Tamura et al. (1983)

[a] 298.2 K, $I = 0.1$.
[b] 298.2 K, $I = 1.0$.
[c] 298.2 K, $I = 0$.

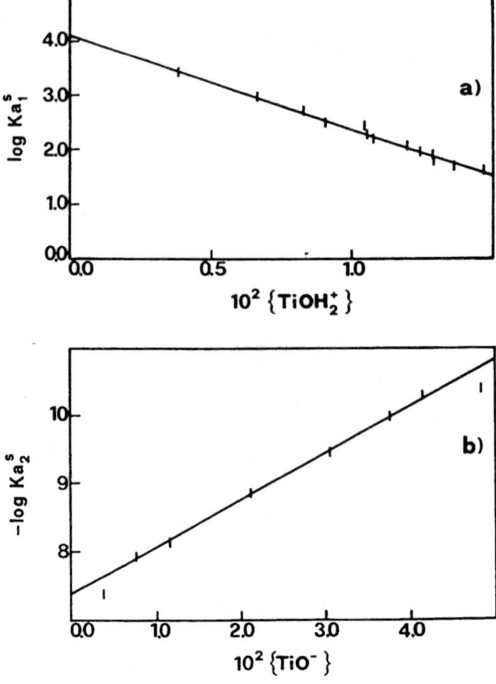

Figure 4.3. Evaluation of intrinsic surface acidity constants from conditional constants. The solid straight lines are calculated with the equations given in Table 4.2. The linear extrapolation made implies that in the low pH range $\{SOH_2^+\} \gg \{SO^-\}$ and in the high pH range $\{SO^-\} \gg \{SOH_2^+\}$.

With charged surface species there is a strong overlap of the first two factors (Sposito, 1983). Our knowledge of surface heterogeneity is meager. No comprehensive theory for calculating activity coefficients of surface species is available. Hence most efforts have been focused on the effect of surface potentials. The observable change in Gibbs free energy related to the adsorption process is thus split according to

$$\Delta G_{\text{adsorption}} = \Delta G_{\text{intrinsic}} + \Delta G_{\text{coulombic}} \tag{4.11}$$

where

$$\Delta G_{\text{adsorption}} = -RT\,(\ln 10)\log K^s \tag{4.12}$$

and

$$\Delta G_{\text{intrinsic}} = -RT\,(\ln 10)\log K^s_{(\text{int})} \tag{4.13}$$

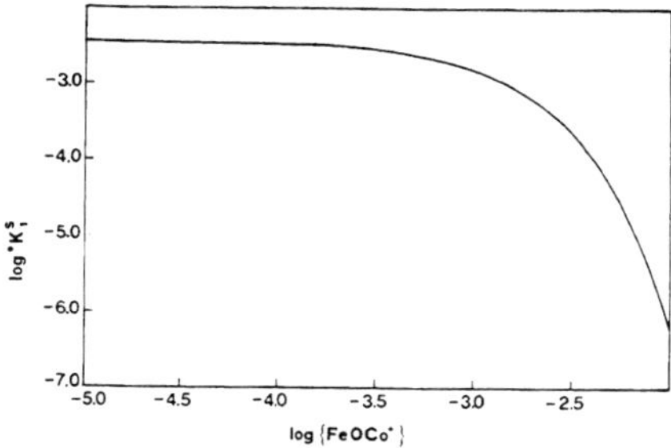

Figure 4.4. The conditional stability constant of the complex $>FeOCo^+$ on spherical magnetite is a function of its surface concentration. At surface concentrations lower than 3×10^{-4} mole kg^{-1}, this dependence is vanishing and the conditional constant accepts the value of the intrinsic constant. (Calculated from Tamura et al., 1983.)

$\Delta G_{\text{coulombic}}$ is the energy required to bring an ion from the bulk of the solution to a surface site at a potential ψ_0. Hence

$$\Delta G_{\text{coulombic}} = zF\psi_0 \tag{4.14}$$

where z is the charge number of the ion under consideration. Combining Eqs. 4.11–4.14 results in

$$\log K^s = \log K^s_{(\text{int})} - \frac{zF}{RT \ln 10} \psi_0 \tag{4.15}$$

The obvious difficulty lies in the fact that ψ_0 values of suspended oxide particles are experimentally inaccessible. [For qualifications of this statement, see

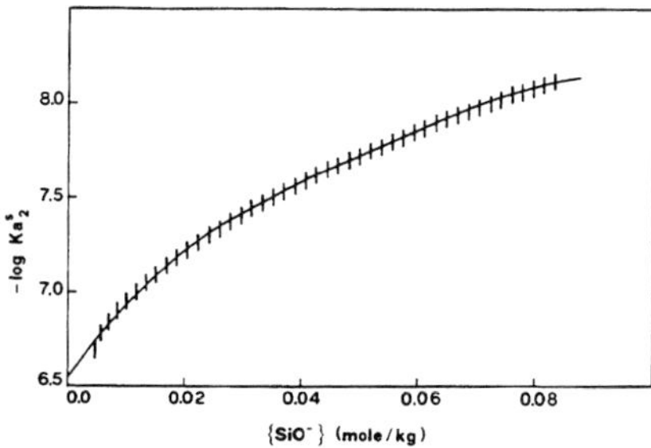

Figure 4.5. Evaluation of the intrinsic acidity constant of surface silanol groups (Aerosil 200, 298.2 K, 0.2 M KNO_3). The solid curve was calculated with the equation $\log K^s_{a2} = -6.527 - 48.74\{SiO^-\} + 9.29.3\{SiO^-\}^2 - 11236\{SiO^-\}^3 + 52443\{SiO^-\}^4$.

Westall (Chapter 1, this volume).] Earlier attempts to estimate surface potentials assuming ideal Nernstian behavior

$$\psi_0 = \frac{RT \ln 10}{F} (\mathrm{pH_{zpc}} - \mathrm{pH})$$

have been proved to be inadequate. We are thus left with attempts to compute the surface potential from the surface charge using a double-layer model. Some of the current models are depicted in Figure 4.6. Model I is the Helmholtz or constant-capacitance model

$$\psi_0 = \frac{\sigma}{\kappa} \quad (\mathrm{V}) \tag{4.16}$$

where σ is the specific surface charge in coulombs per square meter and κ is the specific capacitance in farads per square meter. Model II is the diffuse-layer model as applied by Stumm et al. (1970). It differs from the familiar Gouy–Chapman model insofar as a fixed number of surface sites is implied. Model III is the triple-layer model as introduced by Yates et al. (1975) and applied by Davis et al. (1978). It is essentially an extended Stern model that differs from its precursor in that the compact layer is split into two parts. A modified version has been proposed by Barrow et al. (1981).

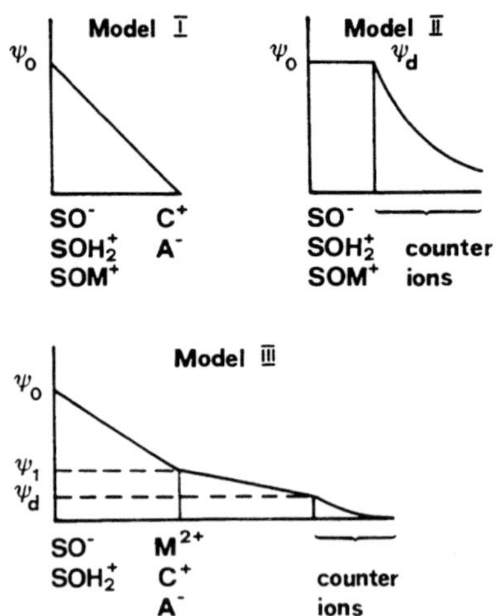

Figure 4.6. Schematic presentation of the surface–solution interface for the case of an oxide suspended in a solution containing the divalent metal ion M^{2+} and the supporting electrolyte C^+A^-. For each model the potential is shown as a function of distance from the surface; the assumed positions of the adsorbed ions are indicated. (Reproduced by permission from Schindler, 1985.)

Incidentally, the constant-capacitance model is formally consistent with the empirical equation (4.10a): For a reaction that increases the surface charge by formation of a charged surface species X^z, Eq. 4.16 takes the form

$$\psi_0 = \frac{zF\{X^z\}}{\kappa S} \quad (V) \tag{4.17}$$

where S is the specific surface area in square meters per kilogram. Combining Eqs. 4.15 and 4.17 results in

$$\log K^s = \log K^s_{(\text{int})} - \frac{(zF)^2}{RT(\ln 10)\kappa S}\{X^s\} \tag{4.18}$$

From Eq. 4.10a and 4.18 it is seen that

$$\alpha = \frac{(zF)^2}{RT\kappa S} \quad (\text{kg mole}^{-1}) \tag{4.19}$$

As shown by Westall and Hohl (1980) and by Morel et al. (1981), the three models are equivalent in fitting the experimental data. They differ, however, in their perception of how $\Delta G_{\text{adsorption}}$ is split into $\Delta G_{\text{intrinsic}}$ and $\Delta G_{\text{coulombic}}$; they thus produce somewhat different values of $K^s_{(\text{int})}$ (Table 4.3) and sometimes even different surface species. This means that published values of stability constants of surface species are not yet as reliable as stability constants of metal–ligand complexes in homogeneous systems. Moreover, the number of values published to date is comparatively small. The subsequent sections will demonstrate, however, that despite the scarcity and imprecision of the available data, characteristic features can be recognized and—which is of utmost important—missing information can often be obtained from solution chemistry.

Table 4.3. Acidity Constants of Surface Al—OH Groups Computed with Different Double-Layer Models[a]

Method/Model	$\log K^s_{a1}$	$\log K^s_{a2}$	Reference
Extrapolation[b]	7.2	−9.5	Hohl and Stumm (1976)
Model I[b]	7.40	−9.24	Westall and Hohl (1980)
Model II	7.66	−8.98	Westall and Hohl (1980)
Model III	7.33	−9.31	Westall and Hohl (1980)

[a]Experimental data from Hohl and Stumm (1976).
[b]The differences originate from the fact that α values observed in linear regressions to obtain K^s_{a1} and K^s_{a2} are usually not identical, whereas fittings on the basis of model I assume the same value of κ (and thus α) for the entire pH range.

3. ACID–BASE CHARACTERISTICS OF SURFACE HYDROXYL GROUPS

Representative values of both $K^s_{a1(\text{int})}$ and $K^s_{a2(\text{int})}$ are collected in Table 4.4. In order to facilitate comparisons, the selection is restricted to values obtained from extrapolation or computed on the basis of the constant-capacitance model. Note the striking effect of the ionic strength exhibited in the system TiO_2—LiCl. It strongly suggests the formation of ion pairs with the components (M^+, A^-) of the background electrolyte:

$$S{-}OH_2^+ + A^- \rightleftharpoons S{-}OH_2^+A^-$$

$$S{-}O^- + C^+ \rightleftharpoons S{-}O^-C^+$$

On the basis of the triple-layer model or by a double-extrapolation technique, it is possible to split $K^s_{a(\text{int})}$ values as defined by Eq. 4.10 into parts related to true proton transfer and ion pairing, respectively. For details, the reader is referred to the lucid review by James and Parks (1982).

As suggested by Parks (1965), the constants K^s related to surface acidity should correlate with the corresponding constants K in solution. Specifically, the equilibria to be compared are

$$S{-}OH_2^+ \rightleftharpoons S{-}O^- + H^+ \qquad K^s = 1/K^s_{a1(\text{int})} \quad (\text{mole dm}^{-3})$$

$$S(OH)^+_{z-1}(aq) \rightleftharpoons S(OH)^0_z(aq) + H^+ \qquad K = {}^*K_z \quad (\text{mole dm}^{-3})$$

and

$$S{-}OH \rightleftharpoons S{-}O^- + H^+ \qquad K^s = K^s_{a2(\text{int})} \quad (\text{mole dm}^{-3})$$

$$S(OH)^0_z(aq) \rightleftharpoons S(OH)^-_{z+1}(aq) + H^+ \qquad K = {}^*K_{z+1} \quad (\text{mole dm}^{-3})$$

respectively. For a rigorous comparison, statistical factors should be included, since the surface stability constants are microscopic constants whereas the constants K are macroscopic constants. The data shown in Figure 4.7 suggest the correlation

$$\log K^s_{(\text{int})} = \log K \pm 1$$

A noticeable and so far unexplained exception is the acidity of surface silanol groups. For H_4SiO_4 (298.2 K, $I = 1$, $H_4SiO_4 \rightleftharpoons H_3SiO_4^- + H^+$), the intrinsic microscopic constant $\log K_{a_{\text{micr}}}$ is $\log(K_{a_1}/4) = -10.07$, whereas the related surface acidity constant ($\log K^s_{a2(\text{int})} = -6.71$) is considerably larger.

Table 4.4. Acidity Constants of Surface Hydroxyl Groups (298.2 K)

Group	Solid	Ionic Medium	$\log K^s_{a1(\text{int})}$	$\log K^s_{a2(\text{int})}$	Reference
Al—OH	γ-Al_2O_3	0.1 *M* $NaClO_4$	7.2	−9.5	Hohl and Stumm (1976)
	δ-Al_2O_3[a]	0.1 *M* $NaClO_4$	7.4	−10.0	Kummert and Stumm (1980)
$Al(OH)(OH_2)$	γ-$Al(OH)_3$	1 *M* KNO_3	5.24[a]	−8.08[b]	Pulfer et al. (1984)
Si—OH	SiO_2(am)	0.1 *M* $NaClO_4$		−6.8	Schindler and Kamber (1968)
		0.2 *M* KNO_3		−6.53	See text.
		1.0 *M* $LiClO_4$		−6.57	Sigg (1973)
		1.0 *M* $NaClO_4$		−6.71	Sigg (1973)
		1.0 *M* CsCl		−5.71	Sigg (1973)
Ti—OH	TiO_2				
	Anatase	3.0 *M* $NaClO_4$	4.98	−7.80	Schindler and Gamsjäger (1972)
	Rutile	1.0 *M* $NaClO_4$	4.13	−7.39	Gisler (1980)
		10^{-3} *M* LiCl	2.75[a]	−9.1[b]	Davis et al. (1978)
		10^{-2} *M* LiCl	3.25	−8.9	Davis et al. (1978)
		0.1 *M* LiCl	3.6	−8.4	Davis et al. (1978)
Zr—OH	ZrO_2	1.0 *M* KNO_3	5.67[a]	−7.91	Huang (1981)
Th—OH	ThO_2	1.0 *M* $NaClO_4$	5.15[a]	−7.90[b]	Schindler et al. (1976a)
Fe—OH	$Fe(OH)_3$(am)	I = 0.1	6.6	−9.1	Farley et al. (1985)
	α-FeOOH	0.1 *M* $NaClO_4$	6.4[a]	−9.25[b]	Stumm et al. (1980)

[a] Material similar to that classified earlier as δ-Al_2O_3.
[b] Values shown in Figure 4.7.

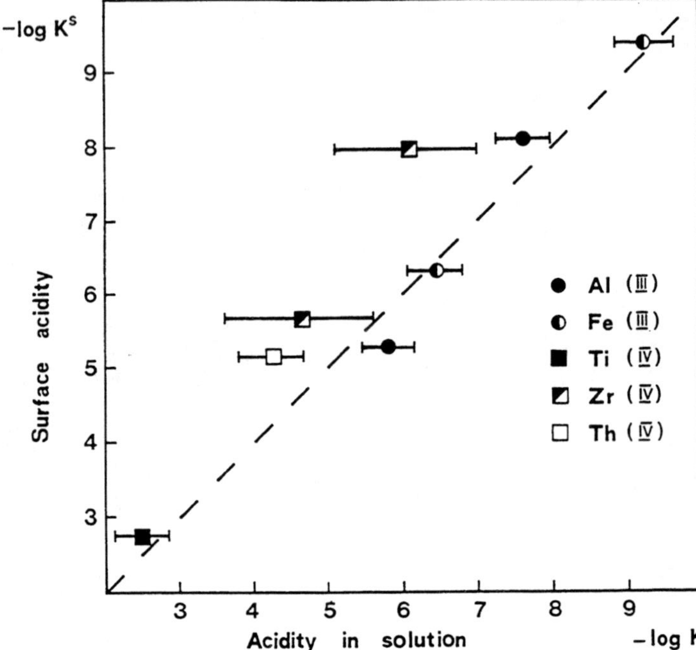

Figure 4.7. Correlation of surface acidity and acidity in solution. Values for surface acidity are taken from Table 4.4. Acidity constants in solution and their relative errors were taken from Baes and Mesmer (1976).

4. METAL COMPLEXES

The assumption that metal ion adsorption is based on competitive complex formation (Eqs. 4.3 and 4.4) justifies the use of the pH value as the master variable that governs the extent of adsorption (Fig. 4.8). For a given system consisting of metal ions M^{z+} and an adsorbing (hydr)oxide, there is a range of 1–2 pH units where the extent of adsorption rises from 0% to almost 100%. Although these general features are confirmed by numerous studies, an unified interpretation has not yet been accepted. Some authors prefer to present their data in terms of Langmuir or Freundlich isotherms. Within the scope of the surface complexation model, both the species obtained and their stability constants are dependent on the choice of double-layer model. Whereas the constant-capacitance model assumes the formation of a surface chelate VI (Eq. 4.4), the triple-layer model leads to ternary surface complexes **VII**.

```
S—O
    \
     M          S—OM—OH
    /
S—O
 VI               VII
```

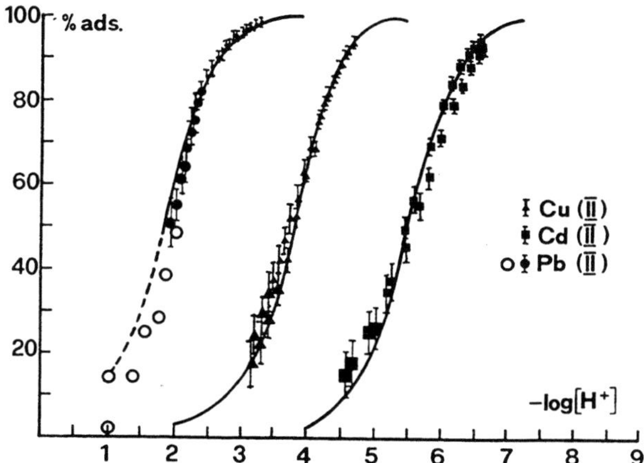

Figure 4.8. Adsorption of some divalent metal ions on TiO_2 (rutile) (Fürst, 1976). For each metal ion there is an interval of 1–2 pH units where the extent of adsorption rises from zero to almost 100%. The solid lines are calculated with the constants given in Table 4.5.

Again for internal consistency, the data collected in Table 4.5 are obtained from the constant-capacitance model or from work at low surface coverage where conditional and intrinsic constants become identical.

Comparison with solution chemistry can be made by relating

$$\text{S—OH} + M^{z+} \rightleftharpoons \text{S—OM}^{(z-1)+} + H^+ \qquad K^s = {}^*K^s_{1(\text{int})}$$

$$\text{H—OH} + M^{z+} \rightleftharpoons \text{H—OM}^{(z-1)+} + H^+ \qquad K = {}^*K_1$$

and

$$2\text{S—OH} + M^{z+} \rightleftharpoons (\text{SO})_2\text{M}^{(z-2)+} + 2H^+ \qquad K^s = {}^*\beta^s_{2(\text{int})}$$

$$2\text{H—OH} + M^{z+} \rightleftharpoons (\text{HO})_2\text{M}^{(z-2)+} + 2H^+ \qquad K^s = {}^*\beta_2$$

Figure 4.9 shows the excellent correlation for the case of silica as an adsorbing solid. The correlation is less pronounced for TiO_2.

Significant correlation has again been found (on the basis of the triple-layer model) for γ-Al_2O_3 (Hachiya et al., 1984a) as well as for α-FeOOH and amorphous $Fe(OH)_3$ (Balistrieri et al., 1981). These correlations are the basis for the well-known (James and Healy, 1972) coincidence of adsorption and hydrolysis and the controversy about the "hydrolysis + adsorption" scheme (Eqs. 4.20 and 4.21) and the surface complexation model (Eq. 4.22).

$$(H_2O)_5M(OH_2)^{z+} + H_2O \rightleftharpoons (H_2O)_5MOH^{(z-1)+} + H_3O^+ \qquad (4.20)$$

$$(H_2O)_5MOH^{(z-1)+} + SOH \rightleftharpoons SOM(H_2O)_5^{(z-1)} + H_2O \qquad (4.21)$$

$$\overline{\qquad\qquad\qquad\qquad\qquad\qquad\qquad\qquad\qquad\qquad}$$

$$(H_2O)_5M(OH_2)^{z+} + SOH \rightleftharpoons SOM(H_2O)_5^{(z-1)} + H_3O^+ \qquad (4.22)$$

Table 4.5. Stability Constants of Metal Complexes (298.2 K)

Group	Solid	M^{z+}	Ionic Medium	$\log {}^*K^s_{1(\text{int})}$	$\log {}^*\beta^s_{2(\text{int})}$	Reference
Al—OH	γ-Al_2O_3	Ca^{2+}	0.1 *M* $NaNO_3$	−6.1	—	Huang and Stumm (1973)
		Mg^{2+}	0.1 *M* $NaNO_3$	−5.4	—	Huang and Stumm (1973)
		Ba^{2+}	0.1 *M* $NaNO_3$	−6.6	—	Huang and Stumm (1973)
		Pb^{2+}	0.1 *M* $NaClO_4$	−2.2	−8.1	Hohl and Stumm (1976)
		Cu^{2+}	0.1 *M* $NaClO_4$	−2.1	−7.0	Hohl and Stumm (1976)
Si—OH	SiO_2(am)	Mg^{2+}	1 *M* $NaClO_4$	−7.7	−17.15	Gisler (1980)
		Fe^{3+}	3 *M* $NaClO_4$	−1.77	−4.22	Schindler et al. (1976a)
		Cu^{2+}	1 *M* $NaClO_4$	−5.52	−11.19	Schindler et al. (1976a)
		Cd^{2+}	1 *M* $NaClO_4$	−6.09	−14.20	Schindler et al. (1976a)
		Pb^{2+}	1 *M* $NaClO_4$	−5.09	−10.68	Schindler et al. (1976a)
Ti—OH	TiO_2 (rutile)	Mg^{2+}	1 *M* $NaClO_4$	−5.90	−13.13	Gisler (1980)
		Co^{2+}	1 *M* $NaClO_4$	−4.30	−10.60	Gisler (1980)
		Cu^{2+}	1 *M* $NaClO_4$	−1.43	−5.04	Fürst (1976)
		Cd^{2+}	1 *M* $NaClO_4$	−3.32	−9.00	Fürst (1976)
		Pb^{2+}	1 *M* $NaClO_4$	0.44	−1.95	Fürst (1976)
Mn—OH	δ-MnO_2	Ca^{2+}	0.1 *M* $NaNO_3$	−5.5	—	Stumm et al. (1970)
Fe—OH	Fe_3O_4	Co^{2+}	$I = 0$	−2.44	−6.71	Tamura et al. (1983)

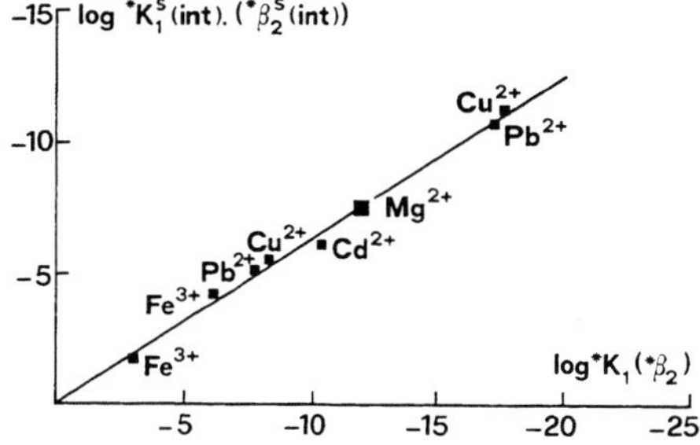

Figure 4.9. Correlation of stability constants log $^*K^s_{1(int)}$ ($^*\beta^s_{2(int)}$) of surface complexes at amorphous silica with stability constants log $^*K_1$ ($^*\beta_2$) of hydroxo complexes. The solid line represents the equation log $^*K^s_{1(int)}$ ($^*\beta^s_{2(int)}$) = $-0.09 + 0.62$ log $^*K_1$ ($^*\beta_2$). Reproduced by permission from Schindler, 1985.

Although the two mechanisms are indistinguishable on the basis of stoichiometric and thermodynamic observations, kinetic studies (Hachiya et al., 1984b) clearly support the surface complexation model.

5. ADSORPTION OF ANIONS

As with metal ions, the extent of adsorption of anions is strongly governed by the pH of the solution. Since adsorption of anions is coupled with a release of OH^- ions, adsorption is favored by low pH values (Fig. 4.10). The equilibrium constants observed to date are given in Table 4.6. Again the tendency

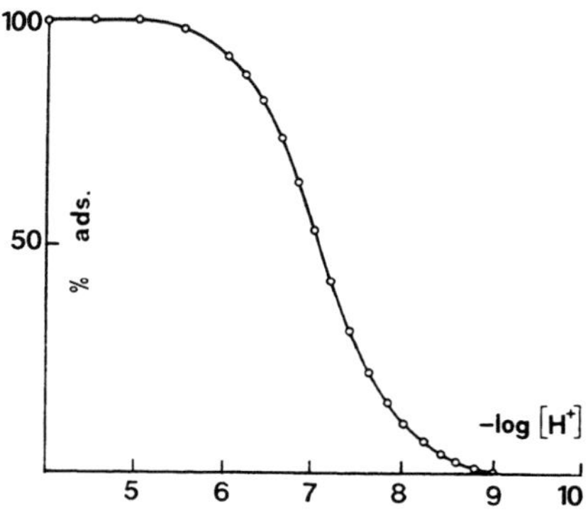

Figure 4.10. Adsorption of F^- on γ-$Al(OH)_3$ (calculated from data given by Pulfer et al., 1984). Since adsorption of anions is coupled with a release of OH^- ions, the extent of adsorption increases with decreasing pH.

Table 4.6. Intrinsic Stability Constants of Anion Complexes (295 K, ionic strength not specified)

Equilibrium		$\log K^s_{(int)}$
	1. α-FeOOH	
$>FeOH + F^-$	$\rightleftharpoons >FeF + OH^-$	−4.8
$>FeOH + SO_4^{2-}$	$\rightleftharpoons >FeSO_4^- + OH^-$	−5.8
$2{>}FeOH + SO_4^{2-}$	$\rightleftharpoons (>Fe)_2SO_4 + 2OH^-$	−13.5
$>FeOH + HAc$	$\rightleftharpoons >FeAc + H_2O$	2.9
$>FeOH + H_4SiO_4$	$\rightleftharpoons >FeSiO_4H_3 + H_2O$	4.1
$>FeOH + H_4SiO_4$	$\rightleftharpoons >FeSiO_4H_2^- + H_3O^+$	−3.3
$>FeOH + H_3PO_4$	$\rightleftharpoons >FePO_4H_2 + H_2O$	9.5
$>FeOH + H_3PO_4$	$\rightleftharpoons >FePO_4H^- + H_3O^+$	5.1
$>FeOH + H_3PO_4 + H_2O$	$\rightleftharpoons >FePO_4^{2-} + 2H_3O^+$	−1.5
$2{>}FeOH + H_3PO_4$	$\rightleftharpoons (>Fe)_2PO_4H + 2H_2O$	8.5
$2{>}FeOH + H_3PO_4$	$\rightleftharpoons (>Fe)_2PO_4^- + H_2O + H_3O^+$	4.5
	2. γ-Al_2O_3	
Benzoic acid		
	$>AlOH + HA \rightleftharpoons >AlA + H_2O$	3.7
Catechol		
	$>AlOH + H_2A \rightleftharpoons >AlAH + H_2O$	3.7
	$>AlOH + H_2A \rightleftharpoons >AlA^- + H_3O^+$	−5
Phthalic acid		
	$>AlOH + H_2A \rightleftharpoons >AlAH + H_2O$	7.3
	$>AlOH + H_2A \rightleftharpoons >AlA^- + H_3O^+$	2.4
Salicyclic acid		
	$>AlOH + H_2A \rightleftharpoons >AlAH + H_2O$	6.0
	$>AlOH + H_2A \rightleftharpoons >AlA^- + H_3O^+$	−0.6

Source: Reprinted by permission from Stumm et al. (1980).

to form surface complexes is correlated with the tendency to form complexes in solution. This can be exemplified by comparing the surface reaction

$$>FeOH + H_2A \rightleftharpoons >FeHA + H_2O \qquad {}^*K_1^s$$

with the corresponding reaction in solution

$$FeOH^{2+} + H_2A \rightleftharpoons FeHA^{2+} + H_2O \qquad {}^*K_1$$

The results of this comparison are shown in Figure 4.11.

Besides having this support from thermodynamics, the proposed model is in agreement with investigations on the kinetics of adsorption of acetic acid

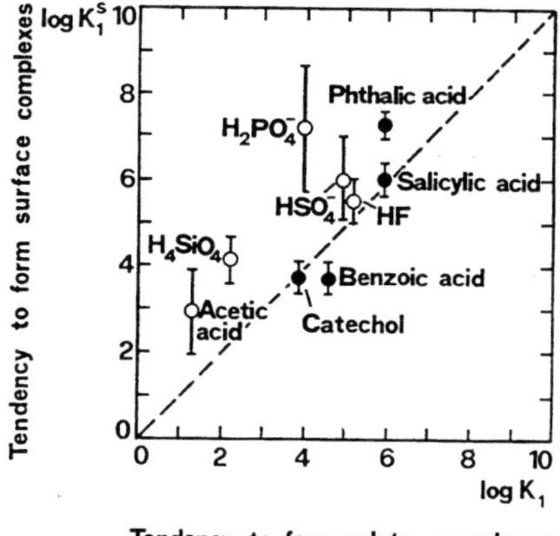

Figure 4.11. Comparison of the tendency to form surface complexes (expressed by log $^*K_1^s$) with the tendency to form solute complexes (log $^*K_1$). (○) with >Fe–OH and $FeOH^{2+}$, respectively. (●) with >Al–OH and $AlOH^{2+}$, respectively. Reproduced by permission from Stumm et al., 1980.

(HAc) on silica–alumina surfaces (Ikeda et al., 1982) that demonstrate that the important step consists of

$$>\mathrm{Al{-}OH_2^+} + \mathrm{Ac^-} \rightleftharpoons >\mathrm{AlAc} + \mathrm{H_2O}$$

6. TERNARY SURFACE COMPLEXES

The studies on metal ion adsorption mentioned so far were carried out in the presence of noncomplexing anions such as ClO_4^- and NO_3^-. Similarly, adsorption of anions was performed with counterions (Na^+, K^+) that exhibit little Lewis acidity. Natural aquatic systems contain both Lewis acids and Lewis bases, and the question of how complex formation in solution affects adsorption (and thus transport properties) is of some relevance. Qualitatively one can say that dissolved ligands will as a rule impede the adsorption of metal ions and that dissolved metal ions will impede the adsorption of ligands. Both effects are based on competition,

$$\mathrm{S{-}OM^{(z-1)+}} + \mathrm{HL} \rightleftharpoons \mathrm{S{-}OH} + \mathrm{ML^{(z-1)+}}$$

$$\mathrm{SL} + \mathrm{M^{z+}} + \mathrm{H_2O} \rightleftharpoons \mathrm{S{-}OH} + \mathrm{ML^{(z-1)+}} + \mathrm{H^+}$$

and can thus quantitatively be predicted if the composition of the system and the pertinent stability constants are known. The realiability of such calculations is based on the condition that no ternary complexes (Eqs. 4.5 and 4.8) are formed.

In 1978, Bourg and Schindler found evidence for formation of type A ternary surface complexes $>\mathrm{SiOCu(en)^+}$ and $(>\mathrm{SiO})_2\mathrm{SiOCu(en)^0}$ (en =

Table 4.7. Stabilities of Ternary Surface Complexes (298.2 K)

System[a]	Ionic Medium	$R_{1,1}$	$R_{1,2}$	$R_{2,1}$	Reference
SiO_2—Cu(II)en	1 *M* $NaClO_4$	2.0	0	4.2×10^{-2}	Bourg and Schindler (1978)
SiO_2—Cu(II)Gly	1 *M* KNO_3	0.38	0	8.9×10^{-4}	Basak et al. (1987)
SiO_2—Cu(II)Ox	1 *M* KNO_3	0.13	0	5.1×10^{-3}	Basak et al. (1987)
SiO_2—Cu(II)bipy	0.1 *M* $NaClO_4$	4.9×10^{2}	4.57×10^{3}	0	Basak et al. (1987)
SiO_2—Mg(II)Gly	1 *M* $NaClO_4$	0.68	0	0	Basak et al. (1987)
SiO_2—Mg(II)Ala	1 *M* $NaClO_4$	1	0	0	Basak et al. (1987)
TiO_2—Mg(II)Gly	1 *M* $NaClO_4$	0.94	0	0	Basak et al. (1987)
TiO_2—Co(II)Gly	1 *M* $NaClO_4$	1.15	0	4.6×10^{-3}	Basak et al. (1987)

[a]en, ethylenediamine; Gly, glycine; Ox, oxalate; bipy, 2,2′-bipyridine; Ala, alanine.

ethylenediamine). Simultaneously, Davis and Leckie (1978) presented evidence for the existence of the type B ternary complex $>FeOSO_2SAg$. Since then, the number of identified type A ternary surface complexes has increased. As in solution chemistry, the stability of ternary surface complexes can conveniently be expressed by the ratio

$$R_{n,e} = \frac{{}^*\beta_n^s(ML_l)}{{}^*\beta_n^s(M)} \tag{4.23}$$

The values of $R_{n,e}$ obtained so far are collected in Table 4.7.

The laws that govern the stabilities of type A ternary surface complexes are not well understood. From statistical considerations [including uncertainties related to the coordination spheres of Co(II) and Cu(II), respectively], one would expect $0.33 \leqq R_{1,1} \leqq 0.67$; $0.08 \leqq R_{2,1} \leqq 0.42$; $0 \leqq R_{1,2} \leqq 0.33$. It seems that $R_{1,1}$ is partially controlled by statistics. In addition, charge effects may be involved and may even play a major role. The negatively charged silica surface shows a slight preference for positively charged complexes, explaining the sequence $R_{1,1}(Cuen^{2+}) > R_{1,1}(CuGly^{+}) > R_{1,1}(CuOx^{0})$. For unknown reasons the observed values for both $R_{2,1}$ and $R_{1,2}$ are far from statistical predictions. The overall contribution of formation of ternary surface complexes is that it lessens the effect of ligand competition. Nevertheless, the addition of dissolved ligands shifts the region of metal adsorption to higher pH values even in systems where ternary surface complexes are formed (Fig. 4.12).

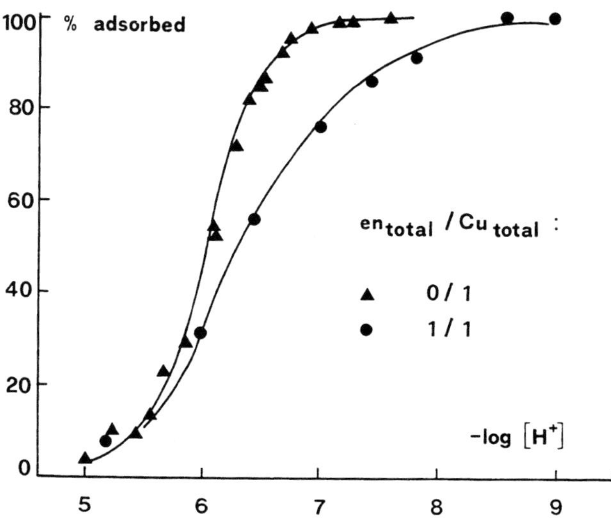

Figure 4.12. Effect of ethylenediamine upon the adsorption of Cu(II) on amorphous silica. The solid lines have been calculated on the basis of ligand competition including formation of ternary surface complexes. (Reproduced by permission from Bourg and Schindler, 1978.)

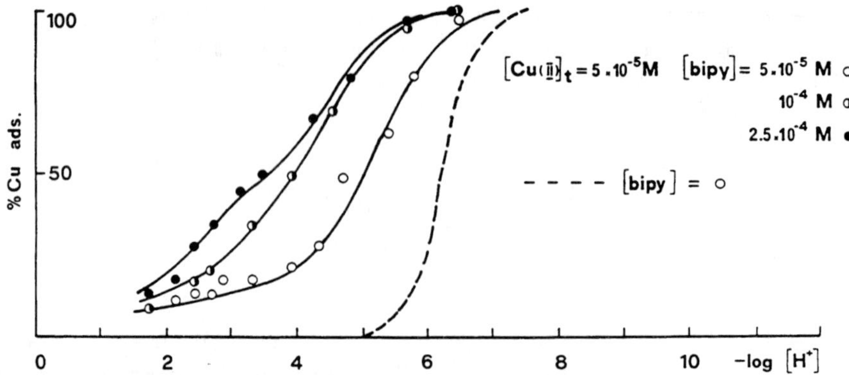

Figure 4.13. Effect of 2,2′-bipyridine upon the adsorption of Cu(II) on amorphous silica. The solid lines are calculated with the constants given in Table 4.7 including the equilibria $\geqslant$SiOH + $Cubipy^{2+}$ $\rightleftharpoons$ $\geqslant$$SiOHCubipy^{2+}$, log K_1^s = 0.8; and $\geqslant$SiOH + $Cu(bipy)_2^{2+}$ $\rightleftharpoons$ $\geqslant$$SiOHCu(bipy)_2^{2+}$, log K_2^s = 1.63.

A striking effect is exhibited by 2,2′-bipyridine. The addition of this ligand to systems consisting of Cu(II) and amorphous silica greatly favors Cu(II) adsorption (Fig. 4.13). The effect is presumably based on hydrophobic interaction of the coordinated ligand with the silica surface. This presumption is sustained by the observation that in addition to the ternary surface complexes mentioned in Table 4.7 two surface species, $SiOHCubipy^{2+}$ and $SiOHCu(bipy)_2^{2+}$, are formed. On the other hand, adsorption of uncoordinated 2,2′-bipyridine on SiO_2 surfaces has so far not been observed. The effect is not restricted to 2,2′-bipyridine; similar promotion of Cu(II) adsorption has been observed with 1,10-phenanthroline and 2,2′,6′,2″-terpyridine (von Zelewsky and Bemtgen, 1982). Moreover, 2,2′-bipyridine also promotes adsorption of Zn(II) at silica (Basak et al., 1987).

7. CONCLUSIONS

The interaction of hydrous oxide surfaces with solute H^+, OH^-, metal ions, and ligands is interpreted in terms of coordination chemistry. Thus, the adsorption equilibria of H^+ (OH^-), metal ions, ligands, and metal–ligand complexes with the surface can be handled in the same way as equilibria in solutions. Experimentally accessible equilibrium constants often yield intrinsic constants by extrapolation to low surface coverage. Surface acidity constants can usually be correlated with the corresponding constants in solution. Similarly, the tendency to form surface complexes (metal ions with surface oxygen donor atoms and ligands with the Lewis acid–metallic centers, respectively) can be correlated with solute equilibrium constants, that is, metal ions to form hydroxo complexes, and ligand exchange equilibria, respectively.

Treating the adsorption of metal ions and anions (ligands), respectively, as competitive complex formation (metal ion versus H^+ of surface OH^- group) and competitive ligand exchange (anion versus surface OH^-) establishes the pH value as a master variable that governs the extent of adsorption (increasing or decreasing with pH, respectively, for metal ions and ligands). Adsorption of both Lewis acids and Lewis bases often yields ternary surface complexes.

A better insight into the mechanism of surface binding has been gained by combining thermodynamic information with that on the structure of both binary and ternary surface complexes obtained by molecular methods (spectroscopy) and from kinetic studies on the mechanism of adsorption rate constants. Surface complexes are of importance also in geochemical kinetics, for example, in the dissolution and formation (heteronucleation) of mineral phases, where surface processes are often the rate-controlling steps.

Acknowledgment

Research grants to Paul W. Schindler and to Werner Stumm, respectively, from the Swiss National Science Foundation have supported our research on surface chemistry for the past decade.

REFERENCES

Baes, C. F., Mesmer, R. E. (1976), The hydrolysis of cations, Wiley Interscience.

Balistrieri, L. S., Brewer, P. G., and Murray, J. W. (1981), Scavenging residence times of trace metals and surface chemistry of sinking particles in the deep ocean, *Deep Sea Res.* **28,** 101–121.

Barrow, N. J., Bowden, J. W., Posner, A. M., and Quirk, J. P. (1981), Describing the adsorption of copper, zinc, and lead on a variable charge mineral surface, *Aust. J. Soil Res.* **19,** 309–321.

Basak, M., Bourg, A. C. M., Cornell, R. M., Gisler, A., Schindler, P. W., Stettler, E., and Trusch, B. (1987), The effect of dissolved ligands upon the adsorption of metal ions at oxide–water interfaces. (to be published)

Bleam, W. F., and McBride, M. B. (1985), Cluster formation versus isolated-site adsorption. A study of Mn(II) and Mg(II) adsorption on boehmite and goethite, *J. Colloid Interface Sci.* **103,** 124–132.

Bourg, A. C. M., and Schindler, P. W. (1978), Ternary surface complexes 1. Complex formation in the system silica–Cu(II)–ethylenediamine, *Chimia* **32,** 166–168.

Davis, J. A., James, R. O., and Leckie, J. O. (1978), Surface ionisation and complexation at the oxide/water interface. I. Computation of electrical double layer properties in simple electrolytes, *J. Colloid Interface Sci.* **63,** 480–499.

Davis, J. A., and Leckie, J. O. (1978), Effect of adsorbed complexing ligands on trace metal uptake by hydrous oxides, *Environ. Sci. Technol.* **2,** 1309–1315.

Farley, K. J., Dzombak, D. A., and Morel, F. M. M. (1985), A surface precipitation model for the sorption of cations on metal oxides, *J. Colloid Interface Sci.* **106,** 226–242.

Gisler, A. (1980), Die Adsorption von Aminosäuren an Grenzflächen Oxid-Wasser. Ph.D. thesis, University of Bern, Bern, Switzerland.

Hachiya, K., Sasaki, M., Saruta, Y., Mikami, N., and Yasunaga, T. (1984a), Static and kinetic studies of adsorption–desorption of metal ions on a γ-Al_2O_3 surface. 1. Static study of adsorption–desorption, *J. Phys. Chem.* **88,** 23–27.

Hachiya, K., Sasaki, M., Ikeda, T., Mikami, N., and Yasunaga, T. (1984b), Static and kinetic studies of adsorption–desorption of metal ions on a γ-Al_2O_3 surface. 2. Kinetic study by means of pressure jump technique, *J. Phys. Chem.* **88,** 27–31.

Hingston, F. (1981), "A Review of Anion Adsorption," in M. A. Anderson and A. J. Rubin, Eds., *Adsorption of Inorganics at Solid–Liquid Interfaces,* Ann Arbor Science, Ann Arbor, MI, pp. 51–90.

Hohl, H., and Stumm, W. (1976), Interaction of Pb^{2+} with hydrous γ-Al_2O_3, *J. Colloid Interface Sci.* **55,** 281–288.

Hohl, H., Werth, E., Giovanoli, R., and Posch, E. (1985), Heterogeneous nucleation. I. Nucleation of calcium fluoride on cerium(IV) oxide. Unpublished report as quoted in P. W. Schindler (1985).

Huang, C. P. (1981), "The Surface Acidity of Hydrous Solids," in M. A. Anderson and A. J. Rubin, Eds., *Adsorption of Inorganics at Solid–Liquid Interfaces,* Ann Arbor Science, Ann Arbor, MI, pp. 183–217.

Huang, C. P., and Stumm, W. (1973), Specific adsorption of cations on hydrous γ-Al_2O_3, *J. Colloid Interface Sci.* **43,** 409–420.

Ikeda, T., Sasaki, M., Hachiya, K., Astumian, R. D., Yasunaga, T., and Schelly, Z. A. (1982), Adsorption-desorption kinetics of acetic acid on silica-alumina particles in aqueous suspensions, using the pressure-jump relaxation method, *J. Phys. Chem.* **86,** 3861–3866.

James, R. O., and Healy, T. W. (1972), Adsorption of hydrolyzable metal ions at the oxide–water interface, I, II and III, *J. Colloid Interface Sci.* **40**(1), 42–81.

James, R. O., and Parks, G. A. (1982), "Characterization of Aqueous Colloids by Their Electrical Double Layer and Intrinsic Surface Chemical Properties," in E. Matijevic, Ed., *Surface and Colloid Science,* Vol. 12, Plenum, New York, pp. 119–215.

Kawakami, H., and Yoshida, S. (1985), Quantum-chemical studies of alumina. Part 1: Brønsted acidity and basicity, *J. Chem. Soc. Faraday Trans. 2* **81,** 1117–1127.

Kinniburgh, D. G., and Jackson, M. L. (1981), "Cation Adsorption by Hydrous Metal Oxides and Clay," in M. A. Anderson and A. J. Rubin, Eds., *Adsorption of Inorganics at Solid–Liquid Interfaces,* Ann Arbor Science, Ann Arbor, MI, pp. 91–160.

Kiselev, A. V. (1971), The effect of the geometrical structure and the chemistry of oxide surfaces on their adsorption properties, *Discuss. Faraday Soc.* **52,** 14–32.

Kummert, R., and Stumm, W. (1980), Surface complexation of organic acids on hydrous γ-Al_2O_3, *J. Colloid Interface Sci.* **75,** 373–385.

Morel, F. M. M., Westall, J. C., and Yeasted, J. G. (1981), "Adsorption Models: A Mathematical Analysis in the Framework of General Equilibrium Calculations," in M. A. Anderson and A. J. Rubin, Eds., *Adsorption of Inorganics at Solid–Liquid Interfaces,* Ann Arbor Science, Ann Arbor, MI, pp. 263–294.

Motschi, H. (1987), "Aspects of the Molecular Structure in Surface Complexes: Spectroscopic Investigations," this volume, Chapter 5.

O'Melia, C. R. (1987), "Particle–Particle Interactions," this volume, Chapter 14.

Parks, G. A. (1965), The isoelectric points of solid oxides, solid hydroxides and aqueous hydroxo complex systems, *Chem. Rev.* **65,** 177.

Peri, J. B. (1965), A model for the surface of γ-alumina, *J. Phys. Chem.* **69,** 220–230.

Pulfer, K., Schindler, P. W., Westall, J. C., and Grauer, R. (1984), Kinetics and mechanism of dissolution of bayerite (γ-$Al(OH)_3$) in HNO_3–HF solutions at 298.2 K, *J. Colloid Interface Sci.* **101,** 554–564.

Schindler, P. W. (1985), Grenzflächenchemie oxidischer Mineralien, *Oester. Chem. Z.* **86,** 141–147.

Schindler, P. W., Fürst, B., Dick, R., and Wolf, P. U. (1976b), Ligand properties of surface silanol groups. I. Surface complex formation with Fe^{3+}, Cu^{2+}, Cd^{2+} and Pb^{2+}, *J. Colloid Interface Sci.* **55,** 469–475.

Schindler, P. W., and Gamsjäger, H. (1972), Acid–base reactions of the TiO_2 (anatase)–water interface and the point of zero charge of TiO_2 suspensions, *Kolloid Z. Polymere* **250,** 759–763.

Schindler, P. W., and Kamber, H. R. (1968), Die Acidität von Silanolgruppen, *Helv. Chim. Acta* **51,** 1781–1786.

Schindler, P. W., Wälti, E., and Fürst, B. (1976a), The role of surface hydroxyl groups in the surface chemistry of metal oxides, *Chimia* **30,** 107–109.

Schott, J., and Petit, J. C. (1987), New evidence for the mechanisms of dissolution of silicate minerals, this volume, Chapter 11.

Sigg, L. (1973), "Untersuchungen über Protolyse und Komplexbildung mit zweiwertigen Kationen von Silikageloberflächen," M. Sc. thesis, University of Bern, Bern, Switzerland.

Sigg, L. (1987), "Surface Chemical Aspects of the Distribution and Fate of Metal Ions in Lakes," this volume, Chapter 12.

Sposito, G. (1983), On the surface complexation model of the oxide–aqueous solution interface, *J. Colloid Interface Sci.* **91,** 329–340.

Stumm, W., Ed. (1967), *Equilibrium Concepts in Natural Water Systems* (Adv. Chem. Ser., No. 67), American Chemical Society, Washington, DC.

Stumm, W., and Furrer, G. (1987), "The Dissolution of Oxides and Aluminum Silicates; Examples of Surface-Coordination-Controlled Kinetics," this volume, Chapter 8.

Stumm, W., Huang, C. P., and Jenkins, S. R. (1970), Specific chemical interaction affecting the stability of dispersed systems, *Croat. Chem. Acta* **42,** 223–245.

Stumm, W., Kummert, R., and Sigg, L. (1980), A ligand exchange model for the adsorption of inorganic and organic ligands at hydrous oxide interfaces, *Croat. Chem. Acta* **53,** 291–312.

Tamura, H., Matijevic, E., and Meites, L. (1983), Adsorption of Co^{2+} ions on spherical magnetite particles, *J. Colloid Interface Sci.* **92,** 303–314.

Westall, J. C., and Hohl, H. (1980), A comparison of electrostatic models for the oxide/solution interface, *Adv. Colloid Interface Sci.* **12,** 265–294.

Whitfield, M., and Turner, D. R. (1987), "The Role of Particles in Regulating the Composition of Seawater," this volume, Chapter 17.

Yates, D. E. (1975), "The Structure of the Oxide–Aqueous Electrolyte Interface," Ph.D. thesis, University of Melbourne, Melbourne, Australia.

Zelewsky, A. von, and Bemtgen, J. M. (1982), Formation of ternary copper(II) complexes at the surface of silica gel as studied by esr spectroscopy, *Inorg. Chem.* **21,** 1771–1777.

5

ASPECTS OF THE MOLECULAR STRUCTURE IN SURFACE COMPLEXES; SPECTROSCOPIC INVESTIGATIONS

Herbert Motschi
Ciba-Geigy AG, Basel, Switzerland

Abstract

Hydrous oxide surfaces of the abundant elements Al, Si, Ti, and Fe and hydrous surfaces with organic functional groups (e.g., cellulose derivatives, cells of bacteria) interact with metal ions and ligands by distinct but not always easily distinguishable mechanisms. Magnetic resonance methods (EPR, ENDOR, ESEEM, NMR) can be applied to obtain information about structural fragments and dynamic behavior of adsorbate interactions with functional groups in aqueous suspensions. Static EPR parameters of adsorbed Cu^{2+}*(aq) are a sensitive measure to discriminate between inner-sphere and outer-sphere complex formation with functional surface groups. Resolution of hyperfine interactions with neighboring magnetic nuclei is required in order to obtain the same standard of information when* VO^{2+}*(aq) ion is adsorbed which can be achieved by applying ENDOR and ESEEM spectroscopy. Vibrational methods (IR/Raman) bear the potential of a diagnostic tool although an analysis of the dynamic state of an adsorbed species is beyond its time scale. Generally, the most fertile strategy to attack the problems involved in aqueous surface chemical structure and exchange properties is to select a combination of sensitive methods together with a sound and imaginative understanding of colloid and coordination chemistry.*

1. INTRODUCTION

Chemical processes in natural water systems tend to take place at the solid–solution interface. The partition of a chemical species between solution and solid phase can be treated by thermodynamic modeling (Westall and Hohl, 1980; Sposito, 1983). The challenge of proving or falsifying the surface coordination model of hydrous oxides as presented by Schindler and Stumm cannot be met by thermodynamic and double-layer considerations only. Therefore, information is required to characterize the molecular structure of the adsorbed species. This can be achieved by applying spectroscopic methods suitable to investigations of particles which are in a hydrated state, spatially random distributed and have considerable lack of structural organization. Hydrous oxide minerals of relevance (Si, Al, Fe, Ti) and particles coated with organic functional groups (e.g., functionalized cellulose derivatives as model surfaces and bacterial cells) are subject to these investigations.

Spectroscopic methods accessible to such information have to be classified according to their characteristic frequency domain which is to be compared with molecular motions (vibrational, diffusional; see Fig. 5.1). If the mobility of a molecule is rapid compared to the time scale of a particular type of spectroscopy, an isotropic spectrum is observed; conversely, if the frequency of the spectroscopy is high compared to molecular motion, an anisotropic spectrum is observed. Molecular dynamics can be interpreted for those spectroscopic methods which overlap with molecular motions. In an attempt to discriminate spectroscopically between inner-sphere and outer-sphere surface coordination, for example, static parameters sensitive to the interaction of adsorbed molecules with occasionally only slightly different functional groups have to be identified.

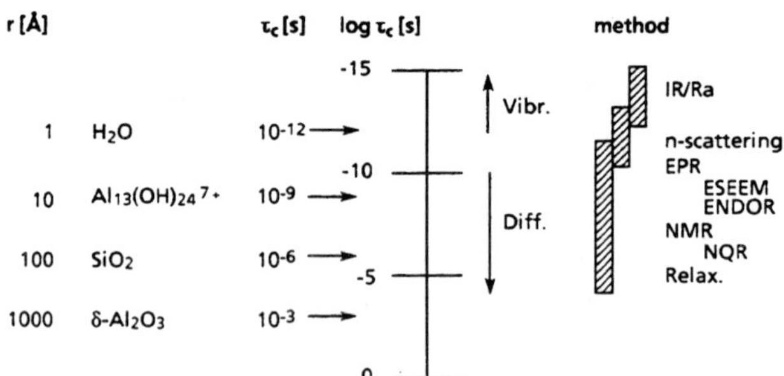

Figure 5.1 Comparison of the time domains of molecular dynamics (characteristic parameters: molecular radius r and correlation time τ_c) with frequency windows of various spectroscopic methods.

2. CATION ADSORPTION

2.1. Cu(II) EPR to Probe Surface Complexation

Copper(II) belongs to the strongest complex formers according to the Irving–Williams series, and hence complex formation is expected to be an important factor on hydrous oxide and organic surfaces. This characteristic should be a prominent feature of this element. Furthermore, in the electronic configuration of Cu^{2+} in its dominantly tetragonal local environment, the unpaired electron occupying a $d_{x^2-y^2}$-orbital ground state has a maximum spin density directed toward the ligand places which renders the EPR parameters (g, A^{Cu}) sensitive to ligand variations. This behavior is most clearly reflected in the EPR parameter $g_\parallel$ for which a correlation with the overall stability constant ($\log K_{tot}$) has been established (Motschi, 1985) and which is shown for a number of hydrous surfaces in Figure 5.2.

The appearance of an anisotropic spectrum (slow motion: $\tau_c \geq 10^{-9}$ s) has to be expected when particle mobility is time-limiting (compare Figure 5.1; kinetics of ligand exchange for $Cu^{2+} + L \rightleftharpoons CuL$ is also typically on the order of $k \leq 10^8\ s^{-1}$). However, Bassetti et al. (1979) have demonstrated in an investigation of Cu^{2+} sorption into porous silica and alumina that the mobility of a $Cu^{2+}(aq)$ ion can be reduced by narrow pores to such an extent that it appears as a rigid limit spectrum on the EPR time scale. Hence, the features of an anisotropic spectrum appear not to be a sufficient criterion to postulate chemisorption. Static EPR parameters ($g_\parallel$, $A_\parallel$) are a sensitive measure of surface complex formation of a Cu(II) ion as indicated by a decrease in the $g_\parallel$ value compared to the free aquo ion ($g_\parallel = 2.44$). Consecutive complex formation

$$Cu^{2+}(aq) + L_1 + L_2 + \cdots + L_n \rightleftharpoons Cu(L_1, L_2, \cdots, L_n) \qquad (\log K_{tot} = \sum \log K_i) \quad (5.1)$$

is manifested by an incrementally additive decrease of $g_\parallel$ which can be associated with a concomitant gain in stability of the complex formed. Yet, EPR spectra do not always unequivocally prove complex formation with surface functional groups; for example, hydrolysis of the $Cu^{2+}(aq)$ ion leads to the same effect as adsorption on silica [specified as $(\equiv SiO)_2$ in Fig. 5.2 at $g_\parallel = 2.38$]. The OH^- ion, being the strongest of all monodentate oxygen donor ligands, represents a marker in Figure 5.2, and there is now independent evidence that terminal monocarboxylates, sulfonates, and phosphates bind the $Cu^{2+}(aq)$ or $CuOH^+(aq)$ ion by hydrogen bonds (outer-sphere). Carboxylic acid functional groups of chelate-forming ligands (charcoal and fulvic acids) do form strong inner-sphere surface complexes.

Extremely high stability constants are postulated for $Cu^{2+}(aq)$ adsorbed

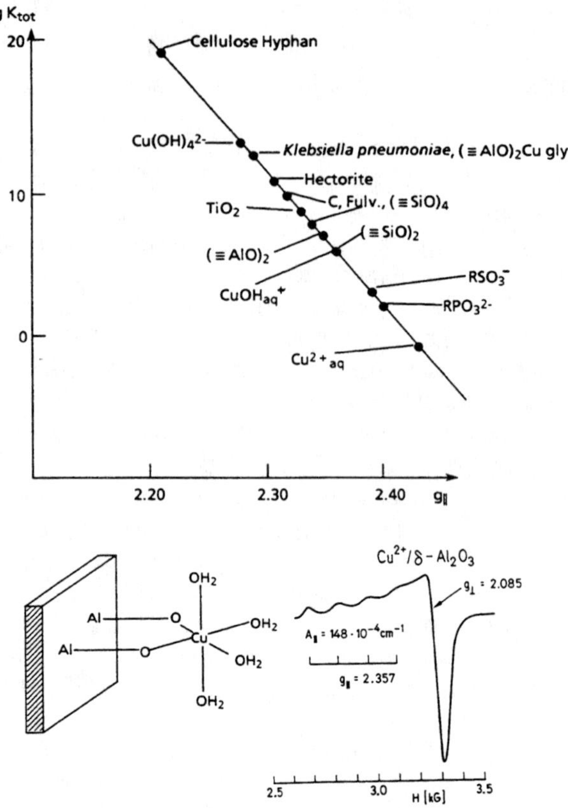

Figure 5.2. Stability constants of Cu(II) surface complexes for hydrous oxide and organic surfaces estimated from EPR ($g_{\|}$). Oxide surfaces: TiO_2, δ-Al_2O_3 = $(\equiv AlO)_2$, silica $(\equiv SiO)_2$, and $(\equiv SiO)_4$, hectorite, ternary surface complex $(\equiv AlO)_2$ Cu(glycine). Organic surfaces: Cellulose hyphan (azo dyestuff), bacteria cells (*Klebsiella pneumoniae*), charcoal (C), fulvic acid (Fulv.), sulfonate ion exchange resin (RSO_3-), cellulose phosphate (RPO_3^{2-}).

on a cellulose dyestuff functional surface and on the surface of the bacterium *Klebsiella pneumoniae* (amino acid chelate groups, probably containing histidine, acting as ligands). An indirect way to determine the number of surface functional groups coordinated to Cu(II) is by the formation of ternary complexes; for example, for

$$(\equiv SiO)_2 + Cu(H_2O)_4 + L_1L_2 \rightleftharpoons (\equiv SiO)_2CuL_1L_2 \qquad (5.2)$$

equilibria can be studied preferentially by EPR (Zelewsky and Bemtgen, 1982).

On a number of surfaces, such as anatase-TiO_2, Aerosil 300-SiO_2, and cellulose derivatives, there is evidence of Cu^{2+}–Cu^{2+} interactions in the EPR spectrum (line broadening, loss of sensitivity) which may be termed cluster formation (Bleam and McBride, 1985). Such interactions are far-reaching (10–20 Å), and the assumption of surface precipitation of $Cu(OH)_2$ is not always justified; diamagnetic dilution with Pb^{2+} or Cu^{2+} adsorbed on cellulose phosphate did not reduce these interactions (Möhl et al., 1986a). It seems more adequate to postulate an uneven distribution of functional groups on the surface of a bulk support. This view is in agreement with the well-known behavior of cellulose fibers in which a large part of a crystalline body is inactive toward adsorption whereas all the active adsorption parts are located in the amorphous domain. Lineshape analysis of ^{1}H-NMR measurements (second moments) on different fibrous silicas have demonstrated the cluster arrangement of surface hydroxyl functional groups (Fripiat, 1982).

2.2. VO(IV) EPR: An Insufficient Probe

The vanadyl ion VO^{2+} forms complexes of comparable strength to Cu(II) with a strong preference to oxygen donors over nitrogen functional groups (A-metal behavior). In the d^1 ground state the single electron occupies a d_{xy} orbital which now has its maximum electron density between ligand positions, and only minor changes are observed in the EPR parameters (g, A) upon ligand substitution. These effects are not strong enough to discriminate between surface complexation and hydrolysis. More selective methods are required for a characterization of the adsorbed species.

ENDOR (electron nuclear double resonance) and ESEEM (electron spin echo envelope modulation) are resonance techniques which allow resolution of interactions of the unpaired electron with neighboring magnetic nuclei. [For an introduction to ENDOR and ESEEM spectroscopy, the reader is referred to Schweiger (1982) and Mims (1972).]

The situation is pictorially represented in Figure 5.3, where it is shown

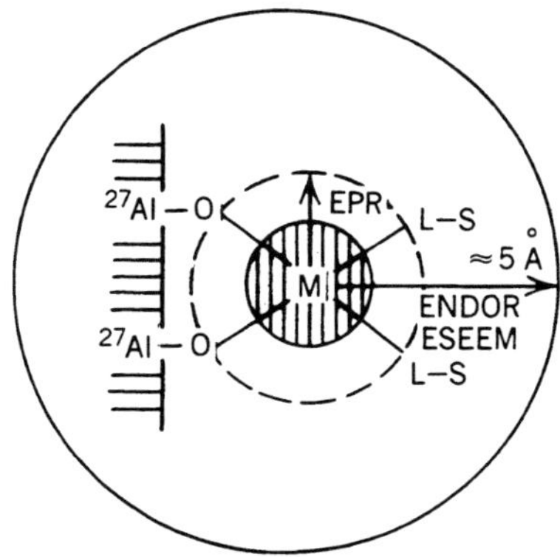

Figure 5.3. Approximate ranges of EPR, ENDOR, and ESEEM.

that EPR measurements on transition metal complexes only probe the closest vicinity of the paramagnetic center, which in favorable cases may include hyperfine interactions with nuclei of the first coordination sphere (L) or reflect changes in the ligand field as in the case of Cu(II). ENDOR/ESEEM spectroscopy allows the resolution of weak interactions between the unpaired electron and nuclei within a distance of about 5Å (L and substituent S).

2.3. ENDOR and ESEEM of Adsorbed Metal Ions

The magnetic interactions of a paramagnetic metal ion in a magnetic field can be described by the spin Hamiltonian (Wertz and Bolton, 1972).

$$\mathcal{H} = \text{H (el. Zeeman)} + \text{H (zero field splitting)} + \text{H (hyperfine)} + \text{H (nuclear Zeeman)} + \text{H (quadrupole)} \tag{5.3}$$

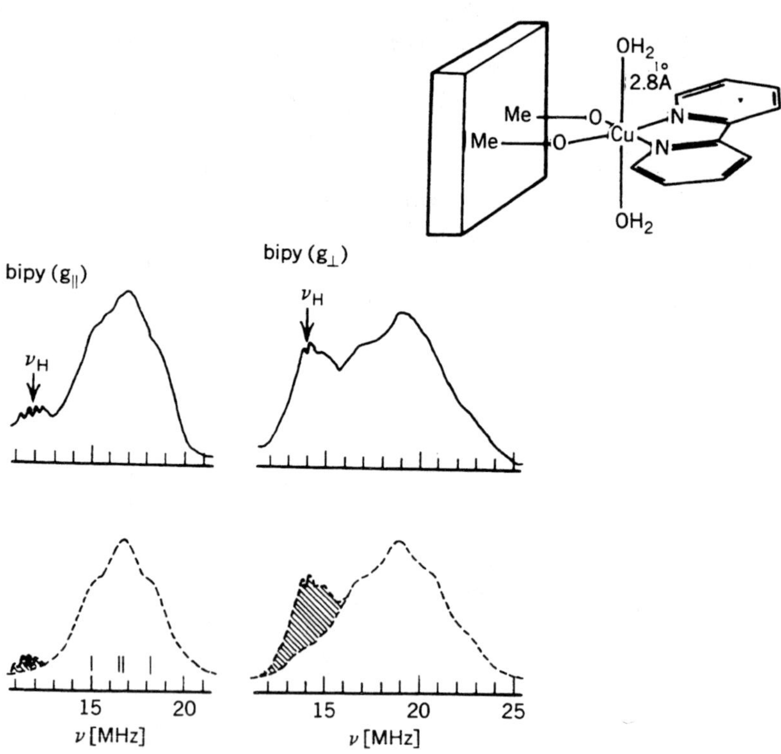

Figure 5.4. ENDOR spectrum of the ternary complex [(≡AlO)$_2$Cu(bipy), with the EPR observer parallel and perpendicular to the molecular axis] and derived molecular structure.

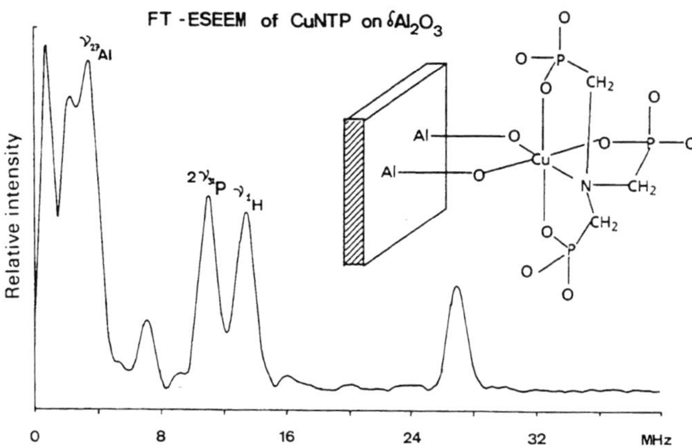

Figure 5.5. Fourier transformed ESEEM spectrum of the ternary complex $(\equiv AlO)_2Cu$ NTP.

Whereas in an EPR spectrum the electron Zeeman and zero field splitting ($S \geq 1$) and in favorable cases hyperfine coupling can be resolved, ENDOR and ESEEM allow the resolution of ligand hyperfine, nuclear Zeeman, and, in favorable cases, nuclear quadrupole interactions ($I \geq 1$).

ENDOR measurements allow the determination of bond distances if the interaction between the magnetic moments of the electron and the nucleus is of a dipolar nature. Electron delocalization through chemical bonding can lead to isotropic couplings, in which case the analysis of the spectrum is no longer straightforward. Spectrum simulation is considerably facilitated by the input of the parameters from model compounds. The advantage of a higher resolution in the ENDOR and ESEEM experiment is gained at the expense of a loss in detection sensitivity compared to the EPR experiment. For example, detailed insight into the molecular structure has been obtained from the analysis of ENDOR spectra of ternary Cu(II) complexes with well-characterized ligands (see Fig. 5.4) (Rudin and Motschi, 1984). Because there are no molecular model complexes resembling the surface functional group and no ^{27}Al signal has been observed in the ENDOR experiment, the structural assignment of the bonding with the surface remains a black box.

Neither the number of Al centers involved in the binding nor their geometric arrangements (bond angles, bond distances) can be determined. Thus far it was possible in only one case to identify the coupling of a paramagnetic metal ion (VO^{2+} on δ-Al_2O_3) with the surface Lewis center (^{27}Al) in the ENDOR experiment (Motschi and Rudin, 1984). Very recently, Möhl et al. (1986b) by applying the ESEEM technique, have identified the close proximity of the ^{27}Al surface in a series of ternary surface complexes of the type $(\equiv AlO)_2CuL_n$ (Fig. 5.5).

In the spin echo envelope function

$$I(\tau) = \sum a_v F_v(\tau) \cos(\omega_v \tau) \tag{5.4}$$

where ω_v is the linear combination of Larmor frequencies, α_v is the amplitude coefficient, and $F_v(\tau)$ is the decay function, important parameters such as

Number of interacting nuclei from the surface
Spin delocalization
Bond distances, bond angles
Electric field gradient (quadrupole interaction of ^{27}Al)

are not known and may vary considerably in disordered systems. For this reason a definite determination of the magnetic parameters in Eq. 5.3 cannot be achieved because many more input parameters are required for spectrum simulation than are experimentally available.

Combining all the pieces of complementary evidence obtained so far, it can be stated that Cu(II)) strongly binds to bacteria by the formation of inner-sphere complexes with mainly peptide functions and to certain hydrous oxides, but ion association is dominant for the interaction with single RSO_3^-, RPO_3^{2-}, and $RCOO^-$ groups. Also, VO^{2+} forms inner-sphere complexes with hydrous oxide surfaces. In comparison, there is EPR evidence that Mn^{2+} binds through ion association to hydrous oxide surfaces and polycarboxylates (Burlamacchi et al., 1983), which also reflects the weaker tendency of this metal ion to form complexes in solution.

The results obtained thus far are of considerable diagnostic value, but there remains a need for a refined experimental technique and an improved mathematical analysis in the future to obtain the same quality of information on surface-adsorbed species as has been possible for bulk single crystals in the past.

3. ANION AND ORGANIC LIGAND ADSORPTION

3.1. Ligand Exchange Mechanisms

The characterization of direct interactions between the surface and an adsorbed ligand by vibrational spectroscopy (IR and Raman) is fraught with difficulties. Bands of metal–ligand interactions are generally weak, they occur at low frequencies in a range where many deformation bands overlap, and lattice vibrations of the surface support obscure most of the transitions. However, in a number of cases characteristic adsorption bands can be observed within substituents of functional groups of the ligand. An example is given in Figure 5.6, where symmetric and antisymmetric stretching frequencies

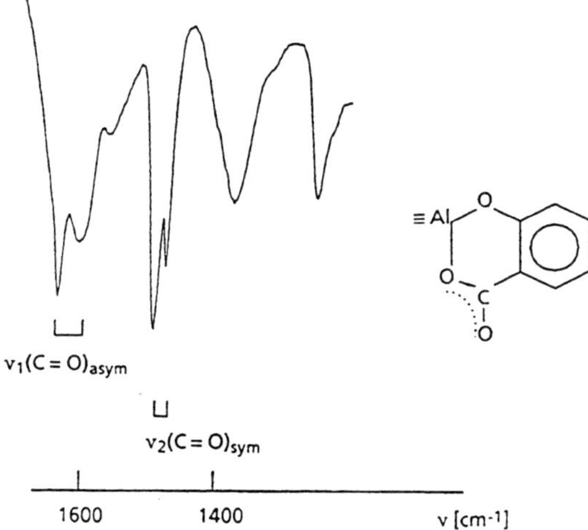

Figure 5.6. Diffuse reflection IR spectrum of salicylate adsorbed on δ-Al_2O_3. (Experimental assistance by Bruker GmbH, Karlsruhe BRD, is gratefully acknowledged.)

of the salicylate carboxylate chelate bond to the Al ion can be diagnostically identified by comparison with the spectrum of the $Al(sal)_3^{3-}$ ion (Table 5.1).

While the formation of a chelate complex with an Al ion may appear established, the IR time scale (see Fig. 5.1) cannot discriminate between a mobile surface complex and an effective complex formation with an "anchored" surface Al center. ^{1}H- and ^{13}C-NMR measurements indicate that the salicylate surface complex is immobile on the NMR time scale (no signals are observable); that is, the molecular mobility might be limited by the particle motion of δ-Al_2O_3 (cf. Fig. 5.1). Such circumstantial evidence, however, is not sufficient to prove this hypothesis. Because of the limited sensitivity of the method, positive observation of NMR signals is rarely achieved for most nonporous surfaces except for the highly active internal surfaces of zeolites (Pfeifer, 1976).

Table 5.1. IR bands of Salicylates

	v_1 (cm^{-1})	v_2 (cm^{-1})
KHsal	1605	1490
	1592	1470
$Al(sal)_3^{3-}$	1615	1478
	1590	1464
≡Al sal	1613	1478
	1585	1464

The characterization of the dynamic behavior of adsorbed ligands is a sensitive measure to discriminate between elementary exchange mechanisms at the surface, such as between the formation of an outer-sphere association complex (k_{ass}) versus the formation of an inner-sphere complex by ligand exchange (k_{exch}).

$$\equiv\text{Al—OH}_2^+ + \text{L}^- \begin{cases} \xrightarrow{k_{ass} \approx 10^9\text{–}10^{11}\ \text{s}^{-1}} \equiv\text{Al—OH}_2^+\text{L}^- \\ \xrightarrow[k_{exch} \approx 1\ \text{s}^{-1}]{} \equiv\text{Al—L} \end{cases} \tag{5.5}$$

An alternative way to investigate the dynamic behavior of adsorbed ligands is to employ so-called spin labeling techniques. A stable nitroxide radical is attached to an organic ligand L (Fig. 5.7), which renders this molecule amenable to the paramagnetic resonance techniques mentioned earlier.

Studies with "carboxy proxyl" (L = COO^-) on porous silica have shown that the mobility is controlled by the size of the pores (Martini, 1984). On hydrous aluminum oxide surfaces, fast motion limit spectra ($\tau_c \leq 10^{-11}$ s) and slow motion limit spectra ($\tau_c \geq 10^{-9}$ s) have been identified (McBride, 1982). On hydrous δ-Al_2O_3, the fast motion regime prevails, whereas on hydrous TiO_2 (anatase), strong spin pairing is observed, an effect similar to the "clusters" formed when Cu^{2+}(aq) is adsorbed (Motschi, unpublished results).

In all cases the exchange rate of the carboxy proxyl on aluminum oxide surfaces far exceeds the rate for an inner-sphere ligand exchange process. Jump diffusion of the outer-sphere complexes (ion association), on the other hand, may indeed be very fast ($\tau_c \approx 10^{-10}$ s). If a carboxylate ligand were rigidly held by a direct bond to the surface Lewis center (≡Al—OOC—R), the correlation time (τ_c) would become dependent on the internal rotation of the molecule around the R—COO bond which can be expected to fall between 10^{-4} s (E_a = 10 kcal/mol, activation energy for internal rotation) and 10^{-7} s (E_a = 5 kcal/mol). Such processes are too slow to be detected by conventional EPR methods, but they can be determined by other techniques such as STEPR (saturation transfer EPR) or spin echo experiments. Parfitt et al. (1977) found IR evidence that carboxylates and phosphates particularly form strong bonds to hydrous oxide surfaces if they can form a binuclear bridge.

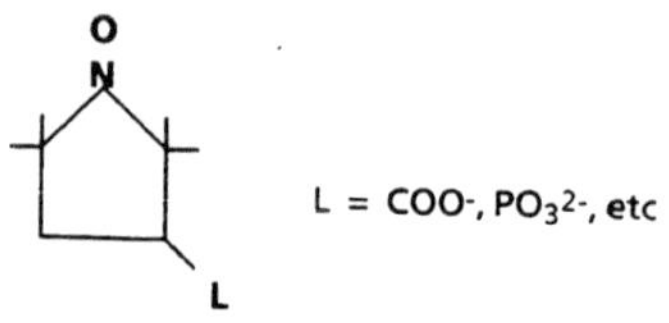

O
N
L = COO^-, PO_3^{2-}, etc
L

Figure 5.7. Spin label of the "proxyl" type with different functional groups.

The combination of all these possible spectroscopies together with titration experiments will not only allow the identification of the nature of the chemical bond but also deliver answers to such important questions as site topology (e.g., isolated sites on δ-Al_2O_3, clustered sites on TiO_2, pore adsorption in many silicas, and patches of sites in fibrous cellulose or silica) and surface diffusional processes.

3.2. Linear Alkylbenzene Sulfonate (LAS) Association as Influenced by Metal Surface Complexes

Linear alkylbenzene sulfonates represent an important and versatile class of surfactants, mainly due to their amphiphilic properties, which are involved in a number of adsorption processes with both mineral and organic hydrous surfaces. This is also reflected in a high content of LAS in sewage sludge ($\geq$ 1% of dry matter). Soluble and insoluble micelle formation critically depends on the nature and concentration of the metal ion present in the solution which is also reflected in the complicated behavior of the octanol–water partition coefficient for C_{10}-benzene sulfonates (C_{10}BS; K_{ow} ranges from $\approx$ 4 to 100). It has been shown (Motschi and McEvoy, 1985) that porous surfaces (charcoal, differently functionalized cellulose derivatives) bind C_{10}BS by a nonspecific mechanism, that is, one independent of pH. Hydrous oxides (TiO_2, δ-Al_2O_3), on the other hand, show a strong pH dependence for the C_{10}BS adsorption which is dominantly influenced by a charge mechanism: increasing adsorption toward decreasing pH values and decreasing adsorption with increasing pH. In the presence of a surface complexed metal ion (e.g., Cu^{2+}) the adsorption curve is displaced to higher pH values (parallel to the electrophoretic mobility). The situation is schematically represented in Figure 5.8.

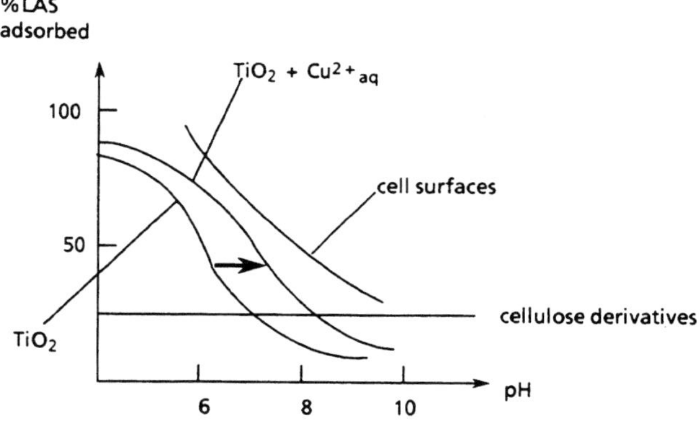

Figure 5.8. Schematic representation of LAS adsorption as a function of pH for different classes of surfaces.

EPR spectra of the Cu^{2+} surface complexes are not influenced by the adsorbed $C_{10}BS$, which is consistent with an ion-association mechanism.

Metal ions adsorbed on functionalized cellulose derivatives show three distinct mechanisms of secondary effects of LAS adsorption (Fig. 5.9):

1. Cu^{2+} forms a cationic inner-sphere complex capable of forming an ion pair with LAS (Fig. 5.9*a*).
2. Cu^{2+} forms a strongly anionic outer-sphere complex with, for example, a surface phosphate, which prevents additional interaction with the anionic surfactant and hence shows no influence upon its adsorption (Fig. 5.9*b*).
3. Surface complex formation is not a prominent feature for a particular functional group (e.g., acetyl cellulose), and hence surface precipitation of $Cu(C_{10}BS)_2$ prevails (Fig. 5.9*c*).

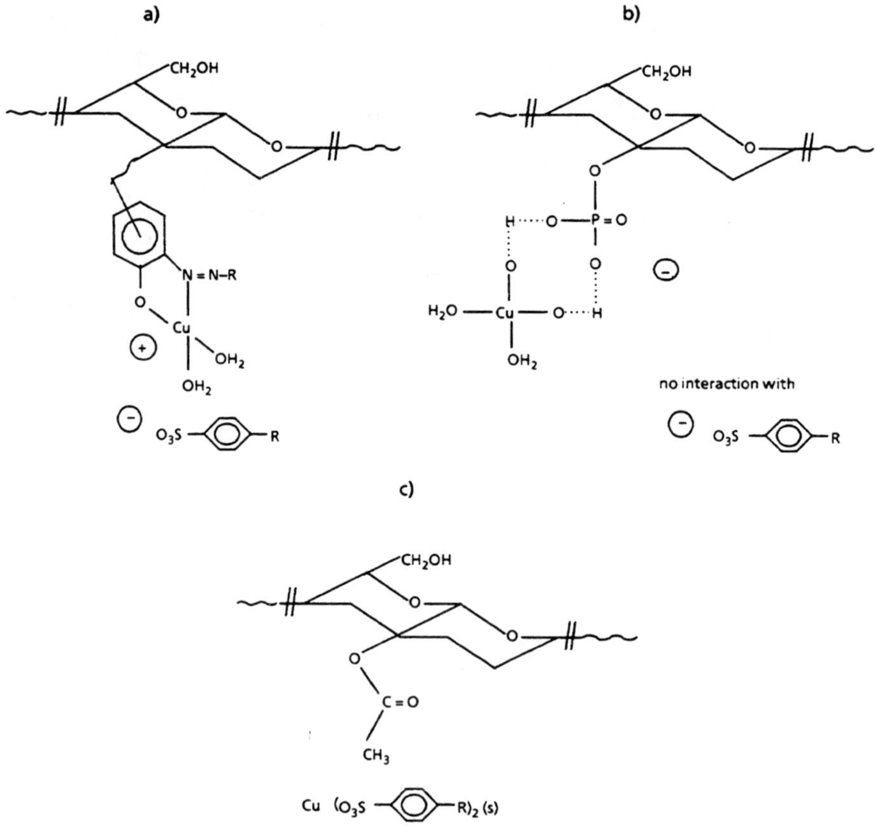

Figure 5.9. Effects of functional groups of cellulose surface upon Cu^{2+} and C_{10}-benzene sulfonate adsorption.

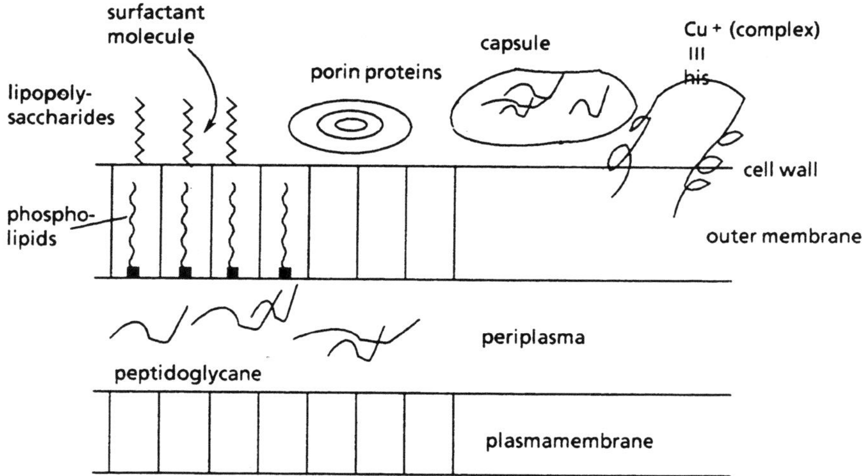

Figure 5.10. Biological cell surfaces (*Klebsiella pneumoniae*) providing separate adsorption sites for LAS adsorption (lipophilic interaction) and Cu^{2+} uptake (complex formation on a peptide chain).

Lipophilic interactions, presumably with the phospholipid layer, appear to be responsible for the strong uptake of LAS by the surface of bacteria (*Klebsiella pneumoniae*). This surface strongly binds Cu^{2+} at sites presumably containing a peptide chelate with histidine as a ligand (stability constant derived from EPR, log $K_{tot} \geq 14$). The adsorption of surfactants is not influenced by surface complexation to the cell membrane, because lipophilic interactions are much stronger than effects of ion association (Fig. 5.10).

4. FINAL CONSIDERATIONS

Despite the heterogeneous nature of the surfaces encountered in aquatic systems, these interactions can be reconstructed from a basic set of fundamental building principles. To approach these equilibria (and even more often metastable pseudo-equilibria), the concepts of coordination chemistry are remarkably fertile, bearing the potential of unifying thermodynamic and bonding principles. To obtain an insight into the aspects of chemical bonding in aquatic surface chemistry, a combination of selected spectroscopic techniques sensitive to static and dynamic states of the "surface complex" is the most promising strategy. Model surfaces, many of which are still far from being uniform, may serve to identify the fundamental interactions including complex formation, ion-pair interaction, cluster formation, Donnan potentials, lipophilic interactions, and micelle formation.

All these interactions may take place at different topological sites and therefore render a particular surface selective for adsorption, dissolution, or nucleation processes; this situation differs remarkably from the diffusionally averaged uniform solution processes of small solute complexes. Cell surfaces of living organisms such as those of bacteria have more highly organized functional principles which cannot be understood simply by a reductionist model. It will therefore become increasingly important to study these systems.

Acknowledgments

The encouragement of Werner Stumm for this project is appreciated. The author is grateful for stimulating discussions with and experimental assistance from Markus Rudin, Arthur Schweiger, Wolfgang Möhl, and Jean-Michel Fauth. Financial support by the Swiss National Science Foundation is acknowledged.

REFERENCES

Bassetti, V., Burlamacchi, L., and Martini, G. (1979), Use of paramagnetic probes for the study of liquid adsorbed on porous supports. Cu(II) in water solution, *J. Amer. Chem. Soc.* **101,** 5471.

Bleam, W. F., and McBride, M. B. (1985), Cluster formation versus isolated-site adsorption. A study of Mn(II) and Mg(II) adsorption on boehmite and goethite, *J. Colloid Interface Sci.* **103,** 24.

Burlamacchi, L., Ottaviani, M. F., Ceresa, E. M., and Visca, M. (1983), Stability of colloidal TiO_2 in the presence of polyelectrolytes and divalent metal ions, *Colloids and Surfaces* **7,** 165.

Fripiat, J. J. (1982), "Silanol Groups and Properties of Silica Surfaces," in J. S. Falcone, Jr., Ed. *Soluble Silicates,* ACS Symp. Ser. no. 194 p. 165.

McBride, M. B. (1982), Organic anion adsorption on aluminium hydroxides: Spin probe studies, *Clays and Clay Min.* **6,** 438.

Martini, G. (1984), ESR study of a negative spin probe adsorbed from a silica/water dispersion, *Colloids and Surfaces* **11,** 409.

Mims, W. B. (1972), "Electron Spin Echos," in S. Geschwind, Ed., *Electron Paramagnetic Resonance,* Plenum, New York, Chap. 4.

Möhl, W., Schweiger, A., Fauth, J.-M., and Motschi, H. (1986a), unpublished results.

Möhl, W., Schweiger, A., and Motschi, H. (1986b), unpublished results.

Motschi, H. (1985), Cu(II) EPR: A complementary method for the thermodynamic description of surface complexation, *Adsorption Sci. Technol.* **2,** 39.

Motschi, H., and McEvoy, J. (1985), Influence of metal/adsorbate interactions on the adsorption of linear alkylbenzenesulphonates to hydrous surfaces, *Naturwissenschaften* **72,** 654.

Motschi, H., and Rudin, M. (1984), ^{27}Al ENDOR study of VO^{2+} adsorbed on δ-alumina, *Colloid and Polymer Sci.* **262,** 579.

Parfitt, R. L., Fraser, A. F., Russell, J. D., and Farmer, V. C. (1977), Adsorption on hydrous oxides, II. Oxalate, benzoate and phosphate on gibbsite, *J. Soil Sci.* **28,** 40.

Pfeifer, H. (1976), Surface phenomena investigated by nuclear magnetic resonance, *Physics Report* (Sec. C *Phys. Lett.)* **26**(7), 293.

Rudin, M., and Motschi, H. (1984), A molecular model for the structure of copper complexes on hydrous oxide surfaces: An ENDOR study of ternary Cu(II) complexes on δ-alumina, *J. Colloid Interface Sci.* **98,** 385.

Schindler, P., and Stumm, W. (1987), this volume, Chapter 4.

Schweiger, A. (1982), "Electron Nuclear Double Resonance of Transition Metal Complexes with Organic Ligands," in M. J. Clarke, J. B. Goodenough, et al. (Eds.) *Structure and Bonding,* Vol. 51, Springer Verlag, Berlin-Heidelberg-New York.

Sposito, G. (1983), On the surface complexation model of the oxide–aqueous solution interface, *J. Colloid Interface Sci.* **91,** 329.

Wertz, J. E., and Bolton, J. R. (1972), *Electron Spin Resonance,* McGraw-Hill, New York, p. 279.

Westall, J., and Hohl, H. (1980), A comparison of electrostatic models for the oxide/solution interface, *Advan. Colloid Interface Sci.* **12,** 265.

Zelewsky, A. von, and Bemtgen, J. M. (1982) Formation of ternary copper (II) complexes at the surface of silica gel as studied by EPR spectroscopy, *Inorg. Chem.* **21,** 1771.

6

SURFACE CHEMICAL PROCESSES IN SOIL

Gerard H. Bolt and Willem H. Van Riemsdijk

Department of Soil Science and Plant Nutrition, Agricultural University, Wageningen, The Netherlands

Abstract

The surface chemical processes governing solute transport in soil must be interpreted using available information about the nature of the soil solid phase. Broadly speaking, this concerns mainly the clay fraction, comprising clay minerals and soil (hydr)oxides, and the soil organic matter. Rather detailed knowledge about the clay minerals is available in the existing literature; this information is summarized here, culminating in the schematic presented as Table 6.1, together with fairly precise referencing to standard texts. The nature of the surface groups to be expected in clays and (hydr)oxides is discussed at some length, leading to the expectation that surface protonation of oxides should preferably be described in terms of a one-step reaction. The information on soil organic matter, though extensive, is much less precise with respect to the structure of the active surface groups. A review of the available multilayer models describing surface complexation and adsorption reactions that occur in soil leads to a schematic presentation of the electrochemical control system involved in Figures 6.9 and 6.10. Here attention is focused on the distribution of the adsorption energies to be expected for "natural" surfaces exhibiting site heterogeneity. In any discussion of the application of adsorption modeling to the soil–water system, heterogeneity and variability in space and time must be a main concern. Several scales of heterogeneity may be discerned, ranging from the surface–site heterogeneity mentioned before, via the solid phase as an intimate mixture of different constituents, to field and formation scale inhomogeneities. This leads to sobering thoughts as to the amount of detail that can be taken into account in the submodels to be used for the description of transport and accumulation phenomena in the soil–water system.

1. INTRODUCTION

A large fraction of the water on its way from the atmosphere to open water (lakes, rivers, etc.) is in contact with soil during a longer or shorter time period. During this contact period the interaction between water and the soil solid phase becomes the determining factor for changes in water quality. Although no doubt redundant for readers of the present book, it is stressed that this interaction between water and soil has always been viewed as predominantly beneficial for the quality of the water. Filtering action is the key aspect here, which carries with it the notion of a "finite capacity unless . . . ." Omitting a discussion of sifting out coarse materials (and subsequent plugging of the filter), permanence of the capacity of the soil to filter water implies either secondary breakdown (particularly microbial, hopefully leading to eventual disappearance via the gas phase as CO_2 and H_2O, in some situations as N_2 following denitrification) or "shunting" of the filtered compound to an inactive position via precipitation (the fate of phosphate comes to mind in this respect). As the second inactivation process implies the unlimited availability of a coprecipitant (Al ions in the case of PO_4) if it is to be a permanent system property, it is clear that the transition between temporary, finite-capacity removal by soil and "permanent inactivation" is a gradual one.

Within the context of the above remarks, the reverse action of soil—adding compounds to the water passing through—should be mentioned. Basically, the origin of life on earth derives from this feature, as the "nearly distilled" water of the unpolluted atmosphere will not support life as it exists on this planet. In turn, this thought sends one back to early geological history, when the sorting out of highly soluble and/or weatherable materials of the earth's crust determined the composition of natural waters. It is of interest to note that while the ensuing geographic distribution of water composition on earth provided for a highly diverse "miraculously well adjusted" spectrum of forms of life, even under fully natural conditions rather hostile extremes of water composition may be found. The Dead Sea provides a contemporary example.

Human interference in the form of irrigation has often induced disturbances known as salinization and/or sodication. Somewhat oversimplifying, these disturbances derive from short-cycling the limited local supply of water, leading to a residue of locally derived and rather soluble salts. Once these salts have accumulated in the root zone, crop yields decrease and agricultural areas are on their way to becoming abandoned land. Increasing water application in order to leach the salts down below the root zone often amounts to postponing the problem as deeper layers become salinized, and eventually a rising water table brings the salts back into this zone. Also, increased leaching may push the local salts toward neighboring areas. Finally, increased leaching of the root zone almost invariably leads to additional dissolution of salts from the subsoil. When these are collected in a drainage system, it depends on the fate of the drainwater whether or not the (farming) system may be considered as a satisfactory solution to the problem of removal of the leached-out salts. As an example, the salinization of the lower Colorado River following intensive irrigation in Arizona appears to include Arizona subsoil salts that one better should have left in place. Modern irrigation practices are directed toward developing systems that will not touch salts present in the deeper layers.

In the twentieth century chemical loading of the soil has increased sharply at certain locations (in particular, at chemical waste dump sites) and more gradually over large areas (for example, in cases of atmospheric fallout). While the chemicals concerned are often totally foreign to natural soil, a third aspect of the increase in chemical loading relates to the increased use of agricultural chemicals. Meant, in part, to help increase agricultural production, these fertilizers are certainly not foreign to soil. If, however, they are incompletely consumed by the crop, the excess amounts tend to leach to deeper layers of the soil, where they eventually may become unwanted constituents of aquifers. Agricultural pesticides are, obviously, often foreign to natural soil and are always unwanted once they move outside the zone where the cidal action was sought. A densely populated and heavily industrialized world eventually leads to (local) pollution of soil and the water passing through it, unless appropriate control is exercised.

Within the context of the present chapter, meant to summarize expectations as to the interaction between the soil solid phase and the water passing through the soil, we first present an overview of the nature of the soil solid phase, stressing the diversity to be expected in general. It is pointed out here that the appearance in the last decade of several good and also fairly complete textbooks in the field of soil chemistry makes it difficult to avoid repetition of some arguments. In particular, we have found it useful to include a number of page references to the following textbooks, for which complete bibliographic information will be found in references at the end of the chapter:

(A) Weaver, C. E. and Pollard, L. D. (1975), *The Chemistry of Clay Minerals.*

(B) Dixon, J. B., and Weed, S. B., Eds. (1977), *Minerals in Soil Environments.*

(C) Greenland, D. J., and Hayes, M. H. B., Eds., (1978), *The Chemistry of Soil Constituents.*

(D) Bolt, G. H., Ed. (1982), *Soil Chemistry,* Part B.

(E) Sposito, G. (1984), *The Surface Chemistry of Soils.*

(F) Gieseking, J. E., Ed. (1975), *Soil Components,* Vol. I, *Organic Components.*

To simplify text citation of these works, they will be referred to by letter.

2. SOIL MATERIALS

There are several ways in which the soil solid phase can be subdivided into groups of materials. Here the division into inorganic and organic components is chosen as a first step. Within the inorganic fraction, a second division is made on the basis of the specific surface area, because the interaction between the solid and liquid phases is critically dependent on the interfacial area between the two. It is easily shown that the round number of 1 m^2/g is a very convenient division line. On the one hand, for isodiametric particles this number corresponds to a Stokes diameter of about 2 μm (for the solid-phase density of oxidic materials at around 3×10^3 kg/m^3), so one may assign the name "soil clay fraction" to those materials that exhibit a specific surface area of 1 m^2/g or larger. On the other hand, it is also of interest to note that for a (negative) charge density corresponding to less than two sites per square nanometer (a fair estimate for soil materials), the maximum amount of cations held by these charges would be of the order of 3 μeq per gram of solid phase. At a field moisture content of 30% and a concentration of the major cations around 0.01 eq/L, this implies that for these cations the distribution ratio is then close to unity, which is far below the level of interest when exploring adsorption and the ensuing retardation phenomena in soil.

Admitting that in the case of highly specific binding of trace constituents in soil, materials with a specific surface area less then 1 m^2/g could also play a role, it would still be unlikely that these coarser materials would be the only ones present in soil. Accordingly, it appears warranted to limit a discussion of soil materials causing adsorption to those exhibiting a specific surface area in excess of 1 m^2/g, that is, the soil clay fraction. This choice is much less realistic if dissolution reactions are to be discussed (cf. Chapters 8–11). Briefly, the high surface area materials will have a low solubility, since they would have been leached out if this were not the case. Thus, dissolution reactions are likely to be dominated by the coarser fractions containing the primary minerals.

2.1. The Clay Fraction: Composition

2.1.1. Clay Minerals. Within the clay fraction the actual clay minerals occupy a dominant position. Ever since the early X-ray work of Hendricks and Fry (1930) and Hoffmann et al. (1933), which established the preponderance of crystalline phyllosilicates in this size fraction of the soil solid phase, a considerable amount of research effort has been devoted to these materials. The existence of an international clay minerals society (AIPEA) and its many national counterparts and of a separate journal *Clays and Clay Minerals,* now in its 33rd year, may be indicative of the difficulty of summarizing the state of the art in this field in the few pages available within this chapter. Moreover, the textbooks listed in Section 1 cointain some good summarizing efforts—500 pages in Dixon and Weed (1977), 150 pages in Weaver and Pollard (1975), and 90 pages in Greenland and Hayes (1978). Sposito (1984) devotes a considerable part of the first chapter to a comparative discussion of soil clay minerals and their surface properties. This fairly short overview is particularly recommended as a source of information to the reader of this chapter.

Reiterating the main lines of the information needed here, it is pointed out that the phyllosilicates called "clay minerals" may be regarded as polycondensates of $SiO_{4/2}$ tetrahedra with $M(OH)_{6/2}$ octahedra, the denominator of the fractions indicating the coordination number of the O or OH surrounding the central metal ion. Basic to these structures is the lateral condensation within the same group, giving $SiO_{3/2}OH$ sheets and $Al(OH)_{6/2}$ sheets, for example, supplemented by normal condensation between these sheets to give either 2:1 or 1:1 layer structures. A two-dimensional presentation of these reactions is given in Figure 6.1.

Whereas in Figure 6.1 the octahedral layer is based on the dioctahedral (i.e., two-thirds of the interstices being filled) Gibbsite structure (and composition), a corollary set of structures may be based on the presence of a trioctahedral layer derived from the Brucite structure. This information may be pictured in the form of Ping-Pong ball structures, which is probably most

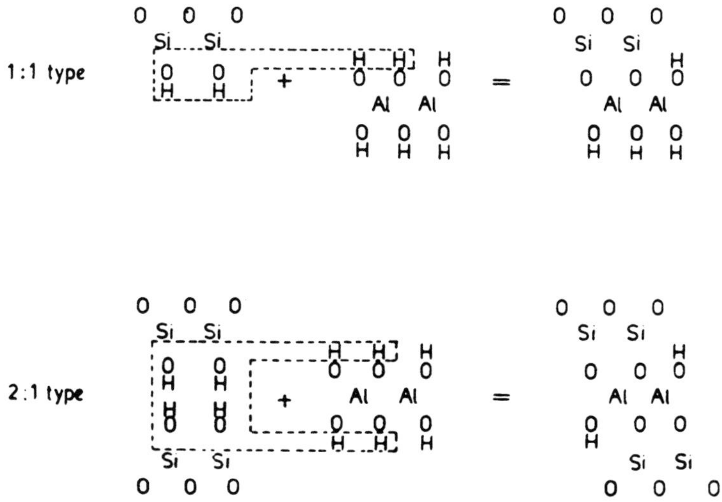

Figure 6.1. Clay minerals as condensates of $Si_2O_3(OH)_2$ and $Al_2(OH)_6$. Taken from Bolt and Bruggenwert (1976).

realistic from the standpoint of spatial distribution of the atoms. The dominant position of the O (or OH) "balls", however, obscures the interstitial characteristic metal cations. Examples of this type of picture may be found in Marshall (1949) and in the slide collection issued by SSSA (1974). Easier to read, but somewhat less to scale, are the "spoke-and-ball" constructions used extensively throughout textbooks [see, for example, Brown et al. (1978) in C, pp. 45–55]. An example of this type taken from Brown et al. has been reproduced in Figures 6.2–6.4 (which show also the corresponding tetrahedra and octahedra in a schematic fashion).

Aside from the above structural characeristics typical of the large majority

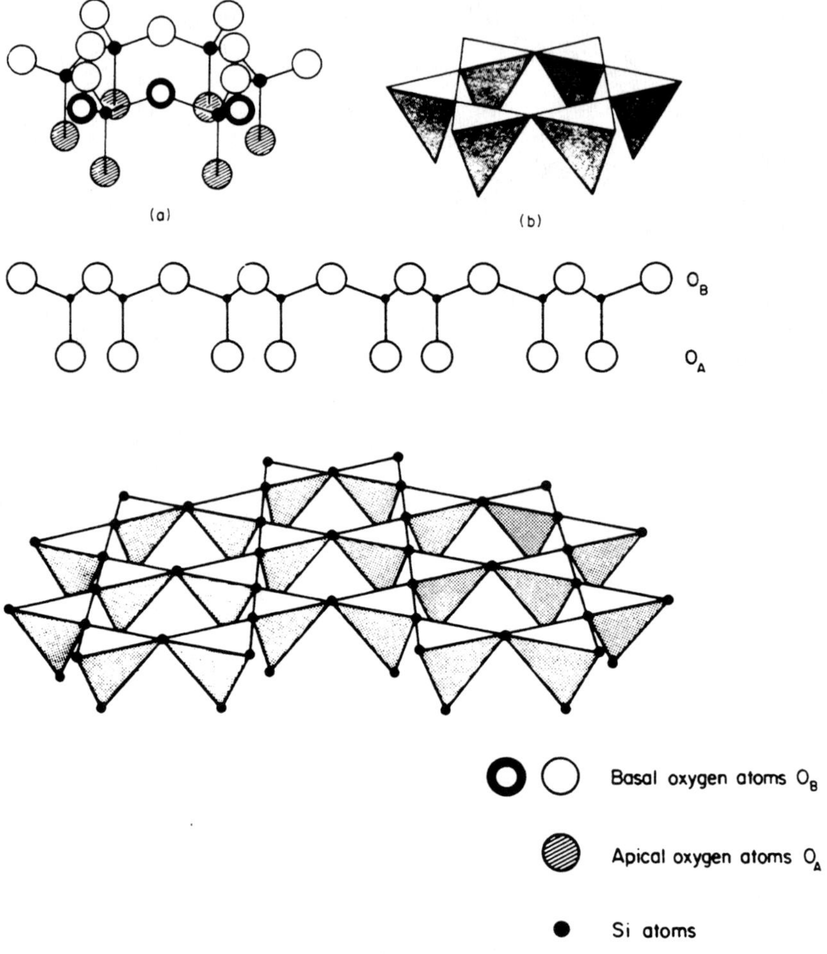

Figure 6.2. $SiO_{4/2}$ groups linked into Si_6O_6 rings forming the tetrahedral sheet. Ball-and-spoke model (*a*) and linked tetrahedra (*b*) are shown. Taken from Brown et al. (1978).

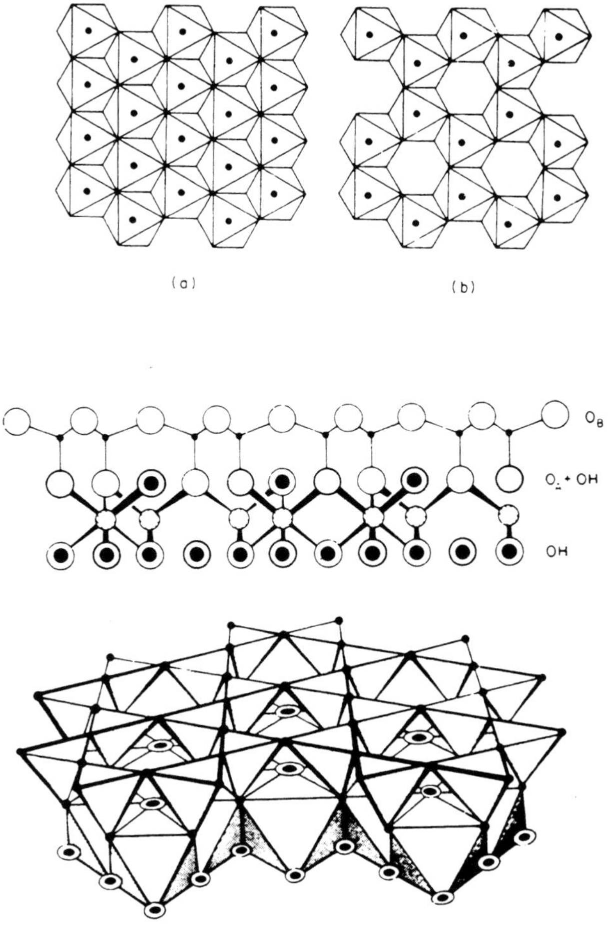

Figure 6.3. Dioctahedral sheet in a 1:1 clay. Taken from Brown et al. (1978).

of clay minerals, there are several other features that lead to a further subdivision of the main groups of 2:1 clays and, to a lesser extent, the 1:1 clays. Looking first at the 2:1 group, one finds that central with regard to the adsorptive properties are the magnitude and location of the layer charge due to substitution of the major cations Si and Al [or sometimes Fe(III)] by cations of lower valence. In the special case of a precise match at 2:1 between these major cations, one finds the end-member named pyrophyllite (with Al in two-thirds of the octahedral positions). The corresponding dioctahedral member is talc (with Mg in all octahedral positions). Aside from these special cases the 2:1 clays generally possess a layer charge due to replacement of Si by Al and/or Al by, for example, Mg. The magnitude of the layer charge is usually specified per unit cell of the crystal, here taken to contain four Al (or six

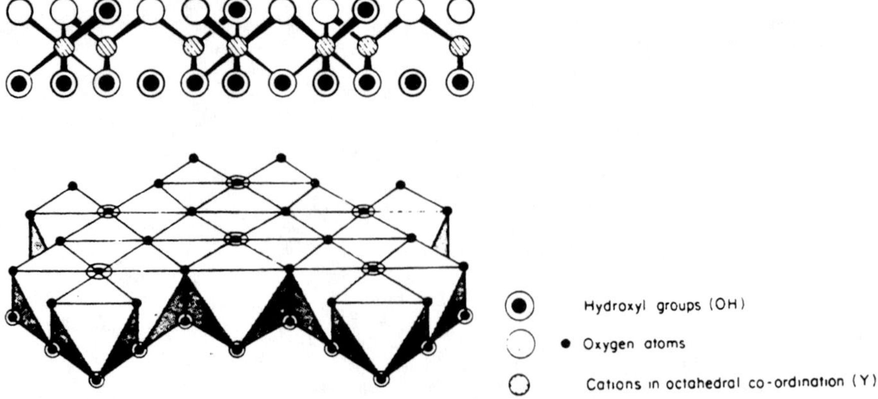

Figure 6.4. Trioctahedral (*a*) and dioctahedral sheets (*b*). The ideal structure of a dioctahedral 1:1 clay is shown at the bottom. Taken from Brown et al. (1978).

Mg) atoms in the octahedral layer. Closely associated with the amount and location of the substitution charge in the clay lattice are other characteristics such as the type of stacking of the unit layers and the binding and spacing between them.

Pointing to the various ways of distinguishing between clay minerals as described above, it is clear that several tens of differently named species have been recognized. While referring in particular to Dixon and Weed (1977) for a much more complete account, it will be attempted here to list the more dominant types in a somewhat generic fashion. As a starter in the 2:1 group the *micas* are mentioned. These are almost entirely primary minerals and are themselves a source of other 2:1 clays. For the most common members of this group, the dioctahedral muscovite and the trioctahedral biotite, the (permanent) layer charge is high (2 per unit cell of four Al atoms) and located entirely in the tetrahedral layer. It is very typical of the micas that this substitution charge is compensated by dehydrated K ions forming inner-sphere complexes (see Section 3) situated in the "hexagonal" cavities of the tetrahedral layer. Presumably the micas may upon weathering lose part of their interlayer K. Thus one finds in many soils of the great river deltas clay minerals that are often referred to as *illites* (or hydrous micas), ranging from high-K illites to low-K illites depending on the history of the soil concerned. Some difficulty remains in explaining why these illites have a layer charge close to 1.5, as it is not quite clear how the layer charge would decrease during weathering reactions. An overview of the layer composition and charge (here given per two Al atoms) of some 30 illite specimens may be found in Weaver and Pollard (1975, in A, pp. 10 and 11). Upon taking out all interlayer K, one attains the composition of the *vermiculites,* of which both di- and trioctahedral members are known. The interlayer charge is then taken care of by the common cations Na and/or Ca, which remain hydrated. It should be noted here that again the substitution charge is considerably below 2, ranging from 1.2 to 1.8 (Fanning and Keramidas, 1977, in B, p. 200), but is still located in the tetrahedral layer. In fact, the total analysis often indicates some excess positive charge in the octahedral layer (Douglas, 1977, in B, pp. 266 and 267). The presence of the substitution charge in the tetrahedral layer indicates rather strong interlayer forces, and accordingly the interlayer spacing shows rather distinct values for most interlayer cations (Brown et al., 1978, in C, pp. 82–85). Tabulated compositions of 11 soil vermiculites are given in Weaver and Pollard (1975, in A, p. 103).

Quite well known are the *smectites* in soil, of which the Wyoming montmorillonites have been used for many research efforts. They belong also to the 2:1 group, and both dioctahedral (montmorillonite, beidellite, nontronite) and trioctahedral (e.g., hectorite, saponite) forms are known, the latter occurring only to a small extent in soil. The dioctahedral members have a layer charge usually below 1 per four Al atoms and for the best-known member, montmorillonite, located (almost) entirely in the octahedral layer. This represents the key to the enormous tendency toward swelling if the interlayer is saturated with a high fraction of monovalent cations. Na-montmorillonite has thus served as the material par excellence for measurements related to the presence of fully extended diffuse double layers like anion exclusion and swelling pressures (cf. Bolt and De Haan, 1979, in D, pp. 243–255). It is believed that the smectites may be formed by weathering processes acting on micas, although here again the considerable reduction of the layer charge requires an explanation. New formation as a secondary mineral via a solution-phase reaction is also assumed to occur. For a discussion of the various possibilities envisaged, reference is made to Borchardt, 1977 (B, pp. 303–312). Clay transformations involving smectites are also discussed in Brinkman, 1979 (D, pp. 433–455). Tabulated composition data are to be found in Weaver and Pollard (1975) (A, pp. 64, 65).

The discussion of the above groups of 2:1 types of the clay minerals concentrated on both the sequence of potassium removal and charge reduction acting on micaceous materials. Another quite common alteration process is the aluminization of clays. Aluminum ions are easily liberated from various sources in the soil system, particularly under acidic conditions. As a first reaction, such Al ions become adsorbed on available clay surfaces, exchanging against cations of lower valence. Eventually such aluminization may extend to more or less complete interlayering of clay particles with gibbsite-like structures. The end point in this process is named chloritization because the ensuing "pseudo" 2:2 clays are *chlorites.* It is stressed that the interlayer is a fully developed gibbsite layer, whereas the octahedral gibbsitic layer has been condensed with the tetrahedral layers. Removal of the interlayer thus remains much easier than dissolving out the octahedral layer, although at a sufficiently low value of the pH this is also possible. Weaver and Pollard (1975, in A, p. 97) give some composition data.

In contrast to the often fairly involved compositions of the group of 2:1

clays discussed above, the 1:1 group tends to be dominated by more uniformly composed members corresponding closely to the simple combination of the prototype arrangement pictured in Figure 6.4. Taking the most abundant member of the 1:1 clays, the dioctahedral mineral *kaolinite,* as an example, one may infer immediately from stability diagrams like the one shown in Figure 6.5 that excessive leaching of aluminum silicate materials in the presence of some silicate brings one into the stability field of kaolinite. Indeed, all over the earth's surface the high leaching conditions associated with the humid tropics have in general led to large deposits of kaolinitic soils. It appears that most kaolinites are newly formed during the weathering process. Only complete desilication will destabilize the kaolinite to form bauxite, the "dead-end" of this weathering process. In addition to the mentioned Al dioctahedral member kaolinite, there exist the minerals dickite and nacrite, which exhibit slightly different stacking patterns but have the same chemical composition.

Substitution charges are presumably absent from kaolinite. Together with the low value of the specific surface area resulting from the hydrogen bonding between the gibbsite sheet on one side of the crystal and the siloxane surface of the neighboring one, kaolinites have a comparatively low adsorption ca-

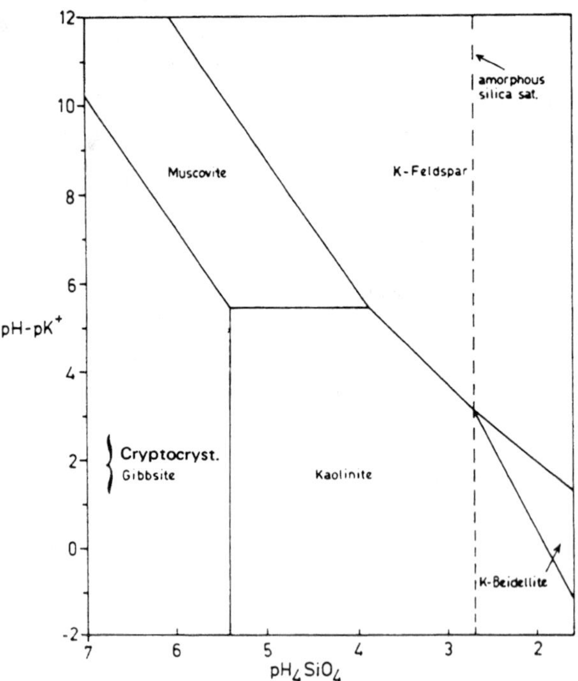

Figure 6.5. Stability diagram of minerals in the system Al_2O_3—SiO_2—K_2O—H_2O at 25°C and 1 bar total P. Taken from Van Breemen and Brinkman (1976) and after Garrels and Christ (1965).

Table 6.1. Charge Properties of Common Clay Mineral Groups

PYROPHYLLITE: $SI_8\ AL_4\ O_{20}\ (OH)_4$

CLAYS	EXT. SURF.	SI_4 / AL	HHHH OOOO AL_4 / MG	SI_4 / AL	INTER LAYER	σ C / M²
		α_T	α_O	α_T		
MICA	(M)	1 →		← 1	K	.35
ILLITE	M	3/4 ------→		←-------3/4	K, H	.28
VERMICULITE	M	----→ (1^+	----TO----	$^-2$) ←----	MG, K	.2 –.3
SMECTITE	M	—	(½ TO 1)	—	(M)	.1 –.2
CHLORITE	(M)	(1 TO ½) ←------		------→ (½ TO 1)	$AL(OH)_{5/2}(OH_2)_{1/2}$ + CHARGE !	??

[a]For comparison, the composition of the prototype layer-silicate *pyrophillite* is given. The α (fractional substitution per unit cell comprising four octahedral Al) is given in rounded-off numbers, specifying tetrahedral (T) and octahedral (O) position, arrows indicating range of variation. The interlayer cations are specified if typical; the ensuing approximate value of the surface density of charge, σ, is indicated.

pacity for cations. A well-known Mg trioctahedral member of the 1:1 group of clay minerals, *serpentine*, is found only in rocks, where it serves as a source of weathering products. Again substitution charge appears to be absent; a slight misfit between the octahedral and tetrahedral layers leads to the tubular structure found in the fibrous serpentine *chrysotile,* an asbestos mineral.

The above rather sketchy description of the major clay minerals is most easily summarized in the form of a table specifying the stoichiometry in terms of the main constituents of the tetrahedral and octahedral layers. Table 6.1 gives an indication of the range of lattice substitutions and the ensuing permanent charge of the lattice specified per unit cell containing four octahedral units. In order to relate the fractional substitution to α to the corresponding surface density of charge (areic* charge), it is pointed out that the basal

*The expression "areic" for an amount per unit area is the officially accepted, but very new, choice of the CGPM and its international committee, CIPM (areic, volumic, massic).

surface area of such a unit cell equals roughly 0.9 nm^2, counting both sides. The maximum charge density mentioned earlier and corresponding to about two sites per square nanometer (cf the mica in Table 6.1 with two sites per unit cell, i.e., about 20–25% substitution in the tetrahedral layer) thus gives a permanent charge of 0.3 C/m^2.

Within the context of the present description of surface reactions in soil, it suffices to merely point out that in addition to the clay minerals discussed so far, "mixed layer" types have also been found. This mixing of layers has been named *interstratification*; obviously, one of the causes of this phenomenon may be sought in the transformation processes indicated in the discussion of certain sequences following, for example, the removal of K from micas (during weathering) and the chloritization process. Examples of interstratified forms found are mica-vermiculite, mica-smectite, chlorite-vermiculite, and chlorite-smectite. Occasional mention has been made of kaolinite-smectite interlayering. Interstratification may be in a very regular pattern as well as in random arrangements.

A brief mention is made here of some "amorphous" members of the aluminosilicates, the *allophanes*. Characteristic of the weathering sequences in volcanic ashes, they have been described as (hollow) spherical or tubular structures comprising exterior Al hydroxide octahedra condensed with Si (hydr)oxide tetrahedra protruding into the interior. Although the entire group of allophanes comprises amorphous compounds of widely varying composition, one type, *imogolite,* has been defined fairly well. Its overall chemical composition corresponds fairly closely to that of a 1:1 clay; a striking difference is the presence of SiOH (silanol) groups in the interior of the spherule. More information on the allophanes may be found in Wada (1977, in B, pp. 603–638) and Brown et al. (1978, in C, pp. 152–163).

2.1.2. (Hydr)Oxides. A rather different group within the inorganic clay fraction in soil is formed by the (hydr)oxides. Silicon dioxides such as quartz and tridymite are mentioned here only in passing, as either the surface area falls far below the chosen limit of 1 m^2/g or they are present as amorphous constituents of which the surface silanol groups, though of interest, tend to be few in number compared with the ones on the edges of clay minerals. Thus the main (hydr)oxides of concern in soil are those of aluminum, iron, and manganese.

For the present purpose it suffices to mention *gibbsite* as the leading $Al(OH)_3$ compound. Figure 6.3 illustrates the structure of the gibbsite sheet; the gibbsite crystal consists of hydrogen-bonded stacks of these sheets. The partly dehydrated form boehmite, AlO(OH), a (minor) constituent in only a few soils, will be treated here as having the same type of active groups on its surface as gibbsite (see below). Its crystal structure is quite different from that of gibbsite, the composing sheets being positioned perpendicular to the *b* axis.

Quite different is the situation with regard to iron (hydr)oxides: Several crystalline forms are of importance in soils, while the transition between the trivalent and divalent states occurring under natural conditions in soil favors changeover between different forms (see Figure 6.6). For a rather complete account, see Schwertmann and Taylor (1977, in B, pp. 145–175), from which Figure 6.6 has been reproduced. The most abundant form of the iron oxides mentioned is *goethite,* α-FeO(OH). For the present purpose its structure is described as double bands of FeO(OH) octahedra sharing edges or corners, the bands bonded in part by H bonds. The ensuing macroscopic structure is often needle-shaped, with grooves and ridges (Brown et al., 1978, in C, pp. 135–137). The other Fe oxide of widespread occurrence is the fully dehydrated *hematite,* α-Fe_2O_3. Its structure somewhat resembles that of gibbsite, although in the hematite crystal the octahedral sheets are bonded fully by additional Fe atoms, thus providing for O atoms in fourfold coordination with Fe. The third common compound in this group is the somewhat weakly defined *ferrihydrite,* a highly hydrated, semicrystalline material that forms upon rapid hydrolysis of Fe(III) solutions and also upon the fast oxidation of ferrous iron brought into solution following reducing conditions. Summarizing, it is in particular the comparatively high solubility of Fe(II) compounds that causes the iron in soil to remain subject to transitions between the different forms if periods with low redox potentials occur regularly. Because of such transitions between iron compounds via the solution phase, the inclusion of other ions present during the "dissolved" state, in particular Al ions, appears possible.

Other oxides found in soil comprise, for example, those of Mn (birnessite,

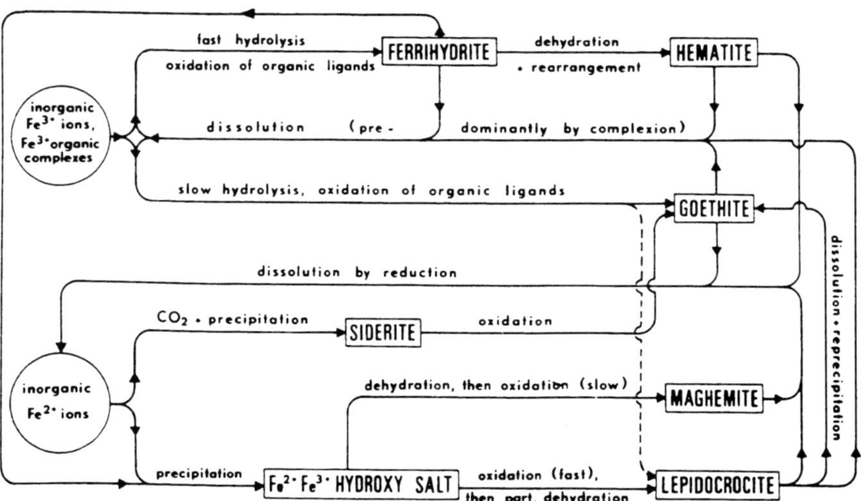

Figure 6.6. Possible pathways of iron oxide formation under near pedogenetic conditions. Adapted from Schwertmann and Taylor (1977) by permission of Soil Science Society of America.

and less commonly, todorokite, pyrolusite, etc.) and of Ti (anatase, ilmenite, pseudo-rutile). Their share in contributing to the surface area of soil is rather small because of the minor amounts present and the often large crystals. As with the iron oxides, the manganese oxides are subject to pϵ-dependent (pϵ = $-\log \{e^-\}$) dissolution and reprecipitation; their ability to accommodate heavy metals within the crystal lattice is probably of greater importance than the adsorptive properties of the surface.

2.2. The Clay Fraction: Surface Properties

Of prime importance are the (presumed) surface properties of the soil materials described above. At the onset of this discussion the reader is referred to Sposito (1984), who (E, pp. 12–23) presented his view on this topic in a concise and very interesting manner.

Referring once more to Figures 6.2 and 6.3, note that the planar surface of the inverted SiO_4 tetrahedron is chemically quite inert. Apparently the full balance of charge on the O atoms in twofold coordination with the *tetra*valent Si effectively prevents further association of a proton even at very low pH values. This fact is borne out by the absence of any swelling or intercalation of pyrophyllite with either water or organic compounds. Also, the known softness of powders of this mineral and its trioctahedral counterpart, talc, indicate that only very weak forces are holding such layers together. Free apical O atoms of silica tetrahedra protruding from quartz surfaces, for example, remain protonated to high pH values, because they are only singly coordinated with Si. In amorphous silica the number of open-ended SiOH groups per unit surface area is much higher than in quartz. In view of the combination of low charge density with low surface area, it appears that the bulk of the silica present in soil contributes very little to its adsorptive capacities.

Returning to the clay minerals, the siloxane surface of these presents two features of interest for the adsorption behavior, the ditrigonal holes within the Si_6O_6 rings shown in Figure 6.2 and (negative) substitution charges located either in the Si plane or, at greater depth, in the plane of the octahedral Al atoms. In Figures 6.7*a*, *b*, an attempt has been made to picture the situation (using a 2:1 clay as an example) with the help of a packed-spheres model. Obviously, any attempt to use a two-dimensional image of a three-dimensional array precludes a complete representation of the actual situation. For comparison, see the spoke-and-ball presentation used by Sposito (1984, in E, pp. 16 and 17).

The total charge density due to isomorphic substitution in the 2:1 soil clays ranges from about 0.1 to 0.3 C/m^2 (see Table 6.1). This corresponds to a permanent cation-binding capacity of roughly 1–3 $\mu eq/m^2$. It seems rather important that the ditrigonal holes are too small for the entry of hydrated cations (anions will seldom, if at all, penetrate in this region of rather high negative electric potentials). Thus, although the substitution charge must be fully neutralized by externally adsorbed cations, the formation of inner-sphere complexes with these seems an exception, the well-known one arising in the case of so-called potassium fixation. Taking the original mica as model for this situation, it appears that the combination of (1) a high charge density in tetrahedral positions—close to 0.4 C/m^2—with (2) the 12-fold coordination of two Si_6O_6 rings of adjacent plates, and (3) the K ion with a medium energy of hydration is needed to provide for this rather unique form of potassium storage in nature. The smectites, with their predominantly octahedral substitution at about 0.1–0.2 C/m^2, do not exhibit this phenomenon. The illites, with about 0.3 C/m^2, predominantly in the tetrahedral layer, will still exhibit potassium-fixation behavior.

It is of interest to point out here that apparently the situation in the case of illites is such that the force fields in and around a single ditrigonal hole are not yet able to cause complete dehydration and subsequent *ab*sorption of the naked K ion inside the hole. Thus, for the external faces of the clay crystals, ionic selectivity for K with respect to Na ions is less than a factor of 10 (Bolt et al., 1963), while very much higher values are found for those ions that become situated in an interlayer position of the same clay.

If the cations do not penetrate into the hexagonal holes, their adsorption must be dominated by simple coulombic forces. The ion selectivity still observed has been interpreted as an indication of surface complexation, albeit as outer-sphere complexes. It remains somewhat doubtful whether the term "complexation" in this case implies more than the expectation that in the rather strong electric fields of the highly charged surface (around 10^8 V/m) any difference in a distance of closest approach could cause cation selectivities of the magnitudes commonly observed.

Switching to the gibbsite surface exposed on the 1:1 clays, one finds again a surface that is basically rather inert, although this surface is fully protonated. The OH group doubly coordinated to Al ions in sixfold coordination is again fully neutralized, and the pH range of zero charge appears to stretch out on both sides of pH 7. In summary, though undoubtedly opening more possibilities for surface reactions than the siloxane surface discussed above, it should still be regarded as rather inert in the pH range prevailing in soil.

Much greater reactivity is exhibited by singly coordinated M—OH groups situated along the edges of the clay minerals and the hydroxides of Al and Fe(III). In Figure 6.7 the different types of groups occurring are shown schematically. The edge-standing —SiOH group (occurring along the edges of all clay minerals and also on SiO_2 particularly the amorphous forms) exhibits an (intrinsic) p*K* value certainly below 9.8, the p*K* value given for monosilicic acid. Using the data given by Bolt (1957), one might venture a fair guess at the intrinsic p*K* value of —SiOH on the surface of colloidal silica particles. Using appropriate approximation equations (cf. Bolt, 1982, in D, p. 24), it would seem permissible to guess at a surface potential of less than 10 mV.

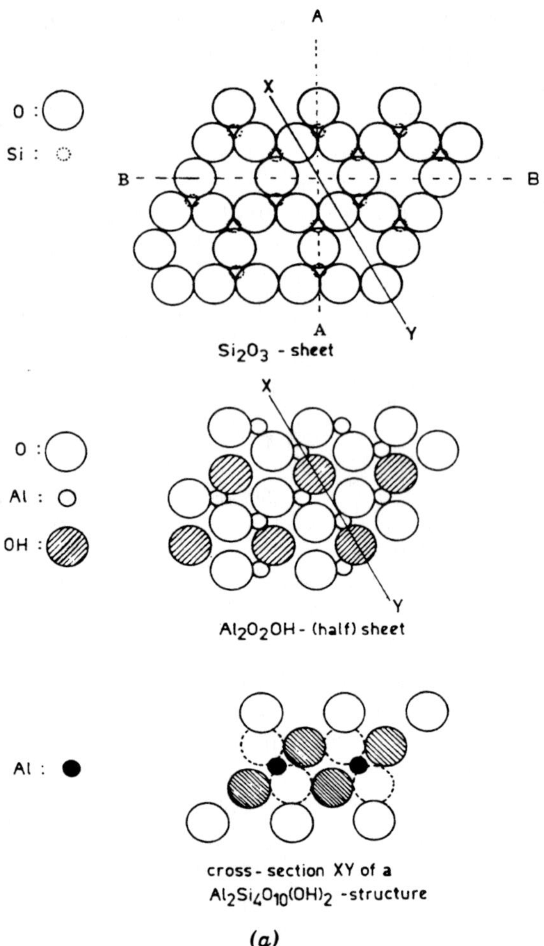

Figure 6.7. (*a*) Scale drawing of a 2:1 clay pictured as a packed array of O and OH with interstitial cations. Upper and middle part viewed along *c*-axis, lower is cross section. After Bolt and Bruggenwert (1976). (*b*) Various cross sections based on Figure 6.7*a*, showing reactive groups. *Upper*: Section of ditrigonal cavity along line *B*—*B*, indicating position of negative charge if a trivalent cation replaces Si (tetrahedral substitution) or a divalent cation replaces Al (octahedral substitution). *Middle*: Same, along the line *A*—*A* of Figure 6.7*a*, showing edge groups Si—OH (1) and $Si—O(Al)H^{1/2+}$ (2). *Lower*: Section along line *X*—*Y* of the octahedral sheet with the gibbsite composition $Al(OH)_{6/2}$, showing edge group of $Al—(OH)H^{1/2+}$ (3).

For a total site density on amorphous silica of about 10 $\mu mol/m^2$, the ensuing degree of dissociation of 0.02 would then suggest an intrinsic pK value somewhat below 8.

Possibly this might be used as a general indication that the proximity of many tetravalent Si atoms close to the surface SiOH tends to increase the proton dissociation of these groups compared to that of monomeric $Si(OH)_4$, thus lowering

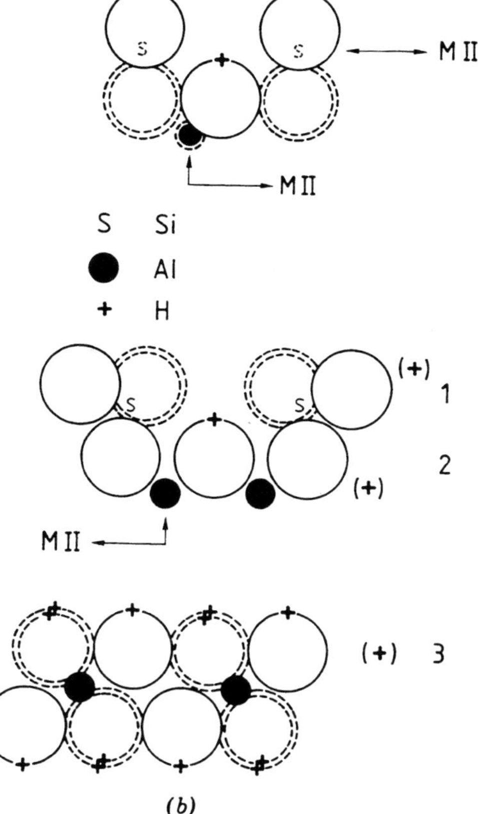

Figure 6.7. *(Continued)*

the intrinsic pK value. Obviously, the negative value of the surface potential arising upon the increase of the negative charge of the surface will soon check further dissociation; thus the surface charge is only about 1 μeq/m^2 at pH 8. Similar reasoning appears to apply when comparing the values of the consecutive pK values for Al monomers and polymers. Thus pK_1 of the monomer $Al(H_2O)_6$ is very close to 5. The pK_2 of the second deprotonation should be some pH units higher. As polymerization now sets in simultaneously, the second proton apparently dissociates much more easily, so the effective pK_2 value is also around 5.

Turning to the octahedral (hydr)oxides of trivalent Fe and Al, one might defend that a singly coordinated OH group as present on the edge surfaces of the common soil solid-phase constituents such as gibbsite and hematite will contribute locally a $\frac{1}{2}-$ charge to the surface. Upon association of a proton, the local charge of this group will be $\frac{1}{2}+$. The pK value of this proton association, or dissociation, reaction could then be found for a particular surface from the experimentally determined point of zero charge, PZC. While admittedly two adjoining singly coordinated OH groups could also be described in terms of two consecutive pK values, it should be questioned whether such a distinction is sensible in view of the use of a smeared-out surface charge and the ensuing surface potential as a controlling factor on the effective pK value. Referring to Bolt and Van Riemsdijk (1982) for details, it is pointed out here that the use of one or two pK values gives roughly the same type of charging curve as long as the difference ΔpK remains less than 2. In Figure 6.7 it has thus been assumed that the surface groups of interest may be typified by one pK value. Relating the situation at the edge surfaces of the oxides to that of the octahedral layer exposed on the edge surface of clay minerals, one finds a fair similarity between gibbsite and most of the other common clay minerals we have discussed. This is particularly so in the 1:1 clays, where the absence of a negative electric field arising from substitution charges combines with the presence of one fully protonated side of the gibbsite layer. There remains the difference of the Si-O-Al groups present on the other side. It seems difficult to predict the pK value of a single proton associated with this structure (see Fig. 6.7.). Using arguments similar to those given in relation to the surface SiOH group, it appears that this proton should be more acidic than the one on the —AlOH group.

The comparable groups on the edges of the 2:1 clays should experience the influence of the close proximity of a fixed negative surface charge, causing increased pK values, particularly at low salt levels. It might seem somewhat disappointing to observe that expectations on pK values of characteristic groups on oxides and clay minerals that appear to be quite well characterized mineralogically are limited to vague statements about their magnitude. As is known from other compounds of much simpler structure (e.g., AgI), however, experimental information provides the only source of such data. In fact, it appears that the sensitivity of physicochemical parameters, like pK values, to the details of local atomic structural arrangements is so great that even such experimental values are subject to change. In this respect it is of interest to mention that PZC values of synthetically prepared hematite could be shifted by several pH units by heating the sample to 500°C, without any observable change in the X-ray diffraction pattern. This observation stands separate from the other aspect of very profound influence of the presence of specifically adsorbed cations and/or anions on the position of the PZC. Against this background it seems futile to hope to be able to produce fairly accurate values for certain system parameters like pK values and PZC for a *particular* soil on the basis of a complete inventory of mineralogical composition. It will have to be characterized with some physicochemical information obtained experimentally on that specific soil if the information is to be used for predictive purposes.

2.3. Soil Organic Matter

Depending on the soil type, the organic soil fraction may be second (after the inorganic clay fraction discussed above) or first in supplying surface sites for interaction with solutes. Again, an impressive amount of literature has been produced on the composition and properties of this material, and it is of interest here to point out a few summarizing texts that have appeared in the past ten years or so. Referring in particular to the "humic compounds" and the polysaccharides, one finds about 250 pages in Gieseking (1975) and 150 pages in Greenland and Hayes (1978). A rather brief discussion may be fond in Sposito (1984).

Understandably, but in contrast to the clay fraction discussed before, the outcome of all research on soil organic matter has not yet produced a very precise description of its composition. Understandably, because soil serves as the "burial place of the plant kingdom"—aside from some animals—and thus the input to the system producing the soil organic matter is already very diverse. Separating off those parts of the fraction that still betray some morphological characteristics of their origin, the remainder is in part resynthesized, in part a breakdown product of organic origin. For lack of crystallinity of these organic polymeric substances, identification and taxonomy are based mainly on the recognition of characteristic groups. The question then arises as to whether or not such a group is indeed a characteristic of the compound studied or only an (accidental) heritage from the input compound. Also the question must often be posed whether this group has possibly entered as an artifact during the analysis phase.

Prior to a brief discussion of the main characteristics of humic substances in soil, the broad division of the whole of the soil organic matter fraction based on Oden (1922), or on even earlier work still in use, is given in Figure 6.8. This figure, taken from Hayes and Swift (1978, in C, p. 190), indicates the relation between polymer size and properties of the material. Obviously, although the fulvic acid fraction lends itself best for identification studies, it should be understood that such information is hardly to be considered representative for the whole fraction of humic materials present in soil.

Referring to the extensive discussion by Hayes and Swift (1978, in C, pp. 180–280), it is pointed out that many functional groups have been identified. Listing these, one finds in order of prominence with regard to adsorption phenomena the carboxyl, phenolic OH, and alcoholic OH as the dominant ones of the oxygen-containing groups, and amino acids as dominant in the nitrogen-containing groups. Attempts at a more detailed identification of functional groups usually involve chemical degradation procedures. Referring again to Hayes and Swift (1978), the number of products identified following several degradation methods amounts to 100 or more.

The problem of describing the nature of soil organic matter becomes clear from the above. As the pK value of a simple monomeric substance, such as the different Cl-substituted acetic acids, varies widely depending on the number of Cl's present, it seems hopeless to attempt to predict, for example, pK values to be expected for the many different carboxylic OH groups anchored to the organic matter polymers present in soil. Furthermore, it is in reality not so important to predict the pH titration curve of the soil organic matter fraction, as it is rather easily determined and in addition must be summed

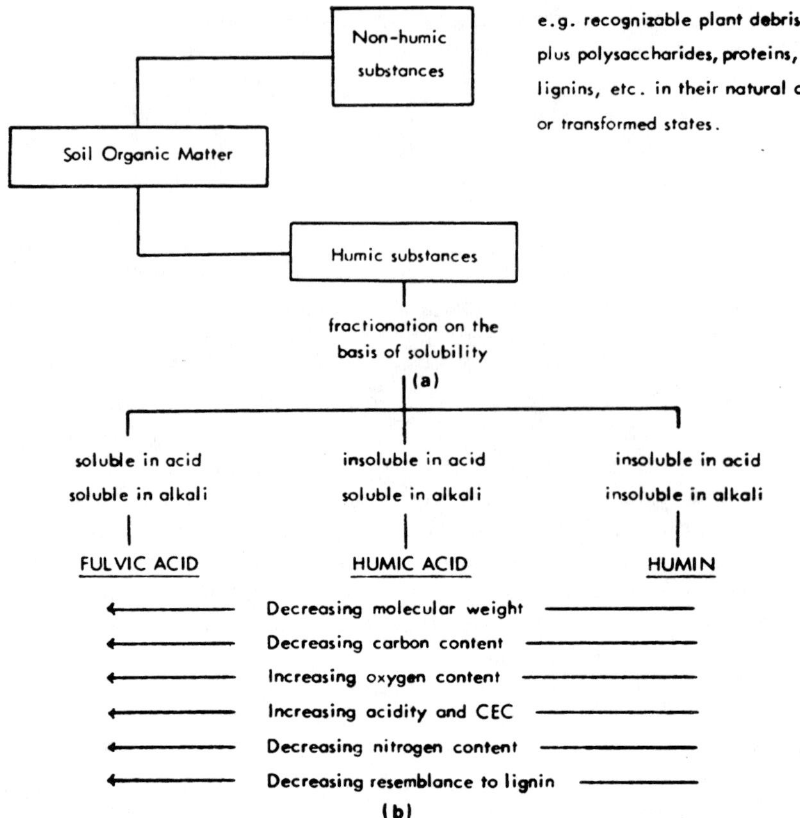

Figure 6.8. Common method of fractionation of soil organic matter, using solubility as a criterion (*a*). General tendency of variation in some properties shown in (*b*). Taken from Hayes and Swift (1978).

up with the contribution of the other proton-buffering compounds like the clay minerals. Of much greater importance are the complexation characteristics of different functional groups for different metal cations, in particular the transition and/or heavy metals. It appears impossible to be more specific in predictions than a generalizing statement to the effect that Cu^{2+} is known to tie up rather well with organic constituents usually present in soil. Similarly, it is well known *in general* that Al^{3+} and particularly Fe^{3+} are easily kept in the solution phase in the presence of ligands often supplied by decomposing soil organic matter.

Superimposed on the variability in sources for the soil organic matter fraction are the variability in breakdown and resynthesis processes occurring in soil, the sum total leading to a high degree of variability in the momentary composition to be expected, which in turn is subject to changes over relatively short time periods following, for example, the addition of the yearly crop residues to soil. Isolation for analytic studies must also be seen in the light of possible changes incurred during analysis. If, then, as pointed out by Sposito (1984), the different characteristic groups are likely to exhibit partly random positioning on the backbone of the organic matter polymers, it might even be impossible to pick up the physicochemical characteristics such as pK values and there remains only the possibility of using the picture of "smeared-out" constants. This somewhat disappointing conclusion will be commented on again in the discussion of the use of models for the description of the adsorption properties of soils.

3. AVAILABLE MODELS

3.1. Electrochemical Control of the Multilayer System

Referring to Chapter 1 for more information on derivations, the models now commonly used for the description of an adsorbed phase residing on solid surfaces are summarized for the present purpose in terms of the schematic drawing shown in Figure 6.9. Thinking primarily in terms of oxidic surfaces,

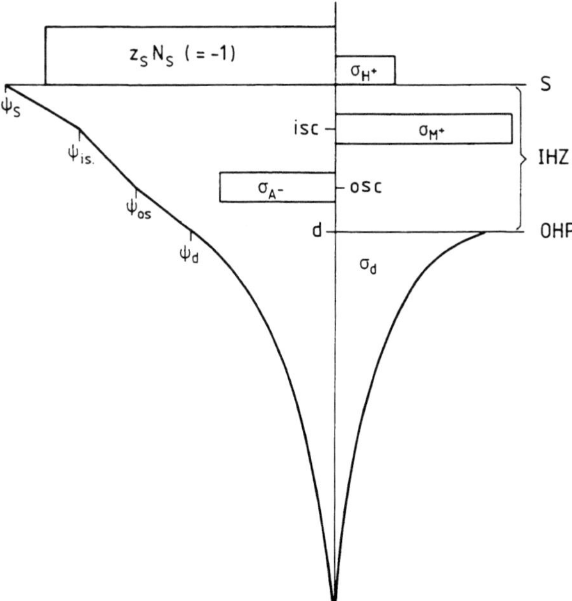

Figure 6.9. Generalized distribution of charge and potential in quadruple layer arrangement serving as a prototype for oxide surfaces. The proton charge is placed on the surface S; two types of surface complexes with ions (inner-sphere, isc, and outer-sphere, osc) are located in the inner Helmholtz zone (IHZ); diffuse charge is situated outward from the outer Helmholtz plane (OHP).

the total number of sites available for adsorption, $N_S m^{-2}$, multiplied with an assigned negative valence z_S, is represented as a unit surface area of negative charge. Electroneutrality is provided for by the adsorption of cations and anions, which are grouped into several different types according to:

1. The primary (see Bolt and Van Riemsdijk, 1982) potential-determining ions (ppdi), in this case H^+.
2. Surface-complex-forming cations M_k^+ and anions A_l^-, both species subdivided into those forming inner-sphere complexes, (isc) and the ones forming outer-sphere complexes (osc).
3. Inert cations and anions, accumulating in the diffuse tail of the electric double layer (DDL).

As shown in Figure 6.9, the ppdi H^+ have been located within the solid surface S such that the surface charge density σ_S is given by

$$\sigma_S = z_S N_S + \sigma_{H^+} \qquad (6.1)$$

The charge densities associated with the ionic complexes, σ_{is} and σ_{os}, are located in the inner Helmholtz zone (IHZ), where the positioning of M_k and A_l is an arbitrary choice. The diffuse part of the countercharge, σ_d, is then located beyond the outer Helmholtz plane (OHP). The ensuing trajectory of the electric potential ψ is sketched in.

In Figure 6.10, a diagram is presented of the "electrochemical control system." The electrical part of this is governed by the capacitances C_0 (between S and the isc's), C_1 (between isc and osc), C_2 (between osc and OHP), and the diffuse layer capacitance, C_d. Obviously, the potential increments between the locations distinguished are controlled by the product of the reciprocal capacitance C^{-1} and the cumulative charge density $\Sigma\sigma$. The chemical part of the control system is determined by the complexation reactions. For the present system the following reaction equations are used:

$$SOH^{(1/2)-} + H^+ \rightleftharpoons SOH_2^{(1/2)+} \qquad K_H \qquad (6.2)$$

$$SOH^{(1/2)-} + M^{z+} \rightleftharpoons SOHM^{(z-1/2)+} \qquad K_M \qquad (6.3)$$

$$SOH_2^{(1/2)+} + A^{z'-} \rightleftharpoons SOH_2A^{(z'-1/2)-} \qquad K_A \qquad (6.4)$$

The reaction constants K above will be used as "association" or pair-forming constants. Applying these equations to the inner Helmholtz zone (IHZ) leads to a multicomponent Langmuir adsorption isotherm for each of the adsorbates involved.

In the isotherm equations deriving from the specified reactions, the concentration of the adsorbate must obviously be specified for the layer involved, so the concentrations in the equilibrium solution must be multiplied by the

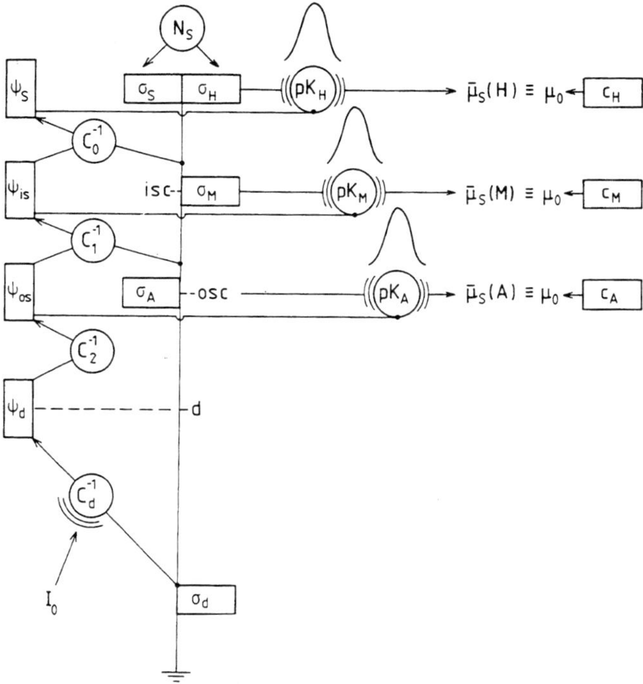

Figure 6.10. Electrochemical control system regulating the surface charge σ_S, and potential, ψ_S, in Figure 6.9. Input variables are the solution concentrations of the primary potential-determining ion H^+ (ppdi), the ions M and A forming surface complexes, and the ionic strength, I_0. The intrinsic complexation constants are expressed as pK values assumed to exhibit a distribution of values.

appropriate Boltzmann factor. In turn the latter is determined by the value of the electric potential in the layer considered, leading to the electrochemical control system pictured in Figure 6.10. Also the adsorption maximum follows from the reactions specified, as $z_S N_S = -\frac{1}{2}N_s$. The use of a "one-p$K$" dissociation reaction to describe the interaction between protons and the (hydr)oxide surface has been discussed on pages 144 and 145.

It should be stressed that the distinction between inner-sphere and outer-sphere complexes will be reflected in general by the larger value of the association constant, K, for the former. Also, as a rule, the isc will be located closer to the surface (no interposed water molecules), implying that in the absence of charge reversal by the adsorption of the isc $|\psi_{is}|$ tends to exceed $|\psi_{os}|$.

Expressing the above values of the association constants in terms of the corresponding pK_H, pK_M, and pK_A, one finds the overall control as pictured. For a given set of concentrations c_H, c_M, and c_A and the corresponding values of $\mu_0(H)$, etc., in the solution phase, the surface adsorbs these three species in Langmuir-type competition for the sites. The ensuing buildup of the electric potential serves as a moderator, and equilibrium is reached upon equality of the electrochemical potential at the surface, $\bar{\mu}_S$, with the chemical potential in solution, μ_0, for all ions involved.

While it may seem that the above "quadruple" arrangement of particle charge and countercharge is the simplest account that could be given of the actual configuration to be expected, it must be admitted that at the same time it constitutes probably the maximum complexity that can be handled when constructing a model of the adsorption behavior of the soil solid phase. Such limitation in handling facilities is obviously not dictated by computational problems but by the inability to assess dependable values for the system parameters involved. The "simple" model described above thus involves three different capacitance values in addition to the (presumably known) functional relationship between I_0, the ionic strength, and C_d, as well as the intrinsic pK values for *all* ions involved in complexation reactions. Including the total number of sites available per unit mass of the solid phase, the system described formally in Figure 6.9 thus comprises seven system parameters plus the needed relation $C_d(I_0)$.

In fact, the system described above, which contains one salt in addition to the ppdi H^+, comprises the *three* capacitances separating the *two* IHZ planes from the particle surface S and the DDL commencing at position d together with *one* value for the total number of sites N_S available to all three ions involved. If it were considered that the ion species involved are all adsorbed at adsorbate-peculiar sites present in an amount N_k, the one multicomponent Langmuir equation would be replaced by three separate equations, thus increasing the number of system parameters to three capacitances, three site-density parameters, and the three complexation constants K. Extending this reasoning to a multicomponent system with n ion species would eventually require one more decision as to whether one is satisfied with only two different locations for the positioning of *all* isc's and osc's. Clearly, some restraint is necessary in this situation, as $3n$ system parameters (K_k, N_k, the position parameter x_k) seems far beyond a possible span of control for modeling purposes.

Indeed, many models in use for the description of adsorption isotherms are simplified versions of Figure 6.10, as shown in Figure 6.11. Quite clearly, the dominance of a "permanent" substitution charge, present in many typical clay minerals, effectively limits the possible role of a deprotonation-induced CEC to marginal values. Accordingly, as shown in Figure 6.11*a*, the value of σ_S is taken as fixed. Using the cation pair M_{1-2}, and accepting that full dehydration of either one is not likely, one is left with the selection of appropriate values for pK_1 and pK_2 and also C_2. An attempt to describe cation exchange in binary cation–clay systems using a model of the above type was made first by Heald et al. (1964), who used C_2 approaching infinity, that is, they placed the osc ions subject to K_1 and K_2 in a plane coinciding with the OHP. Further attempts in this direction were made by Shainberg and Kemper

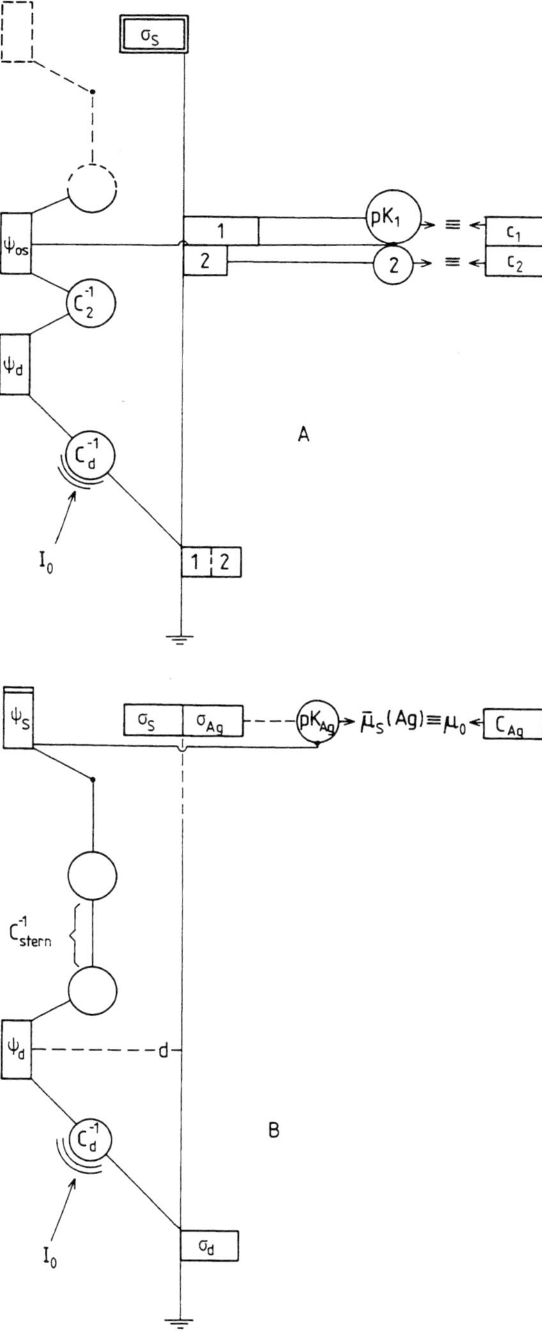

Figure 6.11. Two simplified situations of the system pictured in Figures 6.9 and 6.10. (*a*) The constant-charge clay system in equilibrium with a solution containing two cation species subject to surface complexation and accumulation in the nonselective diffuse layer. (*b*) The classical AgI system in the presence of the ppdi Ag and inert electrolyte only. The chemical part of $\bar{\mu}_S(Ag)$ is taken constant, giving Nernstian behavior of the surface potential ψ_S.

(1967) (cf. also Bolt, 1982b, in D, pp. 74, 75) and Neal and Cooper (1983). It may be shown that for a fixed value of the total surface charge (and thus the CEC), one may satisfactorily describe "mildly" preferential cation adsorption in terms of a "fitted" value for K_1/K_2 and a not very sensitive choice of (1) the absolute value of either K_1 or K_2 and (2) the value of C_2. In fact, selecting a rather high value for K_1 (and K_2) effectively eliminates the role of the diffuse part of the double layer and brings one to the earliest models used for describing cation exchange on clays. These [named after Vanselow (1932), Gapon (1933), and others] were to be seen as either straightforward or adapted versions of the solid solution model that could be termed "condensed double-layer" models within the present context. Much attention was eventually paid to an enhanced statistical treatment of such a condensed first layer allowing for the modes of placement of polyvalent ions on a fixed array of charged lattice sites characterized by certain prescribed geometrical patterns (cf. Harmsen, 1979). At the other extreme end of the scale, one may deny the notion of complex formation if it involves hydrated cations in the proximity of a surface "charged permanently at *some depth*" (in the octahedral layer). The observed ionic preference is then attributed to differences in distance of closest approach of cations of rather different hydrated sizes. Because of the quite steep increase in the electric potential close to the charged surface, differences of 0.1–0.2 nm produce a sizable cation selectivity. In view of the above it appears that there is some freedom to select either the ratio of complexation constants or of ionic radii as a guideline for observed "mild" exchange preferences on constant-charge surfaces like the planar surfaces of many clays.

Mentioning in passing the AgI surface in contact with inert electrolyte as a primordial example, it is pointed out that here the original Stern–Grahame extension of the Gouy–Chapman diffuse layer was brought to quantitative usage, calculating the capacitance of the presumably empty Stern layer from the experimental σ_S(pAg) curve. Pointing to the corresponding Figure 6.11*b*, it is also noted that in this particular case it was almost always accepted that the abundance of Ag ions throughout the crystal would allow the assumption that the chemical part of $\bar{\mu}_S$(Ag) could be taken as a constant. In the figure this is indicated by a dashed line connecting σ_{Ag} and $\bar{\mu}_S$(Ag), the latter thus being influenced solely by ψ_S. This condition obviously leads to perfect Nernstian behavior of the solid surface in response to a change in c(Ag) in the solution phase, as was generally assumed in the early treatment of the AgI surface. Referring to the extensive literature on the adsorption behavior of AgI (for an excellent summary up to 1978, see Bijsterbosch and Lyklema, 1978), it is of interest within the present context to note that the back-calculated value of $(1/C_1 + 1/C_2)^{-1}$ found for AgI in inert electrolyte tends to be larger than expected. This may be interpreted to indicate that even "inert" ions tend to approach the surface rather closely. In turn this may serve as a warning not to over-estimate the significance of a too detailed subdivision of the Stern layer.

Returning to the oxides that served as model for Figures 6.9 and 6.10, one finds various simplified versions in the literature. Briefly, such simplifications differ in their choice of the capacitances distinguished formally in Figure 6.10: C_0, C_1, and C_2. Using infinitely large values for these (such that their reciprocals vanish) implies that the bordering planes are taken to coincide. The ground state in this process is to condense the IHZ by placing the surface complexes (not distinguishing between inner-sphere and outer-sphere complexes) directly on the solid surface S. In this manner, C_0^{-1} and C_1^{-1} vanish, while now the surface complexes are competing with and subject to the same electric potential as the ppdi, in the present case the proton. Because of this choice, the name "secondary potential-determining ions" (spdi) might be appropriate. This model could be seen as a constant-capacity model, as introduced and used by Stumm et al. (1970), Schindler and Gamsjaeger (1972), and Huang and Stumm (1973) for modeling oxides and described as such by Westall and Hohl (1980). When used over a range of concentrations of the inert background electrolyte, it actually comprises a combination of a constant (adjustable) capacitance C_S and a (calculable) DDL capacitance.

The above maximally simplified Gouy–Stern model is probably limited to systems comprising at the most one strongly bound spdi in addition to inert electrolyte (cf. the AgI model shown in Figure 6.11*b*). A slightly more developed model accepts the presence of Langmuir–Stern complexes outside the surface S but places these at the d plane, thus letting C_2^{-1} vanish. In this manner the number of adjustable capacitances is still limited to one (C_1), while the basic features of the classical Stern arrangement are introduced. Westall and Hohl (1980) refer to this model as the "basic Stern model." They also infer that the model used by Bowden et al. (1977) to describe the adsorption of phosphate by soil oxides belongs to this type, but Bowden et al. bring in one more system parameter by assigning a separate set of sites to the adsorption of phosphate. Referring to the "complete" model discussed in the small print section following Figure 6.10, the Bowden et al. version amounts to setting $C_0^{-1} = C_2^{-1} = 0$ but using a site-density paramter N_P in addition to the site density of the surface with respect to the adsorption of the ppdi, H^+. Referring to the earlier simplified model, it seems logical to include with the basic Stern model the feature of strongly adsorbed spdi assigned to the actual surface S. In that case these spdi could be identified with inner-sphere complexes, while the Stern ions placed at the d plane would be considered outer-sphere complexes.

Actually the intermediate between the constant-capacity model above and the triple-layer models to be discussed below appears to be a fair choice. Thus, in the presence of inert ions only, one may derive the value of the overall Stern capacitance according to $C_S^{-1} = C_0^{-1} + C_1^{-1} + C_2^{-1}$ from a curve-fitting procedure applied to the pH–titration curve of the oxide. Doing so, one often finds an indication that the distance d corresponding to the value of C_S is only a few tenths of a nanometer. It then seems hardly warranted to use the titration curves in the presence of surface-complexing ions both to obtain the value of an appropriate complexation constant and to assign a complex-peculiar position to such an ion situated within the small distance *d*. In fact, in view of the rather small values found for the full distance *d*, it might be questioned whether it serves any purpose to distinguish between outer-sphere and inner-sphere complexes of the surface. Perhaps the only sensible question is whether any surface complex—which in principle will have its own characteristic distance to the surface—should be considered as located with the ppdi H^+ *at* the surface *S* or whether it should be grouped with other complexes at some distance from the surface, which for lack of accurate information is approximated by the distance *d*, also used as the starting point of the DDL.

The well-known triple-layer models are an obvious extension of the above description. Suggested by Grahame as early as 1947, the triple layer was the materialization of a "filled Stern layer": surface complexes (i.e., ions subject to a complexation *K* and Langmuir statistics) were placed within the assigned IHZ at a position in between the surface *S* and the OHP located at *d*. Obviously, this design involves the splitting of the (reciprocal) Stern capacitance C_S^{-1} into its two parts C_1^{-1} and C_2^{-1}.

Although perfectly all right on paper, as mentioned earlier, one can hardly expect to acquire the accurate experimental information needed for this purpose. Furthermore, if it is recognized that in reality every complex should have its own peculiar distance to the surface—even if this is subsequently approximated by a group distinction in terms of inner-sphere and outer-sphere complexes (see Figs. 6.9 and 6.10)—it appears wisest to stay with the simpler models. Certainly in the case of rather poorly defined mixtures of solid-phase components such as are present in soils, one should probably forgo any attempt to specify precise locations, either expressed as a set of fitted constants *C* or as a priori guesses based on an assumed set of distances between adsorbates and surface. Instead, the use of one fitted capacitance value for the location of all complexes with respect to the surface *S* seems the best one can do in this case. If then, as inferred above, because of the rather small value of *d* found for test systems in inert electrolyte, the location of the complexes is taken to coincide with the OHP, one is back to the basic Stern model discussed above.

3.2. Effect of Surface Heterogeneity

Somewhat complementary to the uncertainty about the precise structure of the adsorbed phase when it is viewed in a direction normal to the surface is the question of lateral homogeneity of this layer. The surface of even a pure crystalline metal oxide will probably consist of reactive sites that are characterized by a range of p*K* values for the interaction of such sites with a particular ion species. It has been suggested that it is necessary to take such surface heterogeneity explicitly into account in modeling the adsorption of trace metals on iron oxide (Benjamin and Leckie, 1981). In Figure 6.10 this has been accounted for by picturing a distribution of log K, typified in principle by a "mean" value and the "width" of a distribution curve. A fair amount of literature has been produced on this aspect (Sips, 1948, 1950; Jaroniec, 1983). Recently a thorough analysis was made of the effect of site heterogeneity, as characterized by a continuous distribution of pK values, on the models discussed above (Van Riemsdijk et al., 1986, 1987). Aside from the well-known fact that such a distribution leads in principle to the conversion of the basic Langmuir isotherm into a Freundlich-type isotherm, it could also be shown that for multiadsorbate systems the ensuing isotherm exhibits the appearance of a finite slope at vanishing values of the adsorbate concentration. A relevant equation covering the situation of multicomponent adsorption subject to a distribution of pK values may be written as (Van Riemsdijk et al., 1986):

$$\theta_i = \frac{K_i c_i}{\left(1 + \sum K_i c_i\right)^m \left(\sum K_i c_i\right)^{1-m}} \tag{6.5}$$

in which θ_i refers to the fraction of available sites occupied by i, as summed up over the totality of sites exhibiting a distribution of the value of pK_i of a chosen type (see Van Riemsdijk et al., 1986), characterized by a mean value K_i and the width of the distribution curve as specified by way of m. It is noted here that for $m = 1$ the above equation reverts to the standard multicomponent Langmuir equation. Again, the concentration c_i refers to the concentration present at the plane of adsorption, so in the present case an appropriate Boltzmann factor must be brought in.

The above-mentioned transition of the standard Langmuir isotherm into a finite-slope Freundlich-type isotherm, upon replacing a chosen pK value by a corresponding distributed one, is in practice difficult to ascertain when scrutinizing experimental data (cf. Van Riemsdijk et al., 1986, 1987). Granted that the introduction of a distributed pK value *does* affect the position of the PZC (as compared to the one found for a single pK situated at the mean value of the distributed pK), it is also clear that the latter effect remains unnoticed, because in practice the PZC is *derived* from the experiment.

The above effect could possibly explain the observed shift of the PZC of hematite after heating the sample to 500°C (Penners, 1985), by assuming that the distribution of the pK_H changed upon heating, which seems likely indeed.

The above considerations give rise to the convenient conclusion that in spite of the likely presence of (some) surface heterogeneity, the use of models pertaining to laterally homogeneous surfaces is acceptable if the experimentally determined PZC is treated as an equivalent value rather than one with absolute significance.

4. APPLICABILITY OF THE MODELS TO THE SOIL WATER SYSTEM

4.1. Field Heterogeneity; Need for Relatively Simple Models

Heterogeneity and variability in space and in time, characteristic features of soil, obviously interfere with the possibilities for developing successful modeling efforts. Viewing a unit volume of soil as the basic system to be described, one finds that the phases of primary interest for the processes of concern here—the solution phase, the adsorbed phase, and the solid phase(s)—are subject to continual and rapid variation in the liquid content of the system. Although in aquatic systems also the concentrations of dissolved species are subject to the processes of evaporation and dilution, the rates of variation in time and space tend to be much smaller. In addition, the solvent water comprises a minor share of the totality of the soil system compared to its role in aquatic systems. Thus solid and adsorbed phases are often dominant in determining the course of events: distribution ratios between a rapidly accessible adsorbed phase and the liquid phase of the order of 100 or larger are the rule rather than the exception.

Narrowing the discussion to solid and adsorbed phases present in the soil, here again the heterogeneity of the solid-phase constituents described in the previous section is usually reflected even within a rather small unit of volume of soil. As a rule, clay minerals (often of more than one type), a sand fraction containing several types of primary minerals including quartz, (hydr)oxides in part present as coatings, and organic matter fractions are all present simultaneously in a rather intimate mixture. Superimposed upon this heterogeneity on a micro scale is the variability of the soil composition with its location, as defined in both the vertical and horizontal directions. In actuality one should perhaps distinguish here a heterogeneity on a meso scale, that is, within a certain field, and the macroscale heterogeneity reflected by the different geologic formations found when mapping the route of groundwater along aquifers.

If it is necessary for the object of investigation to take field-scale heterogeneity into account (e.g., the prediction of phosphate transport in the field, Van der Zee and Van Riemsdijk, 1986a), one must also obtain an insight into the variability of surface chemical properties of concern in the field. In such a case, many samples of the soil have to be taken and analyzed. In turn this implies, because of the number of samples involved, that the reactive properties of the soil with respect to the chemical species involved must be characterized by relatively simple procedures. In practice the selection of the appropriate experimental procedure depends critically on the insight one has attained with regard to the processes dominating the interaction between the species under consideration and the constituents of the soil type of interest.

The need for relatively simple descriptions of surface chemical processes in soil will now be illustrated with some examples involving the investigation of excess phosphate, certain heavy metal ions, and the influx of protons into soil. These examples are based largely on our own experience with the investigation of problems of contamination on a field scale.

It has been well established that for the reaction of excess (ortho)phosphate with acid sandy soils in temperate regions, (amorphous) iron and aluminum (hydr)oxides are the dominant soil constituents. That is to say, at concentrations of orthophosphate that prevail in the equilibrium solution of phosphate-saturated soil samples (i.e., around 3 mmol/L), the sorption capacity for phosphate is determined predominantly by the content of the soil of such reactive Fe and Al compounds. This content is then established by an extraction procedure. An estimate of the relevant specific surface area of these oxides may be obtained by establishing the phosphate *adsorption* maximum of the samples in short-term adsorption studies in the laboratory.

It can be shown (Van Riemsdijk, Boumans, and De Haan, 1984; Van Riemsdijk, Van Der Linden, and Boumans, 1984) that the reaction of phosphate with metal oxides at the concentration levels that occur in soil immediately after application of fertilizer (manure) is a conversion of the metal oxide particle into a metal oxide coated with metal phosphate. One thus has to distinguish (for our present purposes) between phosphate that has been adsorbed reversibly at the solid–liquid interface and the much less reactive metal phosphate formed at the interface and progressing toward the interior of the particle.

The actual *adsorption* behavior of phosphate in soil has recently been described by Goldberg and Sposito (1984), using the constant-capacity model discussed earlier. In this approach three surface phosphate species are presumably involved, each with its own intrinsic pK value. The advantage of this model is that it describes well the pH dependence of the phosphate sorption. In practice, however, the pH of a given soil in agricultural use is kept constant within one pH unit or less. Moreover, for the soils treated with excess phosphate from manure as studied here, the pH dependence of phosphate adsorption was rather weak within the range of concern. Finally it was found that the isotherm describing the reversible adsorption of phosphate for the acid sandy soil could be expressed very accurately with a simple Langmuir adsorption isotherm involving only one pK value (Van der Zee and Van Riemsdijk, 1986b).

In contrast to the situation with phosphate, the adsorption of heavy metals on metal oxides is strongly dependent on pH. In soil, however, both the metal oxide content and the content of organic matter are important factors in the binding of heavy metals. Thus it could be shown that for a wide variety of field soils (Lexmond, 1980; Chardon, 1984), the adsorption of the ions of copper and cadmium could be described quite satisfactorily with an empirical equation according to

$$S_{\mathrm{M}} = K[\mathrm{M}]^{a}[\mathrm{H}]^{b} \tag{6.6}$$

In this equation the exponents a and b appeared to be relatively constant for a wide variety of soils (in the Netherlands). The coefficient a differed for Cd (0.8) and Cu (0.5). The factor K was understandably very different for the soils investigated, as it contains a capacity factor. At constant pH, Eq. 6.6 reverts to a simple Freundlich equation. An obvious disadvantage of such an extremely simple equation is the absence of allowance for competition between different metal ions for the same sites. Such competition may be of particular importance when yearly heavy doses of sewage sludge, containing many different heavy metals, are applied to soil.

The acid–base titration curve of a soil is central to the problem of pH control of soil, for example, by liming. Lately this aspect of soil chemical behavior has attracted renewed attention in the light of the suspected problems with "acid rain." In agriculture the practice of liming to maintain the balance of protons and other cations following the removal of crops is quite old indeed. The acuteness of the renewed interest in pH control of soil may be traced to the combination of a widespread and diffuse influx of protons from the atmosphere into soils, including those with very small buffer capacity, and the absence of the practice of yearly fertilizer additions to many of these soils.

4.2. pH–Titration Curves of Solid-Phase Mixtures

Titration curves are the principal means of establishing the properties of oxide surfaces. For a soil this implies titrating a mixture of metal oxides, organic matter, and clay minerals. The PPZCs ("pristine point of zero charge", that is, the PZC in the absence of specifically adsorbed ions aside from the proton, see the relevant discussion in Bolt and Van Riemsdijk, 1982, in D, p. 471), of these different types of surfaces will be quite far apart. In fact, the clay minerals (illites, etc.), at least, will not exhibit a PZC for the basal surface, as the substitution charges remain balanced by dissociated counterions even at very low pH values.

Interpretation of titration curves of real soil is difficult for various reasons. Apart from kinetic effects (such as delayed equilibration because of difficult accessibility of the solid surfaces), dissolution reactions during titration may obscure the information sought. This concerns both the high-pH dissolution of humic compounds and the low-pH liberation of Al^{3+} from many primary and secondary soil minerals.

In order to obtain some insight into the shape of σ_S–pH curves to be expected for mixtures of the above type, one may simulate them using some simplifying assumptions. To this purpose a "standard" curve was chosen representing a two-pK surface in the presence of inert electrolyte only, using an empty Stern layer with a "reasonable" electric capacity (cf. the discussion of this prototype in the previous section). For a chosen value of the PPZC at pH 6, the corresponding single-component curve is shown in Figure 6.12a. Shifting this curve to a PPZC at pH 2 gives the branch corresponding to the

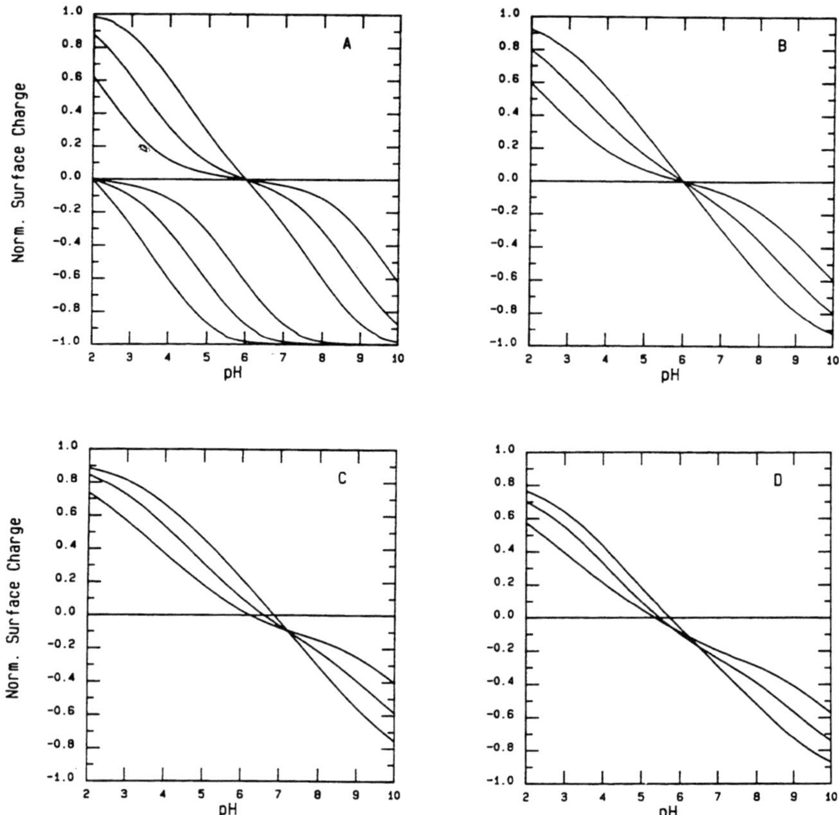

Figure 6.12. Simulated σ–pH curves for mixtures of surfaces with different PZC, ϕ_i signifying the fractional area with PZC at pH = i; see text. All curves given for inert electrolyte levels at 0.001, 0.01, and 0.1 M monovalent salt. (*a*) Homogeneous surface with PZC at pH = 6, so $\phi_6 = 1$; second set showing same with $\phi_2 = 1$. (*b*) "Symmetric" mixture with $\phi_4 = 0.115, \phi_5 = 0.19, \phi_6 = 0.38, \phi_7 = 0.19, \phi_8 = 0.115$. (*c*) "Asymmetric" mixture with $\phi_2 = 0.1, \phi_5 = 0.1, \phi_{6.5} = 0.4, \phi_8 = 0.2, \phi_9 = 0.2$. (*d*) "Asymmetric" mixture with $\phi_2 = 0.2, \phi_4 = 0.05, \phi_5 = 0.1, \phi_6 = 0.15, \phi_{6.5} = 0.2, \phi_7 = 0.05, \phi_{7.5} = 0.15, \phi_9 = 0.1$. Note the absence of a *c*ommon *i*ntersection *p*oint.

negative charge above pH 2 also shown in the figure. Both curves were scaled to yield a maximum value for the surface charge relative to the chargeable sites, thus reaching from −1 to +1 in the graph. All curves were calculated for three different levels of inert electrolyte—0.001, 0.01, 0.1 M uniunivalent salt giving the pH at the common intersection point (CIP), if present, relative to the PZC. In Figures 6.12*a–d*, the charging curves are shown for mixtures comprising different fractions of materials, all following charging curves akin to the standard curve but each component having a different PPZC. Such curves are obtained by summing the amounts of charge present per component, using the abundance (the fraction of the total surface area provided by each component) as the weighting factor. It is easy to prove that for a symmetric distribution (of the weighting factors for the different components plotted as a function of the relevant PPZC values), the mixture will exhibit a PZC equal to the weighted mean value of the individual PZC values. In Figure 6.12*b*, the corresponding charge–pH curve is shown for such a symmetric mixture, characterized in terms of the fractions ϕ_i of the components with a PPZC equal to pH $= i$. The CIP duly reflects the PPZC of the mixture, as expected. In Figure 6.12*c*, an asymmetric distribution is shown, where now a CIP appears to correspond to a net charge at about -0.1 of the unit value. This is obviously caused by the fact that the CIP corresponding to a pH around 7 is so far above the PPZC of one of the constituents (at pH 2) that this component is fully charged at all salt levels, thus giving a constant contribution equal to -0.1, also at the CIP found for the remainder of the components. Two remarks are of interest here: (1) this situation is common in soils containing surfaces with substitution charges like the clay minerals, and (2) the CIP, as expected, is present only within certain limits. In the present case it may be shown that the PZC as applicable to the mixture without the component with a PZC at pH 2, corresponds to pH values at 7.20, 7.23, and 7.23 for the three salt levels, and thus is indistinguishable from a true CIP. Finally the situation found for an arbitrary mixture is shown in Figure 6.12*d*, where now there is no indication of a CIP. From the above it should be clear that for mixtures actually occurring in natural soil the presence or absence of a clearly defined CIP is possible, depending on the particular distribution of different types of surfaces; moreover the apparent presence of a CIP gives no indication with respect to the value of the PZC, aside from the fact that the significance of a PZC in a mixture of components with different PPZCs remains uncertain.

Referring once more to the recent concern about the action of acid rain on soil water composition, it is pointed out in passing that knowledge of the σ–pH curve of the solid phase is obviously needed if buffering of the soil against the addition of protons is to be studied. Several other buffer mechanisms are then of concern, however, because in this case the rate of addition of protons tends to be small but persistent. This implies also that primary minerals will have time to contribute to soil buffering, and chances for breakthrough (into open water) of the protons deriving from the rain appear to be small. In fact, the Al ions produced in dissolution reactions of many soil minerals tend to move only over short distances: once pH values along the route have risen to 5 or above, the Al will cease to move except when organic ligands are around to prevent the formation of hydroxy polymers. Together the above remarks may indicate that breakthrough of Al^{3+} and H^+ into open water would be probable only if the connecting pathway is quite short and consists of coarse-grained sediments with very low buffering capacity. Shallow acidic soils in mountain regions are suspect in this case.

4.3. Information Needed for Modeling Solute Transport in Soil

As may have been shining through the text of this chapter, the authors are not too hopeful that problems arising from prolonged contact between soil and water on its way to open water are to be solved by methodical computer simulation of the process of solute transport through soil without further inquiry. The diversity and variability of the transporting medium requires much more than just local sampling to obtain some suitable parameter values that could then be entered in a somewhat standard simulation program of "solute transport subject to retardation and dispersion phenomena" (cf. Bolt, 1985). Briefly, it appears that the approach should consist of several phases. First, the probable pathway of the water will have to be determined, presumably with the help of soil physicists and hydrologists. Next, the local residence time must be estimated, from which the (soil) chemists might be able to decide what types of reactions are to be considered. Obviously, at this stage one should take into account the nature of the solute to be followed through the soil. Depending then on the specific questions posed (e.g., Is the interest mainly in the quality of the water leaving the system or rather in what is retained? Are the long-term effects of primary interest?, etc.), one might be able to specify the precision needed. Only at that stage should one decide what types of interaction models are to be used and what type of parameter analysis should be warranted. Given then the time requirement as to delivery of an estimate and the means available to obtain the experimental information, further consultation with those knowledgeable about the nature of the formation involved should lead to decisions as to the scope of the total model to be developed, in particular with regard to the details still to be taken into account. Quite often the simpler types of submodels described here will be the best choice, granted that an insight into the processes taking place in the soil will have to provide the proper guidelines for making such decisions.

Acknowledgment

The authors gratefully acknowledge U. Schwertmann, Soils Department, Technische Universität, München, BRD, who read and commented on the section covering soil oxides.

REFERENCES

Benjamin, M. M., and Leckie, J. O. (1981), Multiple-site adsorption of Cd, Cu, Zn and Pb on amorphous iron oxihydroxide, *J. Colloid Interface Sci.* **79,** 209–221.

Bijsterbosch, B. H., and Lyklema, J. (1978), Interfacial electrochemistry of silver iodide, *Adv. Colloid Interface Sci.* **9,** 147–251.

Bolt, G. H. (1957), Determination of the charge density of silica sols, *J. Phys. Chem.* **61,** 1166–1169; **62,** 1308.

Bolt, G. H., Ed. (1982a), *Soil Chemistry, B. Physico-chemical Models,* 2nd ed., Elsevier, Amsterdam.

Bolt, G. H. (1982b), "Theories of Cation Adsorption by Soil Constituents: Distribution Equilibrium in Electrostatic Fields," in G. H. Bolt (1982a), pp. 47–76.

Bolt, G. H. (1985), Transport of solutes in soil: Basic features of front retardation, *Water Sci. Technol.* **17,** 87–99.

Bolt, G. H., and Bruggenwert, M. G. M. (1976), "Composition of the Soil," in G. H. Bolt and M. G. M. Bruggenwert, Eds., *Soil Chemistry, A. Basic Elements,* Elsevier, Amsterdam, pp. 1–12.

Bolt, G. H., and De Haan, F. A. M. (1979), "Anion Exclusion in Soil," in G. H. Bolt (1982a), pp. 233–257.

Bolt, G. H., Sumner, M. E., and Kamphorst, A. (1963), A study of the equilibria between three categories of potassium in an illitic soil, *Soil Sci. Soc. Am. Proc.* **27,** 294–299.

Bolt, G. H., and Van Riemsdijk, W. H. (1982), "Ion Adsorption on Inorganic Variable Charge Constituents," in G. H. Bolt (1982a), pp. 459–504.

Borchardt, G. A. (1977), "Montmorillonite and Other Smectite Minerals," in J. B. Dixon, and S. B. Weed, Eds., *Minerals in Soil Environments,* Soil Science Society of America, Madison, WI, pp. 293–330.

Bowden, J. W., Posner, A. M., and Quirk, J. P. (1977), Ionic adsorption on variable charge mineral surfaces. Theoretical charge development and titration curves, *Aust. J. Soil Res.* **15,** 121–136.

Brinkman, R. (1979), "Clay Transformations: Aspects of Equilibrium and Kinetics," in Bolt (1982a), pp. 233–257.

Brown, G., Newman, A. C. D., Rayner, J. H., and Weir, A. H. (1978), "The Structure and Chemistry of Soil Clay Minerals," in Greenland and Hayes (1978), pp. 29–178.

Chardon, W. J. (1984), Mobiliteit van Cadmium in de bodem, series "Bodembescherming," vol. 36, Government Publishing Co., The Hague.

Dixon, J. B., and Weed, S. B., Eds. (1977), *Minerals in Soil Environments,* Soil Science Society of America, Madison, WI.

Douglas, L. A. (1977), "Vermiculites," in Dixon and Weed, (1977), pp. 293–330.

Fanning, D. S., and Keramidas, V. Z. (1977), "Micas," in Dixon and Weed (1977), pp. 195–258.

Gapon, E. N. (1933), On the theory of exchange adsorption in soil, *J. Gen. Chem. USSR* **3,** 144–152.

Garrels, R. M., and Christ, C. L. (1965), *Solutions, Minerals and Equilibria,* Freeman, Cooper, San Francisco.

Gieseking, J. E., Ed. (1975), *Soil Components,* Vol. I, *Organic Components,* Springer-Verlag, Berlin.

Goldberg, S., and Sposito, G. (1984), A chemical model of phosphate adsorption by soils. II. Non-calcareous soils, *Soil Sci. Soc. Am. J.* **48,** 779–783.

Grahame, D. C. (1947), The electrical double layer and the theory of electro-capillarity, *Chem. Rev.* **41,** 441–501.

Greenland, D. J., and Hayes, M. H. B. (1978), *The Chemistry of Soil Constituents,* Wiley-Interscience, Chichester.

Harmsen, K. (1979), "Theories of Cation Adsorption by Soil Constituents: Discrete-Site Models," in Bolt (1982a), pp. 77–139.

Hayes, M. H. B., and Swift, R. S. (1978), "The Chemistry of Soil Organic Colloids," in Greenland and Hayes (1978), pp. 179–320.

Heald, W. R., Frere, M. H., and De Wit, C. T. (1964), Ion adsorption on charged surfaces, *Soil Sci. Soc. Am. Proc.* **28,** 642–647.

Hendricks, S. B., and Fry, W. H. (1930), The results of X-ray and microscopical examinations of soil colloids, *Soil Sci.* **29,** 457–465.

Hofmann, U., Endell, K., and Wilm, D. (1933), Kristallstruktur und Quellung von Montmorillonit, *Z. Krist.* **86A,** 340–352.

Huang, C. P., and Stumm, W. (1973), Specific adsorption of cations on hydrous γ-Al_2O_3, *J. Colloid Interface Sci.* **43,** 409–420.

Jaroniec, M. (1983), Physical adsorption on heterogeneous solids, *Adv. Colloid Interface Sci.* **18,** 149–225.

Lexmond, T. M. (1980), The effect of soil pH on copper toxicity to forage maize growth under field conditions, *Neth. J. Agr. Sci.* **28,** 164–183.

Marshall, C. E. (1949), *The Colloid Chemistry of the Silicate Minerals,* Academic Press, New York.

Neal, C., and Cooper, D. M. (1983), Extended version of Gouy–Chapman electrostatic theory as applied to the exchange behavior of clay in natural waters, *Clays Clay Min.* **31,** 367–376.

Oden, S. (1922), *Die Huminsäuren,* Verlag Steinkopff, Leipzig.

Penners, N. H. G. (1985), "The Preparation and Stability of Homodisperse Colloidal Haematite," Ph. D. thesis, Agricultura Univ. of Wageningen, The Netherlands.

Schindler, P. W., and Gamsjäger, H. (1972), Acid–base reactions of the TiO_2 (anatase)–water interface at the point of zero charge of TiO_2 suspensions, *Koll. Z, Z. Polymere* **250,** 759–763.

Schwertmann, U., and Taylor, R. M. (1977), "Iron oxide," in Dixon and Weed (1977), pp. 145–180.

Shainberg, I., and Kemper, W. D. (1967), Ion exchange equilibria on montmorillonite, *Soil Sci.* **103,** 4–9.

Sips, R. (1948), On the structure of a catalyst surface I, *J. Chem. Phys.* **16,** 490–495.

Sips, R. (1950), On the structure of a catalyst surface II, *J. Chem. Phys.* **18,** 1024–1026.

Sposito, G. (1984), *The Surface Chemistry of Soils,* Oxford University Press, New York.

Stumm, W., Huang, C. P., and Jenkins, S. R. (1970), Specific chemical interaction affecting the stability of dispersed systems, *Croat. Chem. Acta* **42,** 223–245.

Van Breemen, N., and Brinkman, R. (1976), Chemical equilibria and soil formation, in Bolt and Bruggenwert (1976), pp. 141–170.

Van der Zee, S. E. A. T. M., and Van Riemsdijk, W. H. (1986a), Transport of phosphate in a heterogeneous field, *TPIM* **1,** 339–359.

Van der Zee, S. E. A. T. M., and Van Riemsdijk, W. H. (1986b), Sorption kinetics and transport of phosphate in sandy soil, *Geoderma* **38,** 293–309.

Van Riemsdijk, W. H., Bolt, G. H., Koopal, L. K., and Blaakmeer, J. (1986), Electrolyte adsorption on heterogeneous surfaces: Adsorption models, *J. Colloid Interface Sci.* **109,** pp. 210–228.

Van Riemsdijk, W. H., Boumans, L. J. M., and De Haan, F. A. M. (1984), Phosphate sorption by soils I: A model for phosphate reaction with metal oxides in soil, *Soil Sci. Soc. Am. J.* **48,** 537–541.

Van Riemsdijk, W. H., De Wit, J. C. M., Koopal, L. K., and Bolt, G. H. (1987), Metal ion adsorption on heterogeneous surfaces: Adsorption models, *J. Colloid Interface Sci.* in press.

Van Riemsdijk, W. H., Van Der Linden, A. M. A., and Boumans, L. J. M. (1984), Phosphate sorption by soils III: The P diffusion-precipitation model tested for three acid sandy soils, *Soil Sci. Soc. Am. J.* **48,** 545–548.

Vanselow, A. P. (1932), Equilibria of the base-exchange reactions of bentonites, permutites, soil colloids and zeolites, *Soil Sci.* **33,** 95–113.

Wada, K. (1977), "Allophane and Immogolite," in Dixon and Weed (1977), pp. 603–638.

Weaver, C. E., and Pollard, L. D. (1975), *The Chemistry of Clay Minerals,* Elsevier, Amsterdam.

Westall, T., and Hohl, H. (1980), A comparison of electrostatic models for the oxide/solution interface, *Adv. Colloid Interface Sci.* **12,** 265–294.

PART TWO

THE FORMATION AND DISSOLUTION OF SOLID PHASES

7

THE HYDROLYSIS OF IRON IN SYNTHETIC, BIOLOGICAL, AND AQUATIC MEDIA

Walter Schneider and Bernhard Schwyn*

Laboratory for Inorganic Chemistry, Swiss Federal Institute of Technology (ETH), Zurich, Switzerland

Abstract

At moderate temperatures, a variety of sols, gels, and amorphous or crystalline precipitates emerge from hydrolysis of iron(III) in aqueous solution. The morphological properties of the products reflect the pathways of formation in any specific system. The prime variables are the rate of change of pH as a function of time and space and the chemical components. It is a useful concept to consider first the relevant mononuclear iron species from which primary polynuclear cores $Fe_pO_r(OH)_s$ are formed and which maintain growth processes. Starting from hydrated ferric ions in acid solution, positively charged polynuclears are obtained which undergo general and specific anion interactions. Starting from anionic polyalcohol complexes in very alkaline solution, rather small polynuclears can be obtained which undergo interactions with ligand groups competing for monodentate H_2O/OH^- at the peripheral Fe(III) sites of polynuclears. In both types of systems, precipitation of solid products can either be inhibited or induced by chemical interactions at the polynuclear–solution interface. The types of products which are found, or are expected to be found, in biological and aquatic systems, are discussed in terms of the prototype behavior as verified in synthetic media.

*Present address: Department of Biochemistry, University of California, Berkeley, California.

1. INTRODUCTION

The principal iron oxide (hydroxide) minerals of sedimentary origin are hematite α-Fe_2O_3 and goethite α-FeO(OH), whereas lepidocrocite γ-FeO(OH) and maghemite γ-Fe_2O_3 are moderately common (Wedepohl, 1969). The rather rare akaganeite β-FeO(OH) was detected in the late 1950s (Mackay, 1962). It is worthwhile to remember that magnetite Fe_3O_4, one of the principal iron ores, is found in igneous rock. The geochemist becomes aware that each of these minerals could form only within a restricted range of conditions such as temperature, oxygen pressure or redox potential, pH, activity of water, and further chemical components. Thermodynamic factors are important (Garrels and Christ, 1965), but kinetics cannot be disregarded. The prehistory of all iron oxides (hydroxides) is wiped out when they are heated in air at temperatures around 500°C to form α-Fe_2O_3. A variety of products are found in settling particles in lakes, on the one hand, and in deposits in cells and tissues on the other hand. These specimens emerge at a rather narrow range of temperature, 0–40°C, and pH 5–9. Their properties reveal that subtle effects of many variables are involved in the pathways of formation of iron oxides or hydroxides. Actually, the latter term refers to a broad range of phenomena rather than to a set of well-defined compounds. "Iron hydroxides" may differ widely with regard to chemical composition, structure, particle size, size distribution, shape of particles, and interfacial properties, and consequently with regard to chemical reactivity. It is highly pertinent for our health status whether the oxidic iron stores in our body are easily mobilized or chemically inert (Jacobs, 1980). The awareness of aquatic chemists is very much advanced in emphasizing the role of Fe(III)—Fe(II) in the network of chemical processes in lakes (Stumm and Morgan, 1981; Baccini, 1984).

In experimental studies, hydrated ions Fe^{3+} and Fe^{2+} are widely used as well-defined educts in the investigation of iron(III) hydrolysis, which implies addition of a base or an oxidant, respectively. From the point of view of chemical processes occurring in the environment including biological systems, the hydrated ferric ion is far less a realistic source of iron(III) hydroxides than the hydrated ferrous ion. It is all the more important to evaluate the results obtained in synthetic media with respect to genuine properties of the species appearing in well-behaved systems. It is useful to define the media as considered in this chapter.

1.1. Synthetic Media

Electrolyte solutions containing inert salts such as alkali metal perchlorates, nitrates, and chlorides. We have to consider weak complexes FeX_i^{3-i} with $\log K_i$ (individual stability constants) < 1 for $X^- = Cl^-$, but not for ClO_4^- or NO_3^-. We should expect more severe interactions at the level of mononuclear Fe(III) species in solutions with $[SO_4^{2-}] > 0.1\ M$. As an allusion to this we may recall the mineral jarosite, $KFe_3(OH)_3(SO_4)_2$.

Strong alkali hydroxide solutions containing polyalcohols such as xylitol or sorbitol. In these media, mononuclear complexes as described in Section 4.3 are formed. They are the best approximation to $Fe(OH)_6^{3-}$, which does not exist in aqueous solution. It is interesting to approach pH $\approx$ 7 from the alkaline side using a suitable mononuclear, that is, an analog of the hydrated Fe^{3+}, in moving from pH < 1 to pH 7.

1.2. Biological Media

Although iron is an essential bioelement for all organisms, we focus attention on vertebrates where the interplay of blood and intracellular fluids dominates the iron cycle. The prime chemical components of fluid domains are macromolecular—proteins, glycoproteins, carbohydrates—and low molecular weight predecessors. There is no point in conceptually separating solution areas from macromolecular boundaries. It is more appropriate to think of a biomolecular matrix extending over an aqueous space, with the matrix and space providing a diversity of local chemical environments having pH in the range 5–9 and ionic strength up to 0.15 M. The gastrointestinal fluids in humans and mammals therefore provide the link between the gastric juice (with pH as low as 2 at rest) and the intracellular media of the mucosal cells in the duodenum.

1.3. Aquatic Media

There is a striking contrast to biological media, with regard to total concentration levels of both iron and organic carbon which have upper limits around $10^{-6}\ M$ (Fe) and $10^{-4}\ M$ (C). We focus attention on metal transfer mechanisms in lakes (Sigg, 1985), where the turnover of iron reaches the highest rate in conjunction with redox processes Fe(III) $\rightleftarrows$ Fe(II). It is useful to consider first of all the hydrated ferrous ion Fe^{2+} in a homogeneous solution with pH 5–9 and low ionic strength. Oxidation by O_2 will induce hydrolysis of Fe(III). We then have to admit that perturbations by particulate matter could occur at all stages of the hydrolysis reaction series ranging from adsorption of Fe^{2+} to surface precipitation of iron hydroxides.

This chapter aims at an understanding of the morphologic variety of hydrolysis products in terms of the molecular processes that occur at the onset of the formation of the polynuclears. Prime attention is given to the (hydr)oxide cores $Fe_pO_r(OH)_s$ involved where $2 < p < 10^{+5}$.

The guidelines are provided by the results of experiments in which the educt solutions were either strongly acid (Fe^{3+}), neutral (Fe^{2+}), or strongly alkaline (polyalcoholato complexes); the mononuclear sources of iron(III) are indicated in parentheses.

2. MONONUCLEAR AQUAHYDROXO COMPLEXES OF IRON(III) AND THE PRIMARY STEPS OF POLYNUCLEAR GROWTH

It is useful to consider the imaginary situation of the series of mononuclear species in Table 7.1. Actually, the first three members of this series were experimentally verified. The hypothetical mononuclear mn-$Fe(OH)_3$ is thermodynamically unstable with respect to solid phases, whereas the anionic $Fe(OH)_4^-$ was shown to be responsible for the solubility of iron hydroxides at pH > 10 (Schindler et al., 1963).

For equilibrium data up to 1974 the reader is referred to an excellent monograph (Baes and Mesmer, 1976), where selected constants for equilibria 7.1–7.4 are compiled and discussed.

$$Fe^{3+} + H_2O \rightleftharpoons FeOH^{2+} + H^+ \qquad \log {}^*K_1 = -3 \qquad (7.1)$$

$$FeOH^{2+} + H_2O \rightleftharpoons Fe(OH)_2^+ + H^+ \qquad \log {}^*K_2 = -3.3 \qquad (7.2)$$

$$FeOH^{2+} + FeOH^{2+} \rightleftharpoons Fe_2(OH)_2^{4+} \qquad \log K_D = 3 \qquad (7.3)$$

$$Fe(OH)_3(s) \rightleftharpoons Fe^{3+} + 3OH^- \qquad -39 \geq \log K_{s0} \geq -41 \qquad (7.4)$$

$$Fe(OH)_3(s) + OH^- \rightleftharpoons Fe(OH)_4^- \qquad 4 < \log K_{s4} < 7 \qquad (7.5)$$

Table 7.1. The Series of Mononuclear Complexes $Fe(OH)_i(H_2O)_x^{3-i}$

Mononuclear	pK	$\log k_{ex}^{H_2O}$ (s)
$Fe(H_2O)_6^{3+}$	3[a]	2[b]
$FeOH(H_2O)_5^{2+}$	3.3[a]	5[b]
$Fe(OH)_2(H_2O)_4^+$	6[c]	≥7[c]
$Fe(OH)_3(H_2O)_x$	9[c]	—[d]
$Fe(OH)_4(H_2O)_x^-$		—[d]

[a]pK values from Hedström (1953).
[b]$k_{ex}^{H_2O}$ from Grant and Jordan (1981).
[c]Estimated.
[d]x may be lower than $6 - i$, and rates of water exchange may be so high as to provoke rates of substitution approaching the diffusion-controlled limit.

The approximate values in Eqs. 7.1–7.4 are typical for 25°C and ionic strength $1\ M < I_c < 3\ M$ ($NaClO_4$). The conventional solubility product K_{s0} spans the range from amorphous ("active") $Fe(OH)_3$ to crystalline α-FeO(OH)(s), as does K_{s4}. Actually, the estimates of pK_3 and pK_4 in Table 7.1 are based on the limit to the equilibrium quotient K_{s3} in Eq. 7.6 as obtained from ultracentrifugation of saturated solutions of freshly prepared "active" $Fe(OH)_3$ (Lengweiler et al., 1961).

$$\text{Active } Fe(OH)_3(s) \rightleftharpoons \text{mn-}Fe(OH)_3 \qquad \log K_{s3} \leq -9 \qquad (7.6)$$

From the values of K_{s0} (upper limit), K_{s3}, and K_{s4} (lower limit), we get estimates for β_3 and β_4 as defined by Eqs. 7.7 and 7.8:

$$Fe^{3+} + 3OH^- \rightleftharpoons \text{mn-}Fe(OH)_3 \qquad \log \beta_3 \approx 30 \qquad (7.7)$$

$$Fe^{3+} + 4OH^- \rightleftharpoons Fe(OH)_4^- \qquad \log \beta_4 \approx 35 \qquad (7.8)$$

The individual stability constants of the hydroxo complexes $Fe(OH)_i^{3-i}$ in Table 7.1 were calculated from Eqs. 7.1, 7.2, 7.7, and 7.8 with $pK_w = 14$. Although rather crude, these estimates are highly valuable in the context of primary steps in the hydrolysis of Fe^{3+} (Schneider, 1984). When base is added in experiments, no matter what mixing technique is applied the situation indicated in Figure 7.1 cannot be circumvented. At the base–bulk solution boundary the pH gradients provoke the buildup of $Fe(OH)_i^{3-i}$ distributions which reflect the local pH profiles. The proton transfer reactions

$$Fe(OH)_i(H_2O)_x^{3-i} + B \xrightarrow{K_{i+1}} Fe(OH)_{i+1}(H_2O)_{x-1}^{2-i} + HB^+ \qquad i = 0, 1, 2, 3 \quad (7.9)$$

are extremely fast, reaching the diffusion-controlled limit provided $pK_{HB} \geq pK_i + 2$ (Hague, 1971). This condition is satisfied if $B = OH^-$ or CO_3^{2-} ($i = 0, 1, 2$), which means that the second-order rate constants k_{i+1} are at least $10^{10}\ M^{-1}\ s^{-1}$.

Within the lifetime of domains with $pH(t) \geq pH$ (bulk) the mononuclear species $Fe(OH)_i^{3-i}$ (abbreviation of $Fe(OH)_i(H_2O)_x^{3-i}$) react with each other to form dinuclear and, subsequently, polynuclear complexes.

$$Fe(OH)_i^{3-i} + Fe(OH)_j^{3-i} \xrightarrow[i \leq j]{k_{ij}} Fe_2(OH)_{i+j} \qquad i, j \geqslant 1 \qquad (7.10)$$

The coordination number $(i + x)$ is very likely to be 6, as for all polynuclears (Section 3) as well as for solids FeO(OH) and Fe_2O_3 (Wells, 1984). At any rate, reactions 7.10 involve the expulsion of at least two water molecules. According to established mechanisms of ligand substitution reactions (Hague,

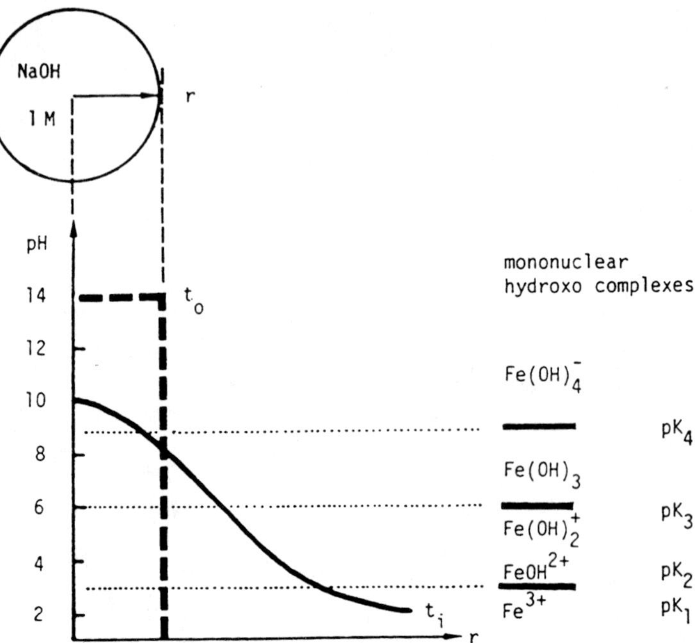

Figure 7.1. The concept of instantaneously formed mononuclears $Fe(OH)_i^{3-i}$ at the boundary between a drop of NaOH solution and bulk solution of hydrated Fe^{3+} with pH < 2. The rates of mixing are slow on the time scale of diffusion-controlled proton transfer reactions. Dinuclear and further polynuclear species appear within milliseconds at the concentration level of 10^{-4} *M* mononuclears at pH around 7. A nonuniform mixture of products emerges from the pH heterogeneity depending on its lifetime. (Used with permission from Schneider, 1984.)

1971), the rate constants k_{ij} will increase roughly in the order of increasing k_i^{ex}. There will be a dramatic increase in the series:

$$\begin{matrix} k_{11} & \ll & k_{22} & \ll & k_{33} \\ \text{ca. } 10^3 & & (10^5) & & (10^7)\ M^{-1}\,s^{-1} \end{matrix} \tag{7.11}$$

Experimentally, k_{11} was determined for a variety of media from decomposition rates and K_D (Wendt, 1962; Lutz and Wendt, 1970). Values in parentheses are estimates compatible with estimates of k_2^{ex} and k_3^{ex} in Table 7.1. Accordingly, the half-time for the reaction of mn-$Fe(OH)_3$ with itself, $t_{1/2} = 0.5\ (c \cdot k_{33})^{-1}$, is about 1 ms at the concentration level 0.1 m*M*. Of course, we have to consider two further types of reactions in the context of Figure 7.1:

1. The buildup of polynuclears, for example,
   At pH < 2:

$$Fe_2(OH)_2^{4+} + FeOH^{2+} + H_2O \longrightarrow Fe_3(OH)_4^{5+} + H^+ \tag{7.12}$$

At pH ≈ 7:

$$Fe_2(OH)_6 + Fe(OH)_3 \longrightarrow Fe_3(OH)_9 \tag{7.13}$$

2. The acid decomposition of dinuclears and polynuclears, which depends on the bulk acidity with respect to the kinetics and the equilibrium limit; for example,

$$Fe_3(OH)_4^{5+} + 4H^+ \longrightarrow 3Fe^{3+} + 4H_2O \tag{7.14}$$

We cannot predict by a priori arguments the composition and structures of species appearing in Eqs. 7.12–7.14. However, it is clear by now that the fundamental properties of mononuclear species $Fe(OH)_i^{3-i}$ provoke an unavoidable variety of intermediate products due to the overlapping time scales of mixing and substitution as well as protonation reactions.

The structural variety which has to be considered for the transition from dinuclears ($p = 2$) to trinuclears ($p = 3$) is shown in Figure 7.2. We write

$$Fe_pO_r(OH)_s^{3p-(2r+s)} \tag{7.15}$$

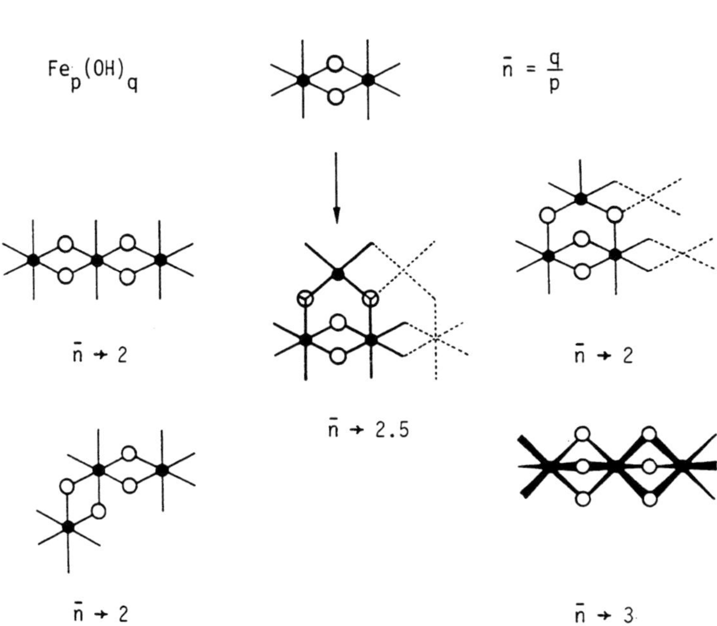

Figure 7.2. The structural variety of trinuclear ferric hydroxo complexes $Fe_3(OH)_q^{9-q}$ that could form from $Fe_2(OH)_2^{4+}$ at low pH. The limiting stoichiometries $Fe(OH)_{\bar{n}}^{3-\bar{n}}$ corresponding to infinite chains ($p \rightarrow \infty$) or twin chains of octahedra have not been observed in any experiment. The arrow indicates a transition ($p = 2 \rightarrow p = 3$) which is realistic as an early step in the growth of polynuclears with β-FeO(OH) type structure discussed in Section 3. The trigonally marked circles are oxide ions in the extended twin chain. The pentanuclear including the dashed part is $Fe_5O(OH)_7^{6+}$.

for the general composition of a polynuclear core which, by definition, contains O^{-2} and OH^- exclusively in bridging positions.

In Figure 7.2 it is indicated how well-defined limiting stoichiometries may result from the growth of one-dimensional crystalline polynuclears. Experimentally, the identification of hydrolysis products is very much confined to rather monodisperse systems, that is, very narrow ranges of (p, r, s).

3. HYDROLYSIS IN SALT SOLUTIONS

3.1. Heterogeneous and Homogeneous Nucleation

Turning to the discussion of macroscopic phenomena, it is useful to consider a specific set of test solutions that are obtained by successive addition of $NaHCO_3$ to a ferric perchlorate solution (0.06 M Fe^{3+}; 2 M $NaClO_4$) with pH ≈ 1. The effect of base addition is shown in Figure 7.3, where a is defined as the molar ratio HCO_3 /Fe(III). It is seen that the pure ferric perchlorate solution (a = 0 by definition) which contains no excessive acid is supersat-

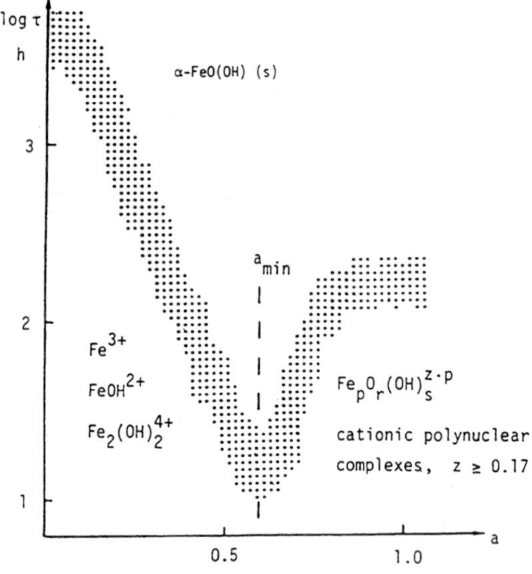

Figure 7.3. The variation of precipitation time τ with molar ration a = HCO_3^-/Fe(III) for 0.06 M iron(III) perchlorate solution; I_c = 2 M ($NaClO_4$); 25°C. (Used with permission from Schneider, 1984; data from Künzi, 1982.) τ is the rather arbitrarily defined time elapsed before a Tyndall effect is observed. The pattern of the dependence of log τ on a refers to characteristic properties of the system, although the numerical values of τ are not reproducible in a strict sense. Crystalline precipitates with α-FeO(OH) structure are confined to the range $a < a_{min}$ within the time τ. The average charge z = 0.17 per Fe(III) in polynuclears was found in potentiometric studies (Ciavatta, 1975).

urated with respect to α-FeO(OH)(s). From the data set (1, 2, 3), pH = 1.8 is calculated and confirmed by measurement. Of course, the classical data of Hedström (1953) were obtained by potentiometric determinations of pH and pFe in similar solutions containing only the species Fe^{3+}, $Fe(OH)^{2+}$, $Fe(OH)_2^+$, and $Fe_2(OH)_2^{4+}$. The pattern as shown by Figure 7.3 was reported by Feitknecht and Michaelis (1962) and later by Knight and Sylva (1974), who also determined precipitation times $\tau(a)$ for nitrate and chloride solutions. There is a common pattern in all cases which refers to the shape of the curve $\tau(a)$ and the type of behavior observed in the ranges $a < a_{min}$ and $a > a_{min}$, respectively.

$a < a_{min}$: The yellow solutions do not contain detectable amounts of polynuclears ($p > 2$) when some visible precipitate has appeared (Künzi, 1982).

$a > a_{min}$: Brown polynuclears are formed immediately in the homogeneous solution.

The stability of sols in the range $1 < a < 2.5$ depends on the type and concentration of anions. Of course, a_{min} also depends on these variables. However, the most important result is the pattern which indicates that nucleation is heterogeneous in the range $a < a_{min}$ and homogeneous beyond a_{min}. The experimental proof for this interpretation would be tedious for perchlorate solutions because the rates of the elementary reactions of the type 7.10, 7.12, 7.14 are slowest in the presence of ClO_4^- but considerably enhanced by Cl^- (Künzi, 1982; Schneider, 1984). The fact that amorphous products precipitate from perchlorate solutions, particularly if the rate of base addition (da/dt) was rather high, is well understood in terms of the arguments discussed in Section 2 and the slow kinetics of acid decomposition of polynuclears. It is important to realize that reactions 7.10 ($i = j = 3$), 7.13, and further growth steps as well as aggregation of uncharged polynuclears are very fast.

3.2. Chloride Solutions: A Case Study

An extended study of iron(III) hydrolysis in chloride solution (Anner, 1975; Künzi, 1982; Jiskra, 1983; Schwyn, 1983) revealed conspicuous effects of Cl^- at all levels, that is, in nucleation, growth, and aging of polynuclears. Actually, we are dealing with a case of crystal growth in solution, since precipitates contain polynuclear rods that formed in the solution. Attention is drawn here to some of the experimental data reviewed recently (Schneider, 1984) which refer to nucleation, size and shape of polynuclears, and interfacial interactions.

It is obvious from Figure 7.3 that the dinuclear $Fe_2(OH)_2^{4+}$ is rather inert with respect to the transition from $p = 2$ to $p = 3$ (in Eq. 7.12) in the region

$a < a_{min}$. As shown by Figure 7.4, the nucleation is still heterogeneous in the lower region $a < a_{min}$, since crystals of NaCl (or $CaCl_2$) induce polynuclear growth far more effectively than dissolved Cl^-. This nonequivalence in lowering the free activation energy of nucleation is removed in the solution, where $a = 0.4$ is slightly larger than $a_{min} \approx 0.3$. No spread in the bottom curves was observed when dust particles had not been removed from the solutions prior to the addition of NaCl solution.

Laser light-scattering methods were used to determine the molecular weight, or rather the polynuclearity coefficient p in Eq. 7.15, and the translational diffusion coefficient of polynuclears (Schwyn, 1983). From these independent data, the length and width of polynuclears were calculated in terms of the approximation as rotational ellipsoids. It turned out that the polynuclears are needle-shaped, indicating one preferential growth direction. This was confirmed by electron micrographs of samples prepared by flocculation with *para*-toluene sulfonate, which merely induces the alignment of the crystallites. The

Induced Nucleation

$$Fe_2(OH)_2^{4+} \xrightarrow[0.2\ M]{NaCl} PN$$

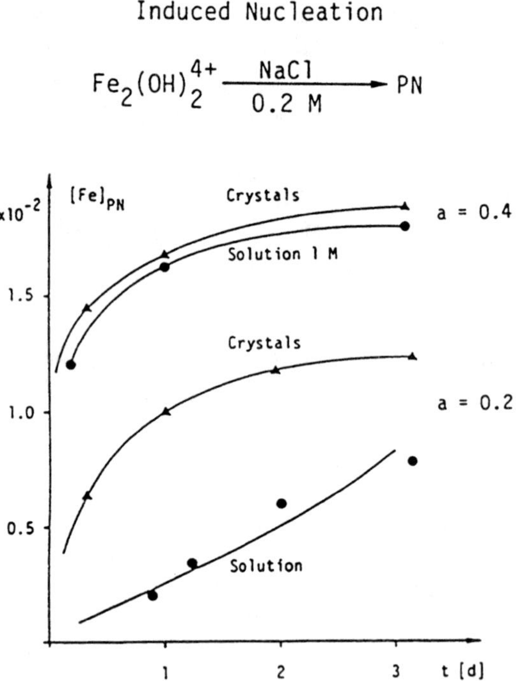

Figure 7.4. Nucleation in supersaturated ferric perchlorate solutions and subsequent evolution of polynuclear complexes (PN). The concentration of chloride (0.2 *M*) was adjusted either with 1 *M* NaCl solution or by adding the corresponding molar quantity of NaCl crystals. Supersaturated solutions were membrane filtered prior to the addition of chloride ($t = 0$). $[Fe]_{PN}$ is the total molar concentration of Fe(III) in the polynuclear fraction. Total concentrations: Fe(III), 0.06 *M*; Cl^-, 0.2 *M*; ClO_4^-, 0.80 *M*. Precipitation times were about 3 days (25°C). Data from Schwyn (1983).

successful determination of two shape parameters by dynamic light scattering (Berne and Pecora, 1976) is restricted to rather monodisperse particle systems. Mössbauer spectra of frozen solutions (Van Steenwijk et al., 1978) and X-ray powder diagrams of flocculated polynuclears (Künzi et al., 1982) confirmed the β-FeO(OH) structure (Szytula, 1970) of products formed all over the range of $a \leq 2.4$, total iron 0.05–0.4 M, and chloride concentration 0.2–1.5 M. This well-defined structure of polynuclears is the only conceivable reason for the constant chemical composition

$$FeO_{2/3}(OH)_{4/3}(H_2O)_xCl_{1/3}, \qquad 0.5 < x < 1.5 \tag{7.16}$$

In this formula, the analytical results are expressed according to the structural hypothesis shown in Figure 7.8.

In view of the crucial role of nucleation, we should expect preformed polynuclears to be effective seeds in supersaturated solutions. This is confirmed by the results (Fig. 7.5*b*) of the experiment sketched in Figure 7.5*a*. The molar concentration of polynuclears persists over time, whereas the size p increases steadily.

From the initial growth period $0 \leq t \leq 6$ h the average rate constant $\bar{k}_i$ of the intrinsic steps

$$PN(p) + FeOH^{2+} + 1.7H_2O \xrightarrow{\bar{k}_i} PN(p + 1) + 1.7H^+ \tag{7.17}$$

was estimated. It is in the order of $10^2\ M^{-1}\ s^{-1}$, which is close to $k_{11} \approx 5 \times 10^2$ for the dimerization of $FeOH^{2+}$ (Eqs. 7.10, 7.11), which does not involve proton transfer to the solvent water. Hence, in acid solution ($1.9 > pH > 1.7$) the growth rate reflects the reactivity of the mononuclear $FeOH^{2+}$ with surface sites of the polynuclear (PN).

The rate of "homogeneous" nucleation ($a > a_{min}$) is not determined simply by the concentration of ferric species and H^+ in the bulk solution, because nuclei originate from the time-dependent inhomogeneity at the inlet of the base. As discussed in Section 2, mn-$Fe(OH)_3$ and $Fe(OH)_4^-$ are formed instantaneously when HCO_3^-/CO_3^{2-}, pH 10, is added. Primary di- and polynuclears are expected to differ from those involved in the nucleation experiment to which Figure 7.4 refers. We cannot identify transient species, but we find the number of nuclei formed as the number of polynuclears in the final solution. It is shown in Figure 7.6 that different concentrations of polynuclears emerged from identical educt solutions when different types of bases were used to adjust a to 0.5 and 2.0, respectively.

In the first pair of solutions ($a = 0.5$) the concentration [PN] differs by a factor of 50. Carbonate induces lower rates of nucleation but more pronounced anisotropic growth. When imidazol was used, the lower pH gradients, obviously, provoked higher nucleation rates but also lower length/width ratios. This suggests that the primary complexes formed at the car-

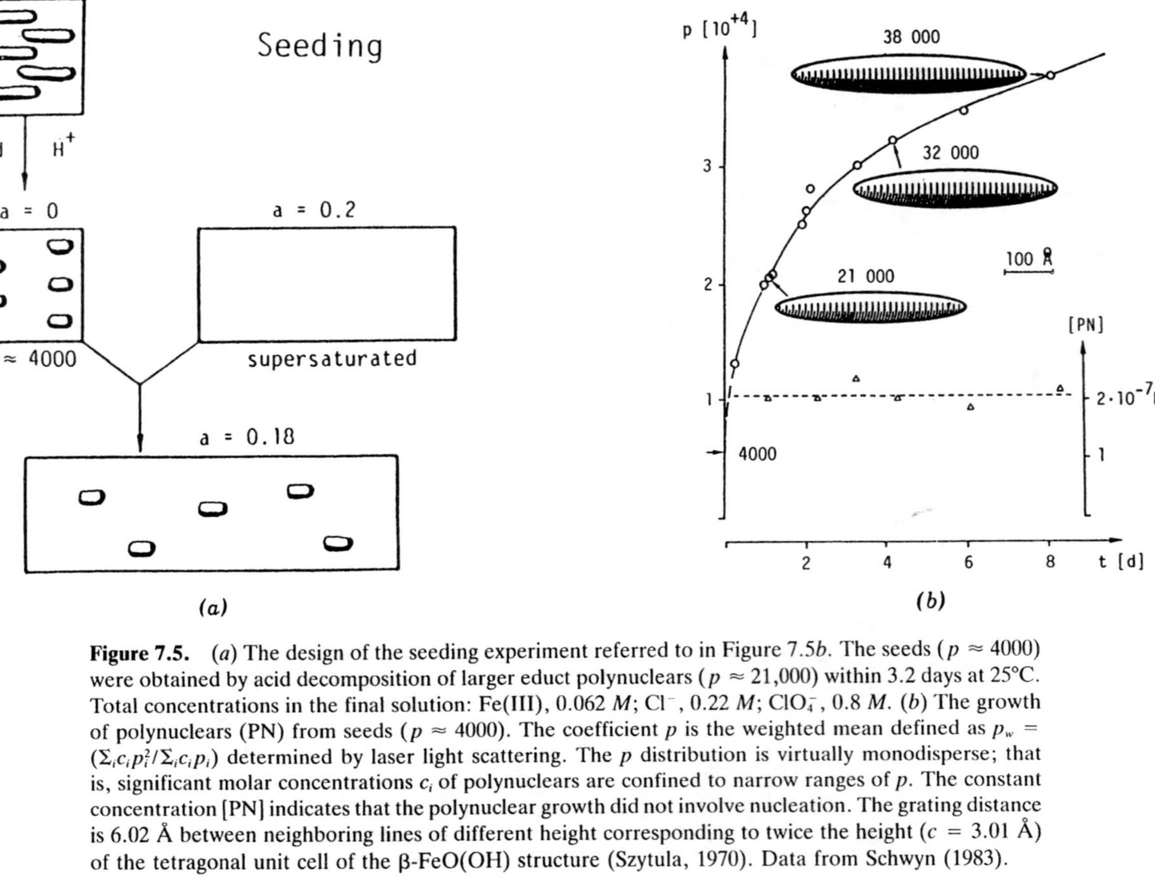

Figure 7.5. (*a*) The design of the seeding experiment referred to in Figure 7.5*b*. The seeds ($p \approx 4000$) were obtained by acid decomposition of larger educt polynuclears ($p \approx 21{,}000$) within 3.2 days at 25°C. Total concentrations in the final solution: Fe(III), 0.062 *M*; Cl^-, 0.22 *M*; ClO_4^-, 0.8 *M*. (*b*) The growth of polynuclears (PN) from seeds ($p \approx 4000$). The coefficient p is the weighted mean defined as $p_w = (\Sigma_i c_i p_i^2 / \Sigma_i c_i p_i)$ determined by laser light scattering. The p distribution is virtually monodisperse; that is, significant molar concentrations c_i of polynuclears are confined to narrow ranges of p. The constant concentration [PN] indicates that the polynuclear growth did not involve nucleation. The grating distance is 6.02 Å between neighboring lines of different height corresponding to twice the height ($c = 3.01$ Å) of the tetragonal unit cell of the β-FeO(OH) structure (Szytula, 1970). Data from Schwyn (1983).

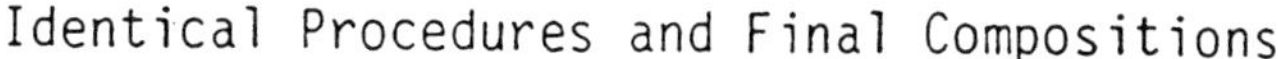

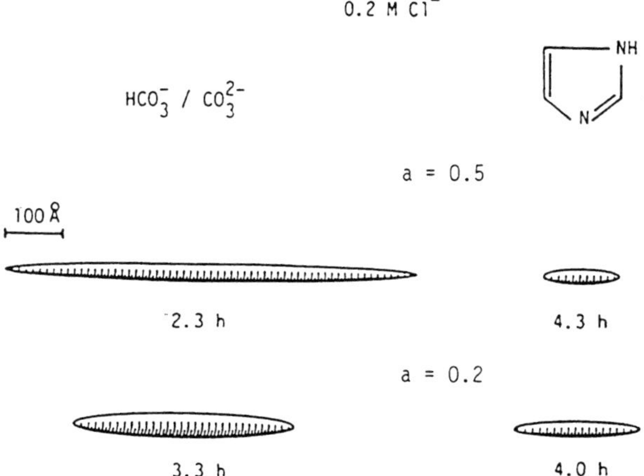

Figure 7.6. The dependence of nucleation rates and subsequent growth on the type of base used to adjust identical $a = 0.5$ and $a = 2.0$, respectively, in identical educt solutions ([Fe(III))], 0.06 M; $[Cl^-]$, 0.2 M; $[ClO_4^-]$, 0.8 M). The base solutions were HCO_3^-/CO_3^{2-} (total $[CO_3^{2-}]$, 0.75 M; pH 10) and ImH^+/Im (total imidazol, 1.2 M; pH 8). The time $t = 0$ refers to the onset of base addition to an acid solution ($a = -0.5$; pH 1.2). The polynuclear sizes are $p = 9400$, 1300 (upper row), and $p = 9900$, 2300 (lower row). Rates of base addition $da/dt \approx 0.9\ h^{-1}$ were identical in all cases. Data from Schwyn (1983).

bonate–bulk solution interface need more chemical modification in order to reach the state of effective nuclei. With regard to experimental procedures, it is more evident by now that the rate of base addition, drop size, rate of mixing, and time are important variables in addition to the chemical components and their concentrations.

In the virtually monodisperse systems with a fixed, aging was shown to promote the increase in width in favor of length when the concentration of polynuclears remained constant (Schwyn, 1983). However, the observation time is limited because of aggregation processes which precede the spontaneous precipitation. This is shown in Figure 7.7 for a solution ($a = 2$; $[Cl^-] = 0.94\ M$), where the precipitation time was about 10 days.

In the macroscopically stationary solution, the weight average p_w of the coefficient p increases in the time range $t = 1$–8 days by a factor of 42. The shape parameters are not compatible with persisting rotational ellipsoids emerging from the continuous growth of cores:

$$[FeO_{2/3}(OH)_{4/3}]_p^{(p/3)+} \tag{7.18}$$

The length/width ratio increases from 8 to ca. 40, which indicates that the aggregation of polynuclears ($t = 0$, $p = 15{,}000$) occurs in this period. Since

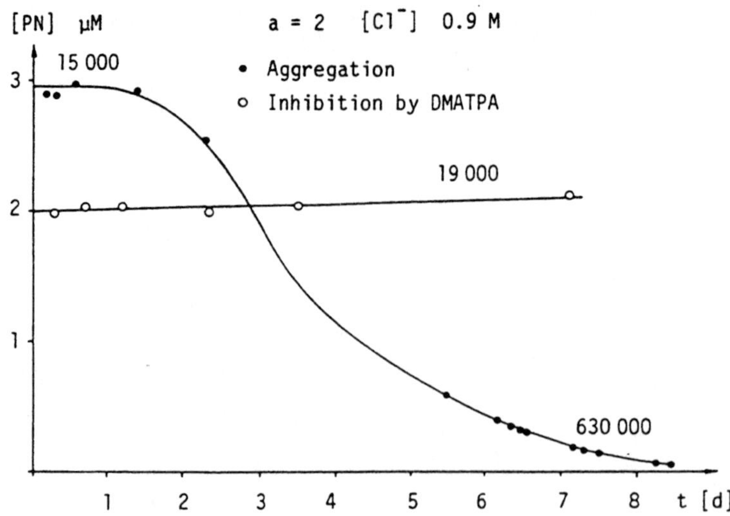

Figure 7.7. Spontaneous aggregation ($t < 4$ days) and coagulation ($t > 5$ days) of cationic polynuclears ($p = 15{,}000$) with charge $+\frac{1}{3}$ per Fe(III) in chloride solution prior to precipitation ($t \approx 10$ days). The same educt solution was used to prepare the polynuclears $p = 19{,}000$ which were protected from aggregation by surface complex formation with *p*-α-(*N,N*-dimethyl)aminotolylphosphonic acid (DMATPA, Eq. 7.22). DMATPA was added immediately after $a = 2$ had been adjusted. The total concentrations of Fe(III) and DMATPA were 5.4×10^{-2} *M* and 9.1×10^{-3} *M*, respectively, corresponding to the molar ratio P/Fe = 0.17. The surface area per P (adsorbed) was in the order of 50 Å^2.

the polynuclear cores (7.18) are polycations with charge $+\frac{1}{3}p$, it is expected that they coagulate in conjunction with charge compensation by chloride ions. Actually, it was shown that this type of coagulation is reversible. Fresh coagulates can be dissolved in perchlorate solution having the same pH ≤ 2 as the educt solution and then be reprecipitated with chloride.

A different type of coagulation will be induced if the pH in the perchlorate solution was raised into the neutral region.

$$\frac{1}{p}\,[\mathrm{FeO_{2/3}(OH)_{4/3}}]_p^{(p/3)+} + x\mathrm{OH}^- + y\mathrm{Cl}^- \longrightarrow \mathrm{FeO_{2/3}(OH)_{4/3+x}Cl_y(s)} \qquad (7.19)$$

$$x + y = \frac{1}{3}; \qquad \frac{1}{3} \geq y \geq 0$$

Equation 7.19 describes the variety of products that can be obtained by "washing out" Cl^- from primary products according to standard procedures (Weiser, 1946). Moreover, it describes thermodynamically favored aging processes in condensed phases which, on the time scale of years at ambient temperature, may yield the limiting product β-FeO(OH). Even in macroscopically stationary solutions ([Fe(III)], 0.05 M; [Cl^-], 0.3 M; [ClO_4^-], 0.7 M; a = 1.1; 25°C) the coefficient $\bar{n}$ in Eq. 7.20 increases from 2.70 to 2.80 over a period of some 2 months whereby the pH decreases from 2.00 to 1.64 (Künzi, 1982).

$$Fe_pO_r(OH)_s^{(3-\bar{n})p+}; \qquad \bar{n} \overset{\text{Def.}}{=\!=} \left\langle \frac{2r + s}{p} \right\rangle_{\text{average}} \qquad (7.20)$$

Conceptually, it is correct to consider the products according to Eq. 7.19 in terms of surface site reactions such as deprotonation and ligand substitution (Westall, 1987; Schindler and Stumm, 1987). In the limit $y \rightarrow 0$, that is, $\bar{n} \rightarrow 3.0$, uncharged cores will undergo aggregation and subsequently coagulation. It is realistic to anticipate that polynuclear–polynuclear interactions may involve reversible hydrogen bonding between surface groups

$$(PN)Fe^{III}\text{—}O(H)\cdots H\text{—}O\text{—}H\cdots O(H)\text{—}Fe^{III}(PN) \qquad (7.21a)$$

or less reversible bridging

$$(PN)Fe^{III}\begin{matrix}\diagup OH \diagdown \\ \diagdown HO \diagup\end{matrix}Fe^{III}(PN) \qquad (7.21b)$$

Structures 7.21a and b are useful in suggesting that polynuclears could be protected from aggregation by surface interaction with, for instance, polyalcohols or polysaccharides. As a matter of fact, this principle is applied in some commercially available preparations for iron therapy (Erni et al., 1984).

A rather transparent case of the inhibition of aggregation or coagulation is shown in Figure 7.7. The molecular size of polynuclears ($p \approx$ 19,000) virtually persists for at least a year in the presence of DMATPA

$$Fe\text{—}O\text{—}P(OH)(O)\text{—}C_6H_4\text{—}CH_2\text{—}NH(CH_3)_2^+ \qquad (7.22)$$

p-α-(N,N-Dimethyl)aminotolylphosphonic acid

at the total concentration level of 0.17 per Fe(III) (mole/mole). The uncharged betain form is the ligand which coordinates to iron(III) surface sites (Schwyn, 1983).

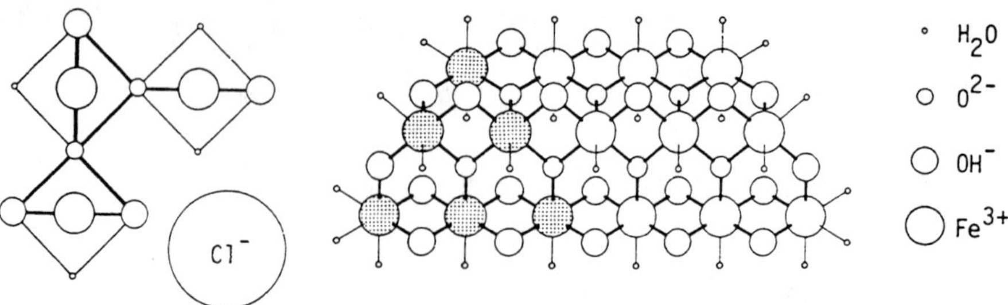

Figure 7.8. The polynuclear $Fe_{15}O_7(OH)_{21}^{10+}$ ($\bar{n}$ = 2.33) and its projection on the plane perpendicular to the chain axis. The limiting composition ($p \rightarrow \infty$) is $[FeO_{2/3}(OH)_{4/3}]_p$; $\bar{n}$ = 2.67. This infinite polynuclear defines a structural embryo. The growth axis corresponds to the crystallographic *c* axis of the tetragonal unit cell indicated in Figure 7.9. Outer-sphere Cl^- is shown in the projection at the minimum energy position. It is suspected that the polynuclear (p = 15) shown here belongs to the ensemble of critical size nuclei.

In this section, it has been shown that the specific structure induced by chloride interactions provides favorable situations for the experimental detection of the hierarchy of factors involved in hydrolysis from the acid medium side. The modelling of the evolution of β-FeO(OH)-type polynuclear structures will be discussed elsewhere (Mathier-Landa, 1986). As an illustration, the structural embryo that accounts for the well-defined stoichiometry $\bar{n} \rightarrow 2.67$; $p \rightarrow \infty$ is shown in Figure 7.8. The projection of a polynuclear composed of such embryos, that is, formally formed by the mere loss of water in joining the embryos, is shown in Figure 7.9. The diameter corresponds to the lowest width that was inferred from light-scattering experiments (see Fig. 7.6, top left ellipsoid). The square in the center of Figure 7.9 is the projection of the idealized (regular octahedra) tetragonal unit cell of the β-FeO(OH) structure (Mackay, 1962; Szytula et al., 1970).

4. IRON HYDROLYSIS IN CELLS

4.1. Iron Storage and Transport

Human iron metabolism (Hughes, 1978) involves the recycling of iron within the body, whereas the daily excretion does not exceed about 1 mg, some .01% of the total body iron (ca. 4000 mg). At least 20% of the iron is stored in cells, a major part of it in the liver. In the present context it is highly interesting that the chemical storage form is iron(III) hydroxide. However, some 2000 Fe atoms (average load) up to 4500 (maximum capacity) are encapsulated as an iron(III) hydroxide core by a protein shell composed of 24 subunits, as a rule. Ferritin is ubiquitous in vertebrates (Crichton, 1985; Harrison, 1983; Theil, 1983). The protein shell, apoferritin, has a typical molecular mass of about 500,000 daltons (24 × 20 kdaltons). Our under-

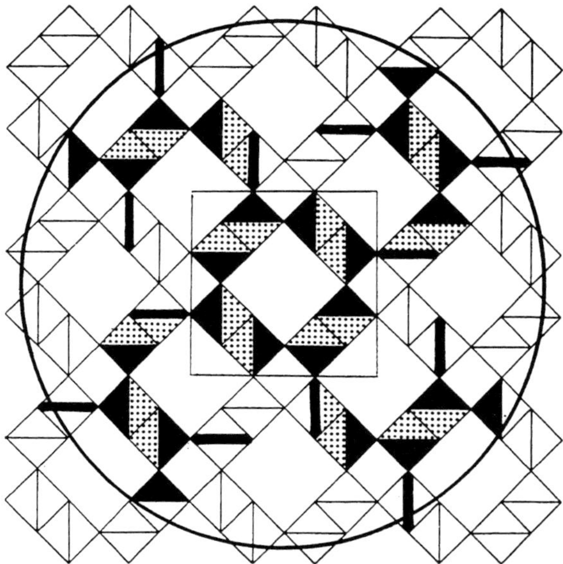

Figure 7.9. The projection of a hypothetical polynuclear composed of structural embryos (Fig. 7.8). The tetragonal unit cell of the β-FeO(OH) structure is indicated in the center. Formally, the polynuclear projected is formed by mere loss of water from embryos. The diameter of the circle is 30 Å, which corresponds to the width of the polynuclear (p = 9400; a = 0.5) shown in Figure 7.6. The structure presented involves maximum electrostatic free energy of ion association compared to isomers with identical cross sections [Used with permission from Schneider (1984).]

standing of the intimate mechanisms of iron deposition on one hand, and of iron mobilization on the other hand, is still rather poor (Romslo, 1980; Crichton, 1985). Iron deposition in ferritin is the growth of intraprotein polynuclears, while mobilization is their dissolution, both occurring in the physiological pH range.

The requirement for iron is particularly obvious in erythropoesis, the production of red cells from the bone marrow (Bessis, 1977; Monnette, 1983). The first stage (amplification) involves a series of four to six cell divisions. Within the lifetime of each cell, typically 20–24 h, the iron required for hemoglobin synthesis is taken up from transferrin, the iron carrier in the bloodstream (Brock, 1985; Aisen and Listowksy, 1980). This protein (80 kdaltons) is an efficient chelator for two iron(III) per protein molecule. Diferric transferrin (Fe_2Tr) is thermodynamically stable with respect to iron(III) hydrolysis in the range $6 < pH < 8$.

The daily flux of iron required for red cell production is about 20 mg (0.36 mmole) in humans. Chemical processing in cells involves uptake and storage of iron followed by consumption by mitochondria for heme synthesis. We are particularly interested here in the iron(III) hydroxide that is confined to the ferritin core area. This defines the concept of an "intracellular transit iron

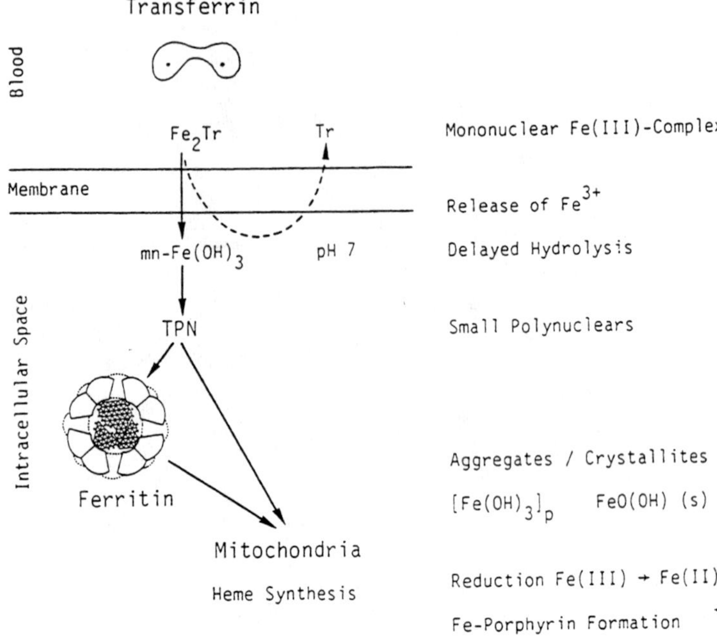

Figure 7.10. Intracellular iron transport: Uptake of diferric transferrin (Fe_2Tr) by specific receptors at the cell membrane is followed by release of Fe^{3+} and subsequent efflux of apotransferrin (Tr). The "transit pool" of iron includes, by definition, all species which undergo iron deposition in ferritin (Ft) and consumption by mitochondria for haem synthesis. The analogous chemical processes apart from cytoplasmic components are indicated to the right according to recent interpretations (Schneider 1982; Funk 1986).

pool" (Jacobs, 1977), intracellular iron species that are transients between release from Fe_2Tr and storage in ferritin or mitochondrial consumption. It is known that there is no direct iron transfer from Fe_2Tr to apoferritin or from Fe_2Tr to mitochondria (Crichton, 1985). Hence, we consider the situation indicated in Figure 7.10.

4.2. A Thermodynamic Paradox in Intracellular Iron Transport

It is useful to consider realistic fluxes and average concentrations of iron in an erythroid cell prior to maturation. As a guide, we use Figure 7.11, which is an abstraction from data on the erythrocytic series (Bessis, 1977). The molar amount of Fe per cell is more reliable than the volume quantities. However, even the increment $\Delta[\mathrm{Fe}]/\Delta t = 10^{-7}\,M^{-1}\,\mathrm{s}^{-1}$ differs by many orders of magnitude from the imaginary concentration $10^{-17}\,M$ of Fe^{3+} required by the solubility product, $\log K_{s0} \approx -38$, at pH 7.

Apart from some endosomic areas that are believed to have relatively low pH $\approx$ 5 (Crichton, 1985), the intracellular aqueous medium is rather neutral.

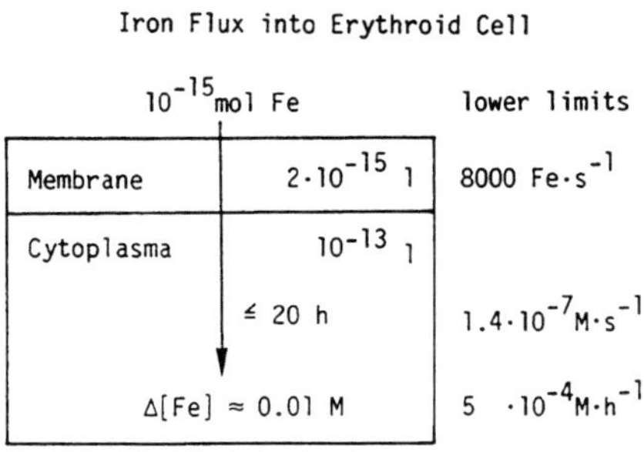

Iron Flux into Erythroid Cell

10^{-15} mol Fe		lower limits
Membrane	$2\cdot10^{-15}$ l	8000 $Fe\cdot s^{-1}$
Cytoplasma	10^{-13} l	
	≤ 20 h	$1.4\cdot10^{-7} M\cdot s^{-1}$
	Δ[Fe] ≈ 0.01 M	5 $\cdot10^{-4} M\cdot h^{-1}$

Figure 7.11. Typical iron fluxes in an erythroid cell. Volume quantities including concentrations are unavoidably crude values which do not affect the reliability of conclusions drawn in the text.

It is quite surprising that Fe(III) hydroxide is not delocalized in cytoplasmic proteins but localized in ferritin. It is correct to anticipate that Fe_2Tr is just a source of mononuclear Fe(III) ions which are released one by one either in the cell membrane or in conjunction with endocytosis of Fe_2Tr. There is no appreciable stationary concentration of apotransferrin in the cell. To date, no specific iron(III) chelator has been identified (Romslo, 1980; Bakkeren, 1985) that could prevent hydrolysis thermodynamically. In fact, this would not be compatible with the function of ferritin. By the same reason, the space and time average of the intracellular redox potential E_H (excluding mitochondrial membranes) cannot be as low as $E_H \leq -0.3$ V, which would provide thermodynamic stability to hydrated Fe^{2+} with respect to FeO(OH)(s). This paradox is removed only if it is assumed that in the cytoplasmic medium, small polynuclears are formed that are unstable with respect to FeO(OH)(s) but kinetically preserved. Nucleation, then, would be required and induced by interaction with apoferritin or partially loaded ferritin. The concept of delayed hydrolysis in cells (Schneider and Erni, 1982) is indicated in Figure 7.12.

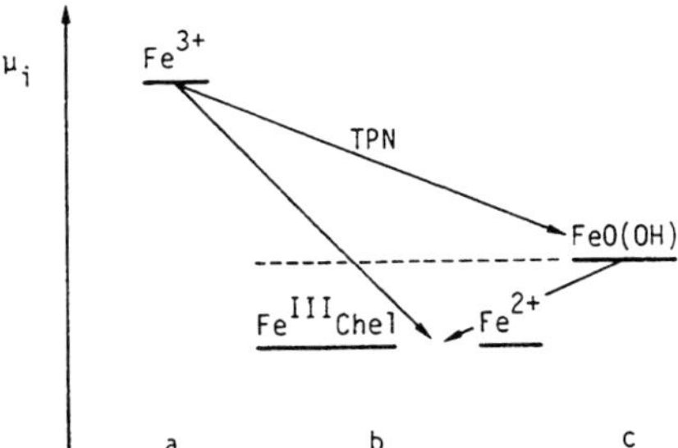

Figure 7.12. The concept of delayed hydrolysis of Fe^{3+} in the cytoplasm prior to iron hydroxide deposition in ferritin. μ_i: Chemical potential (not to scale) of Fe(III) species involved in intracellular transport (Fig. 7.11). (*a*) Fe^{3+} released from Fe_2Tr. (*b*) Chelate complexes (FeChel) which are stable with respect to FeO(OH) around pH 7, and ferrous ion Fe^{2+} stabilized by low redox potentials $E_H \leq -0.3$ V. TPN: transient polynuclears with low nuclearity due to interactions with cytoplasmic components. (*c*) The iron(III) hydroxide core of ferritin.

4.3. Delayed Hydrolysis in Polyalcohol Solutions

Polyalcohols such as sorbitol and xylitol (Fig. 7.13) are weak ligands for Fe^{3+} in acid solution. However, in strongly alkaline sorbitol solutions, pH > 14, tris-bidentate anionic complexes $Fe(SOR)_3^{3-}$ have been identified (Hametner, 1985). We write H_2SOR for sorbitol and use $pK_1 = 13.2$ (1 M KCl; 20°C), as determined for the pair $H_2SOR/HSOR^-$, and a lower limit $pK_2 \geq 15$ for $HSOR^-/SOR^{2-}$ in estimating the conventional stability constants:

$$Fe^{3+} + 3SOR^{2-} \rightleftharpoons Fe(SOR)_3^{3-} \qquad \beta_3 = K_1K_2K_3 \tag{7.23}$$

On the basis of log K versus pK correlations involving well-known complexes with oxygen ligands, the following set of values were estimated, from which the constants K_{s3} (Eqs. 7.25, 7.26) and K'_{s3} (Eq. 7.27) were calculated.

$$\log K_1 \geq 20 \qquad \log K_2 \geq 15 \qquad \log K_3 \geq 10 \qquad \log \beta_3 \geq 45 \tag{7.24}$$

$$FeO(OH)(s) + H_2O + 3H_2SOR \rightleftharpoons Fe(SOR)_3^{3-} + 3H^+ \tag{7.25}$$

$$\log K_{s3} = \log \beta_3 + \log K_{s0} - 3(pK_1 + pK_2 - pK_w) \tag{7.26}$$

$$pH = 7: \quad K'_{s3} = \frac{[Fe(SOR)_3]}{[H_2SOR]^3} = K_{s3}\,[H^+]^{-3} \geq 10^{-15}\,M^{-2} \tag{7.27}$$

With $pK_w = 14$, $\log K_{s0} = -39$, we arrive at $\log K_{s3} \geq -36$ and $\log K'_{s3} \geq -15$ (pH = 7), implying negligible solubility of FeO(OH)(s) even in 1 M sorbitol. As a matter of fact, precipitates can easily be produced if strong

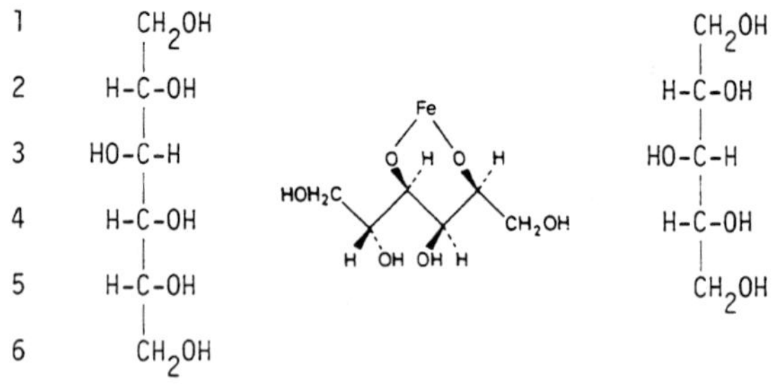

Figure 7.13. Fischer projections of D-sorbitol and D-xylitol. The plausible structural fragment containing twofold deprotonated sorbitol does not involve optimum conformation of free sorbitol.

acid is used to adjust the pH to 7 starting from a solution of $Fe(SOR)_3^{3-}$ at pH >14.

However, when pH gradients are kept as low as possible by using weak acids, the homogeneous solution persists down to pH ≈ 7 even if total Fe(III) is in the order of 0.1 *M* in 1 *M* sorbitol. A variety of physical methods have been applied to elucidate the scheme of Eqs. 7.28–7.30, accounting for hydrolysis, or hydrolytic olation, from the alkaline side.

$$\text{pH} > 12: \quad \underset{\text{yellow}}{2Fe(SOR)_3^{3-}} + 2H_2O \rightleftharpoons \underset{\text{green}}{Fe_2(OH)_2(SOR)_4^{4-}} + 2H_2SOR + 2OH^- \tag{7.28}$$

$$\text{pH} > 10: \quad Fe_2(OH)_2(SOR)_4^{4-} \rightleftharpoons \underset{\text{brown}}{\tfrac{1}{3}Fe_6(OH)_{12}(SOR)_6^{6-}} + 2H_2SOR + 2OH^- \tag{7.29}$$

$$\text{pH} > 7: \quad Fe_6(OH)_{12}(SOR)_6^{6-} \rightleftharpoons \text{gel} \tag{7.30}$$

Obviously, there are striking kinetic preferences which simulate severe constraints on nuclearity. The reversibility of gelation is rather conditional, since it depends particularly on the neutralization procedure. A number of arguments supporting the cyclic structure of the brown hexanuclear are presented elsewhere (Schneider et al., 1986). The core of the hexanuclear is shown in Figure 7.14, where the peripheral sorbitol ligands are not included. They are replaced by pairs of monodentate H_2O, OH^-. This situation is

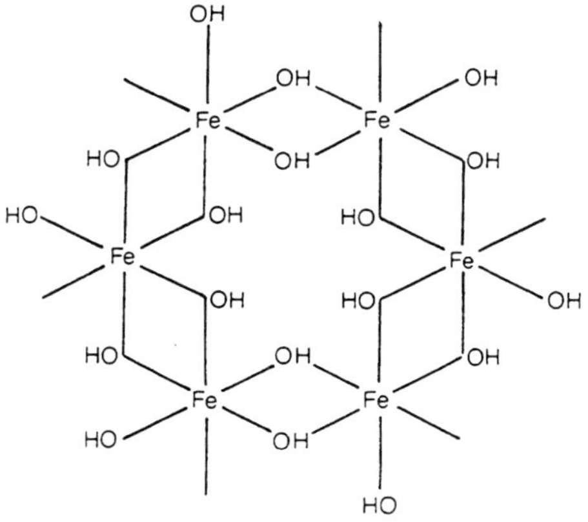

Figure 7.14. The cyclic hexanuclear $[Fe_6(OH)_{12}](OH)_6$ = $[Fe(OH)_3]_6$ related to $Fe_6(OH)_{12}(SOR)_6^{6-}$. The idealized structure has symmetry S_6.

expected to emerge from the dissociation of SOR^{2-} from the Fe(III) sites according to reaction 7.31.

$$\text{Fe}\langle{}^{\text{O}}_{\text{O}}\rangle\text{SOR} + H_2O + H^+ \longrightarrow \text{Fe}\langle{}^{\text{OH}}_{\text{OH}_2} + H_2SOR \qquad (7.31)$$

Hence, the structure in Figure 7.14 describes a hexamer $(Fe(OH)_3)_6$ conforming to pH around 7. The fragment to the right in Eq. 7.31 could undergo a variety of substitutions by weak ligand groups, whereas strong ligands such as EDTA, acetylacetone, and hydroxamic acids invariably react to form stable mononuclear complexes. The gelation (7.30) involves the reaction of the cyclic hexanuclear (Fig. 7.14) with itself. If the cycles were preserved, chemical links between them could only be μ-ol bridges as shown in 7.21*b*. There are two patterns with the highest possible symmetry of olation products (Fig. 7.15*a,b*). It is seen from Figure 7.15*a* that the original hexanuclears lose their identity completely in joining to form the same type of layer structure as in $Al(OH)_3$ (Wells, 1984). Hence, the alternative pattern in Figure 7.15*b* is the more realistic model for the olation involved in gelation as well as for its reversibility. Of course, an enormous variety of arrays emerge from combining the olation principles of Figure 7.15 in a random manner.

The experimental proof for reversibility of gelation is presented in Figure 7.16 (Rich, 1985). There is no nucleation with subsequent growth of inert

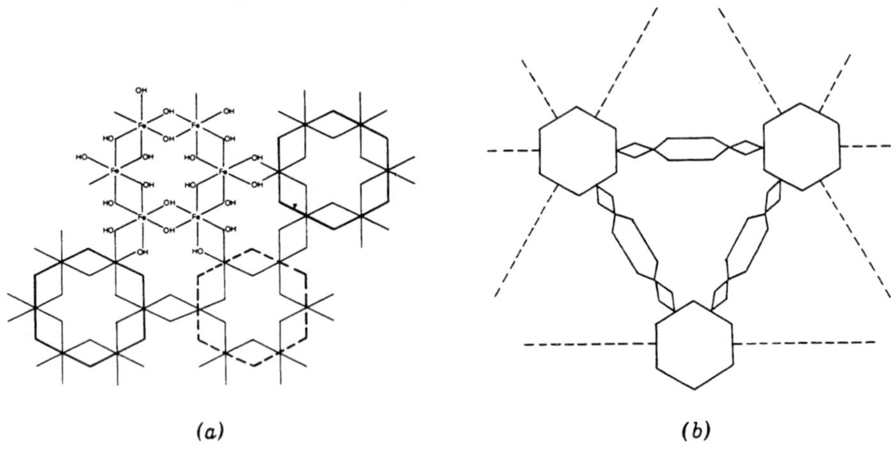

Figure 7.15. (*a*) The $Al(OH)_3$ type of layer structure derived from hexanuclear units in Figure 7.14 representing an unrealistic model of reversible gelation (Fig. 7.16). (*b*) Structural units as shown in Figure 7.14 are linked by di-μ-ol bridges $Fe(OH)_2Fe$. Neighboring hexagons are perpendicular to each other. This type of array is a realistic model of reversible gelation (Fig. 7.16).

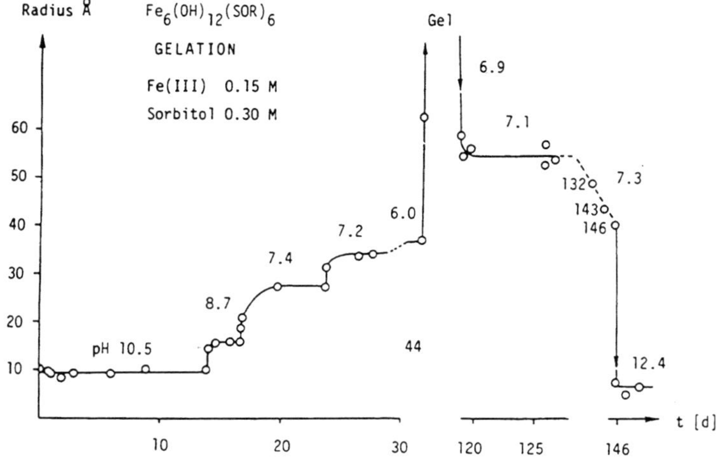

Figure 7.16. The reversible gelation induced by changes of pH in a solution of $Fe_6(OH)_{12}(SOR)_6^{6-}$. The pH was adjusted with imidazolium chloride (10.5 → 6) and NaOH (6 → 12.4). Hydrodynamic radii of polynuclear components were derived from translational diffusion coefficients measured by laser light-scattering methods. The final solution (pH 12.4) is an equilibrium mixture of hexa- and dinuclear species according to Eq. 7.29 (Rich, 1985).

crystalline domains of α-FeO(OH) or α-Fe_2O_3 within 76 days of aging of the gel at room temperature, pH 6, which indicates that the original structural units persist over the entire time scale of Figure 7.16.

4.4. Models of Transients in Intracellular Hydrolysis

It has been postulated that the intracellular medium protects low nuclear hydroxo complexes by interactions with groups such as –CONH–, ROH, $ROPO_3^{2-}$ of proteins, glycoproteins, and a variety of organic phosphates (Schneider and Erni, 1982). Hence, no specific cytosolic iron-binding ligands are required. It was shown that synthetic hexanuclears survive in media containing a high concentration of groups as mentioned above (Jiskra, 1983). However, the hexanuclear core is not expected to be the only model species because it is a particular member of a series of cores $Fe((OH)_2Fe)_{p-1}$. Open-chain members with low p would also meet the requirements discussed in Section 4.2, but isolation procedures would exert strong chemical perturbations. Recently, iron-containing low molecular weight fractions from rat reticulocyte cytosol were analyzed by HPLC methods (Bakkeren, 1985). They contained unspecific oligopeptides and amino acids in widespread portions. The apparent molecular weight of ca. 5000 indicates either one polynuclear with $p < 50$ or, more likely, an aggregate of even lower polynuclears. By oxidation of Fe^{2+} in gluconate (Erni et al., 1985) or albumin solutions at pH

7, low nuclear species were obtained that were readily consumed by yeast mitochondria (in suspension) and used for heme synthesis (Funk et al., 1986). Hence, low nuclear hydroxo complexes could well be important iron carriers in cells quite generally.

5. AQUATIC MEDIA: HYDROLYSIS INDUCED BY OXIDATION OF FERROUS IONS

It is anticipated here that the reader is familiar with the thermodynamics and kinetics of the oxidation of Fe^{2+} by O_2 as presented in a standard text (Stumm and Morgan, 1981). Beyond pH 5, there is a second-order dependence of the rates of oxygenation with respect to OH^-, which indicates that the electron transfer occurs in conjunction with proton transfer from Fe(II) to O_2:

$$Fe^{2+} + O_2 + iH_2O \longrightarrow \text{mn-Fe(OH)}_i^{3-i} + O_2^- + iH^+ \qquad (7.32)$$

Equation 7.32 is an allusion to Table 7.1. The reaction is quite endergonic for $i = 0$, but less so for $i = 3$ at pH 7;

$$O_2 + e^- \rightleftharpoons O_2^- \qquad E_0 = +0.07\text{ V} \qquad (7.33)$$

$$\text{mn-Fe(OH)}_3 + 3H^+ + e^- \rightleftharpoons Fe^{2+} + 3H_2O \qquad E_0' \approx +0.24\text{ V} \qquad (7.34)$$

We do not yet know whether (7.32) is really the first elementary step. However, the concept of instantaneously formed mononuclears is useful in giving rise to the scheme of reactions in Figure 7.17, where equilibrium 7.35 is anticipated.

$$\text{mn-Fe(OH)}_3 + H_2O_2 \rightleftharpoons \text{mn-Fe(OH)}_2(O_2H) + H_2O \qquad (7.35)$$

The substitution reaction 7.35 could also be applied to peripheral sites of polynuclears as well as to surface sites of oxides. It is concluded from Figure 7.17 that the variety of hydrolysis products is strongly involved in the intimate mechanism of oxygenation of Fe^{2+}. Actually, the product-dependent term of the rate law (Eq. 7.36) supports this conclusion (Sung and Morgan, 1980).

$$-\frac{d[\text{Fe(II)}]}{dt} = (K_1 + K_2[\text{Fe(III)}])[\text{Fe(II)}] \qquad (7.36)$$

Moreover, interactions of polynuclears with organic matter are expected to occur according to the principles discussed in Section 4.3. Finally, matrix effects of inorganic particulate matter call for attention in conjunction with adsorption of Fe(III) species. We cannot yet assess the variety of such effects in more detail.

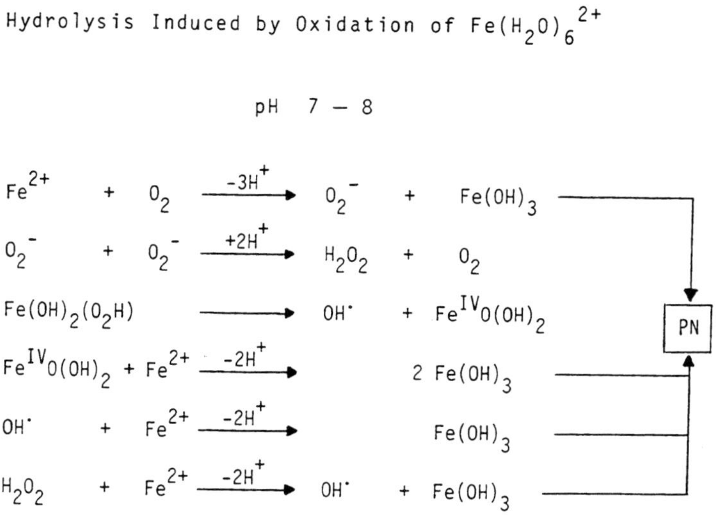

Hydrolysis Induced by Oxidation of $Fe(H_2O)_6^{2+}$

pH 7 – 8

$$Fe^{2+} + O_2 \xrightarrow{-3H^+} O_2^- + Fe(OH)_3$$

$$O_2^- + O_2^- \xrightarrow{+2H^+} H_2O_2 + O_2$$

$$Fe(OH)_2(O_2H) \longrightarrow OH^\cdot + Fe^{IV}O(OH)_2$$

$$Fe^{IV}O(OH)_2 + Fe^{2+} \xrightarrow{-2H^+} 2\,Fe(OH)_3$$

$$OH^\cdot + Fe^{2+} \xrightarrow{-2H^+} Fe(OH)_3$$

$$H_2O_2 + Fe^{2+} \xrightarrow{-2H^+} OH^\cdot + Fe(OH)_3$$

$Fe(OH)_3$ → PN

Figure 7.17. The oxygenation of hydrated ferrous ion in terms of the concept of mononuclear $Fe(OH)_3$ formed at the instant of electron transfer. Hydroperoxide complexes and Fe(IV) intermediates in this scheme are plausible postulates beyond established experimental facts. Polynuclear products (PN) may interact with H_2O_2 to form hydroperoxide surface complexes which decompose in the same way as $Fe(OH)_2(O_2H)$.

It is suggested by the scheme of Figure 7.17 that open-chain and cyclic $[Fe(OH)_3]_p$ could be early products of oxygenation, particularly at very low concentrations of Fe^{2+}. Hence, products resembling aggregates or gels, as discussed with respect to Figure 7.15 could emerge from oxygenation of Fe^{2+} in natural waters.

At present, experimental evidence for low nuclear species as oxygenation products is restricted to Fe^{2+} concentrations as high as 0.01 *M* in Tris buffer (0.5 *M*; pH 7.5). The setup shown in Figure 7.18 was used in laser light-scattering measurements to determine the hydrodynamic radii of polynuclear products (von Gunten, 1985) of the reaction

$$\begin{array}{lcl} 4Fe^{2+} + O_2 + 10H_2O \xrightarrow{\text{pH 7.5}} \frac{4}{p}(Fe(OH)_3)_p + & & 8H^+ \\ \qquad\qquad\qquad (8\ B) & & (8\ HB^+) \\ \qquad\qquad B = H_2NCH_2C(CH_2OH)_3 & & \end{array} \tag{7.37}$$

Although unrealistic with regard to natural waters, the medium chosen is useful as a scavenger of primary products. The radius of polynuclears was in the range $15 < r < 20$ Å. Considering $r \approx 10$ Å for the hexanuclear $Fe_6(OH)_{12}(SOR)_6^{6-}$, the estimate $p \leq 50$ is realistic. It is even more revealing that the polynuclear size was found to remain constant for several months,

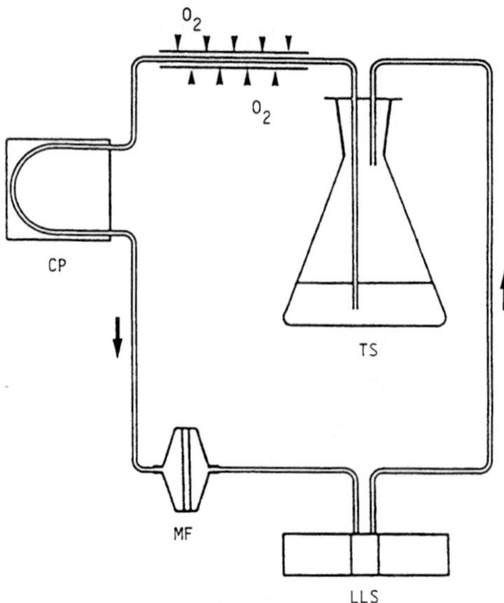

Figure 7.18. The experimental setup for oxygenation of ferrous ion in a closed system (von Gunten, 1985). TS, test solution, educt 0.01 *M* Fe^{2+}; O_2, oxygen admitted by gas-permeable silicon tubing; CP, circulation pump; MF, membrane filter, pores 0.2 μm; LLS, laser light-scattering cell.

at least. Current research is devoted to verifying further the relevance of these phenomena for natural waters.

6. THE MORPHOLOGICAL VARIETY OF HYDROLYSIS PRODUCTS

Applying the principles and phenomena discussed so far, we can now rationalize the potentially unrestricted variety of products that emerge from hydrolysis at moderate temperatures (Fig. 7.19).

In acid solution, the relevant mononuclears are cationic $FeOH^{2+}$ and $Fe(OH)_2^+$ for which

$$[Fe(OH)_2^+] < [FeOH^{2+}] \text{ or } [Fe_2(OH)_2^{4+}] \tag{7.38}$$

is valid under all circumstances. The genuine polynuclear is the positively charged polycation with well-defined preferential structure (Fig. 7.19, top left). Anions may exert specific effects at the level of mononuclear and low nuclear species, but ion association including precipitation (coagulation) is more pronounced at the macromolecular (polyelectrolyte) level. Precipitation can be inhibited by surface ligands without net charge. It is feasible that

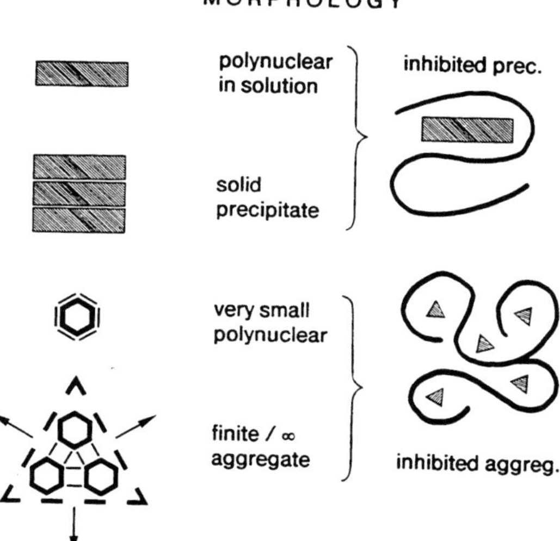

Figure 7.19. The morphological variety of hydrolysis products: prototype behavior as discussed in the text.

macromolecular matrices localized polynuclears in anionic cavities, preventing the alignment required for precipitation (Fig. 7.19, top right). Of course, any system of this type would be unstable with respect to FeO(OH) or Fe_2O_3, but the kinetics of structural transformation are very slow at moderate temperatures.

In alkaline solution, the only interesting type of mononuclears are those with low conditional stability in the neutral region such as sorbitol complexes $Fe(SOR)_3^{3-}$. They are important as spring of restricted size polynuclears such as cyclic $Fe_6(OH)_{12}(SOR)_6^{6-}$ containing a core which persists at pH $\approx$7 if it is protected from further olation by weak peripheral ligands. If this protection is removed, the structural units may be preserved although they are linked to each other by OH^- bridges (Fig. 7.19, lower part).

In neutral solutions, conditions may be such that the relevant mononuclear $Fe(OH)_3$ undergoes olation to open-chain or cyclic units that are subject to delayed hydrolysis by interactions such as those discussed for the prototype cyclic polynuclear $(Fe(OH)_3)_6$. In aquatic systems this possibility is rather restricted to hydrolysis induced by oxygenation of Fe^{2+}.

Preparative procedures involving the addition of base to an acid solution of Fe^{3+} or the reverse, the addition of an acid Fe^{3+} solution to excessive base, are easily complicated by unavoidable time-dependent gradients. That is, primary products typical of alkaline solutions invade an acid solution and vice versa. As a rule, therefore, we should expect nonuniform products with regard to structural and morphological properties.

7. CONCLUSIONS

Hydrolysis products of iron(III) that are uniform with regard to structure, size and shape of particles, interfacial properties, and consequently chemical reactivity are expected to be rather exceptional. The virtually unrestricted variety of products accessible at moderate temperatures can be rationalized in terms of a restricted set of prototypes. It is expected that we can improve the methodology for the analysis of product spectra.

Acknowledgments

This paper summarizes a series of studies carried out during the last 15 years. The senior author (W. Schneider) is indebted to C. Anner, I. Erni, F. Funk, D. Hametner, U. von Gunten, J. Jiskra, W. Künzi, E. Mathier-Landa, N. Oswald, H. W. Rich, B. Rüttimann, B. Schwyn, and F. van Steenwijk for their participation in these studies. Research was supported by Kredite für Unterricht und Forschung, ETHZ; current work on the oxygenation of ferrous ion is being supported by the Swiss National Science Foundation (Project No. 2.382-0.84). We thank Paul Schindler and Werner Stumm for continuous encouragement and fruitful discussions.

REFERENCES

Aisen, P., and Listowsky, I. (1980), Iron transport and storage proteins, *Ann. Rev. Biochem.* **49,** 357.

Anner, C. (1975), "Die Hydrolyse des dreiwertigen Eisens im Chloridmedium," dissertation, Swiss Federal Institute of Technology (ETH), Zurich, No. 5573.

Baccini, P. (1985), "Phosphate Interactions at the Sediment–Water Interface," in W. Stumm, Ed., *Chemical Processes in Lakes,* Wiley, New York, p. 189.

Baes, C. F., Jr., and Mesmer, R. E. (1976), *The Hydrolysis of Cations,* Wiley, New York, pp. 229–237.

Bakkeren, D. L., de Jeu-Jaspers, C. M. H., van der Heul, C., and van Eijk, H. G. (1985), Analysis of iron-binding components in the low molecular weight fraction of rat reticulocyte cytosol, *Int. J. Biochem.* **17,** 925.

Berne, B. J., and Pecora, R. (1976), *Dynamic Light Scattering,* Wiley, New York.

Bessis, M. (1977), *Blood Smears Reinterpreted,* Springer, New York, p. 25.

Brock, J. (1985), "Transferrins," in P. Harrison, Ed., *Metalloproteins,* Part 2, Verlag Chemie, Weinheim, p. 183.

Ciavatta, L., and Grimaldi, M. (1975), On the hydrolysis of the iron(III) ion in perchlorate media, *J. Inorg. Nucl. Chem.* **37,** 165.

Crichton, R. R. (1985), "Intracellular Iron Metabolism," in G. Spik, J. Montreuil, R. R. Crichton, and J. Mazurier, Eds., *Proteins of Iron Storage and Transport,* Elsevier, Amsterdam, p. 99.

Erni, I., Oswald, N., Rich, H. W., and Schneider, W. (1984), Chemical characterization of iron(III) hydroxide–dextrin complexes, *Arzneim.-Forsch./Drug Res.* **34,** 1555.

Erni, I., Oswald, N., Rich, H. W., and Schneider, W. (1985), "The Chemistry Relevant to Oral Iron Preparations," in G. Spik, J. Montreuil, R. R. Crichton, and J. Mazurier, Eds., *Proteins of Iron Storage and Transport,* Elsevier, Amsterdam, p. 305.

Feitknecht, W., and Michaelis, W. (1962), Ueber die Hydrolyse von Eisen(III) perchlorat-Lösungen, *Helv. Chim. Acta* **45,** 212.

Funk, F., Lecrenier, C., Lesuisse, E., Crichton, R. R., and Schneider, W. (1986), A comparative study on iron sources for mitochondrial haem synthesis including ferritin and models of transit pool species, *Eur. J. Biochem.* **157,** 303.

Garrels, R. M., and Christ, C. L. (1965), *Solutions, Minerals and Equilibria,* Harper and Row, New York.

Grant, M., and Jordan, R. B. (1981), Kinetics of solvent water exchange on iron(III), *Inorg. Chem.* **20,** 55.

Gunten, U. von (1985), Swiss Federal Institute of Technology (ETH), Zurich, unpublished results.

Hague, D. N. (1971), *Fast Reactions,* Wiley, New York, p. 80.

Hametner, D. (1985), "Die Hydrolyse von Eisen(III): Die Vernetzung ausgehend von mononuklearen Polyalkoholkomplexen," dissertation, ETH, Zurich, No. 7795.

Harrison, P. (1983), "The Spatial Structure of Horse Spleen Apoferritin," in E. C. Theil, G. L. Eichhorn, and L. G. Marzilli, Eds., *Advances in Inorganic Biochemistry,* Vol. 5, Elsevier, New York, p. 39.

Hedström, B. O. A. (1953), The hydrolysis of the iron(III) ion Fe^{3+}, *Arkiv Kemi* **5,** 1.

Hughes, E. R. (1978), "Human Iron Metabolism," in H. Sigel, Ed., *Metal Ions in Biological Systems,* Vol. 7, Dekker, Basel, p. 352.

Jacobs, A. (1977), "An Intracellular Iron Pool," in *CIBA Foundation Symp.* **51,** *Iron Metabolism,* Elsevier, Amsterdam, p. 91.

Jacobs, A. (1980), "The Pathology of Iron Overload," in A. Jacobs and M. Worwood, Eds., *Iron in Biochemistry and Medicine,* **Vol. II,** Academic Press, New York, p. 428.

Jiskra, J. (1983), "Tracer-Untersuchungen zur hydrolytischen Vernetzung von Eisen(III) in chloridhaltigen und biologischen Medien," dissertation, ETH, Zurich, No. 7394.

Knight, R. J., and Sylva, R. N., Precipitation in hydrolyzed iron(III) solutions, *J. Inorg. Nucl. Chem.,* **36,** p. 591.

Künzi, W. H. (1982), "Die Hydrolyse von Eisen(III). Der Einfluss von Chlorid auf Bildung und Zerfall von Vernetzungsprodukten." dissertation, ETH, Zurich, No. 7016.

Lengweiler, H. Feitknecht, W., and Buser, W. (1961), Die Ermittlung der Löslichkeit von Eisen(III)-hydroxiden mit ^{59}Fe, *Helv. Chim. Acta* **44,** 796.

Lutz, B., and Wendt, H. (1970), Kinetics of the fission and formation of the dimeric $(FeOH)_2^{4+}$ and $(VOOH)_2^{2+}$, *Ber. Bunsenges. Phys. Chem.* **74,** 372.

Mackay, A. L. (1962), "β-Ferric oxyhydroxide—akaganeite, *Mineral. Mag.* **33,** 270.

Mathier-Landa, E. (1986), "Modellierung der Bildung von Beta-Eisen(III)-Oxidhydroxid in wässeriger Lösung," dissertation, Swiss Federal Institute of Technology (ETH), Zürich.

Monnette, F. C. (1983), "Cell Amplification in Erythropoiesis: *In vitro* Perspectives," in C. D. R. Dunn, Ed., *Current Concepts in Erythropoiesis,* Wiley, New York, p. 31.

Rich, H. W. (1985), Swiss Federal Institute of Technology (ETH), Zürich, unpublished results.

Romslo, I. (1980), "Intracellular Iron Transport," in A. Jacobs and M. Worwood, Eds., *Iron in Biochemistry and Medicine,* Vol. II, Academic Press, New York, p. 325.

Schindler, P., Michaelis, W., and Feitknecht, W. (1963), Die Löslichkeit gealterter Eisen(III)-hydroxid-Fällungen, *Helv. Chim. Acta* **46,** 444.

Schindler, P., and Stumm, W. (1987), "The Surface Chemistry of Oxides, Hydroxides, and Oxide Minerals," this volume, Chapter 4.

Schneider, W. (1984), Hydrolysis of iron(III)—Chaotic olation versus nucleation, *Comments Inorg. Chem.* **3,** 205.

Schneider, W., and Erni, I. (1982), "Chemistry of Transit Iron Pools," in P. Saltman and J. Hegenauer, Eds., *The Biochemistry and Physiology of Iron,* Elsevier, New York, p. 121.

Schneider, W., Erni, I., Hametner, D., Magyar, B., and Steenwijk, F. van (1987), Iron(III) hydrolysis in alkaline sorbitol solutions: Evidence for cyclic hexanuclear hydroxocomplexes of iron(III), *Inorg. Chim. Acta, Bioinorg. Chem.*

Schwyn, B. (1983), "Die Hydrolyse von Eisen(III). β-Eisen-Oxidhydroxid: Von der Keimbildung zur Koagulation," dissertation, ETH, Zürich, No. 7404.

Sigg, L. (1985), "Metal Transfer Mechanisms in Lakes; The Role of Settling Particles," in W. Stumm, Eds., *Chemical Processes in Lakes,* Wiley, New York, p. 283.

Steenwijk, F. van, Hametner, D., and Schneider, W. (1978), Mössbauer studies on intermediates in the formation of iron hydroxide, *Proc. Workshop Chem. Appl. Mössbauer Spectry., Seeheim FRG,* p. 17.

Stumm, W., and Morgan, J. J. (1981), *Aquatic Chemistry,* 2nd ed., Wiley, New York, pp. 238, 331, 418.

Sung, W., and Morgan, J. J. (1980), Kinetics and products of ferrous ion oxygenation in aqueous systems, *Environ. Sci. Technol.* **14,** 561.

Szytula, A., Malanda, M., and Dimitrijevič, Z. (1970), Neutron diffraction studies of β-FeO(OH), *Phys. Stat. Sol.* **3,** 1033.

Theil, E. C. (1983), "Ferritin: Structure, Function, and Regulation," in E. C. Theil, G. L. Eichhorn, and L. G. Marzilli, Eds., *Advances in Inorganic Biochemistry,* Vol. 5, Elsevier, New York, p. 1.

Wedepohl, K. H. Ed., (1969), *Handbook of Geochemistry,* Vol. II-3, Springer, New York, p. 26-D-1.

Weiser, H. B., Milligan, W. O., and Cook, E. L. (1946), Beta iron(III) oxide 1-hydrate, *Inorg. Synthesis* **2,** 215.

Wells, A. F. (1984), *Structural Inorganic Chemistry,* 5th ed., Clarendon, Oxford, p. 626.

Wendt, H. (1962), Die Kinetik der Bildung des binuklearen Eisen(III)-hydroxo-komplexes $Fe_2(OH)_2^{4+}$, *Ber. Bunsenges. Phys. Chem.* **66,** 235.

Westall, J. (1987), "Adsorption Mechanisms in Aquatic Surface Chemistry," this volume, Chapter 1.

8

THE DISSOLUTION OF OXIDES AND ALUMINUM SILICATES; EXAMPLES OF SURFACE-COORDINATION-CONTROLLED KINETICS

Werner Stumm and Gerhard Furrer

Institute for Water Resources and Water Pollution Control (EAWAG), Zurich, Switzerland; Swiss Federal Institute of Technology (ETH), Zurich, Switzerland

Abstract

The weathering of rocks and the formation of soils as well as the processes in the formation, alteration, and dissolution of sediments are ruled by surface reactions at the interface between minerals and water. Most of the dissolution reactions are critically dependent on the coordinative interactions taking place on these surfaces, above all on the interaction of hydrous oxide surfaces with H^+, OH^- *(surface protonation), and suitable ligands (anions and weak acids).*

Surface processes rather than transport processes are typically the rate-controlling steps in the dissolution of most hydrous oxides and aluminum silicates; thus, the reaction rates simply depend on the concentration of surface species. Linear dissolution kinetics are observed if the surface protonation and surface ligand concentration remain constant. This indicates that the detachment of the surface metal species is rate-determining. Surface chelates, especially those present as five- and six-membered rings, for example, with oxalate or salicylate, are most efficient in enhancing the dissolution reaction.

1. INTRODUCTION

Almost all aspects of an understanding of the rate processes that control the composition of our environment concern interfaces. Oxides, especially those of Si, Al, Fe, and Mn, are abundant components of the earth's crust. The rates of processes occurring at the hydrous oxide surfaces, such as precipitation (heterogeneous nucleation) of minerals and dissolution of mineral phases, are of importance in the weathering of rocks, in the formation of soils and sediments, and in the corrosion of metals and its inhibition by passive films. Many oxidation–reduction reactions are critically dependent on the reactions occurring on these surfaces. Biota is also very important to most weathering processes, because living matter produces CO_2 as a result of respiration, leading to acidification of surface waters. Organic solutes can (1) interact as complex-forming ligands with the solid surface and the metal ions in solution and (2) serve as reducing agents for reducible oxides and minerals. Both effects enhance the extent and rate of rock dissolution (Furrer and Stumm, 1983).

It is the objective of this chapter to explain with the help of steady-state assumptions the effect of surface coordination on the kinetics of the dissolution of oxides and to extend our considerations to the weathering of aluminum silicates and the effect of biota on the degradation of minerals.

To facilitate the reader's understanding, we will start with very simple models and proceed in a deductive way from the simple to the more complex. Then we will examine the extent to which experimental findings are compatible with the theoretical reasoning.

2. SURFACE-CONTROLLED DISSOLUTION

2.1. The Dissolution Process

The dissolution reaction of a metal oxide implies that the coordinative environment of the metal has changed. For example, in the case of BeO, Be^{2+} in the crystal lattice exchanges its O^{2-} ligands for H_2O or another ligand L, according to the following scheme (Valverde and Wagner, 1976):

$$Be^{2+}(\text{crystal}) + H_2O \rightleftharpoons Be(aq)^{2+} \tag{8.1a}$$

$$Be^{2+}(\text{crystal}) + H_2O + L^{n-} \rightleftharpoons BeL(aq)^{(2-n)^+} \tag{8.1b}$$

$$O^{2-}(\text{crystal}) + 2H^+(aq) \rightleftharpoons H_2O \tag{8.1c}$$

Each one of these partial reactions may consist of smaller reaction steps.

The most important aqueous reactants participating in the dissolution of a solid mineral are H_2O, H^+, OH^-, complex-building ligands, and, in the case of reducible or oxidizable minerals, reductants or oxidants.

However, the dissolution of a solid is a sum of chemical and physical reaction steps. If the chemical reactions at the surface are slow in comparison with the diffusion processes, the dissolution kinetics are controlled by the chemical surface processes, and thus transport of the reactants from the bulk solution to the surface and of products from the surface into the solution can be neglected (Berner and Holdren 1979). The main chemical surface reactions are (1) the attachment of the reactants at the surface sites, where they polarize and weaken the critical metal–oxygen bonds, and (2) the rate-limiting detachment of the surface metal species (Fig. 8.2*a*). When initial sites are regenerated completely after the detachment step by fast protonation reactions, and provided that the concentrations of the reactants are kept constant, steady-state conditions with regard to the oxide surface species are established (Table 8.1). If, furthermore, the system is far from dissolution equilibrium, the back reaction can be neglected and constant dissolution rates occur and can be written as:

$$\text{Rate} \propto (\text{surface concentration of reactants}) \times (\text{surface density of sites})^n \quad (8.1)$$

where n is the number of required (mobile) reactants to activate the central metal ion of one (immobile) surface site.

Table 8.1. Model Assumptions

1. Dissolution of slightly soluble hydroxo oxides.
   Surface process is rate-controlling.
   Back reactions can be neglected if far from equilibrium.
2. The hydrous oxide surface, as a first approximation, is treated like a crosslinked polyhydroxo-oxo acid.
   All functional groups are identical.
3. Steady state of surface phase.
   Constancy of surface area.
   Regeneration of active surface sites.
4. Surface defects, such as steps, kinks, pits, and so on establish surface sites of different activation energy, with different rates of reaction:
   $$\text{Active sites} \xrightarrow{\textit{faster}} \text{Me(aq)} \quad (a)$$
   $$\text{Less active sites} \xrightarrow{\textit{slower}} \text{Me(aq)} \quad (b)$$
   Overall rate is given by (a).
   Steady-state condition can be maintained if a constant mole fraction, x_a, of active sites to total (active and less active) sites is maintained, that is, if active sites are continuously regenerated.
5. Precursor of activated complex:
   Metal centers are bound to surface chelate or surrounded by n protonated functional groups.
   $(C_H^s/S) \ll 1$

2.2. Coordination Reactions at the Hydrous Oxide Surface

The surface chemistry of hydrous oxides, the adsorption mechanisms, and the nature of chemical bonding in surface complexes have been discussed and reviewed in other chapters of this book (Westall; Schindler and Stumm; Motschi). We wish to reemphasize a few aspects:

1. When brought into contact with water, the oxide surfaces are transformed into hydrated oxides or hydroxides. The resulting surface functional groups represent enormous facilities for the adsorption of cations and anions (Fig. 8.1). The pH-dependent charge of a hydrous oxide surface results from proton transfers at the surface. In a general way, hydrous oxides can be thought of as inorganic polymers bearing surface functional groups. Their reactivity determines the properties of the phase boundary between the solid surface and the electrolyte.
2. Energies of interactions include electrostatic and chemical contributions. The electrostatic interaction received early attention and has been integrated into an efficient electric double-layer model useful in describing and predicting many phenomena of colloid stability (dispersion) and conditions of coagulation (particle aggregation). However, the selectivity of interactions of hydrous oxide surfaces with many solute species (metal ions and ligands) can be accounted for only by considering specific chemical interactions with solute species.
3. In recent advances in the surface science of heterogeneous reactions occurring at aquatic interfaces, the interactions of hydrous oxide surfaces with H^+, OH^-, cations and anions or weak acids and the concomitant influences on the surface charge have been interpreted as surface coordination reactions; that is, the surface hydroxyl groups behave potentially toward metal ions as ligands, whereas anions are capable of competing with surface hydroxyl groups (Stumm et al., 1970, 1976; Kummert and Stumm, 1980; Sigg and Stumm, 1981). The extent of surface coordination and its pH dependence can be quantified by mass action equations (Fig. 8.1). The concept of the surface complexation model has been extensively documented in recent reviews (Sposito, 1983; Schindler and Stumm, this volume). As we shall discuss, the nature of the bonding has far-reaching implications for the mechanism of interfacial processes and their kinetics.

3. THE LIGAND-PROMOTED DISSOLUTION OF δ-Al_2O_3

Aqueous ligands tend (1) to become adsorbed and form surface complexes with the Al(III) Lewis acid centers of the hydrous oxide surface and (2) to form complexes with the Al(III) in solution. Complex formation in solution increases the solubility. This has no effect on the dissolution rate, however, if the dissolution is surface-controlled. The enhancement of the dissolution

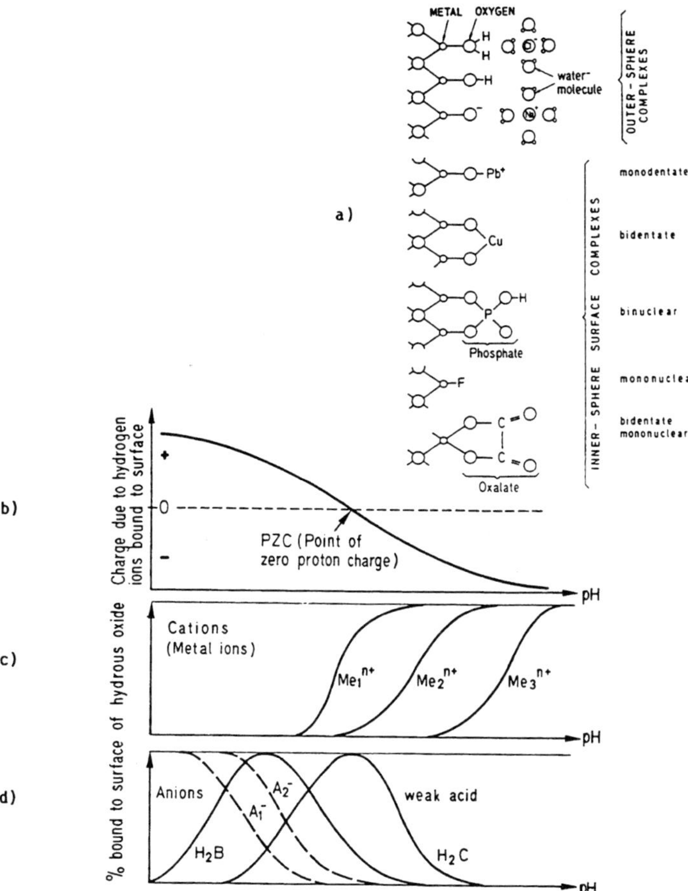

Figure 8.1. (*a*) An oxide surface, covered in the presence of water with amphoteric surface hydroxyl groups, >M-OH, can be looked at as a polymeric oxyacid or base. The surface OH group has a complex-forming O-donor atom that coordinates with H^+ (*b*) and metal ions (*c*). The H^+ bond to the surface can be determined experimentally by alkalimetric titration of an oxide suspension in a given electrolyte solution. The underlying central ion in the surface layer of the oxide acting as a Lewis acid can exchange its structural OH ions against other ligands (anions or weak acids) (*d*). The extent of surface coordination and its pH dependence can be quantified by mass active equations and can be explained by considering the affinity of the surface sites for metal ion or ligand and the pH dependence of the activity of surface sites and ligands. The tendency to form surface complexes may be compared with the tendency to form corresponding solute complexes.

rate by a ligand in a surface-controlled reaction implies that surface complex formation facilitates the release of ions from the surface to the adjacent solution. Ligands form surface complexes by ligand exchange with surface hydroxyl groups. These ligands bring negative charge into the coordination sphere of the surface Al species. This may polarize the critical Al–oxygen bonds, thus facilitating the detachment of the Al from the surface. It has been shown that bidentate ligands (i.e., ligands with two donor atoms) such as dicarboxylates and hydroxycarboxylates can form relatively strong surface chelates, that is, ring-type complexes, such as oxalate:

$$>Al(OH_2)(OH) + C_2O_4^{2-} + H^+ \rightleftharpoons >Al^{\ominus}(O_2C_2O_2) + 2H_2O \qquad (8.2)$$

In Figure 8.2*a,* a simple scheme of reaction steps is proposed. The shorthand representation of a surface site is a simplification that does not take into account either detailed structural aspects of the oxide surface or the oxidation state of the metal ion and its coordination number. It implies (model assumption 2, Table 8.1) that all functional surface groups—such as those in a crosslinked polyhydroxo-oxo acid (Stumm and Morgan, 1981)—are identical.

The scheme in Figure 8.2*a* indicates that the ligand—for example, oxalate—is adsorbed very fast in comparison to the dissolution reaction; thus adsorption equilibrium may be assumed. The surface chelate formed is able to loosen the original Al–oxygen bonds of the crystal lattice. The detachment of the oxalato-aluminum species is the slow and rate-determining step. The negative charge of the surface site after detachment is neutralized by two subsequent fast protonation steps.

The scheme of Figure 8.2*a* corresponds to steady-state conditions (Table 8.1). According to this, the rate of the ligand-promoted dissolution, R_L (mol m^{-2} h^{-1}), is proportional to the surface concentration of ligands, C_L^s (mol m^{-2})

$$R_L = \frac{d[\mathrm{Al(III)(aq)}]}{dt} = k_L C_L^s \qquad (8.3)$$

where k_L is the reaction constant. As shown in Figure 8.2*b,* the experimental results are in accord with Eq. 8.3.

However, we have to reflect on one of our model assumptions. It is certainly not justified to assume a completely uniform oxide surface; we have to consider a more detailed surface configuration. Some of the pertinent features of the surface are given in Figure 10.2 (Blum and Lasaga, this volume, Chapter 10). There are surface defects such as steps, kinks, pits, and corners present. The dissolution is favored at a few localized (active) sites where the reactions have lower activation energy. The overall reaction rate is the sum of the rates of the various types of sites. The reactions occurring at differently active sites are parallel reaction steps occurring at different rates (Table 8.1).

In parallel reactions, the fast reaction is rate-determining (Lasaga, 1981a). We assume that the ratio (mole fraction, x_a) of active sites to total (active +

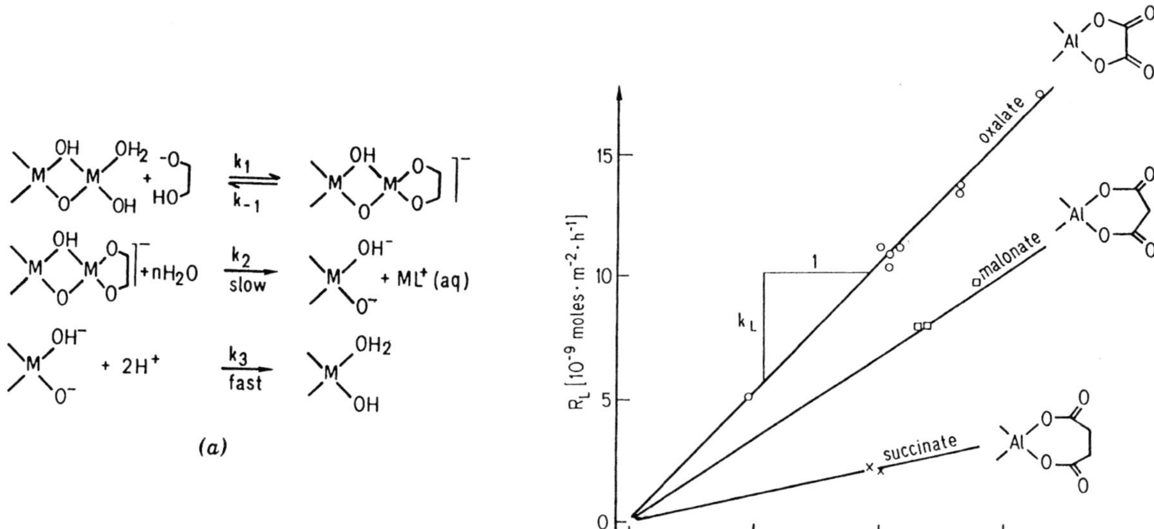

Figure 8.2. (*a*) The ligand-catalyzed dissolution reaction of a M_2O_3 can be described by three elementary steps: A fast ligand adsorption equilibria, a slow detachment process, and a fast protonation after the rate-determining step. (*b*) In accordance with the reaction scheme of (*a*) the rate of ligand-catalyzed dissolution of δ-Al_2O_3 by the aliphatic ligands oxalate, malonate, citrate, and succinate, R_L (nmol m^{-2} h^{-1}), can be interpreted as a linear dependence on the surface concentrations of chelate complexes, C_L^s.

less active) sites remains constant during the dissolution, that is, the active sites are continuously regenerated after Al(III) detachment, and thus steady-state conditions are maintained. The reaction constant k_L in Eq. 8.3 includes x_a, which is a function of the solid and the pretreatment technique.

3.1. Structure-Specific Effects

Figure 8.2*b* illustrates that different ring structures of the surface complex are of different effectiveness in promoting the dissolution. The sequence of k_L for the aliphatic ligands is

$$k_{\text{oxalate}} > k_{\text{malonate}} > k_{\text{succinate}} \tag{8.4}$$

Oxalate, which forms a five-membered chelate ring, is most efficient in promoting the dissolution; malonate, which forms a six-membered ring, and succinate, which forms a seven-membered ring, are progressively less efficient. Similar results were obtained for the aromatic ligands (Furrer and Stumm, 1986):

$$k_{\text{salicylate}} > k_{\text{phthalate}} > k_{\text{benzoate}} \tag{8.5}$$

The catalytic effect of salicylate, which forms a six-membered chelate ring, is approximately four times that of phthalate, which forms a seven-membered ring. Benzoate, although strongly bound as a monodentate surface complex to the Al_2O_3 surface, has no significant catalytic effect on the dissolution.

The results are also of interest if one considers that the rate constants in Eqs. 8.4 and 8.5 have the same sequence as the corresponding stability constants with Al^{3+} in solution, which, in turn, most likely have (see Schindler and Stumm, this volume, chapter 4) the same sequence as the corresponding surface complex formation constants. The promoting effect of salicylate is approximately twice that of malonate, although both ligands form six-membered chelate rings.

Dissolution kinetics depends on the concentration of the individual surface species. The concentration of the specific surface chelate may in some instances be different from the analytically determined total ligand surface concentration. For example, the salicylate surface complex becomes protonated and the ring opens at pH < 5 (see Eq. 8.6). This is accompanied by the

$$\text{Al}\langle\text{O}{-}\text{C}_6\text{H}_4{-}\text{C}({=}\text{O}){-}\text{O}\rangle^{\ominus} + H^+ + H_2O \rightleftharpoons \text{Al}(H_2O)(\text{O}{-}\text{C}({=}\text{O}){-}\text{C}_6\text{H}_4{-}\text{OH}) \tag{8.6}$$

loss of the reaction-promoting effect of surface-bound salicylate. This corresponds to the findings that monodentate coordinated organic ligands such as benzoate have little or no catalytic effect. Thus Eq. 8.3 may be rewritten more exactly by replacing the total (analytically determinable) surface concentration of ligands, C_L^s, by the specific surface concentration of chelate species, $\{M{<}L\}$:

$$R_L = k_L\{M{<}L\} \tag{8.7}$$

4. THE PROTON-PROMOTED DISSOLUTION OF HYDROUS OXIDES

4.1. Nonreductive Dissolution

A simplified scheme for the subsequent steps of the dissolution reaction is given in Figure 8.3*a*. The proton-promoted dissolution is in a certain sense a different mechanism from the ligand-promoted reaction. While the ligands as Lewis bases attack the surface metal centers, which function as Lewis acids, the protons get bound to the surface hydroxyl groups or to the oxide ions closest to the surface of the crystal lattice. At this position, the protons polarize the critical bonds between the oxide and metal ions. The extent of surface protonation can be analytically determined, for example, by alkalimetric titration of the hydrous oxide. The adsorption (binding) of protons onto the surface (steps 1–3) is usually very fast (Hachiya et al., 1984); thus the extent of surface protonation can be assumed to be in equilibrium with the solution. The detachment of the metal ion from the surface (step 4) is the slowest of the consecutive steps. Therefore, the rate of the proton-promoted dissolution, R_H (mol m^{-2} h^{-1}), is proportional to the concentration (or activity) of the surface species D (mol m^{-2})

$$R_H = \frac{d[\mathrm{M(III)(aq)}]}{dt} \propto \{D\} \tag{8.8}$$

4.1.1. How Can We Express {D}? According to the assumptions of our simple model, (Table 8.1), the protons at the solid surface are distributed at random. We may consider the surface a checkerboard and ask such questions as: How does the probability of finding 1, 2, 3, or 4 checkers (excess protons), respectively, in the immediate neighborhood of a black square (metal center) depend on the density of checkers (number of checkers per number of white squares; this corresponds to the fraction of protonated functional groups, Θ_H)? The answer is available from probability theory. At relatively small Θ_H

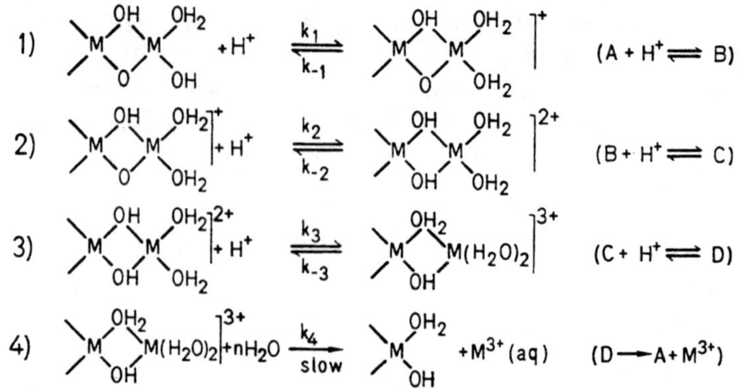

(a)

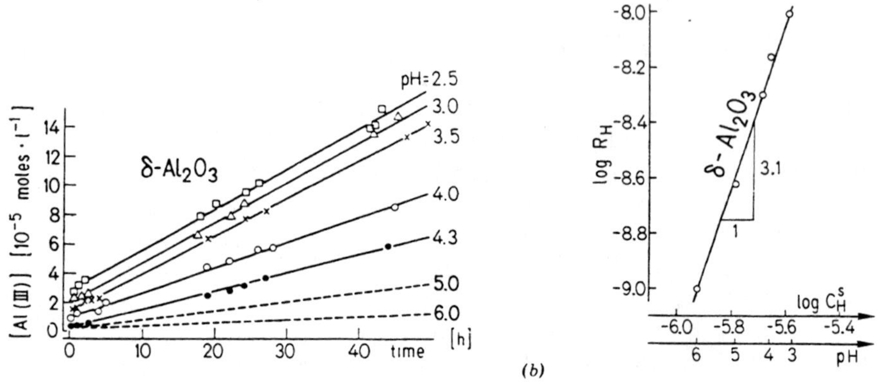

(b)

Figure 8.3. The proton-catalyzed dissolution process at a M_2O_3 surface site. The elementary steps of a reaction mechanism with three preceding protonations of the reactive site (*a*) predict a dependence of the rate of the proton-catalyzed dissolution of δ-Al_2O_3, R_H (mol m^{-2} h^{-1}), on the surface concentration of protons, C_H^s (mol m^{-2}) as shown in (*b*).

values, the fraction of metal centers Θ_n with n nearest-neighbor protonated functional groups* is

$$\Theta_n \approx f_n \Theta_H^n \tag{8.9}$$

where f_n is the appropriate proportionality factor.

By definition, the fraction of the protonated functional groups, Θ_H, and the fraction of surface metal centers with three nearest-neighbor protonated functional groups, Θ_3, are

$$\Theta_H = \frac{C_H^s}{S} \quad \text{and} \quad \Theta_3 = \frac{\{D\}}{S} \tag{8.10}$$

*The authors are indebted to Bernhard Wehrli and Erich Wieland of our laboratory for theoretical elaborations and Monte Carlo simulations (Wehrli and Wieland, 1986).

where C_H^s and S are the surface proton concentration (mol m^{-2}) and the surface density of total crystallographic sites (mol m^{-2}), respectively. Thus the concentration of D is given by

$$\{D\} \approx f_3(C_H^s)^3 \tag{8.11}$$

where f_3 is the proportionality factor.

Combining Eqs. 8.8 and 8.11, we obtain the rate of the proton-promoted dissolution for a metal oxide in the case where D is the precursor of the detachable metal species:

$$R_H = \frac{d[\text{Me(III)aq}]}{dt} = k_H(C_H^s)^3 \tag{8.12}$$

where the reaction constant k_H includes the proportionality factor f_3.

This equation is corroborated experimentally by the dissolution kinetics of δ-Al_2O_3 (Fig. 8.3*b*). If the dissolution mechanism required two preceding surface protonation steps, the rate would be proportional to $(C_H^s)^2$. Experimentally, we found $n = 3.25$ for a α-FeOOH (Zinder et al., 1986) and $n = 2.1$ for BeO (Furrer and Stumm, 1986). More generally, the following rate equation may be postulated:

$$R_H = k_H(C_H^s)^n \tag{8.13}$$

where n is an integer (corresponding to the number of protonation steps prior to detachment) if the dissolution occurs by only one mechanism.

We should remind ourselves that in Eqs. 8.12 and 8.13, k_H includes, as we discussed for Eq. 8.3, a term x_a (mole fraction of active sites in total sites) in order to account for the fact that the dissolution is favored at a relatively small fraction of localized sites that have a lower activation energy.

4.1.2. The Combination of Ligands and Protons. If proton- and ligand-promoted dissolution are independent parallel reactions, ligand promotion becomes superimposed on proton promotion, that is, as has been found experimentally by Furrer and Stumm (1986), the total dissolution rate, R_{tot} (mol m^{-2} h^{-1}) is composed of two or more additive rates:

$$R_{tot} = R_H + R_L \tag{8.14}$$

In the presence of different ligands, a competition between them may result at high ligand concentrations. For example, a very reactive (dissolution-promoting) ligand such as oxalate may become replaced by a more strongly adsorbed but less active ligand such as phosphate or benzoate, in which case the dissolution rate would decrease. This indicates that the specific adsorption of suitable ligands to an oxide film on a metal surface has a critical influence on its passivity behavior. Furthermore, under alkaline conditions, the deprotonated hydroxyl groups can also act as ligands, facilitating the detachment of a metal center and enhancing the dissolution rate.

4.2. Reductive Dissolution

While the rate of proton-promoted dissolution of α-FeOOH under nonreducing conditions is small (10^{-9} mol m^{-2} h^{-1} at pH 3), reducing conditions (in the absence of light) such as those established in the presence of 10^{-3} *M* ascorbate, lead to an acceleration by almost two orders of magnitude (Zinder et al., 1986). A conceivable mechanism assumes that prior to the detachment step the reduction of the metal ion and the protonation of the nearest-neighbor oxide or hydroxide ions must take place.

After it has been transferred to the oxide surface, the electron will move to the metal ion which then becomes the center of the detachable group. After reduction of the central metal ion, the existing coordination sphere of the metal center will be labilized. If the redox potential of the system is sufficiently negative, the general rate equation postulated for a proton-promoted reductive dissolution at a constant redox potential (e.g., 10^{-3} *M* ascorbate) is

$$R_{e,\mathrm{H}} = K_{e,\mathrm{H}}\,(C^s_{\mathrm{H}})^n \tag{8.15}$$

For α-FeOOH, Zinder et al. (1986) have found that $n = 2.9$.

As under nonreducing conditions, chelating ligands (e.g., oxalate) markedly enhance the reductive dissolution. The electron transfer to the surface Fe(III) center and one protonation step precede the detachment, which is noticeably facilitated by surface coordination with bidentate ligands. For a given redox potential, this mechanism leads to a rate law such as

$$R_{e,\mathrm{L}} = k_{e,\mathrm{L}} C^s_{\mathrm{L}} C^s_{\mathrm{H}} \tag{8.16}$$

where $k_{e,\mathrm{L}}$ is the rate constant.

5. KINETIC THEORY

5.1. Promotion or Catalysis

Figure 8.4 illustrates that the surface-bound protons are mobile reactants; surface protonation occurs at random, but the protons may move from one functional group to another and from one surface site to another. Thus two protons may become bound to two functional groups immediately adjacent to one surface metal center (configuration 3). Forward and backward reactions

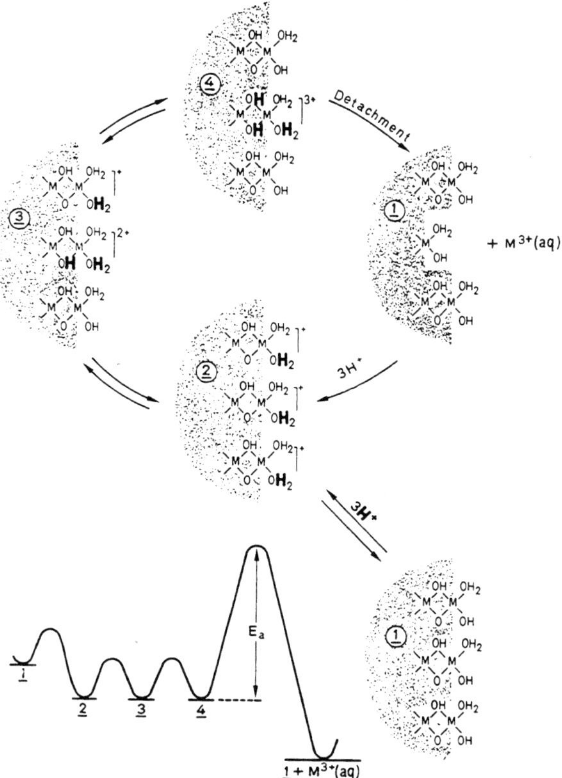

Figure 8.4. The dissolution reaction is illustrated as a catalytic cycle. The migration of surface-bound protons can be interpreted as tautomeric reactions.

characterized by tautometric equilibria may be assumed. Surface species that are characterized by three protonated functional groups on the same metal center (configuration 4 in Fig. 8.4 or species D in Fig. 8.3*b*) can be viewed as the precursors of an activated complex.

The mechanism postulated for the reductive dissolution illustrates that a suitable hydrous oxide surface can act as an efficient catalyst for redox processes. In this context *surface photochemistry* is of great interest; advances have been made in understanding photoassisted reactions. In the presence of light and in the presence of surface complex–forming ligands, oxides or hydroxides of iron(III) and manganese(III,IV) undergo enhanced photoreductive dissolution, presumably because of direct photolysis of inner-sphere surface complexes. Semiconductor surfaces provide special pathways for ligand-promoted photochemical reactions (Faust and Hoffmann, 1986; Waite and Morel, 1984; Sunda et al., 1983; Stone and Morgan, this volume, Chapter 9; Zepp and Wolfe, this volume, Chapter 16).

5.2. Kinetic Dependence on Solute Concentrations; Fractional Reaction Orders

As we have shown, the rate laws of surface-controlled reactions are conveniently expressed in terms of the concentration (activity) of surface *species* (Eq. 8.7). These surface concentrations were experimentally determined in our studies. The relationship between solute concentration and surface concentration is often known and can be characterized by surface coordination equilibrium constants (Schindler and Stumm, this volume, Chapter 4), for example,

$$>\mathrm{MOH} + \mathrm{HL}^{(z-1)-} \rightleftharpoons >\mathrm{ML}^{(z-1)-} + \mathrm{H_2O}$$

$$^{*1}K_1^s = \frac{\{>\mathrm{ML}^{(z-1)-}\}}{\{>\mathrm{MOH}\}[\mathrm{HL}^{(z-1)-}]} \tag{8.17}$$

$$>\mathrm{MOH} + \mathrm{L}^{z-} \rightleftharpoons >\mathrm{ML}^{(z-1)-} + \mathrm{OH}^-$$

$$^{1}K_1^s = \frac{\{>\mathrm{ML}^{(z-1)-}\}[\mathrm{OH}^-]}{\{>\mathrm{MOH}\}[\mathrm{L}^{z-}]} \tag{8.18}$$

$$>\mathrm{MOH_2^+} \rightleftharpoons >\mathrm{MOH} + \mathrm{H}^+$$

$$K_{a1}^s = \frac{\{>\mathrm{MOH}\}[\mathrm{H}^+]}{\{>\mathrm{MOH_2^+}\}} \tag{8.19}$$

Thus, the rate law can also be expressed in terms of the concentration of solutes. As Figure 8.5 illustrates schematically, the surface concentrations of L and H, C_L^s and C_H^s, are proportional to the solute concentrations $[L]^n$ and $[H^+]^m$ with the exponents n, $m < 1$. Therefore, fractional reaction order dependence for dissolution rates is compatible with, or even indicative of, surface-controlled reactions (Grauer and Stumm, 1982).

6. WEATHERING

6.1. Chemical Weathering of Primary Minerals

Chemical weathering, the dissolution of minerals by the action of water and its solutes, is an important feature of the global hydrogeochemical cycle of elements. Rocks and primary minerals become transformed to solutes and soils and eventually to sediments and sedimentary rocks; these processes produce the nutrients for the biota, and the biota in turn influences the weathering processes. Table 8.2 gives chemical dissolution equations for some minerals commonly found in soils; the reactions are listed roughly in decreasing order of dissolution rate. In each reaction, cations Ca^{2+}, Na^+, K^+, and Mg^{2+} are released and alkalinity is produced via OH^- or HCO_3^- production or H^+

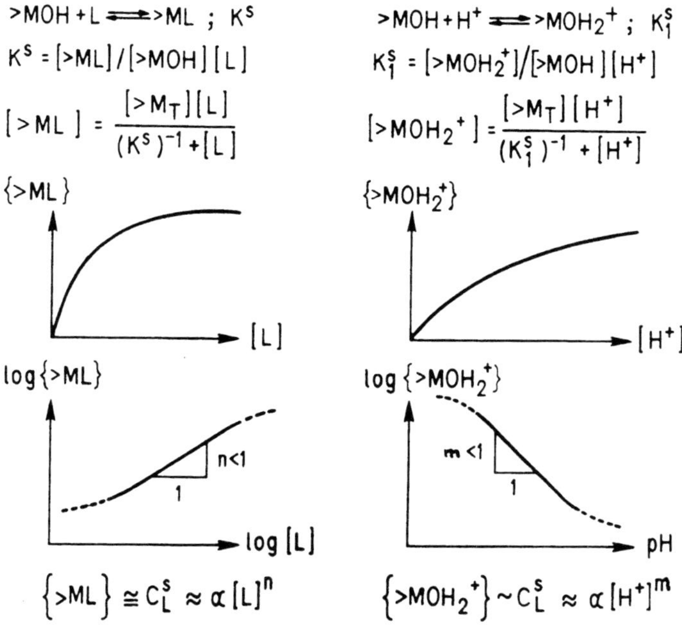

Figure 8.5. Surface complex formation equilibria can be plotted like Langmuir adsorption equations in a double logarithmic plot. A significant part of the curve will be linear, having slopes of $n < 1$ and $m > 1$. Thus a fractional reaction order dependence is often indicative of a reaction rate that is proportional to the concentration of a surface species.

consumption. Many weathering reactions are incongruent, i.e. the dissolution is accompanied (or followed) by the formation (precipitation) of a new phase.

Primary minerals in soils originate from bedrock by mechanical weathering (physical breaking of rocks). They include essentially feldspars (usually most abundant), ferromagnesium silicates, pyroxenes, amphiboles, and quartz. The residues and precipitates of primary minerals—the secondary minerals—include clays, most oxides and hydroxides, and a variety of sulfate and carbonate minerals. It is likely that human activity has increased the overall erosion rate. Of great concern is the increase in soil erosion.

Originally many studies on the dissolution of aluminum silicates carried out in the laboratory reported parabolic dissolution rates (linear functions of the square root of time), a result consistent with diffusional transport-limited kinetics. The works of Berner (Berner and Holdren, 1979) and others (see Blum and Lasaga, this volume, Chapter 10, and Petit and Schott, Chapter 11) have suggested that parabolic dissolution kinetics are an artifact created by surface heterogeneity induced by mineral grinding. Disrupted grain surfaces and ultrafine particles dissolve initially at a high rate. Dissolution rates are actually linear over time. An initial incongruent dissolution may last over a few hours in which a thin cation-depleted layer (2–17 Å) is assumed to exist

Table 8.2. Some Important Dissolution Reactions in Approximate Rank Order for Ease of Weathering[a]

Reaction
$CaCO_3(s) + H_2CO_3 \longrightarrow Ca^{2+} + 2HCO_3^-$ Calcite
$CaMg(CO_3)_2(s) + 2H_2CO_3 \longrightarrow Ca^{+} + Mg^{2+} + 4HCO_3^-$ Dolomite
$FeMgSiO_4(s) + 4H^+ \longrightarrow Fe^{2+} + Mg^{2+} + H_4SiO_4$ Olivine
$CaAl_2Si_2O_8(s) + 2H^+ + H_2O \longrightarrow Ca^{2+} + Al_2Si_2O_5(OH)_4(s)$ Anorthite (Ca feldspar) → kaolinite
$NaAlSi_3O_8(s) + H^+ + \frac{9}{2}H_2O \longrightarrow Na^+ + 2H_4SiO_4 + \frac{1}{2}Al_2Si_2O_5(OH)_4(s)$ Albite (Na feldspar) → kaolinite
$KAlSi_3O_8(s) + H^+ + \frac{9}{2}H_2O \longrightarrow K^+ + 2H_4SiO_4 + \frac{1}{2}Al_2Si_2O_5(OH)_4(s)$ Orthoclase (K feldspar) → kaolinite
$Al_2O_3 \cdot 3H_2O(s) + 6H^+ \longrightarrow 2\,Al^{3+} + 6H_2O$ Gibbsite
$Fe_2O_3 + 6H^+ \longrightarrow 2Fe^{3+} + 3H_2O$ Hematite
$SiO_2(s) + 2H_2O \longrightarrow H_4SiO_4$ Quartz

[a] The reactions could be written with water and dissolved carbon dioxide ($H_2CO_3^*$) or hydrogen ions as reactants; the net result of the reaction is the same. The reactions shown are not elementary reactions; they proceed through a series of reaction steps.
Source: Modified from Schnoor and Stumm (1986).

on the surface of the crystal (Chou and Wollast, 1985). Basic cations are depleted preferentially to Si and Al. This period is followed by a relatively long period of subsequent linear dissolution kinetics.

Table 8.3 gives the reaction orders of the rate laws for some representative minerals: The fractional order dependence on bulk-phase hydrogen ion concentration is indicative of surface-controlled dissolution. The concepts given here for surface-controlled dissolution of hydrous oxides can be extended to the weathering of kaolinite (Wieland and Stumm, 1987; Wehrli and Wieland, 1986). Hydrogen ions and ligands may associate with the octahedral (Al-bearing) or tetrahedral (Si-bearing) sheets. Wieland and Stumm (1986) have found that oxalate and salicylate accelerate this dissolution reaction of ka-

Table 8.3. Reaction Order for the Rate Law for the Dissolution of Minerals

Mineral	Formula	Solution	Reaction Order	Ref.[a]
Dolomite	$(Ca,Mg)CO_3$	HCl	$\{H^+\}^{0.5}$	(1)
Bronzite	$(Mg,Ca)SiO_3$	HCl	$\{H^+\}^{0.5}$	(2)
Enstatite	$MgSiO_3$	HCl	$\{H^+\}^{0.8}$	(3)
Diopside	$CaMgSi_2O_6$	HCl	$\{H^+\}^{0.7}$	(3)
K feldspar	$KAlSi_3O_8$	Buffer	$\{H^+\}^{0.33}$	(4)
Iron hydroxide	$Fe(OH)_3$ gel	Various acids	$\{H^+\}^{0.48}$	(5)
Aluminum oxide	δ-Al_2O_3	HNO_3	$\{H^+\}^{0.4}$	
Beryllium oxide	BeO	HNO_3	$\{H^+\}^{0.17}$	

[a] References:

(1) Busenberg, E., and Plummer, L. (1982), The kinetics of dissolution of dolomite in CO_2–H_2O systems at 1.5 to 65°C and 0 to 1 atm pCO_2, *Amer. J. Sci.* **282,** 45.

(2) Grandstaff, D. (1977), Some kinetics of bronzite orthopyroxene dissolution, *Geochim. Cosmochim. Acta* **41,** 1097.

(3) Schott, J., Berner, R., and Sjöberg, E. (1981), Mechanism of pyroxene and amphibole weathering—I. Experimental studies of iron-free minerals, *Geochim. Cosmochim. Acta* **45,** 2123.

(4) Wollast, R. (1967), Kinetics of alteration of K feldspar in buffered solutions at low temperature, *Geochim. Cosmochim. Acta* **31,** 635.

(5) Furuichi, R., Sato, N., and Okamoto, G. (1969), Kinetics of $Fe(OH)_3$ dissolution, *Chimia* **23,** 455.

Source: Modified from Schnoor and Stumm (1985).

olinite. Their results indicate that these ligands attack the Al centers and facilitate the (slower) detachment of the Al species, which is followed by the (faster) detachment of the silica units.

6.2. Biologically Mediated Weathering

Respiration increases the partial pressure of CO_2 in soil water systems. Furthermore, microorganisms and plants produce a large number of biogenic acids. Oxalic, maleic, acetic, succinic, tartaric, vanillic, ketogluconic, and *p*-hydroxybenzoic acids have been demonstrated in top soils, with oxalic acid the most abundant, occurring in concentrations as high as 10^{-5}–10^{-4} *M*. In addition to complex-forming exudates from algae and plants, these substances are also present in surface waters (Graustein, 1977).

Oxalic and other organic acids can occur in rainwater in concentrations as high as a few micromolar. Many of these organic substances, as we have seen, have a pronounced effect in enhancing the dissolution rates. The complexation in solution tends to raise the concentrations of Al and Fe in solution and to extend the domain of congruent dissolution of minerals; higher concentrations of soluble Al(III) or Fe(III) can be built up before a new phase is formed.

The biogenic solutes aid in the formation of very small particles; surface complex formation of these substances with mineral surfaces imparts negative charge on the surfaces and assists in stabilizing the colloids. Fyfe (this volume, Chapter 18), illustrates colloidal particles with typical crystal sizes down to 100 Å and less; he points out that microorganisms can play a major role in the formation of very fine particulates in aquatic systems, where frequently their walls serve as major complexing agents for the collection of metal ions. On the other hand, the concentration of organic ligands on clays and hydrous oxide particles may have been important in the origination of life.

7. THE ROLE OF CHEMICAL WEATHERING IN THE NEUTRALIZATION OF ACIDS

7.1. Acidic Deposition

Hydrogen ion loading due to acidic deposition occurs in both the terrestrial and aquatic portions of a watershed. Much of the acidity will be neutralized by reactions with soils and bedrocks. In the lake, biogeochemical processes such as the reduction of SO_4^{2-} or NO_3^- can produce alkalinity (Cook et al., 1986). We are here primarily concerned with terrestrial (mineral dissolution) mechanisms of H^+ neutralization.

The worldwide average chemical weathering rate (measured by the rate of hydrogen ion consumption) is estimated to be approximately 3900 eq ha^{-1} yr^{-1}. However, the minimum chemical weathering rate that occurs in small crystalline watersheds is only ca. 200 eq ha^{-1} yr^{-1}. Thus when acid deposition begins to exceed 200 eq ha^{-1} yr^{-1}, one can expect to find some sensitive lakes that will acidify. Such is the case for acid lakes in New England, the Adirondacks of New York, Norway, Sweden, and alpine southern Switzerland.

7.2. Acidification from Biomass Synthesis

It is often not appreciated that the synthesis of terrestrial biomass is accompanied by acidification of the surrounding soil. This acidification results from the fact that plants take up more nutrient cations than anions; thus H^+ ions are released via roots and the acid-neutralizing capacity of the plants increases. The ashes of plants and of wood are alkaline ("potash"). Of course, the production of H^+ ion in photosynthesis is followed by an equivalent H^+ ion consumption when the vegetation is mineralized. But every temporal and spatial imbalance between production and mineralization of biomass (intensive agricultural and seasonal fluctuations) leads to a modification of the H^+ ion balance in the environment (Schnoor and Stumm, 1985). Productivity of crops and forests has been increased by applying fertilizers and by using monocultures. This is another form of eutrophication; the resulting acidification has often caused far-reaching consequences in some soils.

Figure 8.6 illustrates some of the H^+ ion producing and consuming processes in soils with growing vegetation. There is a balance between H^+ ions released from the roots and the H^+ ions consumed by the weathering of minerals. Atmospheric deposition may disturb this delicate balance. Also vegetation produces some organic acids (such as oxalic acids). Decaying humus and litter fall are known to produce organic acids.

In addition to the H^+ ion consumption that occurs in weathering, H^+ may become exchanged on colloidal surfaces. Ion exchange occurs at the surface of clays and organic humus in various soil horizons. The net effect of ion-exchange processes is identical to that observed with chemical weathering;

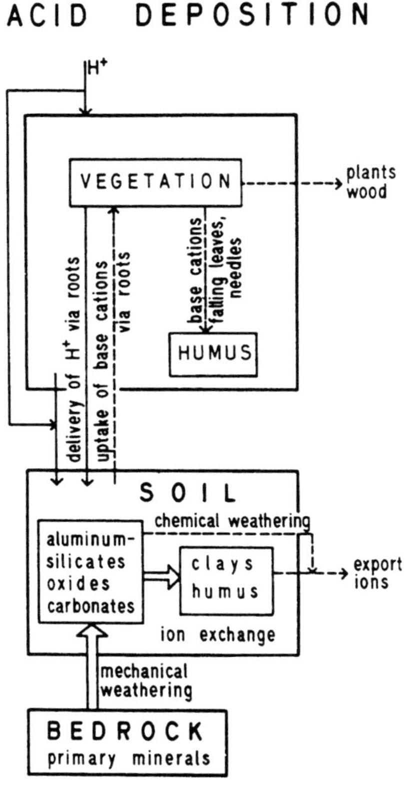

Figure 8.6. Process affecting the acid-neutralizing capacity of soils. H^+ ions from acid precipitation and from the release by vegetation react by weathering carbonates, aluminum silicates, and oxides and by ion exchange on clays and humus. Mechanical weathering resupplies weatherable minerals. The ion exchange surfaces (clays and humus) being produced from the weathering of the primary minerals are a transient reservoir for H^+ ion consumption, but, in the long run, chemical weathering is the rate-limiting step in H^+-ion neutralization and in the supply of base cations to be exported from watersheds. Lines drawn out indicate flux of protons; dashed lines show flux of base cations (alkalinity). The plants of the forest canopy act like a "base pump" whereby base cations taken from lower soil horizons are deposited in a more available form (biological debris, leaves, and needles on the top soil or on the forest floor). (Modified from Zinder and Stumm, 1985, and Schnoor and Stumm, 1986.)

hydrogen ions are consumed, and basic cations (Ca^{2+}, Mg^{2+}, Na^{+}, K^{+}) are released. However, the kinetics of ion exchange are rapid relative to those of chemical weathering (on the order of minutes to hours compared to days). In addition, the pool of exchangeable bases is small compared to the total acid-neutralizing capacity of the soil. Thus, there exist two pools of bases in soils; a small pool of exchangeable bases with relatively rapid kinetics and a large pool of mineral bases with the slow kinetics of chemical weathering. If chemical weathering did not replace exchangeable bases in acid soils of temperate regions receiving acidic deposition, the base exchange capacity would be completely diminished over a period of 50–200 years.

As explained by Schnoor and Stumm (1986), it is important to distinguish between exchange reactions and reactions where the solid or surface phase changes its composition. Both processes modify the composition of the water. The change in alkalinity resulting from weathering and rock alterations is phenomenologically accompanied by a change of ions in solution; but it must not be mistaken for an ion-exchange reaction. The generation of silicic acid in watersheds is one measure of aluminosilicate weathering that is distinctive from ion exchange, although the formation of kaolinite or other clay minerals is a complicating factor which consumes silicic acid and aluminum ions. In

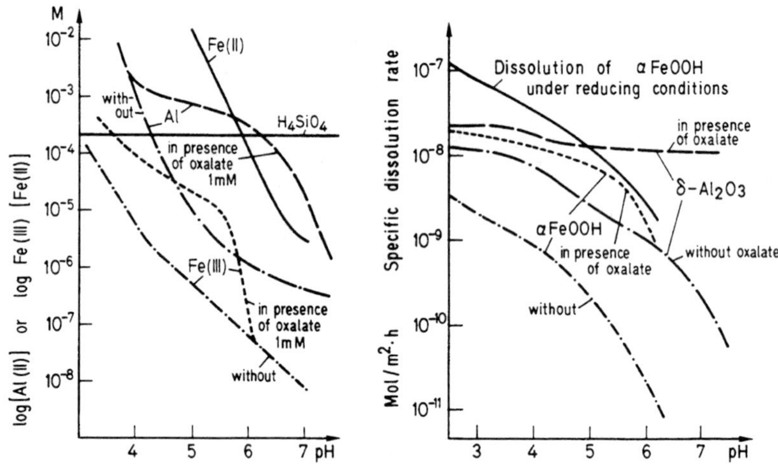

Figure 8.7. Effect of protons and surface complex formers on the dissolution of Al(III), Fe(III), and Si, as it may be of importance in the podzolidation of soils. *Left:* Calculated solubility of Al, Si, and Fe in the presence and absence of oxalate as a function of pH. *Right:* Dissolution rates calculated for different conditions, as a function of pH. For the calculation of the dissolution rate in the presence of oxalate, a constant oxalate surface concentration of 10^{-6} mol m^{-2} was assumed. The calculations should illustrate that the solubility of Al, Si, and Fe in soils is affected differently. In the podzolidation of soils, Al and Fe become dissolved under the influence of low pH and organic complex formers (often released by microorganisms excreted by roots) and displaced into deeper soil horizons. Oxalate serves as a model complex former. (Modified from Zinder and Stumm, 1985.)

the long run, chemical weathering is the rate-limiting step in the supply of basic cations for export from watersheds.

7.3. Podzolization

The effects of H^+ and of complex-forming ligands on the dissolution of minerals are exemplified by the genesis of podzolic acid soils. In a chromatographic effect, bases are depleted from the upper soil profile, and iron and aluminum are mobilized. Soil profiles become stratified into horizons. The organic O horizon is typically a few centimeters deep and contains the humus from leaf and litter decomposition. The A horizon is the uppermost "inorganic" soil horizon. In podzolic soils, the A horizon is depleted of basic cations and appears as a gray-to-white leached layer. The B horizon is classified as "spodic." It appears reddish and contains much of the iron and aluminum that where mobilized from above due to acidic conditions and dissolution of aluminosilicate minerals. Podzolic soils developed over geologic time in response to vegetation and climate in temperate, humid regions. The amount of acid precipitation and vigorously growing vegetation accelerate the leaching of the soil.

As has already been mentioned, organic complex formers, for example, simple organic acids, are formed by microorganisms and released by plants in soils. These complex formers not only increase the solubility of minerals, but are also able to form surface chelates on their hydrous oxide surfaces and thus in turn catalyze the dissolution of oxides and aluminum silicates. The downward vertical displacement of Al and Fe observed in the podsolization of soils can be accounted for by considering the effect of pH and of complex formers on both the solubility equilibria and the dissolution rates. Figure 8.7 illustrates the calculated effects that acidity (pH) and organic complex formers (oxalate is used as an example) exert in enhancing the dissolution of iron(III) and Al(III) oxides.

Acknowledgments

Our research on the surface chemistry of oxide minerals is supported by the Swiss National Science Foundation; it profited from data and ideas given by Bernhard Wehrli, Erich Wieland, and Bettina Zinder and from the influential work on ligand exchange of Robert Kummert and Laura Sigg.

REFERENCES

Berner, R. A., and Holdren, G. R., Jr. (1979), Mechanism of feldspar weathering. II. Observations of feldspars from soils, *Geochim. Cosmochim. Acta* **43,** 1173–1186.

Chou, L., and Wollast, R. (1985), Steady-state kinetics and dissolution mechanisms of albite, *Am. J. Sci.* **285,** 963–993.

Cook, R. B., Kelly, C. A., Schindler, D. W., and Turner, M. A. (1986), Mechanisms of hydrogen ion neutralization in an experimentally acidified lake, *Limnol. Oceanogr.* **31**(1), 134–148.

Faust, B. C., and Hoffmann, M. R. (1986), Photo-induced reductive dissolution of α-Fe_2O_3 by bisulfite, *Environ. Sci. Technol*, **20,** 943–948.

Furrer, G., and Stumm, W. (1983), The role of surface coordination in the dissolution of δ-Al_2O_3 in dilute acids, *Chimia* **37,** 338–341.

Furrer, G., and Stumm, W. (1986), The coordination chemistry of weathering: I. Dissolution kinetics of δ-Al_2O_3 and BeO, *Geochim. Cosmochim. Acta,* **50,** 1847–1860.

Grauer, R., and Stumm, W. (1982), Die Koordinationschemie oxidischer Grenzflächen und ihre Auswirkung auf die Auflösungskinetik oxidischer Festphasen in wässrigen Lösungen, *Colloid Polymer Sci.* **260,** 959–970.

Graustein, W. C. (1977), Calcium oxalate: Occurrence in soils and effect on nutrient and geochemical cycles, *Science* **198,** 1252–1254.

Hachiya, K., Sasaki, M., Saruta, Y., Mikami, N., and Yasunaga, T. (1984), Static and kinetic studies of adsorption–desorption of metal ions on γ-Al_2O_3 surface, *J. Phys. Chem.* **88,** 23–31.

Kummert, R., and Stumm, W. (1980), The surface complexation of organic acids on hydrous γ-Al_2O_3, *J. Colloid Interface Sci.* **75,** 373–385.

Lasaga, A. C. (1981a), "Rate Laws of Chemical Reactions," in A. C. Lasaga and R. J. Kirkpatrick, Eds., *Kinetics of Geochemical Processes* (*Rev. Mineral.,* Vol. 8), Mineralogical Society of America, Washington, DC, pp. 1–68.

Lasaga, A. C. (1981b), "Transition State Theory," in *Kinetics of Geochemical Processes* (*Rev. Mineral.,* Vol. 8), A. C. Lasaga and R. J. Kirkpatrick, Eds., Mineralogical Society of America, Washington, DC, pp. 135–169.

Schnoor, J. L., and Stumm, W. (1985), "Acidification of Aquatic and Terrestrial Systems," in W. Stumm, Ed., *Chemical Processes in Lakes,* Wiley-Interscience, New York.

Schnoor, J. L., and Stumm, W. (1986), The role of chemical weathering in the neutralization of acidic deposition; *Swiss J. Hydrology*, **48,** 171–195.

Sigg, L., and Stumm, W. (1981), The interactions of anions and weak acids with the hydrous goethite (α-FeOOH) surface, *Colloids Surfaces* **2,** 101–117.

Sposito, G. (1983), On the surface complexation model of the oxide–aqueous solution interface, *J. Colloid Interface Sci.* **91,** 329–340.

Stumm, W., Furrer, G., and Kunz, B. (1983), The role of surface coordination in precipitation and dissolution of mineral phases, *Croat. Chem. Acta* **56,** 593–611.

Stumm, W., Hohl, H., and Dalang, F. (1976), Interaction of metal ions with hydrous oxide surfaces, *Croat. Chim. Acta* **48,** 491–504.

Stumm, W., Huang, C. P., and Jenkins, S. R. (1970), Specific chemical interaction affecting the stability of dispersed systems, *Croat. Chem. Acta* **42,** 223–245.

Stumm, W., and Morgan, J. J. (1981), *Aquatic Chemistry,* 2nd ed., Wiley, New York, 780 pp.

Sunda, W. G., Huntsman, S. A., and Harvey, G. R. (1983), Photoreduction of manganese oxides in seawater. Geochemical and biological implications, *Nature* **301,** 234–236.

Valverde, N., and Wagner, C. (1976), Considerations on the kinetics and the mechanism of the dissolution of metal oxides in acidic solutions, *Ber. Bunsenges. physik. Chem.* **80,** 330–333.

Waite, T. D., and Morel, F. M. M. (1984), Photoreductive dissolution of colloidal iron oxides in natural waters, *Environ. Sci. Technol.* **18,** 860–868.

Wehrli, B., and Wieland, E. (1986), The chessboard–A model for two-dimensional distributions of interface species, *EAWAG News* **20/21,** 9–12.

Wieland, E., and Stumm, W. (1987), Dissolution of kaolinite.

Wollast, R. (1967), Kinetics of the alteration of K-feldspar in buffered solutions at low temperature, *Geochim. Cosmochim. Acta* **31,** 635–648.

Zinder, B., Furrer, G. and Stumm, W. (1986), A coordination chemical approach to the kinetics of weathering. II. Dissolution of Fe(III) oxides, *Geochim. Cosmochim. Acta* **50,** 1861–1869.

Zinder, B., and Stumm, W. (1985), Die Auflösung von Eisen(III)-Oxiden; ihre Bedeutung im See und im Boden, *Chimia* **39**(5), 280–288.

9

REDUCTIVE DISSOLUTION OF METAL OXIDES

Alan T. Stone

Department of Geography and Environmental Engineering, The Johns Hopkins University, Baltimore, Maryland

and

James J. Morgan

W. M. Keck Laboratories of Environmental Engineering Science, California Institute of Technology, Pasadena, California

Abstract

Changes in oxidation state have a dramatic impact on the solubility of transition metals. In particular, oxide/hydroxide minerals of iron, manganese, and several other important transition metals are more soluble and dissolve more quickly when surface metal centers are reduced by naturally occurring reductants. This chapter explores a model for the reductive dissolution of metal (hydr)oxide minerals that involves the following surface chemical reactions: (1) precursor complex formation, (2) electron transfer, (3) release of oxidized organic product, and (4) release of reduced metal ion. Surface speciation and rates of surface chemical reactions are intimately associated; their interdynamics controls rates of metal ion release from mineral surfaces. The goal of this discussion is to provide a basis for interpreting experimental results and for predicting rates of reductive dissolution under unexplored conditions.

1. INTRODUCTION

Changes in oxidation state have a dramatic impact on the solubility of transition metals. Reduction of Fe(III) to Fe(II), for example, increases iron solubility with respect to oxide/hydroxide phases by as much as eight orders of magnitude at pH 7 (Stumm and Morgan, 1981). Rates of dissolution reactions can be either transport-controlled or controlled by surface chemical reactions. In either situation, dissolution rates are often strongly influenced by metal oxidation state. Oxidative and reductive dissolution reactions have a major impact on the speciation and transport of metals in the environment and on the stability of oxides and oxide films in engineered systems.

A considerable amount of empirical information is available concerning the effects of mineral structure, reductant reactivity, and medium composition on rates of reductive dissolution reactions. Much recent effort has been directed toward developing a conceptual understanding of this complex class of chemical reactions (Valverde and Wagner, 1976; Gorichev and Kipriyanov, 1984; Segal and Sellers, 1984; Bruyere and Blesa, 1985; Stone, 1986; Waite,

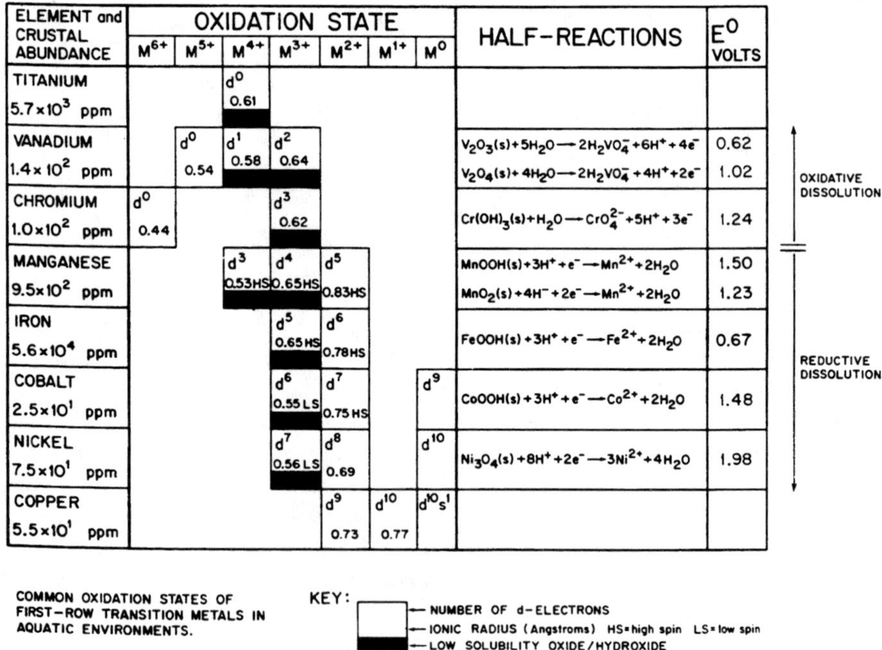

ELEMENT and CRUSTAL ABUNDANCE
OXIDATION STATE
M6+ M5+ M4+ M3+ M2+ M1+ M0
HALF-REACTIONS
E0 VOLTS
TITANIUM 5.7×10³ ppm
d0 0.61
VANADIUM 1.4×10² ppm
d0 0.54
d1 0.58
d2 0.64
V2O3(s)+5H2O—2H2VO4−+6H++4e− 0.62
V2O4(s)+4H2O—2H2VO4−+4H++2e− 1.02
CHROMIUM 1.0×10² ppm
d0 0.44
d3 0.62
Cr(OH)3(s)+H2O—CrO4 2−+5H++3e− 1.24
MANGANESE 9.5×10² ppm
d3 0.53HS
d4 0.65HS
d5 0.83HS
MnOOH(s)+3H++e−—Mn2++2H2O 1.50
MnO2(s)+4H−+2e−—Mn2++2H2O 1.23
IRON 5.6×10⁴ ppm
d5 0.65HS
d6 0.78HS
FeOOH(s)+3H++e−—Fe2++2H2O 0.67
COBALT 2.5×10¹ ppm
d6 0.55LS
d7 0.75HS
d9
CoOOH(s)+3H++e−—Co2++2H2O 1.48
NICKEL 7.5×10¹ ppm
d7 0.56LS
d8 0.69
d10
Ni3O4(s)+8H++2e−—3Ni2++4H2O 1.98
COPPER 5.5×10¹ ppm
d9 0.73
d10 0.77
d10s1
OXIDATIVE DISSOLUTION
REDUCTIVE DISSOLUTION
COMMON OXIDATION STATES OF FIRST-ROW TRANSITION METALS IN AQUATIC ENVIRONMENTS.
KEY:
NUMBER OF d-ELECTRONS
IONIC RADIUS (Angstroms) HS=high spin LS=low spin
LOW SOLUBILITY OXIDE/HYDROXIDE

Figure 9.1. Crustal abundances of first-row transition metals and oxidation states encountered within the redox level domain of natural waters.

1986; Zinder et al., 1986). The goals of this work are to explain differences in behavior among (hydr)oxide minerals and to develop ways of predicting dissolution rates under unexplored conditions.

This chapter focuses primarily on the kinetics and mechanisms of surface chemical reactions that influence overall rates of reductive dissolution. Surface speciation and rates of surface chemical reactions are intimately associated; their interdynamics control rates of metal ion release from mineral surfaces.

2. OXIDATION STATE AND SOLUBILITY

Crustal abundances of first-row transition metals and oxidation states encountered within the redox level domain of natural waters are given in Figure 9.1. Iron, titanium, and manganese are most abundant, but titanium is not an active participant in environmental redox reactions. Changes in oxidation state have a dramatic impact on both the solubility and speciation of transition metals. Sparingly soluble oxides of Mn(III,IV), Fe(III), Co(III), and Ni(III) are thermodynamically stable in oxygenated solutions at neutral pH but are reduced to divalent metal ions under anoxic conditions. Figure 9.2 shows that the Fe(II)-phase amakinite is substantially more soluble than the Fe(III)-phase hematite over a wide pH range. Reduction of surface groups from Fe(III) to Fe(II), therefore, increases iron solubility by several orders of magnitude. Solubility increases of comparable magnitude are observed in the reductive dissolution of Mn(III,IV), Co(III), and Ni(III) (hydr)oxide minerals. V(III,IV) and Cr(III) oxides, in contrast, are solubilized by oxidative dissolution; oxidation to V(V) and Cr(VI) substantially increases solubility.

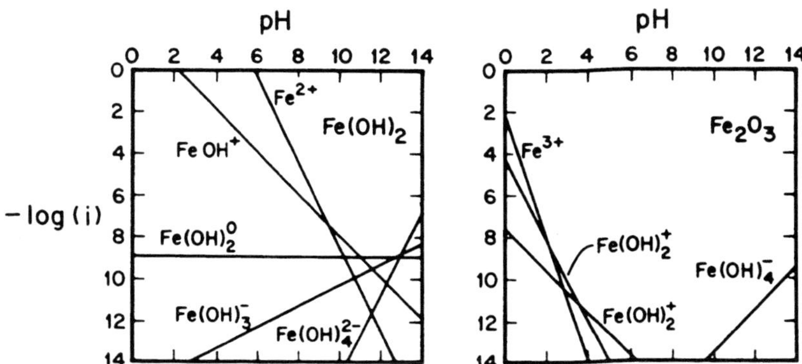

Figure 9.2. Solubilities of the Fe(II)-phase amakinite and the Fe(III)-phase hematite as a function of pH.

3. THE GEOCHEMICAL SETTING

Transition metals are important participants in environmental redox reactions, because changes in oxidation state can take place relatively easily. Direct reaction between important redox species can be quite slow. Many biologically generated organic compounds, for example, are resistant to direct oxidation by molecular oxygen. Transition metals in low oxidation states can be oxidized by molecular oxygen, while those in high oxidation states can be reduced by organic compounds. Both processes can proceed concurrently, at the same location, cycling the transition metal between oxidation states as molecular oxygen and organic compounds are consumed. Oxidation and reduction of transition metals may also be separated in space by a redox boundary that develops as a balance between oxidant and reductant fluxes.

Organic compounds produced by photosynthesis and biological metabolism are the ultimate sources of electrons in natural waters. Alcoholic, carbonyl, carboxylate, and phenolic reductants have been identified in natural organic matter (NOM) collected from a wide variety of environments. Amino and mercapto groups are less abundant and may be important redox participants only in organic-rich or highly reducing environments. Biological processes have a profound influence on the distribution and concentrations of various functional groups and molecular structures. Readily metabolized compounds exhibit high biological turnover rates and relatively low steady-state concentrations. Organic compounds resistant to biological decomposition may persist and accumulate in natural systems. Very likely, organic compounds most important in the reduction of inorganic species are part of the labile organic fraction, which has a high turnover rate.

Bacteria may participate in reductive dissolution reactions by (1) releasing organic reductant metabolites that reduce a stoichiometric amount of metal (hydr)oxide or (2) releasing electron transfer enzymes that couple extracellular redox processes with microbial metabolic processes. There is considerable evidence that bacteria do, in fact, excrete compounds that reduce metal (hydr)oxides (Ghiorse, 1986). Mechanisms of reductive dissolution presented in this chapter are applicable to microbially driven reductive dissolution reactions as well as nonbiological processes.

Chemical and biological processes in anoxic waters also produce several inorganic reductant species that can reduce metal (hydr)oxide minerals. Some inorganic reductants may be important because they react more quickly than available organic species. Inorganic reductants may also alter the course of reductive dissolution reactions by modifying oxide surfaces. Fe^{2+}, for example, may reduce manganese oxide surface sites but inhibit further reaction by formation of an Fe(III) (hydr)oxide coating.

Adsorption:

$$>Mn^{III}OH + Fe^{2+} = (>Mn^{III}OFe^{II})^{+} + H^{+} \qquad (9.1)$$

Electron transfer:

$$(>\mathrm{Mn^{III}OFe^{II}})^{+} + \mathrm{H_2O} = >\mathrm{Mn^{II}OFe^{III}OH} + \mathrm{H^{+}} \qquad (9.2)$$

In a similar manner, reduction of iron (hydr)oxides by sulfide may produce pyrite surface coatings, blocking dissolution and further reaction (D. Canfield, personal communication).

The focus of this chapter is on reductive dissolution by organic reductants because they are present in higher concentrations and are found in a wider variety of aquatic environments than inorganic reductants and therefore have the most significant impact on transition metal geochemistry.

4. DISSOLUTION WITHOUT CHANGES IN OXIDATION STATE

Dissolution of metal (hydr)oxide minerals without changes in oxidation state has been the subject of a considerable amount of research. Recent reviews include those of Diggle (1973), Valverde and Wagner (1976), Gorichev and Kipriyanov (1984), and Stumm et al. (1985). Dissolution reactions may be transport-controlled or controlled by rates of surface chemical reactions (Berner, 1980).

4.1. Surface Chemical Reaction–Controlled Dissolution

Surface chemical reaction–controlled dissolution is limited by the rate of metal ion detachment, replacing lattice bonds with bonds to solvent molecules. Surface features that affect metal-to-lattice bond strength and lability are therefore important in determining dissolution rates. Terrace, ledge, kink, and adatom surface sites have progressively fewer remaining lattice bonds and are released from surfaces at progressively higher rates (Segal and Sellers, 1984). Defects in the crystal lattice and the crystallographic orientation of surface sites can also focus dissolution at particular locations (Gorichev and Kipriyanov, 1984). When "active" sites are only a small fraction of total surface sites, dissolution reactions become particularly sensitive to inhibition by adsorbing species (Berner, 1980).

Protonation of surface sites weakens Me—O lattice bonds, accelerating the rate of detachment (Stumm et al., 1985). Fractional orders with respect to $[\mathrm{H^+}]$ are commonly observed in detachment-controlled dissolution reactions. To explain this observation, it has been postulated that a required number (n) of neighboring surface sites must be protonated before detachment can occur (Stumm et al., 1985):

$$R_{\text{dissolution}} = k_{\mathrm{H}}\{>\mathrm{MeOH_2^+}\}^{n} \qquad (9.3)$$

The value of n is typically 2 for (hydr)oxides of divalent metals, and 3 for (hydr)oxides of tervalent metals.

Specific adsorption of ligands can enhance or inhibit dissolution rates by altering the strength and lability of Me—O lattice bonds. Salicylate, oxalate, and citrate promote the dissolution of alumina (Furrer and Stumm, 1983). In the presence of ligand (L) the dissolution rate becomes

$$R_{\text{dissolution}} = k_H\{{>}\text{MeOH}_2^+\}^n + k_L\{{>}\text{MeL}\} \tag{9.4}$$

Organic ligands without redox reactivity that coordinate metal oxide surface sites have been found to enhance rates of particular reductive dissolution reactions as well (Stumm et al., 1985).

4.2. Transport-Controlled Dissolution

Transport-controlled dissolution reactions represent the upward limit of possible dissolution rates. The outward flux of dissolved products is proportional to the concentration gradient that develops at the mineral surface. In well-stirred suspensions, the concentration gradient can be approximated by finding the difference between the surface concentration (C_0) and bulk solution concentration (C_b) and then dividing by the stagnant layer thickness δ (Bard and Faulkner, 1980). Under transport-controlled conditions, surface chemical reactions are fast and C_0 becomes the metal ion equilibrium solubility. Multiplying the concentration gradient by the diffusion coefficient D gives the metal ion flux from the surface:

$$\text{Flux} = \frac{D}{\delta}(C_0 - C_b) \tag{9.5}$$

When C_b is small, the flux of dissolved metal ions is directly proportional to the equilibrium solubility.

The dissolved iron flux from amakinite [$Fe^{II}(OH)_2(s)$] and from hematite [α-$Fe_2^{III}O_3(s)$] can be estimated using Eq. 9.5 and the solubility data from Figure 9.2. A stagnant layer thickness δ of 0.01 cm, a diffusion coefficient D of 10^{-5} cm^2 s^{-1}, and a negligible dissolved iron concentration in bulk solution will be assumed. Figure 9.3 shows the dissolved iron flux as a function of pH for both iron minerals. Transport-controlled dissolution of the Fe(II)-phase amakinite far exceeds that of the Fe(III)-phase hematite, reflecting a high surface concentration of dissolved iron and a large concentration gradient.

Surface sites of higher-valent, insoluble (hydr)oxides can be reduced by many of the natural reductants discussed earlier. Rates of dissolution are expected to increase as the fraction of reduced surface sites is increased. Whether or not the dissolution rate eventually reaches that of a pure divalent

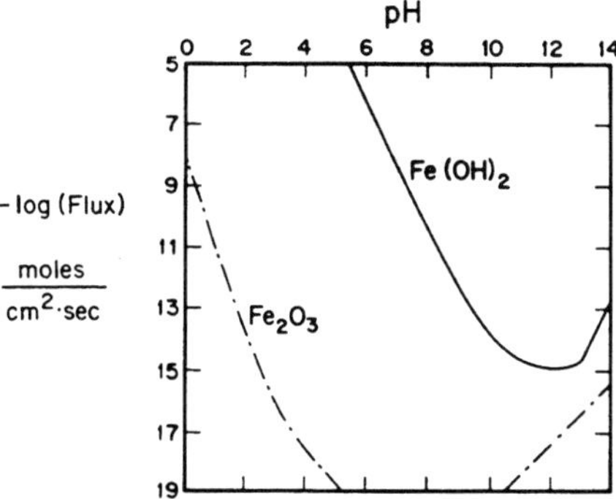

Figure 9.3. Transport-controlled dissolution of the Fe(II)-phase amakinite ($Fe(OH)_2$) and the Fe(III)-phase hematite (Fe_2O_3) as a function of pH.

metal (hydr)oxide is not yet known, but it has been the subject of some speculation (Segal and Sellers, 1980).

5. MECHANISM OF REDUCTIVE DISSOLUTION

Reductive dissolution of metal (hydr)oxide minerals by organic reductants occurs in the following sequence: (1) diffusion of reductant molecules to the oxide surface, (2) surface chemical reaction, and (3) release of reaction products and diffusion away from the oxide surface. In the rare case of transport-controlled reductive dissolution reactions, a decrease in metal oxidation state increases metal ion solubility, the concentration gradient in the stagnant layer, and the outward flux of dissolved metal ions. Most reductive dissolution reactions of environmental and geochemical significance are, however, controlled by surface chemical reactions. Rates and mechanisms of surface chemical reactions will be examined in this section, with particular emphasis on reduction of metal (hydr)oxide surface sites and the subsequent release of reduced metal ion.

Electron transfer between dissolved reductant molecules and metal (hydr)oxide surface sites is an essential component of reductive dissolution. Electrochemical studies have shown that rates of electron transfer to surface sites depend on the density of reductant molecules in the overlying solution and receptor sites in the solid, and upon the activation energy that must be acquired before electron transfer can occur (Morrison, 1980). Chemical speciation is an important determinant of electron transfer rates, affecting the density and energy levels of electron donor and electron acceptor species on the (hydr)oxide surface and in the overlying solution. The stoichiometry and mineralogy of the (hydr)oxide mineral being dissolved also influences electron transfer rates. Electrostatic forces arising from the buildup of electric charge on (hydr)oxide surfaces influence electron transfer rates by affecting the distribution and energy of electron donors and electron receptors (Morrison, 1980).

Dissolved reductant molecules within 15 Å of (hydr)oxide surfaces are available for electron transfer (Morrison, 1980). Electron transfer across such large distances (across several solvent molecules) may not be particularly favorable because of the small degree of orbital overlap between electron donor and electron acceptor. Close-range, specific interactions between reductant molecules and surface sites also occur and have important consequences for electron transfer rates.

A coordination chemistry model for (hydr)oxide surface sites has been developed (Schindler et al., 1976; Stumm et al., 1976) that outlines possible modes of interaction between dissolved species and surface sites. First, recall that hydrated surface sites (>MeOH) undergo protonation–deprotonation reactions, creating a distribution of positive, negative, and neutral surface sites (Stumm et al., 1970):

$$>\mathrm{MeOH} + \mathrm{H}^+ \rightleftharpoons >\mathrm{MeOH}_2^+ \tag{9.6}$$

$$>\mathrm{MeOH} \rightleftharpoons >\mathrm{MeO}^- + \mathrm{H}^+ \tag{9.7}$$

Solute molecules can displace surface-bound OH^- and H_2O via ligand substitution and bind directly to the metal center. An inner-sphere surface complex is formed by this process; this is illustrated in reaction 9.8 for the adsorption of phenol onto a tervalent oxide surface site:

$$>\mathrm{Me^{III}OH} + \mathrm{HO{-}C_6H_5} \rightleftharpoons \underset{\mathbf{I}}{>\mathrm{Me^{III}{-}O{-}C_6H_5}} + \mathrm{H_2O} \tag{9.8}$$

Outer-sphere surface complexes may also form, in which the inner coordination sphere of the metal center remains intact:

$$>\mathrm{Me^{III}OH} + \mathrm{HO{-}C_6H_5} \rightleftharpoons \underset{\mathbf{II}}{\left(>\mathrm{Me^{III}OH},\ \mathrm{HO{-}C_6H_5}\right)} \tag{9.9}$$

Orbital overlap between the adsorbate and the surface metal center is lower in the outer-sphere complex **II** than in the inner-sphere complex **I**, but considerably higher than in the case where no specific interactions occur at all.

Two general mechanisms for electron transfer in homogeneous solution between metal ion complexes and organic compounds have been postulated (Littler, 1970) and are readily extended to reactions at metal (hydr)oxide surface sites. Table 9.1 illustrates both the inner-sphere and outer-sphere mechanisms for the reduction of tervalent metal (hydr)oxide surface sites ($>Me^{III}OH$) by phenol (HA). The two mechanisms have in common the formation of a precursor complex, electron transfer within this complex, and subsequent breakdown of the successor complex. It is assumed that the free radical A· produced by one-equivalent electron transfer is quickly consumed by further reaction with metal oxide surface sites and by radical coupling reactions. Electron transfer via the inner-sphere mechanism is limited by rates of ligand substitution reactions of surface metal centers. Reaction via an outer-sphere mechanism, in contrast, is independent of ligand substitution rates, since the inner coordination spheres of surface metal centers remain intact. These mechanisms operate in parallel, with the overall reaction dominated by the fastest pathway. Substitution-inert surface metal centers must necessarily react via an outer-sphere mechanism.

Inner-sphere and outer-sphere precursor complex formation are both adsorption reactions that increase the density of reductant molecules at the surface, promoting electron transfer. Adsorption influences electron transfer rates in other ways as well. Orbital overlap with surface metal centers is enhanced, particularly in inner-sphere complex formation. Energy levels of reductant molecules change substantially upon adsorption (Morrison, 1980) because of the close proximity of a polarizing metal center and changes in medium composition near the surface. Thus, precursor complex formation can lower the activation energy of reaction. Adsorption may also influence reactions that occur subsequent to electron transfer. The presence of oxidized organic molecules on the surface, for example, may assist or inhibit eventual release of reduced metal ions.

Table 9.1. Inner-Sphere and Outer-Sphere Mechanisms for the Reductive Dissolution of Tervalent Metal Oxide Surface Sites ($>Me^{III}OH$) by Phenol (HA)

Inner-Sphere Mechanism	Outer-Sphere Mechanism
Precursor Complex Formation	
$>Me^{III}OH + HA \underset{k_{-1}}{\overset{k_1}{\rightleftharpoons}} >Me^{III}A + H_2O$	$>Me^{III}OH + HA \underset{k_{-1}}{\overset{k_1}{\rightleftharpoons}} >Me^{III}OH, HA$
Electron Transfer	
$>Me^{III}A \underset{k_{-2}}{\overset{k_2}{\rightleftharpoons}} >Me^{II}{\cdot}A$	$>Me^{III}OH, HA \underset{k_{-2}}{\overset{k_2}{\rightleftharpoons}} >Me^{II}OH^-, HA{\cdot}^+$
Release of Oxidized Organic Product	
$>Me^{II}{\cdot}A + H_2O \underset{k_{-3}}{\overset{k_3}{\rightleftharpoons}} >Me^{II}OH_2 + A{\cdot}$	$>Me^{II}OH^-, HA{\cdot}^+ \underset{k_{-3}}{\overset{k_3}{\rightleftharpoons}} >Me^{II}OH_2 + A{\cdot}$
Release of Reduced Metal Ion	
$>Me^{II}OH_2 + 2H^+ \underset{k_{-4}}{\overset{k_4}{\rightleftharpoons}} >Me^{III}OH + Me^{2+}$	$>Me^{II}OH_2 + 2H^+ \underset{k_{-4}}{\overset{k_4}{\rightleftharpoons}} >Me^{III}OH + Me^{2+}$

6. KINETICS OF REDUCTIVE DISSOLUTION: SIMPLE MODELS

Kinetics and mechanisms of important chemical reactions are studied in detail so that, ultimately, reaction rates can be predicted under unexplored conditions. The understanding of how organic reductant structure, (hydr)oxide mineralogy, medium composition, and temperature affect rates of reductive dissolution is impeded by the apparent complexity of the process; several sequential (and possibly parallel) elementary reactions are involved, as well as a variety of both dissolved and surface chemical species. Simple mechanistic models can, however, be written that capture many of the essential features of complex processes. Applying the principle of mass action to systems of elementary reactions in a simple mechanism yields rate laws that can then be used to explore the dynamics of the system. This section examines how rate constants for reductant adsorption, electron transfer, and product release affect surface speciation and overall reductive dissolution rates for several simple mechanistic models. Conclusions from this study should provide insight into more complex and more realistic mechanistic models for reductive dissolution.

The inner-sphere mechanism for reductive dissolution presented in Table 9.1 will now be examined in detail. To facilitate discussion, some simplifications are made at the outset. The (hydr)oxide surface is assumed to be homogeneous and composed of a small number of distinct chemical species. Electrostatic and neighboring group affects are ignored. A common pH is assumed for all reactions, so that the effects of protonation and/or deprotonation on surface speciation and reaction rate can be ignored. Then, according to the mechanism presented in Table 9.1, a mass balance can be written for S_T, the total moles of surface sites per liter of solution:

$$S_T = [>\mathrm{Me^{III}OH}]_0 = [>\mathrm{Me^{III}OH}] + [>\mathrm{Me^{II}OH_2}] + [>\mathrm{Me^{III}A}] + [>\mathrm{Me^{II}\cdot A}] \quad (9.10)$$

The total number of surface sites remains constant, because a new site is generated whenever a reduced metal ion is released.

Mechanism 1, outlined below, represents the situation where precursor complex formation and electron transfer are slow relative to product release:

Mechanism 1

$$>\mathrm{Me^{III}OH} + \mathrm{HA} \underset{k_{-1}}{\overset{k_1}{\rightleftharpoons}} >\mathrm{Me^{III}A} + \mathrm{H_2O} \quad (9.11)$$

$$>\mathrm{Me^{III}A} \xrightarrow{k_2} >\mathrm{Me^{II}\cdot A} \quad (9.12)$$

$$>\mathrm{Me^{II}\cdot A} + \mathrm{H_2O} \xrightarrow{\text{fast}} >\mathrm{Me^{II}OH_2} + \mathrm{A\cdot} \quad (9.13)$$

$$>\mathrm{Me^{II}OH_2} + 2\mathrm{H^+} \xrightarrow{\text{fast}} >\mathrm{Me^{III}OH} + \mathrm{Me^{2+}} \quad (9.14)$$

Electron transfer is assumed to be irreversible, and subsequent product release is assumed to be instantaneous. Under these conditions, [$>Me^{II}\cdot A$] and [$>Me^{II}OH_2$] are negligible, and the surface mass balance equation becomes

$$S_T = [>\mathrm{Me^{III}OH}] + [>\mathrm{Me^{III}A}] \tag{9.15}$$

According to this mechanism, rates of metal ion release are proportional to the reductant surface coverage:

$$\frac{d[Me^{2+}]}{dt} = k_2[>\mathrm{Me^{III}A}] \tag{9.16}$$

Reductant surface coverage, in turn, is determined by relative rates of adsorption, desorption, and electron transfer.

$$\frac{d[>\mathrm{Me^{III}A}]}{dt} = k_1[\mathrm{HA}][>\mathrm{Me^{III}OH}] - k_{-1}[>\mathrm{Me^{III}A}] - k_2[>\mathrm{Me^{III}A}] \tag{9.17}$$

Holding the dissolved reductant concentration ([HA]) constant, k_1[HA] becomes a pseudo-first-order rate constant. Equation 9.17 can then be integrated with the help of the mass balance equation, yielding

$$[>\mathrm{Me^{III}A}] = S_T\left(\frac{k_1[\mathrm{HA}]}{k_1[\mathrm{HA}] + k_{-1} + k_2}\right) \times (1 - \exp\{-(k_1[\mathrm{HA}] + k_{-1} + k_2)t\}) \tag{9.18}$$

$$\frac{d[\mathrm{Me^{2+}}]}{dt} = k_2S_T\left(\frac{k_1[\mathrm{HA}]}{k_1[\mathrm{HA}] + k_{-1} + k_2}\right) \times (1 - \exp\{-(k_1[\mathrm{HA}] + k_{-1} + k_2)t\}) \tag{9.19}$$

where $(k_1[\mathrm{HA}] + k_{-1} + k_2)^{-1}$ is the characteristic time required for the attainment of steady-state reductant surface coverage.

$$[>\mathrm{Me^{III}A}]_{ss} = S_T\left(\frac{k_1[\mathrm{HA}]}{k_1[\mathrm{HA}] + k_{-1} + k_2}\right) \tag{9.20}$$

$$\left(\frac{d[\mathrm{Me^{2+}}]}{dt}\right)_{ss} = k_2S_T\left(\frac{k_1[\mathrm{HA}]}{k_1[\mathrm{HA}] + k_{-1} + k_2}\right) \tag{9.21}$$

Reductant adsorption is the rate-limiting step when $(k_{-1} + k_2) \gg k_1[\mathrm{HA}]$, and electron transfer is rate-limiting when $(k_{-1} + k_1[\mathrm{HA}]) \gg k_2$. It is interesting to note that when $k_{-1} \gg (k_1[\mathrm{HA}] + k_2)$, both reductant adsorption and electron transfer are rate-limiting (see Denbigh and Turner, 1965). Thus, characteristic times of adsorption, desorption, and electron transfer (k_1, k_{-1}, k_2) are all important in determining reductant surface coverage and overall rates of metal release.

	Elementary Reaction	Characteristic Time	
Adsorption:	$R_1 = k_1[\text{HA}][>\text{Me}^{\text{III}}\text{OH}]$	$(k_1[\text{HA}])^{-1}$	(9.22)
Desorption:	$R_{-1} = k_{-1}[>\text{Me}^{\text{III}}\text{A}]$	$(k_{-1})^{-1}$	(9.23)
Electron transfer:	$R_2 = k_2[>\text{Me}^{\text{III}}\text{A}]$	$(k_2)^{-1}$	(9.24)

Characteristic times for desorption and electron transfer are independent of dissolved reductant concentration, while the characteristic time for adsorption decreases as the reductant concentration is increased.

The dynamics of mechanism 1 will now be explored using numerical examples based on Eqs. 9.18 and 9.19. The following system variable values have been assigned:

$$S_T = [>\text{Me}^{\text{III}}\text{OH}]_0 = 5 \times 10^{-6}\ M$$

$$[>\text{Me}^{\text{III}}\text{A}]_0 = 0$$

$$[\text{Me}^{2+}]_0 = 0$$

$$k_1 = 5 \times 10^2 \quad \text{liters mole}^{-1}\ \text{min}^{-1}$$

$$k_{-1} = 5 \times 10^{-2} \quad \text{min}^{-1}$$

$$k_2 = 1.0 \quad \text{min}^{-1}$$

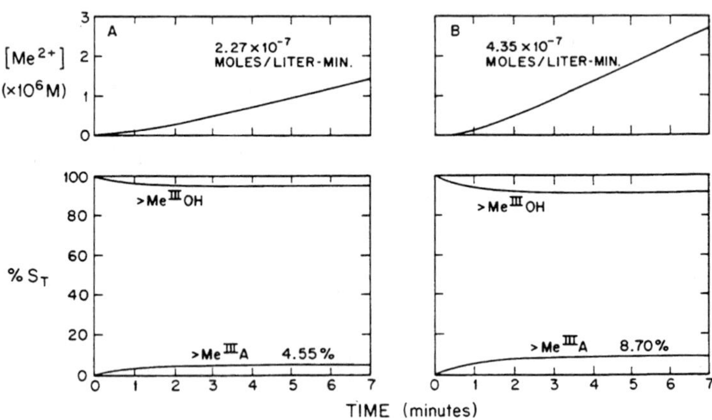

Figure 9.4. Rate-limiting precursor complex formation: surface speciation and dissolved metal ion concentration during reductive dissolution by HA. In case A, $[\text{HA}] = 10^{-4}\ M$, while in case B, $[\text{HA}] = 2 \times 10^{-4}\ M$. ($S_T = 5 \times 10^{-6}\ M$, $k_1 = 5 \times 10^2\ M^{-1}\ \text{min}^{-1}$, $k_{-1} = 5 \times 10^{-2}\ \text{min}^{-1}$, $k_2 = 1.0\ \text{min}^{-1}$).

Figure 9.4 shows reductant surface coverage ([$>$Me^{III}A]) and dissolved metal ion concentration ([Me^{2+}]) as functions of time. Two constant values of [HA], the dissolved reductant concentration, are considered; in both cases, precursor complex formation (adsorption) is slow relative to electron transfer. Another way of stating this is that the characteristic time for adsorption is large relative to the characteristic time for electron transfer. In both cases, steady-state values of [$>$Me^{III}A] and $d[Me^{2+}]/dt$ are achieved within a few minutes. In case B, [HA] is twice that in case A, and as a result, steady-state values of [$>$Me^{III}A] and $d[Me^{2+}]/dt$ increase by almost a factor of 2. As [HA] is increased, k_1 [HA] increases while $(k_{-1} + k_2)$ remains constant; the process forming reductant surface complexes grows in magnitude, while processes breaking them down (desorption and electron transfer) remain constant.

Changes in organic reductant structure influence rates of adsorption, desorption, electron transfer, and product release. Homologous series of organic compounds, in which small, systematic changes in structure have been made, are useful for examining these effects. Consider, for example, the following group of *para*-substituted phenols:

	pK_a	log K_1	$E_{1/2}$ vs. SHE*
p-Methylphenol	10.26	9.26	0.301 V
Phenol	9.98	8.20	0.391 V
p-Chlorophenol	9.42	7.92	0.411 V

pK_a values (from Martell and Smith, 1977) show a decrease in basicity going down the table. log K_1, the stability constant for monodentate complex formation with Fe^{3+}, follows a similar trend.

$$Fe^{3+} + C_6H_5O^- \rightleftharpoons \left(Fe^{III}O{-}C_6H_5\right)^{2+} \quad (9.25)$$

Binding constants for adsorption onto tervalent metal oxide surface sites can also be expected to decrease similarly. $E_{1/2}$ values are polarographic half-wave potentials for reaction with metal electrodes at pH 5.6 (Suatoni et al., 1961). High $E_{1/2}$ values generally signify increased resistance to oxidation.

6.1. Electron Transfer Rate

By inference, both surface coverage and subsequent rates of electron transfer at oxide surface sites should change substantially within a homologous series

*SHE = standard hydrogen electrode.

of organic compounds. Two substituted phenols may have similar adsorption–desorption rate constants but different electron transfer rate constants. It is therefore useful to consider how changes in k_2, the electron transfer rate constant, affect overall rates of reductive dissolution.

In Figure 9.5, k_1 and k_{-1} have the same values as before, but k_2, the electron transfer rate constant, is increased from 10^{-3} min^{-1} to 1.0 min^{-1}. At the lowest value of k_2, electron transfer is the rate-limiting step, while at high values of k_2, reductant adsorption becomes rate-limiting. The effect of the electron transfer reaction on surface speciation depends upon the magnitude of k_2 relative to k_{-1}, the rate constant for reductant desorption. When $k_2 \gg k_{-1}$, electron transfer is the principal pathway for removing adsorbed reductant. Increasing k_2 from 10^{-1} to 1.0 min^{-1} lowers reductant surface coverage by 20%. Once k_2 drops below the value of k_{-1}, reductant is lost from surface sites predominantly by desorption, and changes in k_2 have relatively little effect on speciation. Lowering k_2 from 10^{-2} to 10^{-3} min^{-1} changes reductant surface coverage by only 4%.

At very low values of k_2, $(k_1[\mathrm{HA}] + k_{-1}) \gg k_2$ and reductant adsorption can be described by a quasi-equilibrium step:

$$>\mathrm{Me^{III}OH} + \mathrm{HA} \underset{k_{-1}}{\overset{k_1}{\rightleftharpoons}} >\mathrm{Me^{III}A} + \mathrm{H_2O} \tag{9.26}$$

$$K_1 = \frac{k_1}{k_{-1}} = \frac{[>\mathrm{Me^{III}A}]}{[>\mathrm{Me^{III}OH}][\mathrm{HA}]} \tag{9.27}$$

$$[>\mathrm{Me^{III}A}]_{ss} = S_T\left(\frac{k_1[\mathrm{HA}]}{k_1[\mathrm{HA}] + k_{-1}}\right) \tag{9.28}$$

$$= S_T\left(\frac{K_1[\mathrm{HA}]}{K_1[\mathrm{HA}] + 1}\right) \tag{9.29}$$

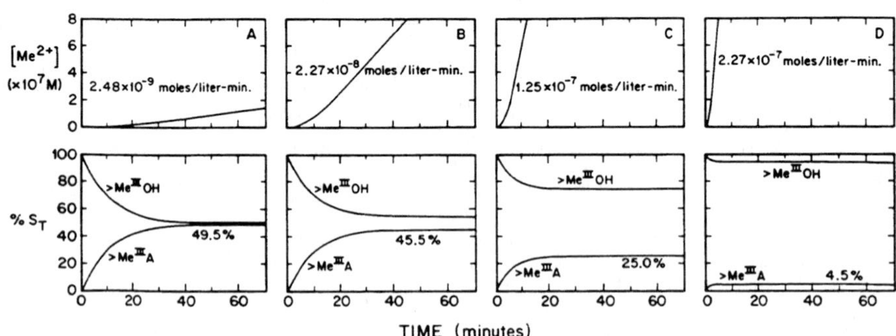

Figure 9.5. The effect of k_2, the electron transfer rate constant, on surface speciation and metal ion release. The value of k_2 is 10^{-3} min^{-1} in case A, 10^{-2} min^{-1} in case B, 10^{-1} min^{-1} in case C, and 1.0 min^{-1} in case D. ($S_T = 5 \times 10^{-6}$ M, $k_1 = 5 \times 10^2$ M^{-1} min^{-1}, and $k_{-1} = 5 \times 10^{-2}$ min^{-1}.)

Using [HA] = 10^{-4} *M*, k_1 = 5×10^2 moles liter^{-1} min^{-1}, and k_{-1} = 5×10^{-2} min^{-1}, the steady-state surface coverage by reductant is 50% of S_T when electron transfer is extremely slow.

6.2. Effect of Reductant Concentration

Increasing [HA] lowers the characteristic time for adsorption relative to characteristic times for desorption and electron transfer (which are both independent of [HA]). As a result, reductant adsorption is the rate-limiting step when reductant concentrations are low, and electron transfer is rate-limiting at high reductant concentrations. Rates of metal ion release plotted against reductant concentration resemble adsorption isotherms (Fig. 9.6). The con-

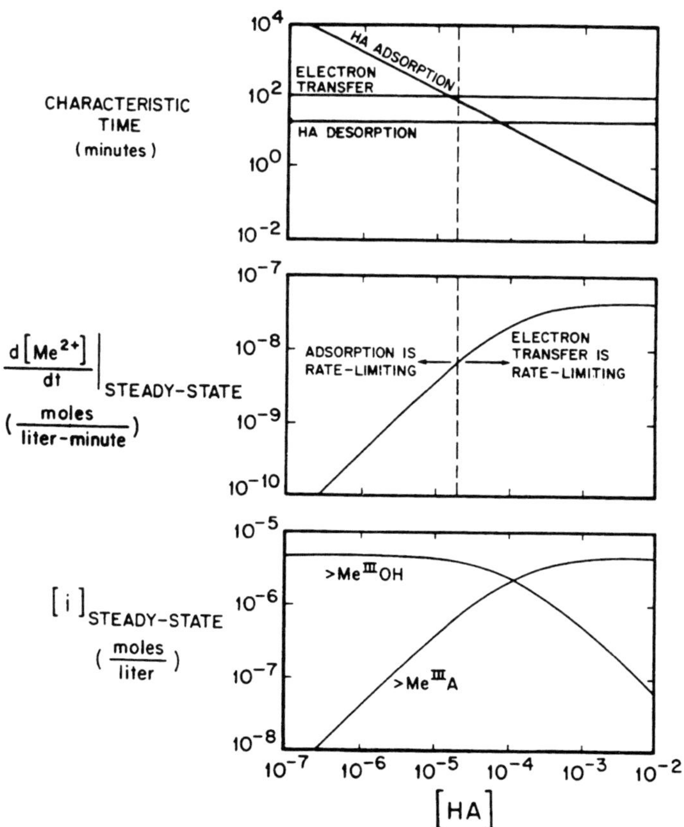

Figure 9.6. The effect of increasing [HA] on steady-state surface speciation and metal ion release rate. Reductant adsorption is rate-limiting at low [HA], and electron transfer is rate-limiting at high [HA]. (S_T = 5×10^{-6} *M*, k_1 = 5×10^2 M^{-1} min^{-1}, k_{-1} = 5×10^{-2} min^{-1}, and k_2 = 1×10^{-2} min^{-1}.)

centration range over which electron transfer is rate-limiting depends upon the magnitude of k_2 relative to k_1 and k_{-1} and does not necessarily depend on reductant surface coverage.

6.3. Release of Reduced Metal Ion as Rate-Limiting Step

Situations may exist where the release of reduced metal ion from the oxide surface is the rate-limiting step, rather than reductant adsorption or electron transfer. To examine this possibility, consider the following mechanism:

Mechanism 2

$$>\mathrm{Me^{III}OH} + \mathrm{HA} \underset{k_{-1}}{\overset{k_1}{\rightleftharpoons}} >\mathrm{Me^{III}A} + \mathrm{H_2O} \tag{9.30}$$

$$>\mathrm{Me^{III}A} \xrightarrow{k_2} >\mathrm{Me^{II}\cdot A} \tag{9.31}$$

$$>\mathrm{Me^{II}\cdot A} + \mathrm{H_2O} \xrightarrow{\text{fast}} >\mathrm{Me^{II}OH_2} + \mathrm{A\cdot} \tag{9.32}$$

$$>\mathrm{Me^{II}OH_2} + 2\mathrm{H^+} \xrightarrow{k_4} >\mathrm{Me^{III}OH} + \mathrm{Me^{2+}} \tag{9.33}$$

A new characteristic time, not considered in mechanism 1, must now be examined.

	Elementary Reaction	Characteristic Time	
Dissolution:	$R_4 = k_4[>\mathrm{Me^{II}OH_2}]$	$(k_4)^{-1}$	(9.34)

Note that the dependence on $[\mathrm{H^+}]$ has been ignored. When k_4, the rate constant for metal ion release, is large relative to k_1 and k_2, the residence time of reduced surface sites and the contribution of $[>\mathrm{Me^{II}OH_2}]$ in the surface mass balance equation are negligible. On the other hand, when k_4 is small relative to k_1 and k_2, a significant fraction of surface sites is found in the reduced form.

$$S_T = [>\mathrm{Me^{III}OH}] + [>\mathrm{Me^{II}OH_2}] + [>\mathrm{Me^{III}A}] \tag{9.35}$$

($[\mathrm{Me^{III}\cdot A}]$ remains negligible because of the high rate of reaction in 9.32.) We have chosen to solve the set of differential equations generated by Eqs. 9.30–9.33 numerically, rather than by seeking an analytical solution. The forward Euler method was used (Forsythe et al., 1977). Figure 9.7 shows surface speciation and metal ion release as a function of time for three values of k_4. In case A, $k_4 \gg (k_1[\mathrm{HA}] + k_{-1} + k_2)$, and less than 1% of the surface sites are found in the reduced form at steady state. As k_4 is decreased, $>\mathrm{Me^{II}OH_2}$

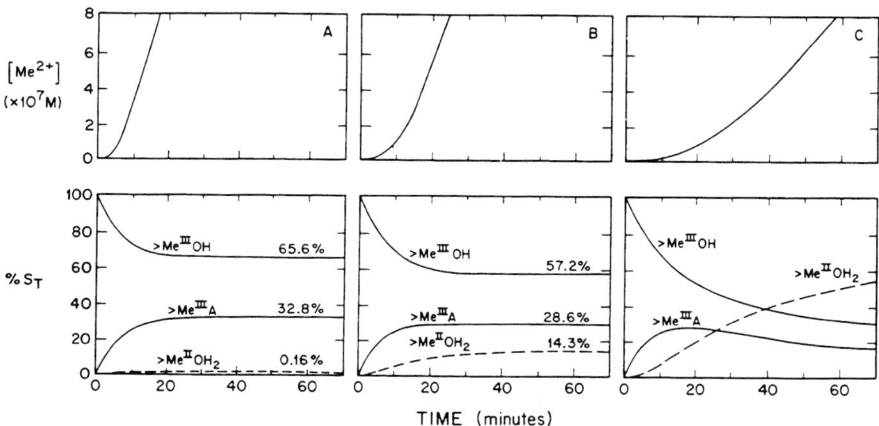

Figure 9.7. The effect of k_4, the rate constant for metal ion release, on surface speciation and dissolved metal ion concentration. The value of k_4 is 1.0 min^{-1} in case A, 10^{-1} min^{-1} in case B, and 10^{-2} min^{-1} in case C. Reductant adsorption and electron transfer are rate-limiting in cases A and B, while metal ion release is rate-limiting in case C. ($S_T = 5 \times 10^{-6}$ M, $k_1 = 5 \times 10^2$ M^{-1} min^{-1}, $k_{-1} = 5 \times 10^{-2}$ min^{-1}, and $k_2 = 5 \times 10^{-2}$ min^{-1}.)

becomes an increasingly important species, eventually covering most of the oxide surface.

6.4. Reversible Electron Transfer

Up to this point, electron transfer and desorption of oxidized organic product have been treated as irreversible reactions. It is quite possible, however, that these steps are reversible and that both oxidized and reduced organic species can adsorb to oxide surfaces and participate in surface electron transfer reactions. To explore this possibility in a simple way, reversible one-equivalent oxidation of the organic reductant yielding a single, stable product will be assumed.

$$A\cdot + H^+ + e^- \longrightarrow HA \tag{9.36}$$

Actual redox reactions involving organic compounds are often considerably more complex and form a mixture of products, often by irreversible reaction pathways. In systems represented by reaction 9.36 where redox equilibrium has been attained, a redox potential can be defined for this half-reaction:

$$E_{HA} = E^0_{HA} - 0.0592\text{pH} + \frac{RT}{F} \ln \frac{[A\cdot]}{[HA]} \tag{9.37}$$

$$E_{HA} = E^{0\prime}_{HA} + \frac{RT}{F} \ln \frac{[A\cdot]}{[HA]} \quad \text{at constant pH} \tag{9.38}$$

One goal of the following discussions is to identify situations where this redox potential is a meaningful and useful system variable.

Reaction mechanism 3 allows for adsorption of oxidized organic species and reversible electron transfer:

Mechanism 3

$$>\mathrm{Me^{III}OH} + \mathrm{HA} \underset{k_{-1}}{\overset{k_1}{\rightleftharpoons}} >\mathrm{Me^{III}A} + \mathrm{H_2O} \tag{9.39}$$

$$>\mathrm{Me^{III}A} \underset{k_{-2}}{\overset{k_2}{\rightleftharpoons}} >\mathrm{Me^{II}\cdot A} \tag{9.40}$$

$$>\mathrm{Me^{II}\cdot A} + \mathrm{H_2O} \underset{k_{-3}}{\overset{k_3}{\rightleftharpoons}} >\mathrm{Me^{II}OH_2} + \mathrm{A\cdot} \tag{9.41}$$

$$>\mathrm{Me^{II}OH_2} + 2\mathrm{H^+} \xrightarrow{k_4} >\mathrm{Me^{III}OH} + \mathrm{Me^{2+}} \tag{9.42}$$

In the present discussion, both [HA] and [A·] are kept constant throughout the course of the reaction. Seven elementary reactions are now being considered, each with its own characteristic time. There are now two characteristic times for adsorption, which depend on concentrations of oxidized and reduced organic species in bulk solution:

	Elementary Reaction	Characteristic Time	
HA adsorption:	$R_1 = k_1[\mathrm{HA}][>\mathrm{Me^{III}OH}]$	$(k_1[\mathrm{HA}])^{-1}$	(9.43)
A· adsorption:	$R_{-3} = k_{-3}[\mathrm{A\cdot}][>\mathrm{Me^{II}OH_2}]$	$(k_{-3}[\mathrm{A\cdot}])^{-1}$	(9.44)

Electron transfer, desorption, and metal ion release reactions are all assumed to be independent of bulk solution concentrations in the present model.

	Elementary Reaction	Characteristic Time	
Forward electron transfer:	$R_2 = k_2[>\mathrm{Me^{III}A}]$	$(k_2)^{-1}$	(9.45)
Reverse electron transfer:	$R_{-2} = k_{-2}[>\mathrm{Me^{II}\cdot A}]$	$(k_{-2})^{-1}$	(9.46)
HA desorption:	$R_{-1} = k_{-1}[>\mathrm{Me^{III}A}]$	$(k_{-1})^{-1}$	(9.47)
A· desorption:	$R_3 = k_3[>\mathrm{Me^{II}\cdot A}]$	$(k_3)^{-1}$	(9.48)
Dissolution:	$R_4 = k_4[>\mathrm{Me^{II}OH_2}]$	$(k_4)^{-1}$	(9.49)

Characteristic times for forward and reverse electron transfer depend only upon the magnitudes of k_2 and k_{-2}. When characteristic times for electron transfer are small relative to characteristic times of all other elementary reactions, the concentrations [$>Me^{III}A$] and [$>Me^{II}\cdot A$] are related to one another through the quasi-equilibrium relationship

$$K_2 = \frac{k_2}{k_{-2}} = \frac{[>Me^{II}\cdot A]}{[>Me^{III}A]} \quad (9.50)$$

When adsorption, desorption, or dissolution reactions are rate-limiting, the ratio [$>Me^{II}\cdot A$]/[$>Me^{III}A$] will approach a constant. When electron transfer reactions are rate-limiting, however, this relationship will not be obeyed.

Four species must now be considered in the surface mass balance equation:

$$S_T = [>Me^{III}OH] + [>Me^{II}OH_2] + [>Me^{III}A] + [>Me^{II}\cdot A] \quad (9.51)$$

Interconversion of surface species during reductive dissolution reactions is shown schematically in Figure 9.8. Increasing bulk solution concentrations of either reduced (HA) or oxidized (A·) organic species transforms unoccupied surface sites ($>Me^{III}OH$ and $>Me^{II}OH_2$) into occupied surface sites ($>Me^{III}A$ and $>Me^{II}\cdot A$). The rate of metal ion release is proportional to [$>Me^{II}OH_2$]. Any perturbation that shifts surface speciation in favor of $>Me^{II}OH_2$ will therefore yield higher rates of reductive dissolution.

The model is now at a point where values of seven individual rate constants must be known or arbitrarily assigned when calculating numerical examples. A wide variety of possible combinations of large and small characteristic times can be chosen; the current discussion will be limited to situations where organic adsorption or metal ion release, rather than electron transfer, is rate-limiting. The forward Euler method of numerical integration will again be

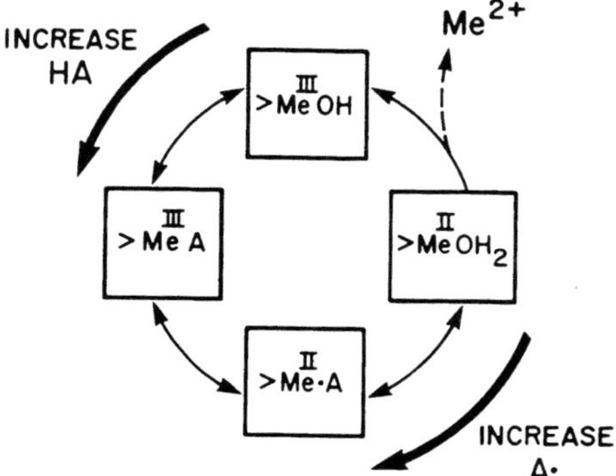

Figure 9.8. Schematic illustration of the surface site mass balance equation. Increasing [HA] promotes conversion of $>Me^{III}OH$ to $>Me^{III}A$, and increasing [A·] promotes conversion of $>Me^{II}OH_2$ to $>Me^{II}\cdot A$.

used. To explore how [HA] and [A·] influence reaction rates, the following values have been assigned:

$$
\begin{array}{llll}
S_T = 5 \times 10^{-6}\ M & \\
k_1 = 5 \times 10^{2}\ \text{L mol}^{-1}\ \text{min}^{-1} & k_{-1} = 5 \times 10^{-2}\ \text{min}^{-1} \\
k_2 = 1.0\ \text{min}^{-1} & k_{-2} = 0.1\ \text{min}^{-1} \\
k_3 = 0.1\ \text{min}^{-1} & k_{-3} = 10^{2}\ \text{L mol}^{-1}\ \text{min}^{-1} \\
k_4 = 1 \times 10^{-2}\ \text{min}^{-1} &
\end{array}
$$

With regard to electron transfer, k_2 is an order of magnitude larger than k_{-2}, and therefore steady-state values of [$>Me^{II}\cdot A$] are close to an order of magnitude larger than [$>Me^{III}A$].

First consider the case where [HA] is increased while [A·] is held constant at a low value (10^{-6} M). This has the effect of enhancing conversion of $>Me^{III}OH$ sites into $>Me^{III}A$ sites, as shown in Figure 9.8. Characteristic times for conversion of $>Me^{III}A$ into $>Me^{II}\cdot A$ and conversion of $>Me^{II}\cdot A$ into $>Me^{II}OH_2$ are unchanged. The relative proportions of $>Me^{III}A$, $>Me^{II}\cdot A$, and $>Me^{II}OH_2$ remain constant, while their sum increases as [HA] is increased. This is shown in Figure 9.9. The rate of metal ion release increases in direct proportion to [$>Me^{II}OH_2$]; both [$>Me^{II}OH_2$] and $d[Me^{2+}]/dt$ eventually reach a limiting value. Figure 9.12 also shows that reductant adsorption is rate-limiting when [HA] is below 2×10^{-5} M. At concentrations above this value, metal ion release becomes rate-limiting.

Letting [HA] = [A·] and increasing both simultaneously has an entirely different effect, as shown in Figure 9.10. At first, $>Me^{III}A$, $>Me^{II}\cdot A$, and $>Me^{II}OH_2$ all increase at the expense of $>Me^{III}OH$. Eventually, however, A· adsorption becomes fast enough that $>Me^{II}\cdot A$ and $>Me^{III}A$ grow at the expense of $>Me^{II}OH_2$. Thus, [$>Me^{II}OH_2$] and $d[Me^{2+}]/dt$ pass through a maximum and then decline as [HA] = [A·] is increased. Conversion of the majority of surface sites into organic-bound species ($>Me^{III}A$ and $>Me^{II}\cdot A$) limits the overall rate of reductive dissolution.

Several authors, including Valverde (1976), have postulated a "redox buffer capacity" poised by the concentration of oxidized and reduced species in overlying solution. In order to examine this possibility, consider the situation where adsorption, desorption, and electron transfer reactions are fast relative to reduced metal ion release. Quasi-equilibrium HA adsorption and electron transfer reactions have already been presented (Eqs. 9.27 and 9.50). In addition, quasi-equilibrium desorption of A· must be considered:

$$K_3 = \frac{k_3}{k_{-3}} = \frac{[>Me^{II}OH_2][A\cdot]}{[>Me^{II}\cdot A]} \tag{9.52}$$

Taken collectively, K_1, K_2, and K_3 represent the overall equilibrium between oxidized and reduced dissolved organic species and oxidized and reduced metal oxide surface sites.

$$>Me^{III}OH + HA = >Me^{II}OH_2 + A\cdot \qquad K_1K_2K_3 \tag{9.53}$$

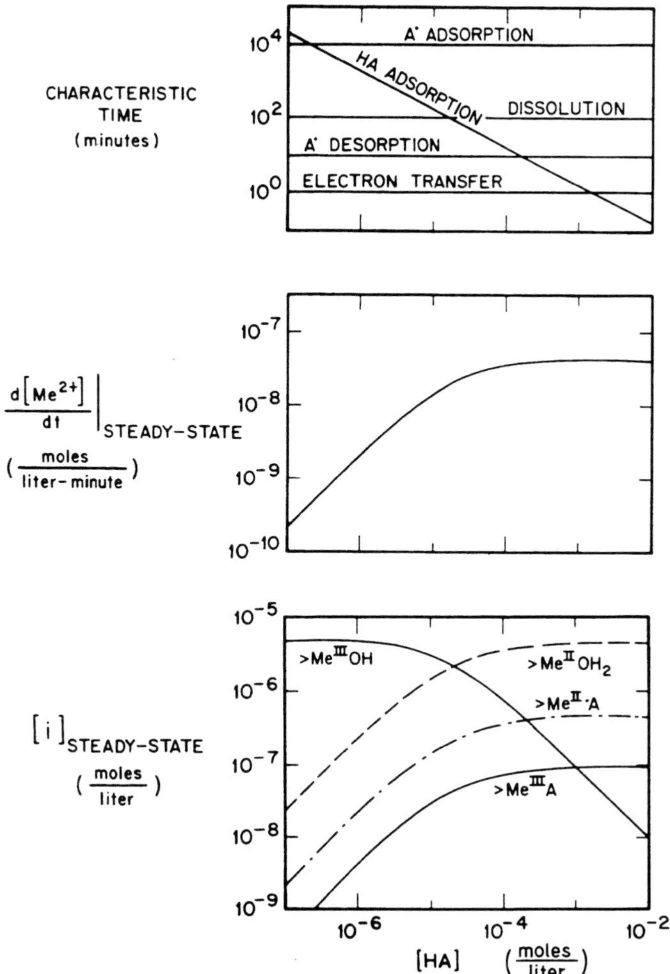

Figure 9.9. Reversible electron transfer: The effect of increasing [HA] (while [A·] is held constant at 10^{-6} M) on steady-state surface speciation and metal ion release rate. ($S_T = 5 \times 10^{-6}$ M, $k_1 = 5 \times 10^2$ M^{-1} min^{-1}, $k_{-1} = 5 \times 10^{-2}$ min^{-1}, $k_2 = 1.0$ min^{-1}, $k_{-2} = 10^{-1}$ min^{-1}, $k_3 = 10^{-1}$ min^{-1}, $k_{-3} = 10^2$ M^{-1} min^{-1}, and $k_4 = 10^{-2}$ min^{-1}.)

Equation 9.53 can also be obtained by adding half-reactions for reduction of surface sites and oxidation of organic reductant.

$$>\text{Me}^{\text{III}}\text{OH} + \text{H}^+ + e^- \longrightarrow >\text{Me}^{\text{II}}\text{OH}_2 \qquad E^0_{\text{Me}} \tag{9.54a}$$

$$\text{HA} \longrightarrow \text{A}\cdot + \text{H}^+ + e^- \qquad E^0_{\text{HA}} \tag{9.54b}$$

$$>\text{Me}^{\text{III}}\text{OH} + \text{HA} \longrightarrow >\text{Me}^{\text{II}}\text{OH}_2 + \text{A}\cdot \tag{9.55}$$

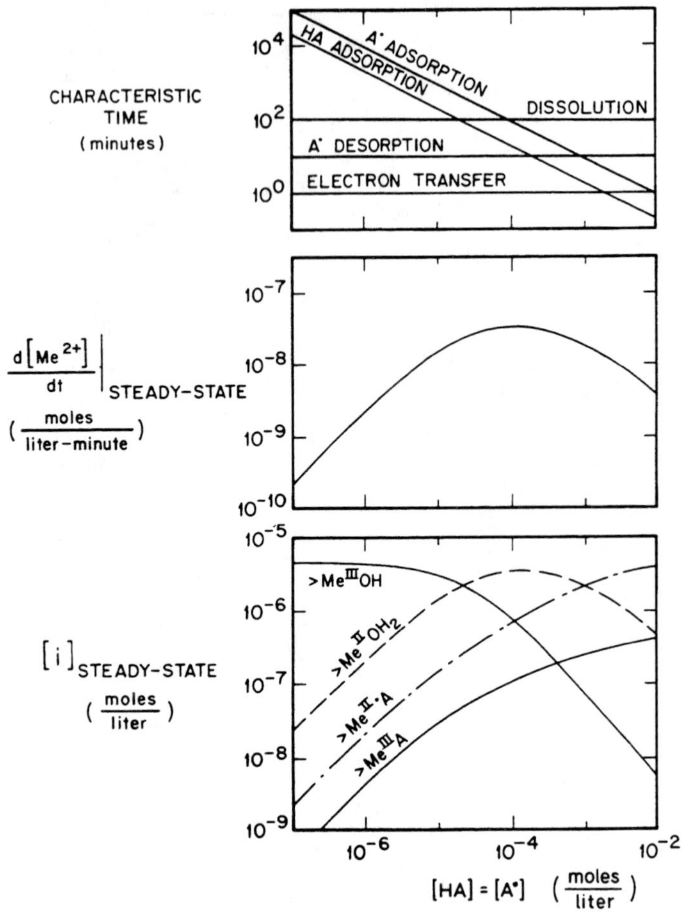

Figure 9.10. Reversible electron transfer: The effect of letting [HA] = [A·] and increasing both simultaneously on steady-state surface speciation and metal ion release rate. ($S_T = 5 \times 10^{-6}$ M, $k_1 = 5 \times 10^2$ M^{-1} min^{-1}, $k_{-1} = 5 \times 10^{-2}$ min^{-1}, $k_2 = 1.0$ min^{-1}, $k_{-2} = 10^{-1}$ min^{-1}, $k_3 = 10^{-1}$ min^{-1}, $k_{-3} = 10^2$ M^{-1} min^{-1}, and $k_4 = 10^{-2}$ min^{-1}.)

The value of $K_1K_2K_3$ can therefore be calculated from the standard electric potentials of reactions 9.54 and 9.55.

$$K_1K_2K_3 = \exp\left[\frac{F}{RT}(E^0_{\mathrm{Me}} - E^0_{\mathrm{HA}})\right] \tag{9.56}$$

Quasi-equilibrium constants can also be used to write an expression for $[{>}\mathrm{Me^{III}OH}]$ in terms of S_T:

$$S_T = [{>}\mathrm{Me^{III}OH}] + [{>}\mathrm{Me^{III}A}] + [{>}\mathrm{Me^{II}{\cdot}A}] + [{>}\mathrm{Me^{II}OH_2}] \tag{9.57}$$

$$S_T = [{>}\mathrm{Me^{III}OH}](1 + K_1[\mathrm{HA}] + K_1K_2[\mathrm{HA}] + K_1K_2K_3([\mathrm{HA}]/[\mathrm{A}\cdot])) \tag{9.58}$$

Equation 9.58, in turn, can be used to find expressions for $[>Me^{II}OH_2]$ and $d[Me^{2+}]/dt$:

$$[>Me^{II}OH_2] = \frac{K_1K_2K_3([HA]/[A\cdot])S_T}{1 + K_1[HA] + K_1K_2[HA] + K_1K_2K_3([HA]/[A\cdot])} \tag{9.59}$$

$$\frac{d[Me^{2+}]}{dt} = k_4\left(\frac{K_1K_2K_3([HA]/[A\cdot])S_T}{1 + K_1[HA] + K_1K_2[HA] + K_1K_2K_3([HA]/[A\cdot])}\right) \tag{9.60}$$

Note that $[HA]/[A\cdot]$ can be written in terms of E_{HA}, the redox potential of the organic half-reaction, using Eq. 9.38:

$$\frac{[HA]}{[A\cdot]} = \exp\left[\frac{-F}{RT}(E_{HA} - E'_{HA})\right] \tag{9.61}$$

Comparison of Eqs. 9.60 and 9.61 indicates that there is no simple relationship between the overall rate of reduction dissolution and the redox potential of the organic half-reaction. The rate of reduced metal ion release depends directly upon $[>Me^{II}OH_2]$, which in turn is a complex function of [HA] and $[A\cdot]$.

Figure 9.11*a* illustrates a situation where adsorption, desorption, and electron transfer are fast relative to reduced metal ion release. To explore the effect of the organic redox couple, the ratio $[HA]/[A\cdot]$ is increased while the sum $[HA] + [A\cdot]$ is held constant. $>Me^{III}OH$ is the dominant surface species at very low $[HA]/[A\cdot]$ ratios but rapidly drops in importance as this ratio is increased. $[>Me^{III}A]$ and $[>Me^{II}\cdot A]$ reach maxima and then gradually diminish at higher $[HA]/[A\cdot]$ ratios. $[>Me^{II}OH_2]$ and $d[Me^{2+}]/dt$ both increase as $[HA]/[A\cdot]$ is increased. In Figure 9.11*b*, the overall rate of reductive dissolution is plotted as a function of E_{HA}, the electric potential of the organic half-reaction. Although $d[Me^{2+}]/dt$ increases as E_{HA} is decreased, the relationship is not a simple one.

A special limiting case of Eq. 9.60 leads to a more direct relationship between redox potential in overlying solution and rates of reductive dissolution. Valverde (1976) stated that dissolution rates are single-valued functions of redox potential when quasi-equilibrium exists between dissolved redox species and surface sites and when adsorption of redox species from overlying solution is small. When $>Me^{III}A$ and $>Me^{II}\cdot A$ are a negligible fraction of total surface sites, Eq. 9.60 can be written as

$$\frac{d[Me^{2+}]}{dt} = k_4\frac{K_1K_2K_3([HA]/[A\cdot])S_T}{1 + K_1K_2K_3([HA]/[A\cdot])} \tag{9.62}$$

Using Eq. 9.61, the rate of metal ion release can be expressed in terms of the organic redox potential E_{HA}:

$$\frac{d[Me^{2+}]}{dt} = k_4S_T\left(\frac{c_1\exp[(-F/RT)E_{HA}]}{1 + c_1\exp[(-F/RT)E_{HA}]}\right) \tag{9.63}$$

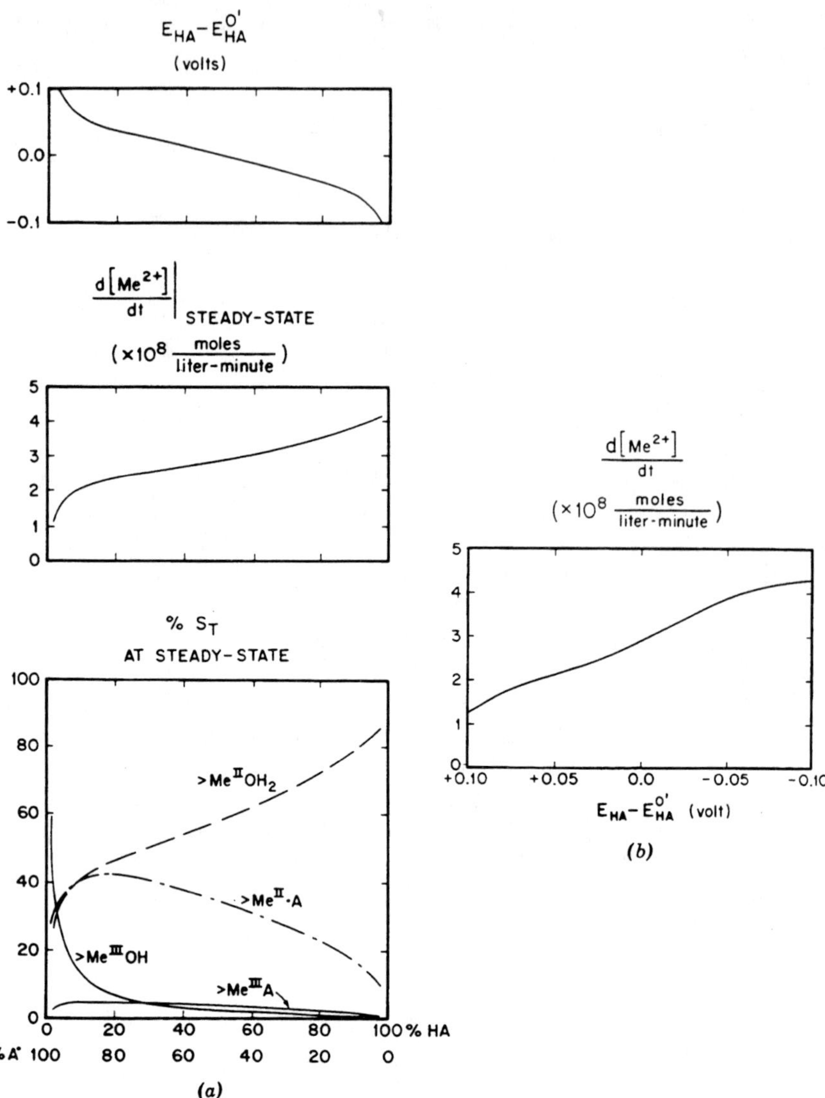

Figure 9.11. The effect of the HA/A· redox potential on steady-state surface speciation and metal ion release rate. ([HA] + [A·]) is held constant while the ratio [HA]/[A·] is increased. In (*a*) [$>Me^{II}OH_2$] and $d[Me^{2+}]/dt$ both increase as the ratio [HA]/[A·] is increased. In (*b*), the rate of metal ion release increases as the redox potential is decreased. ($S_T = 5 \times 10^{-6}\ M$, $k_1 = 5 \times 10^2\ M^{-1}\ \text{min}^{-1}$, $k_{-1} = 5 \times 10^{-2}\ \text{min}^{-1}$, $k_2 = 1.0\ \text{min}^{-1}$, $k_{-2} = 10^{-1}\ \text{min}^{-1}$, $k_3 = 10^{-1}\ \text{min}^{-1}$, $k_{-3} = 10^2\ M^{-1}\ \text{min}^{-1}$, and $k_4 = 10^{-2}\ \text{min}^{-1}$.)

Several constants have been incorporated into c_1 for simplification. When surface coverage by organic species is negligible, lowering E_{HA} shifts surface speciation in favor of $>Me^{II}OH_2$, causing an increase in the rate of reductive dissolution.

In general, rates of reductive dissolution are a complex function of the concentrations of redox-active species in overlying solution. Increases in [HA] and [A·] lessen characteristic times for adsorption relative to characteristic times for desorption, electron transfer, and reduced metal ion release. No simple relationship exists between redox potentials in bulk solution and rates of reductive dissolution when redox-active species occupy a significant fraction of oxide surface sites. When adsorption by redox-active species is negligible, rates of reductive dissolution are related to redox potential through Eq. 9.63.

7. KINETICS OF REDUCTIVE DISSOLUTION: ADDITIONAL CONSIDERATIONS

Several important attributes of reductive dissolution reactions were ignored in the previous section, to facilitate the development of kinetic models. Some additional characteristics of oxide surfaces and surface chemical reactions will now be examined to make the kinetic model more realistic.

7.1. Effect of pH

pH influences rates of reductive dissolution reactions primarily through its effect on chemical speciation. In principle, protonation and deprotonation of dissolved species and surface sites can influence any of the surface chemical reactions outlined in Table 9.1. Reductive dissolution of a tervalent metal oxide by phenol will serve as an illustration.

Precursor complex formation (adsorption) is a ligand exchange reaction. Rates of exchange depend upon the leaving group (H_2O or OH^-) and frequently on the entering group (HA or A^-). Thus, the relative distribution of $>Me^{III}OH_2^+$, $>Me^{III}OH$, and $>Me^{III}O^-$ sites on the oxide surface and of phenol (HA) and phenolate anion (A^-) in overlying solution will influence adsorption rates. An idealized metal oxide with the following surface properties will be considered: $pH_{ZPC} = 8.0$, $pK_{a1}^s(\text{intr}) = 6.5$, and $pK_{a2}^s(\text{intr}) = 9.5$. Phenol has a pK_a (at infinite dilution) of 10.0. Figures 9.12*a* and *b* plot speciation on the oxide surface and in bulk solution as a function of pH at two values of ionic strength.

To fully account for the effect of speciation on rates of precursor complex formation, the following elementary reactions should be considered:

$$>Me^{III}OH_2^+ + HA \xrightarrow{k_1} >Me^{III}AH^+ + H_2O \tag{9.64}$$

$$>Me^{III}OH_2^+ + A^- \xrightarrow{k_1'} >Me^{III}A + H_2O \tag{9.65}$$

$$>Me^{III}OH + HA \xrightarrow{k_1''} >Me^{III}A + H_2O \tag{9.66}$$

$$>Me^{III}OH + A^- \xrightarrow{k_1'''} >Me^{III}A + OH^- \tag{9.67}$$

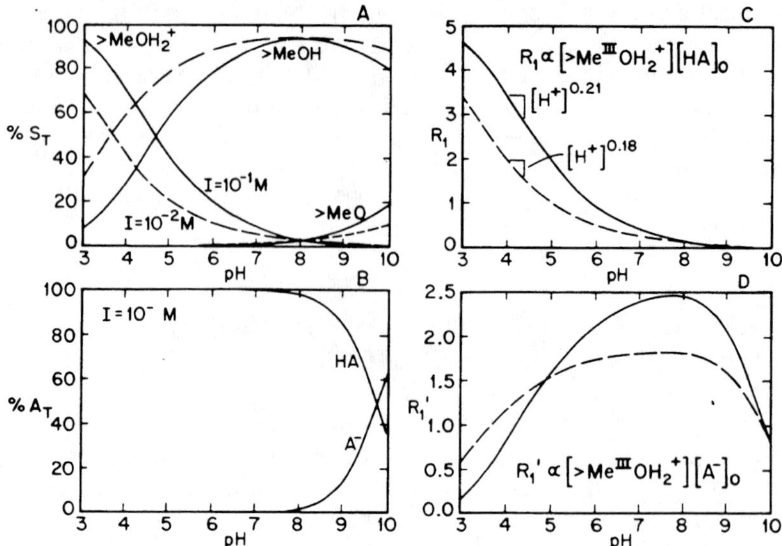

Figure 9.12. The effect of pH on rates of precursor complex formation. (*A,B*) The distribution of surface sites and dissolved reductant species are shown as a function of pH ($S_T = 5 \times 10^{-6}\ M$, $pK^s_{a1}(\text{intr}) = 6.5$, $pK^s_{a2}(\text{intr}) = 9.5$, and $pK_a(\text{phenol}) = 10.0$). (*C,D*) Rates of phenol (HA) and phenolate anion (A^-) adsorption to protonated surface sites.

The overall rate of precursor complex formation (R_1) now reflects changes in speciation with pH.

$$R_1 = k_1[\text{>Me}^{\text{III}}\text{OH}_2^+][\text{HA}] + k_1'[\text{>Me}^{\text{III}}\text{OH}_2^+][\text{A}^-] + k_1''[\text{>Me}^{\text{III}}\text{OH}][\text{HA}] + k_1'''[\text{>Me}^{\text{III}}\text{OH}][\text{A}^-] \quad (9.68)$$

The overall pH dependence of R_1 depends on the relative magnitudes of k_1, k_1', k_1'', and k_1'''. If, for example, precursor complex formation occurs only via reaction 9.64, then R_1 is proportional to the product $[\text{>Me}^{\text{III}}\text{OH}_2^+][\text{HA}]$. Figure 9.12*c* shows the variation in $[\text{>Me}^{\text{III}}\text{OH}_2^+][\text{HA}]$ against pH for two values of ionic strength. R_1 increases with decreasing pH, exhibiting a fractional order with respect to $[\text{H}^+]$ of about 0.20. If, on the other hand, precursor complex formation is dominated by the phenolate anion (A^-) reacting with $\text{>Me}^{\text{III}}\text{OH}_2^+$ (reaction 9.65), the pH dependence looks quite different. $[\text{>Me}^{\text{III}}\text{OH}_2^+]$ increases as the pH is decreased, while $[\text{A}^-]_0$ grows as the pH is increased. The product of these two species, and therefore the reaction rate via reaction 9.65, passes through a maximum at intermediate pH values. This is shown in Figure 9.12*d*. If both reactions 9.64 and 9.65 are important, then the overall pH dependence of precursor complex formation lies somewhere in between Figures 9.12*c* and *d*.

Electrostatic interactions affect the rates of surface chemical reactions in several ways. At equilibrium, the concentration of dissolved ions at a charged oxide interface is given by

$$[i]_0 = [i]_{bulk} \exp(-Z_i F \psi_0 / RT) \quad (9.69)$$

where ψ_0 is the electric potential at the interface, Z_i is the charge of species i, and F/RT has its usual meaning. At pH values below the pH_{ZPC}, the oxide surface potential is positive and $[H^+]_0$ is less than $[H^+]$ bulk. At pH values above the pH_{ZPC}, the oxide surface potential is negative and $[H^+]_0$ is greater than $[H^+]$ bulk. The distribution of $>Me^{III}OH_2^+$, $>Me^{III}OH$, and $>Me^{III}O^-$ groups shown in Figure 9.12*a* reflect this electrostatic effect on $[H^+]_0$. Increasing the ionic strength decreases $|\psi_0|$, lessening the impact on electrostatics.

In Figure 9.12*c* the effect of changes in ionic strength on R_1 is due to changes in the relative amounts of $>Me^{III}OH_2^+$, $>Me^{III}OH$, and $>Me^{III}O^-$. Since phenol is a neutral species, $[HA]_0 = [HA]_{bulk}$, and electrostatics have little effect on its distribution. The situation is different in Figure 9.12*d*. The surface phenolate anion concentration is elevated with respect to bulk concentrations when the pH is less than the pH_{ZPC}, and depressed relative to the bulk concentration when the pH is greater than the pH_{ZPC}. Thus, changes in ionic strength influence both the protonation level of surface sites and the surface phenolate anion concentration.

Changes in species concentrations near charged interfaces can be calculated using one of several available electrostatic models (Westall and Hohl, 1980). Species activities are also perturbed by charged interfaces, but this effect is considerably less understood. The dielectric constant and other medium constants, for example, are perturbed by charged interfaces (Nurnberg, 1974), causing changes in the chemical potentials of reactants, products, and activated complex species. As a result, reaction *rates* are also perturbed by charged interfaces.

In situations where reductant adsorption is fast relative to electron transfer, Me(III) surface sites are in quasi-equilibrium. H^+ and OH^- compete with reductant molecules for surface sites in solution of high or low pH.

Low pH

$$>Me^{III}A + H^+ + H_2O \rightleftharpoons >Me^{III}OH_2^+ + HA \quad (9.70)$$

High pH

$$>Me^{III}A + OH^- \rightleftharpoons >Me^{III}O^- + HA \quad (9.71)$$

When the reductant binding constant ($K_1 = k_1/k_{-1}$) is high, reductant surface coverage is high and relatively constant throughout a broad range of intermediate pH values. When K_1 is low, significant reductant adsorption occurs over a narrower pH range.

Protonation of adsorbed reductant molecules may influence rates of electron transfer by changing the energetics of activated complex formation. Protonation of adsorbed phenol molecules can be postulated:

$$>\mathrm{Me^{III}{-}O{-}C_6H_5} + \mathrm{H^+} \rightleftharpoons >\mathrm{Me^{III}{-}O^{\oplus}(H){-}C_6H_5} \tag{9.72}$$

$>\mathrm{Me^{III}A}$ $\quad$ $>\mathrm{Me^{III}AH^+}$

The net rate of electron transfer (R_2) must take both surface species into account.

$$R_2 = k_2[>\mathrm{Me^{III}A}] + k_2'[>\mathrm{Me^{III}AH^+}] \tag{9.73}$$

If k_2' is much greater than k_2, the second term in Eq. 9.73 may dominate, even when $[>\mathrm{Me^{III}AH^+}]$ is smaller than $[>\mathrm{Me^{III}A}]$. In practice, distinguishing between reaction via $>\mathrm{Me^{III}A}$ and via $>\mathrm{Me^{III}AH^+}$ is quite difficult.

The effect of pH on the reductive dissolution of magnetite (Fe_3O_4) by thioglycolic acid ($HSCH_2COOH$) is quite dramatic; the dissolution rate attains a maximum value at pH 4.5 and drops off sharply at higher or lower pH values (Baumgartner et al., 1982). Protonated surface sites and deprotonated thioglycolate anion are joined together in the precursor complex:

$$>\mathrm{Fe^{III}OH_2^+} + \mathrm{HSCH_2COO^-} \rightleftharpoons \{>\mathrm{Fe^{III}OH_2^+}, \mathrm{HSCH_2COO^-}\} \tag{9.74}$$

$$\{>\mathrm{Fe^{III}OH_2^+}, \mathrm{HSCH_2COO^-}\} \xrightarrow{\text{slow}} \text{products} \tag{9.75}$$

The inner- or outer-sphere characteristics of the precursor complex are not known. The concentration of thioglycolate anion increases with pH, while the surface coverage by $>\mathrm{Fe^{III}OH_2^+}$ decreases with increasing pH. Hence the product $[\mathrm{HSCH_2COO^-}][>\mathrm{Fe^{III}OH_2^+}]$ reaches a maximum value, explaining the observed pH dependence (Baumgartner et al., 1982).

8. ADSORPTION OF REDOX-INERT SPECIES

Redox-inert species can adsorb to metal oxide surfaces, changing the availability and reactivity of surface sites and altering rates of product release. Blocking adsorbates, for example, bind surface sites and inhibit subsequent redox reactions. The effectiveness of blocking adsorbates is improved when rates of ligand exchange at blocked sites are lower than rates at unoccupied sites. On heterogeneous surfaces, blocking adsorbates may bind preferentially to sites that react most quickly with reductants. In this way, the degree of inhibition may be greater than the fraction of total surface sites occupied by blocking adsorbate.

The presence of calcium and phosphate lowers rates of Mn(III,IV) oxide reductive dissolution by hydroquinone (Stone and Morgan, 1984a). Measurements of adsorbed phosphate show that dissolution rates decrease in proportion to increases in phosphate surface coverage.

Adsorption of redox-inert species may also promote electron transfer and reduced metal ion release. In the mechanism outlined below, adsorption of X^- changes the coordinative environment of metal centers on the oxide surface, promoting reaction with the outer-sphere reductant HA:

$$>\mathrm{Me^{III}OH} + \mathrm{HA} \longrightarrow \text{no reaction} \tag{9.76}$$

$$>\mathrm{Me^{III}OH} + \mathrm{X^-} \rightleftharpoons\ >\mathrm{Me^{III}X} + \mathrm{OH^-} \tag{9.77}$$

$$>\mathrm{Me^{III}X} + \mathrm{HA} \rightleftharpoons [>\mathrm{Me^{III}X,HA}] \xrightarrow{\text{fast}} \text{products} \tag{9.78}$$

In this case, rates of reductive dissolution by HA increase as the surface coverage by X^- is increased.

When release of reduced metal ion is rate-limiting, adsorbed ligand molecules can facilitate dissolution through the following mechanism:

$$>\mathrm{Me^{II}OH_2} + 2\mathrm{H^+} \xrightarrow{k_4}\ >\mathrm{Me^{III}OH} + \mathrm{Me^{2+}(aq)} \tag{9.79}$$

$$>\mathrm{Me^{II}OH_2} + \mathrm{X^-} \rightleftharpoons\ >\mathrm{Me^{II}X^-} + \mathrm{H_2O} \tag{9.80}$$

$$>\mathrm{Me^{II}X^-} + 2\mathrm{H^+} \xrightarrow{k_4'}\ >\mathrm{Me^{III}OH} + \mathrm{MeX^+(aq)} \tag{9.81}$$

$$\frac{d[\mathrm{Me^{2+}}]}{dt} = k_4[>\mathrm{Me^{II}OH_2}] + k_4'[>\mathrm{Me^{II}X^-}] \quad \text{(at constant pH)} \tag{9.82}$$

The rate enhancement in the presence of X^- depends on the surface coverage by X^- and the magnitude of k_4' relative to k_4. To prove the existence of such a mechanism, it is important to verify that (1) the auxiliary ligand is not reducing surface sites and (2) the auxiliary ligand is not promoting the release of Me(III) surface groups as well as Me(II) surface groups. It has recently been observed that adsorbed oxalate ligands promote the reductive dissolution of goethite and other iron oxides by ascorbate (Zinder and Stumm, 1985). Reactions 9.76–9.78 or reactions 9.79–9.81 may account for this observation.

9. EFFECTS OF OXIDE MINERALOGY AND SURFACE STRUCTURE

Oxide mineralogy and surface structure affect rates of both nonreductive and reductive dissolution reactions. The effects of surface morphology, crystallographic orientation, nature and density of defects, and presence of impurities on dissolution rates have been addressed in several reviews (Diggle, 1973; Gorichev and Kipriyanov, 1984; Bruyere and Blesa, 1985). Even for a particular mineral, method of preparation and previous history influence dissolution rates. Differences in NiO(s) annealing temperature, for example, cause order-of-magnitude changes in area-normalized dissolution rates in acid solutions (Jones et al., 1977). Oxides with identical stoichiometry but different crystal structure, such as hematite (α-Fe_2O_3) and maghemite (γ-Fe_2O_3), exhibit differences in metal—oxygen bond length and electronic structure (Sherman, 1985). Different rates of ligand substitution and electron transfer can therefore be expected.

When release of reduced metal ions from oxide surfaces is slow, a significant fraction of surface sites may be in a reduced form. Under these conditions, the oxidation state of neighboring metal centers may influence rates of release. As illustration, consider the coordinative environment of the central metal site in each of the following three surface configurations:

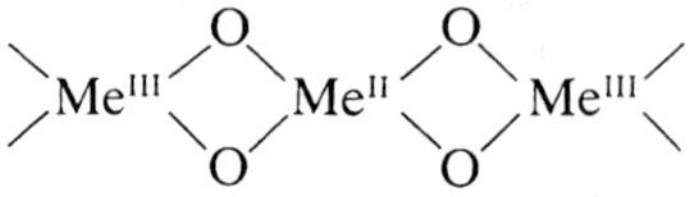

$$>Me^{III}\overset{O}{\underset{O}{\diamond}}Me^{II}\overset{O}{\underset{O}{\diamond}}Me^{III}<$$

I

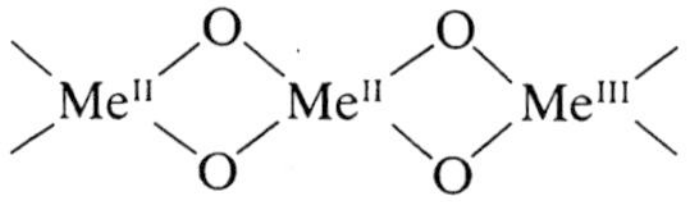

$$>Me^{II}\overset{O}{\underset{O}{\diamond}}Me^{II}\overset{O}{\underset{O}{\diamond}}Me^{III}<$$

II

$$>Me^{II}\overset{O}{\underset{O}{\diamond}}Me^{II}\overset{O}{\underset{O}{\diamond}}Me^{II}<$$

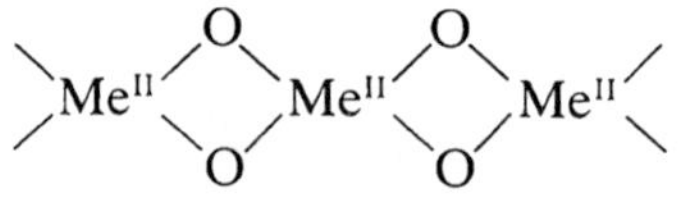

III

As neighboring metal centers are reduced, metal—lattice bond lengths are modified, influencing reactivity. In addition, Me(III)—Me(II) orbital interactions between neighboring metal centers in oxide minerals have been shown to occur that can lead to resonance stabilization (Sherman, 1986). As a consequence, it is possible that on some mineral surfaces, two or three reduced surface groups must be neighboring one another before a metal center can be released into the overlying solution.

10. EXPERIMENTAL VERIFICATION OF THEORY

Rates of surface chemical reactions depend directly upon surface speciation. Experimental determination of surface speciation is, however, quite difficult, and therefore rate expressions are most often expressed in terms of composite variables (such as total surface coverage) or in terms of bulk solution concentrations. This complicates the validation of theories concerning reductive dissolution reactions that were developed in earlier sections.

Small changes in reductant structure often cause substantial changes in rates of reductive dissolution. Measurements of dissolution rate alone do not allow changes in rates of adsorption, desorption, and electron transfer reactions to be distinguished. Measurements of reductant surface coverage add to the understanding of reaction mechanism, by allowing rates of electron transfer and metal ion release to be distinguished from rates of adsorption and desorption. Most methods of measuring surface coverage are based upon physical separation of oxide particles and supernatant before analysis. Problems arise when surface speciation changes during separation or when redox reactions following separation consume the adsorbed reductant. Thus, measurements of surface coverage are meaningful only when the time scale required for measurement is shorter than characteristic times of adsorption, desorption, and electron transfer.

The apparent order of reductive dissolution reactions with respect to $[H^+]$ can arise from a combination of several factors, including (1) changes in rates of ligand exchange and electron transfer following protonation, (2) changes in reductant surface coverage and surface speciation as the pH is changed, and (3) participation of protons in the metal ion release step. Measurements of reductant surface coverage during reaction would indicate the relative importance of factor 2 in determining the pH dependence.

The effect of temperature on rates of reductive dissolution are often reported in terms of an apparent, or overall, activation energy. The apparent activation energy depends on medium composition and surface characteristics as well as on activation energies of individual elementary reactions contributing to the overall mechanism. For reaction steps in quasi-equilibrium, the temperature dependence arises from $\Delta H°$, the enthalpy of reaction.

11. CONCLUSIONS

A model for the reductive dissolution of metal (hydr)oxide minerals has been explored that involves the following surface chemical reactions: (1) precursor complex formation, (2) electron transfer, (3) release of oxidized organic product, and (4) release of reduced metal ion. Surface speciation and rates of surface chemical reactions are intimately associated; their interdynamics control rates of metal ion release from mineral surfaces. Characteristic times for individual surface chemical reactions respond differently to variations in overlying solution composition. As a consequence, increases in pH or reductant concentration can shift the rate-limiting step from one surface chemical reaction to another.

Understanding of reaction kinetics and mechanism is ultimately limited by what is known about surface speciation and rates of surface chemical reactions.

All system attributes that influence dissolution rates must be carefully taken into account when interpreting experimental results. New analytical approaches that enlarge the list of experimentally measurable quantities may substantially improve our understanding of reductive dissolution reactions.

Acknowledgments

This work has been supported by the National Science Foundation (U.S.A.) under Grant No. CEE-04076. Mathematical support by Bofu Yu and Daniel Zwillinger is gratefully acknowledged.

REFERENCES

Bard, A. J., and Faulkner, L. R. (1980), *Electrochemical Methods,* Wiley, New York.

Baumgartner, E., Blesa, M. A., and Maroto, A. J. G. (1982), Kinetics of the dissolution of magnetite in thioglycolic acid solutions, *J. Chem. Soc. Dalton Trans.* **1982,** 1649–1654.

Berner, R. A. (1980), *Early Diagenesis: A Theoretical Approach,* Princeton University Press, Princeton, NJ.

Bruyere, V. I. E., and Blesa, M. A. (1985), Acidic and reductive dissolution of magnetite in aqueous sulfuric acid; site-binding model and experimental results, *J. Electroanal. Interfacial Chem.* **182,** 141–156.

Denbigh, K. G., and Turner, J. C. R. (1965), *Chemical Reactor Theory,* Cambridge University Press, Cambridge, MA.

Diggle, J. W. (1973), "Dissolution of Oxide Phases," in J. W. Diggle, Ed., *Oxides and Oxide Films,* Vol. 2, Marcel Dekker, New York.

Forsythe, G. E., Malcolm, M. A., and Moler, C. B. (1977), *Computer Methods for Mathematical Computations,* Prentice-Hall, Englewood Cliffs, NJ.

Furrer, G., and Stumm, W. (1983), The role of surface coordination in the dissolution of δ-Al_2O_3 in dilute acids, *Chimia* **37,** 338–341.

Furrer, G., and Stumm, W. (1986), A coordination chemical approach to the kinetics of weathering: I. Dissolution of oxides; case studies on δ-Al_2O_3 and BeO, *Geochim. Cosmochim. Acta* **50,** 1847–1860.

Ghiorse, W. C. (1986), "Microbial Reduction of Manganese and Iron," in A. J. B. Zehnder, Ed., *Environmental Microbiology of Anaerobes,* Wiley, New York.

Gorichev, I. G., and Kipriyanov, N. A. (1984), Regular kinetic features of the dissolution of metal oxides in acidic media, *Russ. Chem. Rev.* **53,** 1039–1061.

Jones, C. F., Segall, R. L., Smart, R. S. C., and Turner, P. S. (1977), Semiconducting oxides: The effect of prior annealing temperature on dissolution kinetics of nickel oxide, *Trans. Faraday Soc. I* **73,** 1710–1720.

Littler, J. S. (1970), "The mechanisms of Oxidation of Organic Compounds with One-Equivalent Metal-Ion Oxidants: Bonded and Non-bonded Electron Transfer," in *Essays on Free-Radical Chemistry,* Special Publication, The Royal Chemical Society, No. 24.

Martell, A. E., and Smith, R. M. (1977), *Critical Stability Constants,* Plenum, New York.

Morrison, S. R. (1980), *Electrochemistry at Semiconductor and Oxidized Metal Electrodes,* Plenum, New York.

Nurnberg, H. W. (1974), "The Influence of Double Layer Effects on Chemical Reactions at Charged Interfaces," in V. Zimmerman and J. Dainty, Eds., *Membrane Transport in Plants,* Springer-Verlag, New York.

Schindler, P. W., Fuerst, B., Dick, R., and Wolf, P. U. (1976), Ligand properties of surface silanol groups: I. Surface complex formation with Fe^{3+}, Cu^{2+}, Cd^{2+}, and Pb^{2+}, *J. Colloid Interface Sci.* **55,** 469–475.

Segal, M. G., and Sellers, R. M. (1980), Reactions of solid Fe(III) oxides with aqueous reducing agents, *J. Chem. Soc. Chem. Commun.* **180,** 991–993.

Segal, M. G., and Sellers, R. M. (1984), Redox reactions at solid–liquid interfaces, *Adv. Inorg. Bioinorg. Mechanisms* **3,** 97–129.

Sherman, D. M. (1985), The electronic structures of Fe^{3+} coordination sites in iron oxides; applications to spectra, bonding, and magnetism, *Phys. Chem. Minerals* **12,** 161–175.

Sherman, D. M. (1986), Cluster molecular orbital description of the electronic structures of mixed-valence iron oxides and silicates, *Solid State Commun.*, **58,** 719–723.

Stone, A. T. (1986), "Adsorption of Organic Reductants and Subsequent Electron Transfer on Metal Oxide Surfaces," in J. A. Davis and K. F. Hayes, Eds., *Geochemical Processes at Mineral Surfaces*, American Chemical Society Symposium Series 323, Washington, DC, 446–461.

Stone, A. T., and Morgan, J. J. (1984a), Reduction and dissolution of manganese(III) and manganese(IV) oxides by organics. 1. Reaction with hydroquinone, *Environ. Sci. Technol.* **18,** 450–456.

Stone, A. T., and Morgan, J. J. (1984b), Reduction and dissolution of manganese(III) and manganese(IV) oxides by organics. 2. Survey of the reactivity of organics, *Environ. Sci. Technol.* **18,** 617–624.

Stumm, W., Furrer, G., Wieland, E., and Zinder, B. (1985), "The Effects of Complex-Forming Ligands in the Dissolution of Oxides and Aluminosilicates," in J. Drever, Ed., *The Chemistry of Weathering,* NATO ASI Series, Ser. C, Vol. 149, D. Reidel, Dordrecht, Holland.

Stumm, W., Hohl, H., and Dalang, F. (1976), Interaction of metal ions with hydrous oxide surfaces, *Croat. Chem. Acta* **48,** 491–504.

Stumm, W., Huang, C. P., and Jenkins, S. R. (1970), Specific chemical interaction affecting the stability of dispersed systems, *Croat. Chem. Acta* **42,** 223–245.

Stumm, W., and Morgan, J. J. (1981), *Aquatic Chemistry,* 2nd ed., Wiley-Interscience, New York.

Suatoni, J. C., Snyder, R. E., and Clark, R. O. (1961), Voltammetric studies of phenol and aniline ring substitution, *Anal. Chem.* **33,** 1894–1897.

Valverde, N. (1976), Investigations on the rate of dissolution of metal oxides in acidic solutions with additions of redox couples and complexing agents, *Bunsenges. Phys. Chem.* **80,** 333–340.

Valverde, N., and Wagner, C. (1976), Considerations on the kinetics and the mechanism of the dissolution of metal oxides in acidic solutions, *Bunsenges. Phys. Chem.* **80,** 330–333.

Waite, T. D. (1986), "Photoredox Chemistry of Colloidal Metal Oxides," in J. A. Davis and K. F. Hayes, Eds., *Geochemical Processes at Mineral Surfaces*, American Chemical Society Symposium Series 323, Washington, DC, 426–445.

Westall, J., and Hohl, H. (1980), A comparison of electrostatic models for the oxide/solution interface, *Adv. Colloid Interface Sci.* **12,** 265–294.

Zinder, B., Furrer, G., and Stumm, W. (1986), A coordination chemical approach to the kinetics of weathering: II. Dissolution of Fe(III) oxides, *Geochim. Cosmochim. Acta* **50,** 1861–1869.

Zinder, B., and Stumm, W. (1985), Die auflosung von eisen(III)-oxiden; ihre bedeutung im see und im boden, *Chemia* **39,** 280–288.

10

MONTE CARLO SIMULATIONS OF SURFACE REACTION RATE LAWS

Alex E. Blum and Antonio C. Lasaga

Department of Geology and Geophysics, Kline Geology Laboratory, Yale University, New Haven, Connecticut

Abstract

The common theories and concepts employed in dealing with crystal growth and dissolution are analyzed with detailed Monte Carlo simulations. The Monte Carlo method is presented in a way that can be used to study various aspects of surfaces and surface reaction rates. In particular, the relation between Monte Carlo parameters and observed properties of minerals (e.g., surface free energy, molecular volume) is made, and the role of microscopic reversibility is emphasized. The Monte Carlo results are then used to follow the details of surface reaction rates as the thermodynamic state of the system passes from great undersaturation through equilibrium and into high supersaturation. The usual rate theories are shown to be qualitatively correct but quantitatively inaccurate. The role of defects is emphasized, and the influence of dislocations on the surface rate laws is discussed. In particular, Monte Carlo results are used to predict when etch pits around dislocations should form, what the etch pit morphology should be, and the relative role of dissolution at dislocation sites versus dissolution in dislocation-free regions of surfaces on the overall dissolution of crystals in the laboratory and in nature.

1. INTRODUCTION

The kinetics of mineral dissolution and precipitation are critical to understanding many geochemical processes, particularly diagenesis, weathering, hydrothermal alteration, and the chemical evolution of natural waters. The focal point of dissolution or growth is the mineral *surface* in contact with the fluid phase. Extensive SEM examination of both experimentally dissolved and natural mineral surfaces has revealed numerous and ubiquitous crystallographically controlled etch pits (e.g., Berner and Holdren, 1979; Berner and Schott, 1982; Wilson and McHardy, 1980; Brantley et al., 1986). The presence of etch pits in minerals suggests that defects in the crystal lattice are sites of strong preferential dissolution during geochemical processes. This phenomenon has been utilized extensively in etching techniques for determining line dislocation densities, delineating subgrain boundaries, and counting radioactive fission tracks. It is important to understand the influence of crystal defects on dissolution mechanisms at a molecular level, in order to evaluate the contribution of defects to bulk dissolution and reaction rates in geochemical processes and to use etch pits and other dissolution features to interpret geochemical environments.

Crystal growth and dissolution involve two stages: (1) transport of ions or molecules to and from the crystal surface and (2) surface reactions that actually incorporate or remove ions from the crystal lattice. These two processes are closely linked. The net flux of species away from the surface must equal the rate of addition or release by surface reactions, and the concentration gradient in the solution immediately adjacent to the crystal surface will adjust itself to maintain this balance.

Two extreme cases can be distinguished. If the surface reaction is "slow," then the solution concentration next to the surface will be close to that in the bulk solution. The concentration gradient will be very small so as to also slow diffusion. In this case, the energetics and dynamics of the reactions at the surface will control the behavior of the overall rate. If the surface reaction is "fast," then the solution concentration adjacent to the mineral surface will adjust to be near equilibrium so as to slow the surface reaction and match the diffusion flux induced by the steep concentration gradient. In this case, the rate of diffusion will determine the magnitude of the overall rate. The transport rate may depend on (1) the effective diffusion rate of the species in solution; (2) the advective flux of species due to fluid flow; and (3) the solid-state diffusion of species through any solid precipitate or reaction rim that may armor a crystal.

Diffusion of aqueous species is rapid relative to the low-temperature dissolution rate of most minerals (Berner, 1978) and therefore does not generally influence the dissolution rate except in rare cases. A typical method of determining whether a reaction is controlled by diffusion in solution is to vary the stirring rate. If a reaction is surface-controlled, the rate will not vary with stirring rate. On the other hand, because the stirring rate affects the thickness of the stagnant fluid boundary layer (diffusive layer) around the surface of a mineral, the rate of a diffusion-controlled reaction will increase with stirring rate (Tole et al., 1986). For example, the dissolution of calcite is found to be surface-controlled at 25°C except for $pH < 4$. At low pH, the surface rates speed up enough that transport in solution becomes rate-limiting (Berner and

Morse, 1974). Likewise, nepheline [(Na,K)$AlSiO_4$] dissolution is surface-controlled except for pH ≤ 3 *and* $T \geq 60°C$ in the range $1 \leq pH \leq 10$ and $25°C \leq T \leq 80°C$ (Tole et al., 1986).

Solid-state diffusion is much slower than diffusion of aqueous species, and the presence of any protective surface layer may severely limit the rate of dissolution or growth. Early work by Wollast (1967), Helgeson (1971), and Paces (1973) suggested that dissolution kinetics of feldspars is controlled by solid-state diffusion through a leached layer. However, more recent examination of the surface chemistry of feldspars, amphiboles, and pyroxenes with XPS (Petrovic et al., 1976; Holdren and Berner, 1979; Fung et al., 1980; Berner and Schott, 1982; Schott and Berner, 1983) has failed to find an extensive leached layer, except possibly in Fe^{2+}-bearing minerals under oxidizing conditions. Therefore, the researchers concluded that in common low-temperature geochemical environments, the dissolution rate of most silicates is dominated by chemical reactions at the mineral surface, not by transport processes (Berner, 1978; Lasaga, 1984). The dominance of surface-reaction-controlled dissolution over transport means that local effects such as crystal defects and impurities may have a strong effect on the mechanism and thus on the rate of growth or dissolution. The control of surface morphology and energetics on the dissolution rate is noticeably emphasized by the pervasive existence of crystallographically controlled etch pits (see Fig. 10.1). These etch pits have been proposed by Berner (1978, 1980, 1981) as evidence for surface-controlled reactions. They substantiate the role of dislocation defects in facilitating dissolution, a role that will be elaborated on in subsequent sections.

This chapter will first briefly summarize some simple theories of crystal growth and dissolution and review thermodynamic considerations of etch pit formation in the vicinity of line dislocations. Because analytical solutions to the equations describing complex surface processes are extremely difficult if not intractable, further development of the theory of the surface kinetics of crystal growth and dissolution will proceed by means of Monte Carlo simulations. The remainder of the chapter will describe the use of Monte Carlo simulations in investigating crystal growth and dissolution mechanisms and will present the results of several simulations and their implications for the mechanisms of silicate dissolution and growth.

2. SURFACE REACTION RATE LAWS: SIMPLE MODELS

In this section, some basic theoretical approaches to surface-reaction-controlled crystal growth and dissolution are discussed. All these approaches consider the detailed surface configuration on a molecular scale. Figure 10.2 shows a schematic crystal surface and some of the atomic processes relevant to the kinetic theory. Before a solute molecule can be incorporated into a crystal, it must impinge on and attach itself to the crystal surface. This process

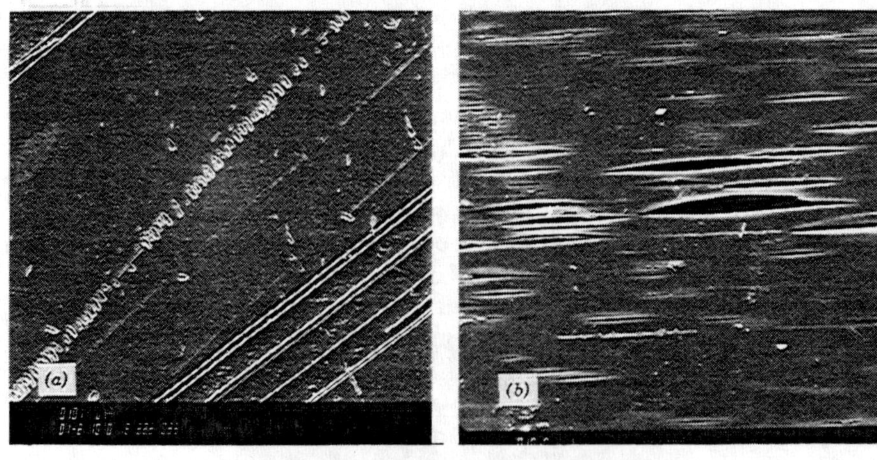

Figure 10.1. (*a*) Etch pits developed during dissolution of (*b*) pyroxene (Berner and Schott, 1982) and (*c*) quartz.

may also involve some loss of water of hydration. When an atom is added to a flat crystal surface (becoming an adatom), it forms only a few bonds. This is an energetically unfavorable situation, and adatoms will have a strong tendency to redissolve into solution. Surface diffusion may enable the adatom to attach itself to a stable step. Surface diffusion has been observed directly in molecular beam experiments on alkali halides (Bjorklund et al., 1977; Bjorklund and Spears, 1977). If the surface diffusion coefficient is D_s, an adatom will on average cover a distance r before redissolving. This distance is given by

$$r \sim (4D_s\tau)^{1/2} \tag{10.1}$$

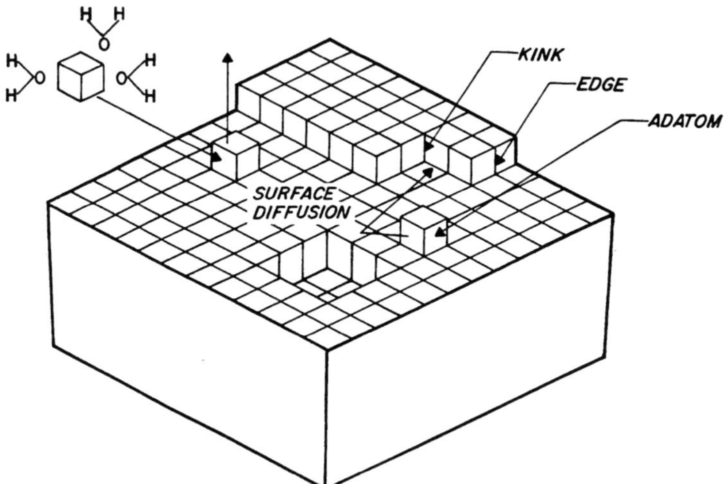

Figure 10.2. Salient features and elementary processes of surfaces.

where τ is the mean lifetime of an adatom between adsorption and desorption. A step is defined by the atoms with four bonds (see Fig. 10.2). At the corner of a step, atoms have three bonds, that is, half the number of bonds in the bulk crystal (assumed here to be simple cubic). These atoms have a special kinetic significance, and the sites are termed "kink" sites. The attachment of an adatom by only one bond allows it to diffuse across the surface easily. The addition of an atom to a step or a kink will form two or three new bonds, respectively, and result in a progressively smaller increase in surface free energy and progressively less tendency to diffuse or redissolve. Steps, and particularly kink sites, will then be energetically more favorable sites for the attachment and retention of new atoms on the surface.

The ultimate evolution of this process of aggregation of surface atoms can be described in several ways, and these distinctions separate the simple kinetic theories. If the reaction rate depends on the formation of a cluster of surface atoms that is stabilized by possessing a large enough size, the two-dimensional nucleation rate theory is obtained. If the surface has sufficient steps (from defects) that only the speed of attachment by diffusion of the unstable adatoms to the steps limits the reaction rate, the Burton–Cabrera–Frank (BCF) theory of screw dislocation growth is obtained (Burton et al., 1951; also see Ohara and Reed, 1973; Christian, 1975; Bennema, 1984).

The surface nucleation theory is based on the free energy change for the formation of a cluster of n atoms,

$$\Delta G_n = -n\,\Delta\mu + A_n\sigma \tag{10.2}$$

where $\Delta\mu$ is the saturation state of the solution ($\Delta\mu = \mu_{\text{liquid}} - \mu_{\text{solid}}$), A_n is the *new* surface area generated by the cluster, and σ is the surface free energy. If the volume of the cluster is set at nv (v = atomic volume) and the cluster is a pancake h units high and r units in diameter,

$$nv = \pi r^2 h \tag{10.3}$$

The new surface area is that along the edge of the pancake,

$$A_n = 2\pi rh$$

Using Eq. 10.3 to solve for r,

$$A_n = 2(\pi vh)^{1/2} n^{1/2} \tag{10.4}$$

Inserting 10.4 into 10.2, we find that the ΔG_n curve initially increases with n and reaches a maximum when

$$n_c = \frac{\pi vh}{\Delta\mu^2}\sigma^2 \tag{10.5}$$

or

$$r_c = \frac{\sigma v}{\Delta\mu} \tag{10.6}$$

the latter equation resulting from combining Eqs. 10.3 and 10.5. For sizes beyond the critical size r_c, ΔG_n decreases. Hence the two-dimensional nucleation theory predicts the reaction rate as the rate of nucleation of clusters greater than r_c. This nucleation rate, R, depends exponentially on the free energy barrier ΔG_n at the critical cluster:

$$\Delta G_{n_c} = \frac{\pi hv\sigma^2}{\Delta\mu} \tag{10.7}$$

$$R \propto \exp\left[-\Delta G_{n_c}/RT\right] \tag{10.8}$$

Equations 10.7 and 10.8 predict too strong a dependence of surface rate on $\Delta\mu$ and are in disagreement with experiment. However, the concept of a critical cluster size is essential to much of crystal growth and dissolution theory and will be useful in later sections.

The BCF theory circumvents the necessity to form critical-size clusters by introducing the presence of dislocation-induced steps on the surface. In particular, if a screw dislocation intersects the surface, a step is produced which, if propagated by growth, will wind into a spiral. The spiral will provide a sequence of steps separated by a certain distance r_{step}. In theory, then, the rate of *lateral* movement of the steps is computed from solution of the surface diffusion equation for adatoms which are attaching themselves to the steps. This step velocity, v_{step}, is found to be proportional to $\Delta\mu$, the supersaturation, for small values of $\Delta\mu$ and independent of $\Delta\mu$ for large values of $\Delta\mu$. The step spacing, r_{step}, also depends on the rate of spiraling of the screw dislocation and thus also depends on $\Delta\mu$. In this case,

$$r_{step} \sim 19r_c \tag{10.9}$$

r_c being the critical radius given by Eq. 10.6 (Ohara and Reed, 1973). The net growth rate is given by

$$R = \frac{hv_{step}}{r_{step}} \tag{10.10}$$

where h is the height of the steps and r_{step}/v_{step} is the time needed to cover the distance r_{step} for each step, thereby filling a whole new crystal layer of height h. Equations 10.6, 10.9, and 10.10 yield

$$R \propto \Delta\mu^2 \tag{10.11a}$$

for small values of $\Delta\mu$. For large values of $\Delta\mu$,

$$R \propto \Delta\mu \tag{10.11b}$$

Equations 10.8, 10.11a, and 10.11b are frequently quoted in the literature. However, it is important to note that they are obtained by making many simplifications [see Ohara and Reed, (1973) or Christian (1975) for further discussion]. In the following sections, extensions of the theory by Monte Carlo results will be taken up.

3. DISLOCATIONS

The effect of a continuously propagating spiral step at the outcrop of a screw dislocation on the surface has already been discussed. However, dislocations have an additional effect. Distortion of the crystal lattice in the vicinity of a dislocation (see Fig. 10.3) strains the crystal bonds, resulting in a field of excess strain energy emanating radially outward from the dislocation. When a dislocation intersects the crystal surface, the stressed region will provide a site for preferential dissolution. The dissolution of the stressed region will release the strain energy, resulting in a greater gain in free energy than dissolution of unstrained material. Note that while the rate laws of the two-

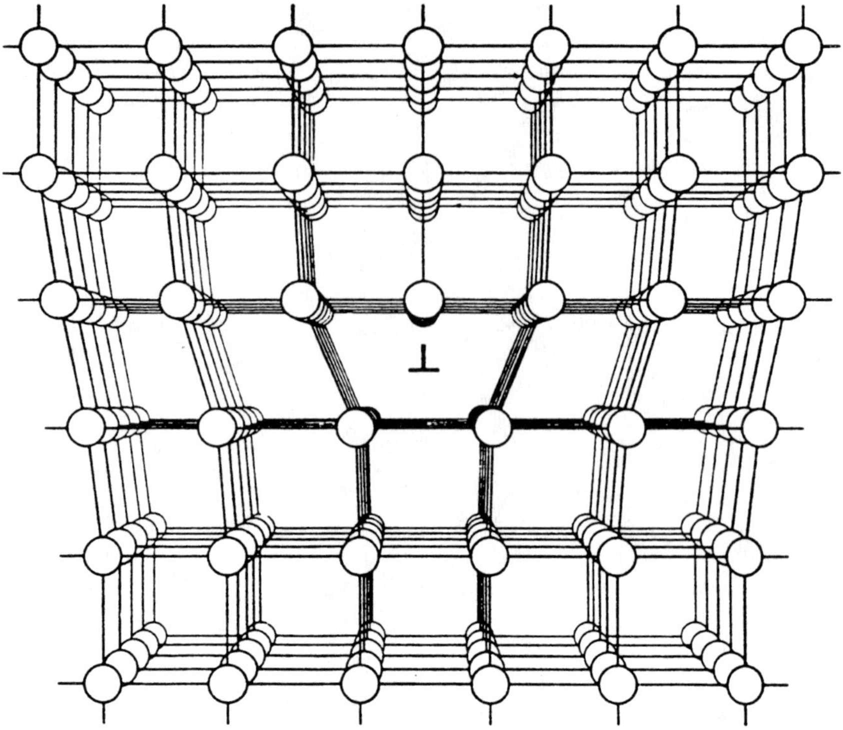

(a)

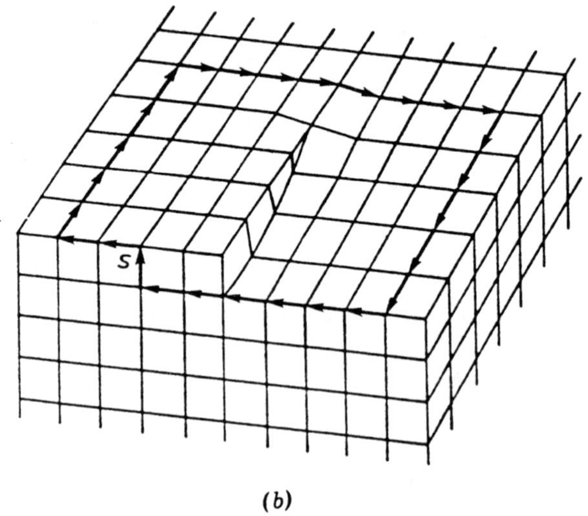

(b)

Figure 10.3. (*a*) Illustration of strain field around an edge dislocation (shown by $\perp$). (*b*) Strain field around a screw dislocation. Burger vector shown by $\vec{s}$.

dimensional nucleation and BCF theories predict that growth and dissolution have the same qualitative dependence on $\Delta\mu$, the effect of a strain field is nonsymmetric. Dissolution will be significantly enhanced in the strained region, and precipitation will be retarded.

Far from the dislocation, where atomic displacements are small (less than the elastic limit), the strain energy density in an isotropic solid can be obtained from elastic theory:

$$u(r) = \frac{\mu b^2}{8\pi^2 K r^2} \tag{10.12}$$

where $u(r)$ is the strain energy density, μ is the shear modulus, b is the magnitude of the Burger vector (i.e., the size of the misfit created by the dislocation), and r is the distance from the dislocation line. K is a constant that describes the relative contribution of screw and edge components in the dislocation strain field and is given by

$$\frac{1}{K} = \frac{\sin^2\psi}{1 - \nu} + \cos^2\psi \tag{10.13}$$

where ψ is the angle between the dislocation line and the Burger vector and ν is the Poisson ratio ($\nu \sim 0.25$ usually). $K = 1$ for a pure screw dislocation, and $K = 1 - \nu$ for an edge dislocation.

Close to a dislocation, the distortion of the lattice is greater than the elastic limit. In this core region, linear elastic theory is no longer applicable. This is evident since Eq. 10.12 approaches infinity as $r \rightarrow 0$. Estimating the nonlinear behavior of the strain energy in the core region has been the topic of several papers, with little conclusive experimental data or agreement as to the form of the nonlinear terms.

Van der Hoek et al. (1982a) have suggested that the strain energy density $u(r)$ be given by

$$u(r) = \frac{\mu b^2/8\pi^2 K}{r_h^2 + r^2} \tag{10.14}$$

where r_h can be thought of as the radius of the core region. If we set r to zero in eqn. (10.14),

$$r_h^2 = \frac{\mu b^2}{8\pi^2 K u(0)} \tag{10.15}$$

where $u(0)$ is the strain energy density at the center of the dislocation. Equation 10.14 is continuous, approaches the approximation of linear elastic theory (Eq. 10.12) as $r \rightarrow \infty$, and attains a finite value as $r \rightarrow 0$. However, the choice of Eq. 10.14 is somewhat arbitrary, and any function meeting the previous criteria would be equally appropriate at our present state of knowledge. The strain energy at the center of the dislocation $u(0)$ can be approximated by ΔH_m, the enthalpy of melting. The rationale behind this approximation is that the strain energy of the bonds associated with thorough disordering of the crystal at the core is similar to the energy needed to disorder the crystal into a melt, ΔH_m.

The presence of dislocations in a crystal will increase the total free energy of a crystal by the amount of strain energy stored in the deformed region around the dislocations. This change in the total free energy will affect the calculated bulk thermodynamic properties of a crystal, including the solubility. Wintsch and Dunning (1985) used this approach to calculate the effect of dislocation density on the solubility of quartz. They calculated that a dislocation density of 10^9, 10^{10}, 10^{11}, and 10^{12} cm of dislocation per cubic centimeter will increase the free energy of quartz by 2.2, 19.7, 170, and 1426 J mole^{-1}, respectively. The effect is significant only at dislocations above about 10^{10}, whereas the dislocation density of quartz rarely exceeds 10^9 cm cm^{-3}. They conclude that dislocations may be an important driving force for recrystallization of highly deformed grains such as in mylonites, where dislocation tangles may locally exceed 10^{11} cm cm^{-3}, but are not important in most geochemical environments. However, this is a thermodynamic argument only.

The strain energy around a dislocation is a highly localized phenomenon. While the strain energy around dislocations may not have a major effect on the bulk thermodynamic properties of a crystal, the local effects may have a substantial influence on the kinetics of the atomic processes and therefore on the rate of surface reactions. Cabrera and Levine (1956) first illustrated that the energy gained by dissolution of the stressed material around dislocations may allow nucleation of dissolution holes with little or no activation energy barrier. The nucleation energy depends on the saturation state of the solution relative to the unstrained crystal and on the magnitude of the strain energy. In highly undersaturated solutions, the steps formed at the strained region near a dislocation may continue to propagate outward across the unstressed crystal with a continuous gain in energy, forming an etch pit. Closer to saturation, the outward propagation of the steps will be restricted by an energy barrier at some critical radius, forming a microscopic hollow core around the dislocation, but not a macroscopic etch pit (Lasaga and Blum, 1986).

The change in free energy required to slightly enlarge a cylindrical hole around a dislocation by an amount dr requires the dissolution of a volume $2\pi rL\,dr$, where r is the radius of the cylindrical hole and L is the length. The energy change will have three components:

1. The release of strain energy $[dG = -2\pi rLu(r)\,dr]$.
2. The dissolution of unstrained crystalline material into solution $(dG = +2\pi rL\Delta\mu/v\,dr)$, where $\Delta\mu$ is the difference in chemical potential between a species in solution and in the unstressed bulk crystal and v is the volume of a molecular unit.

3. The increase in the surface free energy of the crystal resulting from increasing surface area with the enlargement of the hole ($dG = 2\pi\sigma L\,dr$), where σ is the surface free energy.

The change in the free energy is then

$$\frac{dG}{dr} = 2\pi rL\left(\frac{\Delta\mu}{v} + \frac{\sigma}{r} - u(r)\right) \tag{10.16}$$

The total free energy change resulting from the opening of a cylindrical hole of radius r and unit depth is then

$$\Delta G = \int_0^r \frac{dG}{dr}\,dr$$

Integrating Eq. 10.16 and using Eq. 10.14 for $u(r)$:

$$\Delta G = \pi\frac{\Delta\mu}{v}r^2 + 2\pi\sigma r - \frac{\mu b^2}{8\pi K}\ln\left[1 + \left(\frac{r}{r_h}\right)^2\right] \tag{10.17}$$

The maxima of Eq. 10.17 are activation energy barriers to outward expansion of a dissolving step, and any minimum is an energy well which will form a hollow core around the dislocation. Setting $dG/dr = 0$ (Eq. 10.16) and using Eq. 10.14 for $u(r)$, the equation for the extremum points of G is

$$r^3 + r_c r^2 + (r_h^2 - r_o r_c)r + r_c r_h^2 = 0 \tag{10.18}$$

where $r_c = v\sigma/\Delta\mu$ and $r_0 = \mu b^2/8\pi^2 K\sigma$. r_c is the traditional critical radius of two-dimensional nucleation (Eq. 10.6). r_c is positive for supersaturated solutions and negative for undersaturated solutions. A cubic equation has either one or three real roots and the result depends in part on the intensity of the stress field and the saturation of the solution (Van der Hoek et al., 1982a; Lasaga and Blum, 1986). The cases of interest to us are (1) the case with one positive root, which corresponds to a single maximum point, and (2) the case with three positive roots, corresponding to a sequence of maximum, minimum, and maximum points in the curve depicting G versus r. These extrema will be important in the discussions to follow.

4. MONTE CARLO MODEL

Monte Carlo simulations model crystal growth and dissolution as a stochastic process in which microscopic rate constants for individual atomic events are replaced by probabilities (Van der Eerden et al., 1978; Gilmer, 1976, 1977, 1980). In our simulations, there are three possible types of events that may occur at any surface site (e.g., Fig. 10.2): (1) attachment of an atom from solution, (2) dissolution of an atom into solution, and (3) surface diffusion of an atom from one surface site to an adjacent surface site. Crystal growth or dissolution is simulated by choosing surface sites at random and performing the elementary processes of precipitation, dissolution, and surface diffusion with frequencies proportional to the probabilities of each event.

All three possible events are assumed to be activated processes, which means that the rate constant has the form

$$k = A \exp\left(\frac{-E_a}{kT}\right) \tag{10.19}$$

where E_a is the activation energy of the process, k is Boltzmann's constant, T is temperature, and A is the preexponential factor. The activation energies for the various elementary processes are hard to obtain. Surface diffusion activation energies may be obtained from experimental data (Bjorklund and Spears, 1977), and quantum-mechanical calculations may be carried out to obtain the other activation energies (e.g., Lasaga and Gibbs, 1986); but the reliability of these activation energies is still being tested. Because of the uncertainty in activation energies, any constraint on their values will be an important asset to the Monte Carlo simulations. Fortunately, such a constraint can be obtained by use of the kinetic theorem of microscopic reversibility. The principle of microscopic reversibility (Lasaga, 1981; Gilmer, 1977) requires that the rates of the forward and reverse reactions of elementary reactions be equal at equilibrium. This principle can be stated as

$$P(i)p(i \rightarrow j) = P(j)p(j \rightarrow i) \tag{10.20}$$

where $P(i)$ refers to the probability of observing configuration i and $p(i \rightarrow j)$ refers to the probability that the system will make a transition from state i to state j. These $p(i \rightarrow j)$ will correspond to the atomic rate constants needed in the Monte Carlo method. Because the ratio of the p's is related to the ratio of the P's according to Eq. 10.20, and the P's depend exponentially on the configurational energy of the state (e.g., the Boltzmann equation), the activation energies in 10.19 will be related to the difference in energies between states i and j, that is, they will depend on thermodynamic variables. In general terms, taking the ratio of rate constants cancels the preexponential terms in 10.19, and the remaining term, $\exp(-\Delta E_a/KT)$, can be related by 10.20 to the difference in the thermodynamic energies of the reactants and products, $E_j - E_i$. The use of microscopic reversibility will be evident in the discussion of the individual rate constants that follows.

The simulations described here assume a (001) face of an ordered A–B cubic crystal, but *any* face on a general crystallographic system can be modeled using the Monte Carlo method. The simulations discussed here consider only nearest-neighbor interactions between the atoms on the surface of the crystal.

Furthermore, the model divides the crystal into distinct cells of type A or B. Any complex surface configuration of steps, kinks, and surface clusters can be represented, subject to two restrictions: Two solid cells of the same type may not be nearest neighbors (i.e., perfect ordering), and each solid cell must have another solid cell directly below it (i.e., overhangs are excluded). The latter restriction allows any surface configuration to be represented by a rectangular array of integers that specify the height of each column of solid cells perpendicular to the (001) face. The restriction against overhangs is not unrealistic for defect-free crystals at low temperature. However, when etch pits are developed at dislocations, the prohibition against overhangs becomes less realistic. Edge effects are eliminated by using a periodic boundary condition along the directions of the surface. Therefore, the surface topology is assumed to repeat indefinitely along the horizontal axes.

The dissolution rate constant at any surface site with n nearest neighbors is given by

$$k_n^- = \nu \exp\left(\frac{-n\Phi}{kT}\right) \tag{10.21}$$

where k is Boltzmann's constant, T is temperature (K), and ν is a frequency factor. Φ is the energy change in a reaction that breaks a solid–solid and fluid–fluid bond and forms two solid–fluid bonds:

$$\Phi = 2\Phi_{sf} - \Phi_{ss} - \Phi_{ff} \tag{10.22}$$

where Φ_{sf}, Φ_{ss}, and Φ_{ff} are the bond energies of the solid–fluid, solid–solid, and fluid–fluid interactions, respectively. Φ is similar to expressions that arise in regular solution theory. Φ can be approximated by the relation

$$\Phi = 2\frac{\Delta H_{\text{dis}}}{N_b} \tag{10.23}$$

where ΔH_{dis} is the enthalpy of dissolution of the crystal and N_b is the number of bonds per atom in the bulk crystal. Multiplication by 2 in Eq. 10.23 takes into account the fact that each bond is shared by two cells. For a simple cubic crystal, $\Phi = \Delta H_{\text{dis}}/3$.

The precipitation probability at a surface site that is consistent with Eq. 10.21 and microscopic reversibility is

$$k^+ = k_{\text{eq}}^+ \exp\left(\frac{\Delta\mu}{kT}\right) = \nu \exp\left(\frac{-3\Phi}{RT}\right) \exp\left(\frac{\Delta\mu}{RT}\right) \tag{10.24}$$

where k^+ is the precipitation probability at the site, k is Boltzmann's constant, and T is temperature. k_{eq}^+ is the precipitation probability at equilibrium between the crystal and solution and must be equal to the dissolution probability at a kink site [i.e., $k_{eq}^{+} = \nu \exp(-3\Phi/kT)$]. $\Delta\mu$ is the difference in chemical potential between a solid cell in solution and one in the bulk crystal. Its inclusion in Eq. 10.24 is a direct result of microscopic reversibility because $\Delta\mu$ enters in the energy difference between the states with a solid unit on the surface and with the unit in solution. (Note we are comparing k_n^- and k^+.) For dilute solutions and one-component systems (e.g., silica), $\Delta\mu = kT \ln(c/c_e)$, where c is the concentration in solution and c_e is the concentration in equilibrium with the unstrained crystal. At equilibrium, $\Delta\mu = 0$. If $\Delta\mu$ is negative, net dissolution will occur. If $\Delta\mu$ is positive, net growth will occur. Note that the probability of precipitation depends on the concentration of species in solution but is independent of the local surface configuration.

Surface diffusion is internally constrained by microscopic reversibility, but the relationship between the absolute activation energy of surface diffusion and the energetics of dissolution and/or precipitation is still somewhat arbitrary. In our model, the diffusion rate depends on the energetics of the initial state (before the jump). There is an activation energy E_a^{dif} associated with the motion of adatoms (i.e., breaking and forming the basal bond). If, in addition, there are lateral bonds that are broken (non-adatoms), then these energetics are also included. Therefore, for an atom with n total bonds ($n - 1$ lateral bonds), the probability scheme used in these simulations is

$$k_{\text{dif}} = \nu \exp\left(\frac{-E_a^{\text{dif}}}{kT}\right) \exp\left(\frac{-(n-1)\Phi}{kT}\right) \tag{10.25}$$

where k_{dif} is the total probability of surface diffusion and E_a^{dif} is an independent activation energy of surface diffusion. Equation 10.25 assumes that the activation energy of breaking lateral bonds is the same for dissolution and surface diffusion. This scheme for surface diffusion depends only on the number of lateral bonds; it does not discriminate as to whether mobile atoms jump vertically up or down ledges. However, if any single surface diffusion event involves jumping more than six vertical cells, either up or down, it is not allowed.

The unit of time in the Monte Carlo simulations depends on the size of the preexponential terms in Eqs. 10.21, 10.24, and 10.25. In our calculations the unit of time is conveniently given by

$$\Delta t = \frac{1}{\nu} \tag{10.26}$$

In what follows, the actual size of Δt will not be important, only that it depends on $1/\nu$. To put absolute magnitudes on the results of MC we will ultimately need to obtain ν and Δt from experimental data.

Equations 10.21, 10.24, and 10.25 are the heart of a Monte Carlo simulation. Much of the versatility of Monte Carlo simulations stems from the ease with which additional factors such as adsorption of impurities, stress, defects, and different schemes for surface diffusion and surface–solution interactions can be incorporated into the simulation by the modification of only these three equations.

The effect of the stress field around a dislocation is incorporated by modifying Eq. 10.21. The energy change required to dissolve an atom from the surface is decreased by the amount of strain energy introduced into the bonds, so that

$$k_n^- = \nu \exp\left[-n\,\frac{\Phi - u(r)/3}{kT}\right] \tag{10.27}$$

where $u(r)$ is the strain energy density given by Eq. 10.14. $u(r)$ was divided by 3 to partition the strain energy equally among the bonds. By microscopic reversibility, the precipitation probability (Eq. 10.24) remains unchanged.

For very low solubility crystals, such as most rock-forming minerals, Φ/kT will be greater than 2 and k_n^- may vary by several orders of magnitude. Because the surfaces of low-solubility crystals are nearly flat, the majority of events will involve the precipitation, surface diffusion, and dissolution of adatoms that occupy a very small proportion of the surface sites. Thus, a Monte Carlo simulation using the classical procedure of choosing sites at random will spend a large proportion of computing time on null responses at high-n sites, where all the probabilities are very low. This was overcome by varying the length of the time step so that the total probability of any type of event occurring anywhere on the surface during a time step is 1. For the ith time step, we first compute

$$k^{\text{total}} = \nu \sum_{\text{surface}} \left\{\exp\left(-\frac{n\Phi}{kT}\right) + \exp\left(\frac{\Delta\mu}{kT}\right) + \exp\left[-E_a - \frac{(n-1)\Phi}{kT}\right]\right\}$$

Then the size of the time step is given by

$$\Delta t_i = \frac{1}{k_{\text{total}}}$$

The probabilities k^+, k^-, and k^{dif} are now computed (Eqs. 10.21, 10.24, and 10.25) for each site using this Δt_i. One event is then chosen and executed at random from a list of all possible events using these probabilities. Finally, k_{total} is adjusted for the change in surface configuration, and the process is repeated.

The total elapsed time for the simulation is computed from

$$t = \sum_i \Delta t_i$$

where i ranges over all time steps. Growth rates are found by computing the change in volume per unit time.

The input parameters Φ/kT, $\Delta\mu/kT$, μ, E^{dif}, b, and $u(r)$ are all dimensionless (in kT energy units and a length units, where a is the block edge length). The results of simulating one set of dimensionless parameters will then be valid for a series of macroscopic values by scaling a. The surface energy can be related to Φ/kT by dividing the bond strength by the area of a block face:

$$\sigma = \frac{\Phi}{a^2} \tag{10.29}$$

Likewise, the molecular volume in these units is $v = a^3$ for a = unit length, $\sigma = \Phi$, and $v = 1$.

The values of b and r_h are also given in units of a. The shear modulus μ will be in kT energy units per a^3 volume. Lasaga and Blum (1986) show that with these dimensionless units the parameters for quartz can be closely simulated by

$$\frac{\Phi}{kT} = 4.0$$

These relationships will allow us to compare the results of Monte Carlo simulations with the predictions of analytical theory.

5. RESULTS AND DISCUSSION

The Monte Carlo simulations can be used to predict the growth rate and dissolution rate as a function of the deviation from equilibrium, $\Delta\mu$. Figure 10.4*a* gives a plot of total dissolution as a function of time for a typical run. The slope of the curve is the surface reaction rate (Fig. 10.4*b*). Note that a well-defined rate is quickly reached. A plot of surface reaction rates versus $\Delta\mu/kT$ is given in Figure 10.5. The units of growth or dissolution rate are molecular volume per unit area per unit time, or av, where a refers to the length of the atomic block in the simulations and the unit time is given by $1/\nu$, with ν as defined in Eqs. 10.21–10.25. The shape of these reaction rate versus saturation state plots is of great interest not only to our understanding of the theory of crystal growth and dissolution but also to our interpretation of experimental data. The results of the Monte Carlo simulations incorporate many more factors than the early theories could possibly tackle. In particular, the analytical theories have in large part focused on the calculation of thermodynamic quantities such as the critical size of surface clusters and have obtained the growth or dissolution rate from these parameters (e.g., a nucleation rate for surface clusters). Unfortunately, all these theories ignore the

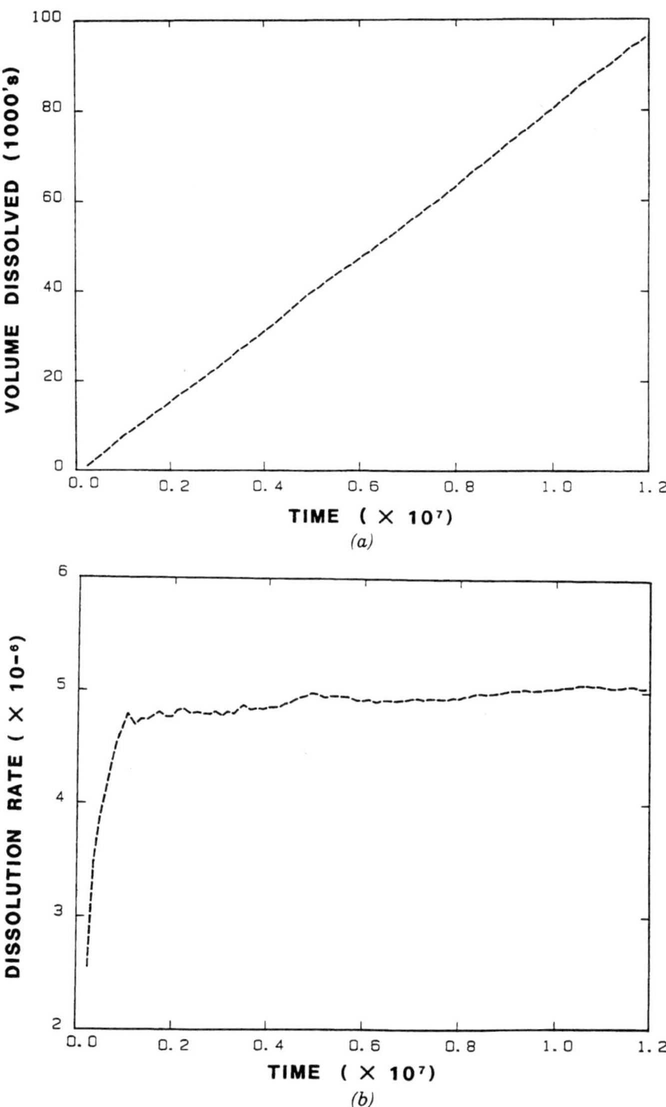

Figure 10.4. (*a*) Plot of total dissolved atomic blocks versus time (in $1/\nu$ units) for a dislocation-free surface. Other parameters: $\Phi/kT = 3$, $\Delta\mu/kT = -2$, $E_a^{\text{dif}}/kT = 100$ (i.e., *no* surface diffusion). (*b*) Slope of curve in (*a*) giving the instantaneous dissolution rate as a function of time.

details of statistical fluctuations in surface morphology and surface energetics during the growth or dissolution process. These fluctuations can be quite significant in the overall kinetics. In addition, most analytical theories treat the crystal as a continuum rather than as a collection of discrete units of finite size.

Monte Carlo simulations avoid making the assumptions discussed above

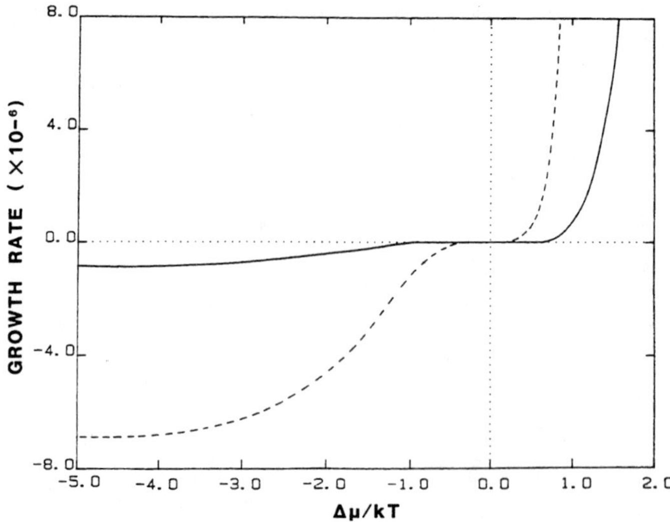

Figure 10.5. Surface reaction rate laws for dislocation-free surfaces. No surface diffusion allowed. Rate in units of $a\nu$. *Solid line,* $\Phi/kT = 3.5$; *dashed line,* $\Phi/kT = 3.0$.

by following the kinetic paths of each discrete molecular unit during dissolution and precipitation. Monte Carlo simulations also yield information on the evolution of crystal surface morphology that is not usually available from analytical models. The major disadvantage of Monte Carlo simulations is the necessity of running numerous simulations to investigate the dependence of the system on several variables, which requires extensive computer resources. However, the extreme flexibility and intuitive simplicity of Monte Carlo simulations make these methods extremely useful in conceptualizing and quantifying the highly complex mechanisms of growth and dissolution of real crystal surfaces.

5.1. Shape of Monte Carlo Rate Laws

Simulations of smooth dislocation-free surfaces or surface steps (i.e., screw dislocations *without* a strain field) used an array of 40 × 40 or 1600 surface sites. Each surface site covers an area of a^2, where a is the unit length defined earlier. Simulations with various array dimensions indicate that there is *no* dependence of growth rate on the size of the array as long as the surface dimensions exceed the size of the critical cluster ($r_c = \sigma v/\Delta\mu$, Eq. 10.6). Surface dimensions smaller than the critical cluster allow the periodic boundary condition to reduce the activation energy of nucleation, increasing growth rates. The chemical potential of A and B species in solution could be varied independently, but all the simulations presented here have equal chemical potentials of A and B species in solution (for an A–B cubic crystal equivalent to equal concentrations or also equivalent to simulating a one-component crystal such as quartz).

The surface reaction rates of dislocation-free surfaces as a function of saturation state for several values of Φ/kT are shown in Figure 10.5. A negative rate indicates dissolution. The growth and dissolution behavior of crystals is extremely asymmetric with respect to solution saturation ($\Delta\mu/kT$). At extremely high undersaturation, the dissolution rate approaches a constant. At high supersaturation, the growth rate increases exponentially. Near saturation, the region of most interest in natural applications, the growth or dissolution rate is a more complex function of the saturation state.

At high undersaturation (as $\Delta\mu \rightarrow -\infty$), the concentration of species in solution approaches zero, and the precipitation of atomic units from solution onto the surface becomes negligible (e.g., Eq. 10.24). As a result, the dissolution rate will be controlled completely by the kinetics of removing atomic units from the surface. In this case, the dissolution rate will cease to vary with variations in $\Delta\mu$ and will approach a constant value that is dependent only on the characteristics of the solid surface. The independence of the dissolution rate on solution composition in very dilute solutions is the basis for the assumption of zeroth-order kinetics that is often used in experimental determinations of dissolution rates (Holdren and Berner, 1979; Tole et al., 1986; Lasaga, 1984). This constant value, however, will be very sensitive to the particular physical and chemical states of surfaces, including number of dislocations and their types, the presence of atomic impurities, and the characteristics of surface diffusion.

Growth from highly supersaturated solutions is dominated by the precipitation of atomic units on the surface, which is an exponential function of the saturation state (Eq. 10.24). Because the precipitation rate is independent of the surface configuration, the Monte Carlo simulations predict that at high supersaturation states, the growth rate will always approach

$$k_{\text{eq}}^{+} \exp\left(\frac{\Delta\mu}{kT}\right)$$

regardless of the variations in the configuration of the surface, such as dislocations, variations in surface diffusion, and impurities. Figure 10.6 exhibits this exponential behavior for high $\Delta\mu/kT$. A plot of ln (rate) versus ln ($\Delta\mu/kT$) becomes linear at high $\Delta\mu/kT$. However, the Monte Carlo simulations assume that dissolution and growth are completely controlled by surface reaction, which means that there is no concentration gradient in the solution adjacent to the crystal surface. When precipitation increases to a point where atomic units cannot be supplied by aqueous diffusion with a negligible concentration gradient, the assumption of surface-reaction control is no longer valid. The growth rate would then become increasingly transport-controlled and approach a linear rate law typical of transport-controlled growth. (Examples of this approach to transport control were given earlier in the

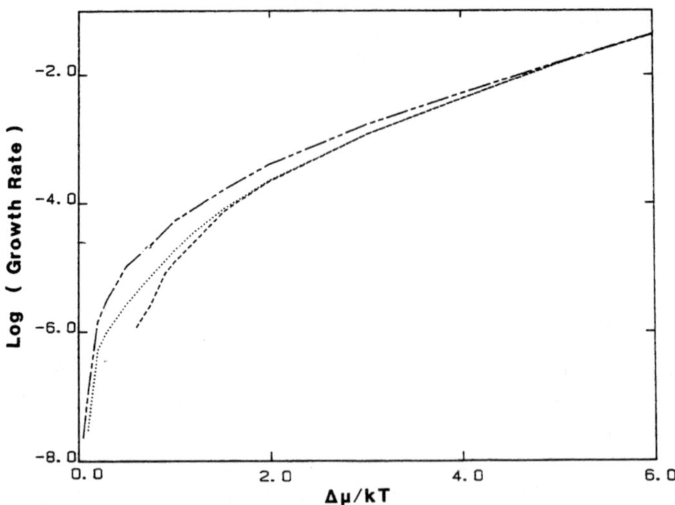

Figure 10.6. Approach to exponential behavior at high $\Delta\mu/kT$ (straight line) for various situations all at $\Phi/kT = 3.0$. (---) Dislocation-free surface, no surface diffusion. (···) Screw dislocation step (no strain), no surface diffusion. (——--) Screw dislocation step (no strain) with $E_a^{\text{dif}}/kT = 0.75$.

chapter.) Therefore, the total rate would have a hybrid shape similar to that given in Figure 10.7. This type of curve should be of wide applicability in natural solutions.

Close to saturation, the growth or dissolution rate will be some complex function of the precipitation and dissolution atomic processes. For dislocation-free surfaces, the surface reaction rate will approach zero at values of $\Delta\mu$ substantially different from zero (See, e.g., Fig. 10.5 or 10.14). Closer to saturation ($\Delta\mu = 0$), the free energy barrier to formation of the critical cluster has increased to the point where nucleation of these clusters in the Monte Carlo simulations is negligible. This is in basic qualitative agreement with the analytical theories of two-dimensional nucleation as discussed earlier (Christian, 1975). In fact, using Eqs. 10.7 and 10.8, the growth rate should depend exponentially on $\Delta\mu$:

$$R \propto \exp\left[-\frac{\pi h v \sigma^2}{\Delta\mu RT}\right] \tag{10.30}$$

Therefore R should be very small (for small $\Delta\mu$) until

$$\frac{\pi h v \sigma^2}{\Delta\mu RT} \sim 1$$

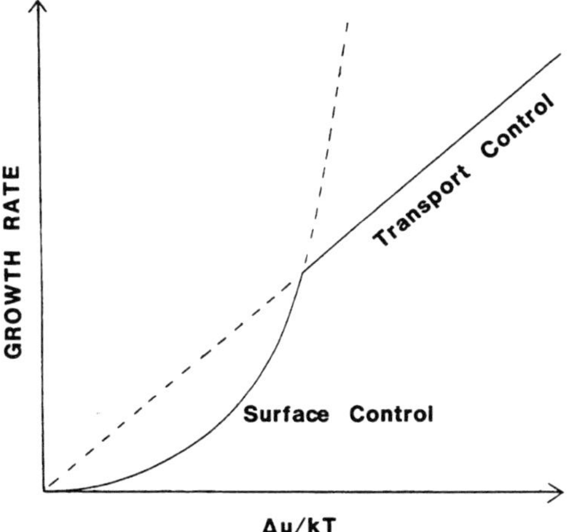

Figure 10.7. General behavior of growth rate laws. At low $\Delta\mu/kT$, surface rates dominate, leading to nonlinear rate laws. At high $\Delta\mu/kT$, the nonlinear curve intersects the transport rate law and the rate follows a linear equation.

Using the formulas for v and σ in the Monte Carlo simulations, we can rewrite this as

$$\frac{\Delta\mu}{kT} \sim \pi \left(\frac{\Phi}{kT}\right)^2 \tag{10.31}$$

The deviation from equilibrium necessary to initiate dissolution or growth is strongly dependent on Φ/kT and on the configuration of the surface. Note, however, that the Monte Carlo simulations predict a near-zero rate until $\Delta\mu/kT$ exceeds a critical value around $(\Delta\mu/kT)_{\text{crit}} = 1\text{–}2$ (Figs. 10.5 and 10.14), but the analytic theory (Eq. 10.31 would yield $(\Delta\mu/kT)_{\text{crit}} = 4\pi\text{–}16\pi$ for the Φ/kT values used here. Obviously the two-dimensional nucleation growth theories predict too great a region of essentially zero rate, a prediction that has also caused disagreement with experiment (Ohara and Reed, 1973). The Monte Carlo results suggest that there may be an intrinsic problem with the approximations in the theory. Further work is continuing along these lines.

The presence of a continuously propagating step at a screw dislocation (excluding the stress field) enhances the dissolution and growth rates in agreement with BCF theory (Fig. 10.8). At large undersaturation ($\Delta\mu/kT = -15$), the presence of a screw dislocation results in only a slight increase in the dissolution rate. For example, if $\Phi/kT = 4$, the rate is 25% higher than the dislocation-free rate, and this increase drops systematically with Φ/kT until at $\Phi/kT = 2.5$ the rate is only 8% higher. Figure 10.9 summarizes the increase in rate due to a screw dislocation step compared to that in dislocation-free

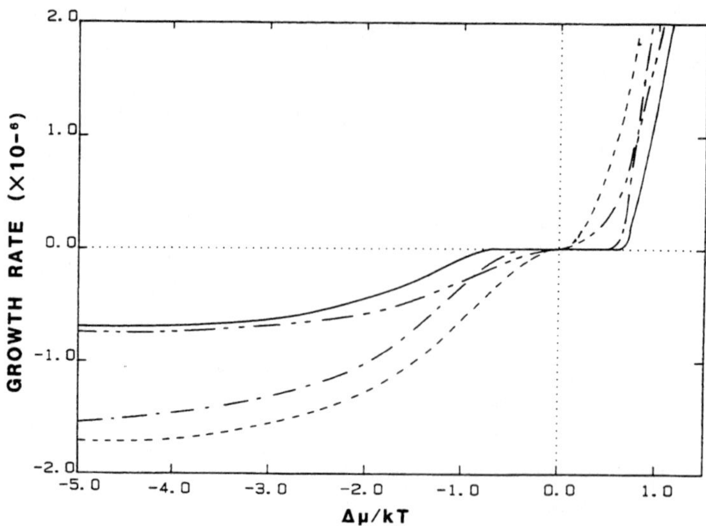

Figure 10.8. Reaction rate ($a\nu$ units) for various surfaces using $\Phi/kT = 3$. (———) Dislocation-free surface, no surface diffusion. (—--) Screw dislocation step (no strain), no surface diffusion. (—·—) Dislocation-free surface (no strain), $E_a^{\text{dif}}/kT = 1.5$. (---) Screw dislocation step (no strain), $E_a^{\text{dif}}/kT = 1.5$.

surfaces. Even more important, the presence of a screw dislocation step allows growth and/or dissolution *without* the formation of a critical cluster. This leads to finite growth and dissolution rates even very close to saturation. In this region, the effect of the dislocation is significant. For example, at $\Phi/kT = 3$ and $\Delta\mu/kT = 1.0$, the dissolution rate in the presence of a dislocation is now 78% higher, and at $\Delta\mu/kT = -0.75$, 270% higher. Below $\Delta\mu/kT = -0.5$, no dissolution occurred on the dislocation-free surface, while the surface with a screw dislocation still dissolved at 14% of its maximum rate.

The presence of these dislocation-induced steps forms the physical basis for the BCF theory of crystal growth discussed earlier (see also Christian, 1975). Close to equilibrium, the BCF theory predicts a rate given by 10.11a:

$$R = A\ \Delta\mu^2$$

The Monte Carlo reaction rates can be compared with the BCF result. Figure 10.10 gives some of the Monte Carlo results for growth in the presence of a screw dislocation step (no strain field) and for $\Phi/kT = 3.0$. When the results in Figure 10.10 are fitted to an exponential function of the type

$$R = A\left(\frac{\Delta\mu}{kT}\right)^n \tag{10.32}$$

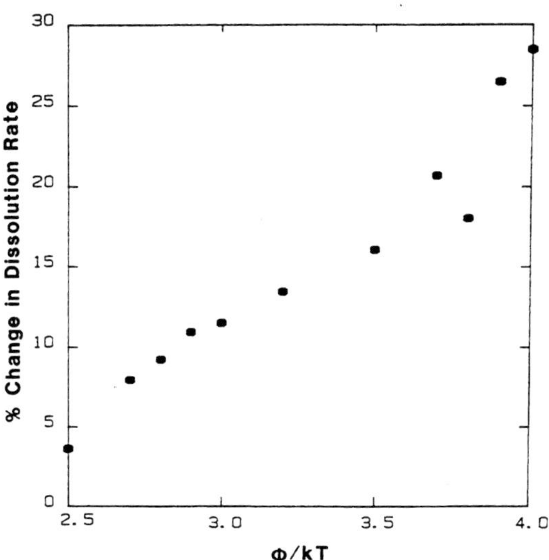

Figure 10.9. Percent increase in dissolution rate for a surface with a screw dislocation step (no strain field included) over that in a dislocation-free surface as a function of Φ/kT. No surface diffusion allowed. $\Delta\mu/kT = -15$ for all cases.

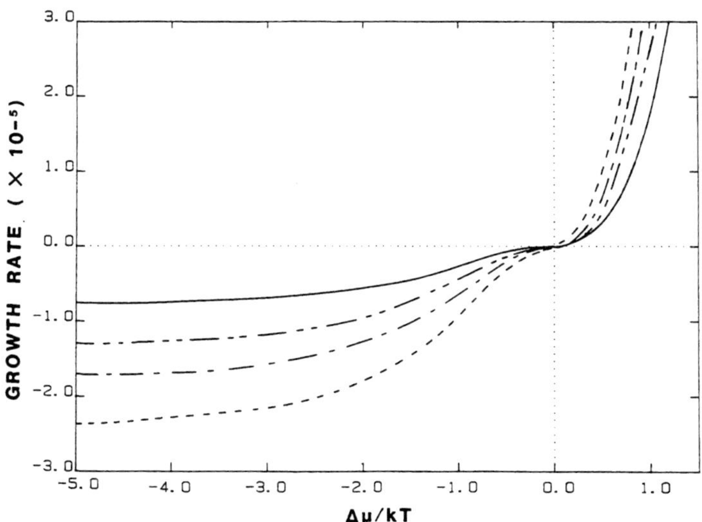

Figure 10.10. Effect of surface diffusion on surface reaction rate laws for surfaces with screw dislocation steps but no strain field included; $\Phi/kT = 3.0$. (———) No surface diffusion; (—-‐) $E_a^{\text{dif}}/kT = 2.25$; (—·—) $E_a^{\text{dif}}/kT = 1.5$; (---) $E_a^{\text{dif}}\ kT = 0.75$.

Table 10.1. Exponential Fits to Growth Data

Φ/kT	E_a^{dif}/kT	A	n
2.0	100.0	1.6×10^{-3}	1.96
3.0	0.75	8.5×10^{-5}	2.76
3.0	1.5	7.0×10^{-5}	2.86
3.0	2.25	6.0×10^{-5}	3.15
3.0	100.0	2.6×10^{-5}	2.78
4.0	100.0	2.3×10^{-7}	2.00

the results are as given in Table 10.1 and plotted in Figure 10.11. The Monte Carlo results validate the BCF theory, although the simplifications inherent in the BCF equation lead to nontrivial differences in the exponents. Note that, in general, the Monte Carlo exponents predict a growth rate that obeys

$$R = A\ \Delta\mu^{2.7\text{–}2.9}$$

which modifies the predictions of Eq. 10.11a. The exponents obtained in the Monte Carlo results are in good agreement with recent experimental data on mineral growth at high temperatures. Schramke et al. (1986) and Lasaga (1986) predict an exponent between 2.0 and 2.7 for the growth of andalusite, Al_2SiO_5, from the dehydration of muscovite, $KAlSi_3O_{10}(OH)_2$.

The mobility of atoms on the surface also has a large effect on the dissolution or growth rate of a crystal surface. Figures 10.8 and 10.10 show the effect of increasing surface diffusion on growth and dissolution of a surface without dislocations and a surface with a screw dislocation.

In a supersaturated solution, the high rate of precipitation of atoms onto a nearly planar surface results in a large steady-state population of adatoms. The adatoms will diffuse across the surface until they either redissolve or reach a site to which they are more strongly bonded, such as a step or kink. The trapping of mobile atoms in high-coordination surface sites results in more efficient retention of atoms that precipitate on the surface and consequently a faster growth rate. Conversely, in undersaturated solutions, the formation rate of adatoms by precipitation is low, and the steady-state population of adatoms will be small. In undersaturated solutions, therefore, surface diffusion of adatoms out of steps and kinks is a mechanism for generating low-coordination surface atoms, which dissolve easily and enhance the dissolution rate. The magnitude and importance of surface diffusion in growth and dissolution mechanisms of complex crystals such as silicates is not well quantified or understood. Clearly, however, surface diffusion is a potentially important process which must eventually be addressed.

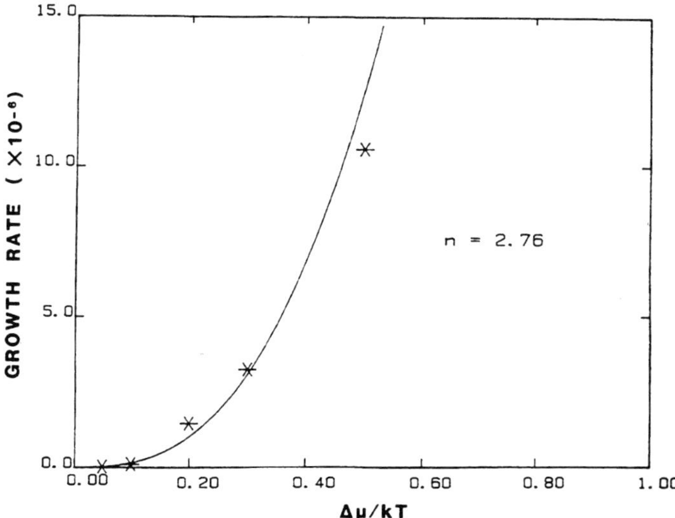

Figure 10.11. Example of nonlinear growth function for the case of a surface with a screw dislocation step (no strain); $\Phi/kT = 3$ and $E_a^{\text{dif}}/kT = 0.75$. (*) Monte Carlo results. The solid line is the fit of Eq. 10.32 with $n = 2.76$.

6. DISLOCATIONS WITH A STRAIN FIELD INCLUDED

In the Monte Carlo simulations, the magnitude of the stress energy is dependent on three dimensionless parameters that are all scaled to the size of the unit cell: b, the magnitude of the Burger vector in unit cell lengths (a); μ, the shear modulus in kT units per cell volume; and $u(0)$, the maximum strain energy density in kT units per cell volume (see Eq. 10.14). The relevance of these parameters to the properties of quartz has been discussed previously. The length of the Burger vector is 2 in all the simulations, the smallest Burger vector allowed by the requirement of perfect ordering of an A–B crystal. Simulations which include a stress field use an 80×80 cell array, or 6400 surface sites.

Including the effects of a dislocation strain field (Eq. 10.27) produces a gradient of increasing probability of dissolution and surface diffusion at sites closer to the dislocation line. There are three major questions to be addressed: (1) What is the effect of stress on the initial dissolution rate of a fresh, flat surface, such as would be measured in most experiments; (2) What is the effect of stress on the steady-state dissolution rate and surface morphology of a crystal, which is pertinent to natural geochemical environments; and (3) What is the effect of solution saturation state and other parameters on etch pit morphology, and can we use etch pit abundance and shape as an indicator of absolute solution saturation state (Lasaga and Blum, 1986).

6.1. Dissolution in the Core Region

The high strain energy near the dislocation core results in strong preferential dissolution in this region, which will form either a microscopic hollow core or a broad etch pit. In either case, there is enhanced removal of material in the immediate vicinity of the dislocation during the initial dissolution of a fresh, flat surface. Figure 10.12 shows the surfaces generated by Monte Carlo simulations of dissolution near a dislocation with a strain field only (no step) at several undersaturation states. Because the etch pits are roughly symmetrical, their morphology can be more concisely represented by a cross section through the center of the surface. Figure 10.13 shows the cross section of the surfaces at various intervals of volume removed.

Figure 10.13 shows that virtually all the dissolution occurs initially at the dislocation cores and may account for a large proportion of the early dissolution. This may be an important factor during dissolution experiments with fresh surfaces. Even after an induction period and the development of a stable surface morphology, the dissolution rate at the core is greater than the steady-state dissolution rate around the perimeter (discussed later) by a factor of 100 at $\Delta\mu/kT = -0.05$ and a factor of 3 at $\Delta\mu/kT = -1.0$. The calculated dissolution rates at the cores are not very reliable for several reasons: (1) At the dislocation core the etch pit is very steep, and a large proportion of the surface area is vertical where precipitation and dissolution are prohibited by the model restriction against overhangs; (2) uncertainies from the choice of the stress function (Eq. 10.14) are at a maximum in the dislocation core; and (3) the restricted geometry of the dislocation core would tend to limit the dissolution rate by transport control.

Despite these limitations, the simulation results are qualitatively reasonable, and very strong preferential dissolution at dislocation cores has been observed (Van der Hoek et. al., 1982, 1983; Tsukamoto and Giling, 1982). Nielson and Foster (1960) and Sears (1960) reported etch channels in synthetic and natural quartz 1 μm in diameter and over 1 cm long. Similar features have been described in lithium fluoride, spinel, and sodium chloride (Heimann, 1982). Such extreme behavior may result from additional stress energy caused by substitutional and interstitial impurities which are strongly concentrated in the distorted sites of the dislocation core (Hirth and Lothe, 1968, p. 462).

6.2. Steady-State Dissolution

The dissolution rate of a simulation with a dislocation and stress field will vary with the array size of the surface. If the surface array size is increased, the proportion of the surface that is appreciably strained will decrease, and the total dissolution rate of the surface will decrease. One way to avoid this dependence of the dissolution rate on the array size is to calculate the growth

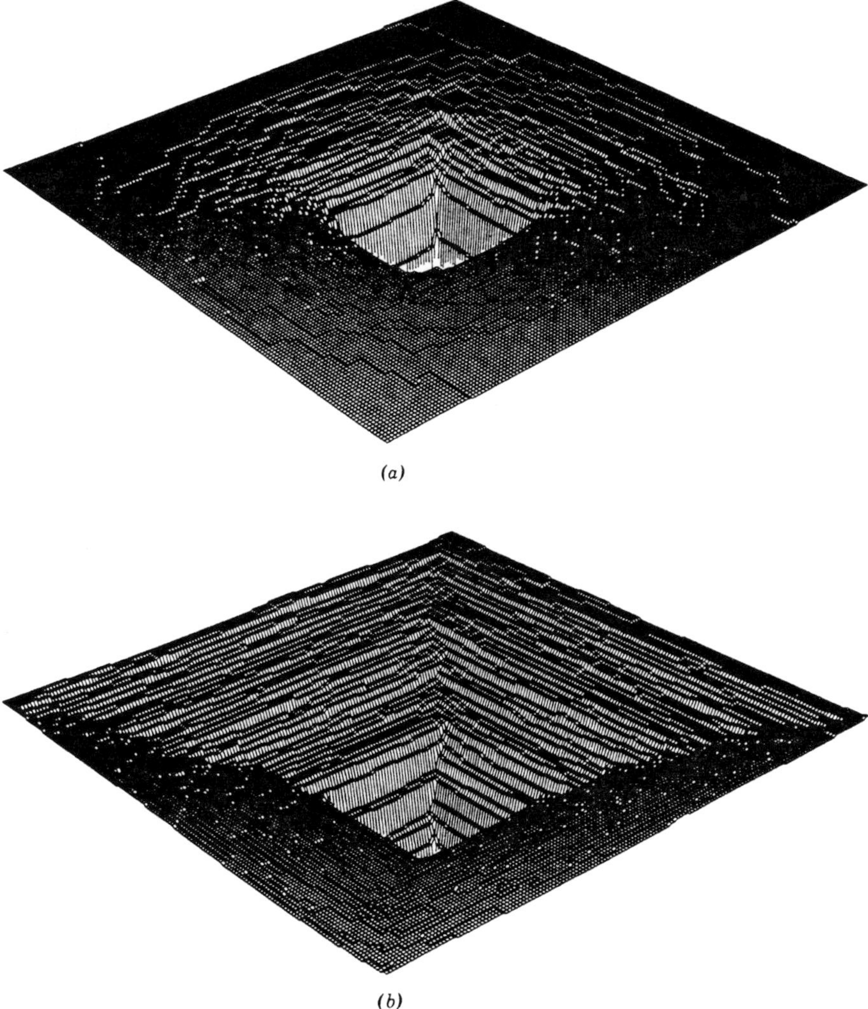

(a)

(b)

Figure 10.12. Example of dissolution surface after dissolution around a screw dislocation (strain energy included); $\Phi/kT = 4.0$. (a) $\Delta\mu/kT = -0.05$; (b) $\Delta\mu/kT = -1.0$.

rate only around the perimeter of an array that exceeds the size of the critical radius. Any dissolving surface layer that expands appreciably beyond the critical radius will continue to expand outward across the unstrained crystal. After an initial induction period for the creation of steps and their outward migration from the dislocation core, the dissolution rate will reach a steady-state value that is independent of array size. This dissolution rate represents the rate at which steps nucleated at the dislocation can expand outward to dissolve the unstrained surface and neglects the effects of dissolution at the dislocation core.

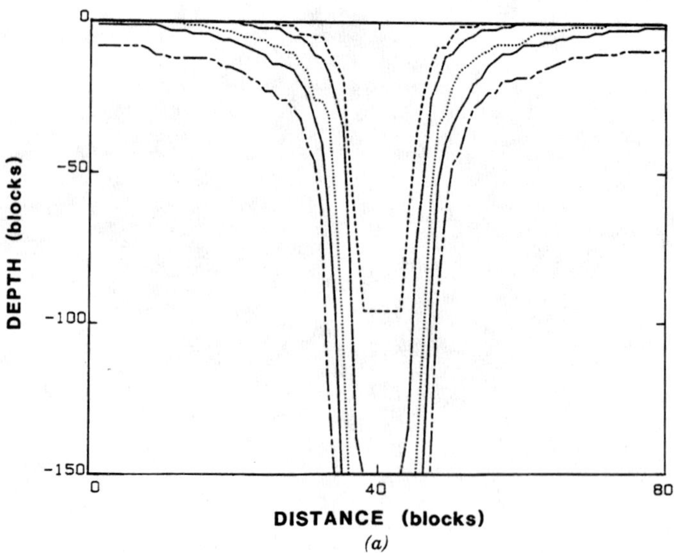

(a)

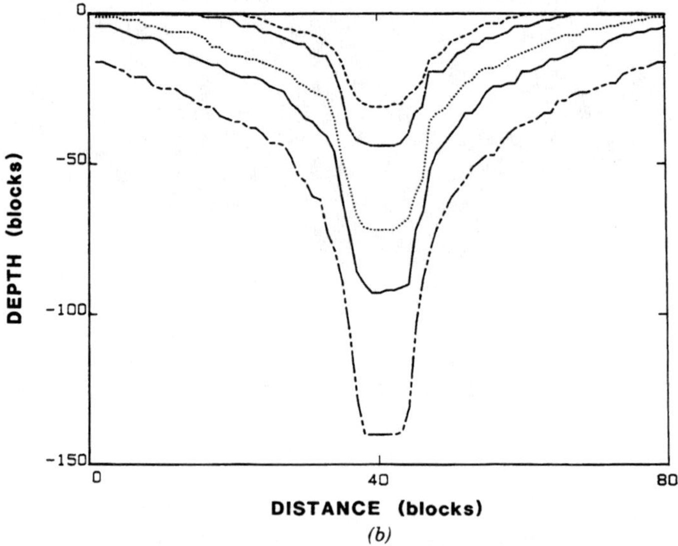

(b)

Figure 10.13. Cross sections of dislocation etch pits after different total dissolution intervals; $\Phi/kT = 4.0$. (———) 10,000 blocks dissolved; (· — ·) 20,000 blocks dissolved; (· · ·) 60,000 blocks dissolved; (———) 100,000 blocks dissolved; (—--) 200,000 blocks dissolved. (*a*) $\Delta\mu/kT = -0.05$; (*b*) $\Delta\mu/kT = -1.0$.

The use of the edge dissolution rate also has the advantage of minimizing the effect of the model restriction against overhangs on the dissolution rate. Dissolution in the region around the critical radius ultimately controls the edge dissolution rate. Since this region is relatively flat, the restriction against overhangs should have a minimal influence on the edge dissolution rate.

Figure 10.14 shows the simulation results for the edge dissolution rate versus the saturation state ($\Delta\mu/kT$) of an undislocated surface, a screw dislocation step, a dislocation stress field, and a dislocation step with stress field. The effect of the stress field is quite large, producing a very rapid increase in the dissolution rate near saturation and increasing the maximum dissolution rate by a factor of about 5. Note that the presence of a screw dislocation step in addition to the stress field has very little additional effect on the dissolution rate, even close to saturation. This means that the continuous production of steps by a pseudo-two-dimensional nucleation mechanism at the dislocation core has overwhelmed the effect of the single spiral step at a screw dislocation. This also implies that enhanced dissolution rates can result from the strain fields around both edge and screw dislocations and are not restricted to the dislocations with a screw component as in BCF theory.

The edge dissolution rate has important implications. Any circular step that is generated at a dislocation and appreciably exceeds the critical radius will continue to expand outward across the unstrained crystal until it either

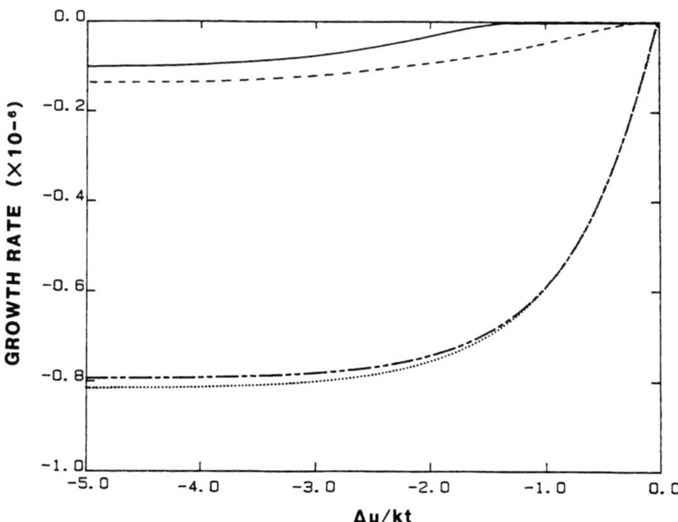

Figure 10.14. Comparison of dissolution rates for various surfaces; $\Phi/kT = 4.0$ and no surface diffusion allowed in all cases. (———) Dislocation-free surface. (---) Screw dislocation step (no strain). (—--) Screw dislocation strain field (no step). (···) Screw dislocation step and strain field.

is annihilated by a step generated at another source or reaches the edge of the crystal. A single dislocation will produce a circular step train radiating outward across the unstrained surface. At steady state, the entire region influenced by this step train will have the "edge" dissolution rate. If the expansion of the dislocation-generated step train is limited by annihilation by steps generated by two-dimensional nucleation on the unstrained surface between the dislocations, the effect of the dislocation on the overall dissolution rate of the crystal will be the product of the area influenced by the dislocation step train and the difference between the "edge" dissolution rate and the two-dimensional nucleation dissolution rate. This would lead to a linear dependence of the overall observed dissolution rate on dislocation density. However, if the dislocations are close enough together that the step trains generated by adjacent dislocations meet and annihilate each other (i.e., the entire surface is covered by dislocation-generated step trains, and two-dimensional nucleation is insignificant), the dependence of the overall rate on the dislocation density will not be so simple. More steps will be generated per unit area, because there are more loci for their formation, but the average distance traveled by a step before annihilation will be proportionately shorter. Obviously, there will be intermediate dislocation densities where the dependence of the overall surface dissolution rate of a surface on the dislocation density will be intermediate between zero- and first-order. Since both the rate of step generation (i.e., the edge dissolution rate) and the horizontal step velocity are strong functions of the solution state, the dependence of the dissolution rate on the dislocation density may also change with the saturation state.

We can carry out a useful simple calculation to compare the effect of dissolution at dislocations to the overall dissolution rate. Just as shown in Figure 10.4, after an initial period there is a nearly constant dissolution rate of total volume with time for a dislocation-induced etch pit. We will label this rate R_d, in units of molecular volumes (a^3) per unit time ($1/\nu$). If n_d is the dislocation density (in numbers per unit area a^2, then the total dissolution rate due to dislocations is given by

$$R^{\text{disloc}} = R_d n_d \tag{10.33}$$

On the other hand, there is a dislocation-free dissolution rate given by R_s (e.g., as computed in Fig. 10.4) in units of volume dissolved (a^3) per unit area (a^2) per unit time ($1/\nu$). The total dissolution rate is given by

$$R_{\text{tot}} = R_d n_d + R_s \tag{10.34}$$

If the first term in Eq. 10.34 dominates, we should expect a linear dependence of experimental and natural solution rates on dislocation density. If the second term dominates, then the dissolution rate should have a zeroth-order dependence on n_d. For a given R_d and R_s, there will be a critical density n_d^c above which the rate will depend on dislocations:

$$n_d^c = \frac{R_s}{R_d} \tag{10.35}$$

Using the data in Figures 10.14 and 10.15, for parameters relevant to quartz dissolution ($\Phi/kT = 4$, $a = 3\text{Å}$) and for $\Delta\mu/kT = -0.1$ we have

$$R_d = 5.6 \times 10^{-4}\, a^3\nu$$
$$R_s = 3.96 \times 10^{-9}\, a\nu$$

so

$$n_d = 7.1 \times 10^{-6}\, a^{-2}$$
$$n_d = 7.8 \times 10^{9} \text{ dislocations cm}^{-2}$$

This value will decrease as we get closer to equilibrium. Note that R_d also will vary with time, that is, as the etch pit opens up. Therefore n_d^c will vary with both dissolution saturation state and with extent of dissolution. More work is needed to extend the simple calculation carried out here.

6.3. Surface Morphology and Etch Pit Development

The Monte Carlo simulation results can be compared with the predictions of the thermodynamic treatment of etch pit formation (Eqs. 10.16–10.18). Figure 10.15 shows, for several saturations, the total free energy change resulting from the dissolution of a circular hole around a dislocation one block deep and with radius r, calculated using Eq. 10.18 and the parameters of the simulations. There is a critical saturation value at $\Delta\mu/kT = -0.23$. At undersaturations greater than -0.23 there is only one free energy maximum (Figs. 10.15*a* and *b*). It is located very close to the dislocation, between 0.5 and 2.5 block units, and the free energy maximum never exceeds 15 kT units. Once a layer has surmounted this small energy barrier, there is no activation barrier to further expansion of the step, and a broad etch pit should form.

At undersaturations less than -0.23, there is an energy minimum surrounded by two maxima (Figs. 10.15*c* and *d*). This minimum corresponds to the radius of a stable hollow core. The radius of the calculated hollow core as a function of saturation state is shown in Figure 10.16. In order for a step to expand outward from the hollow core, it must surmount an activation barrier which is the difference between the minimum and the second maximum in Figures 10.15*c* and *d*. Figure 10.17 shows the magnitude of the activation

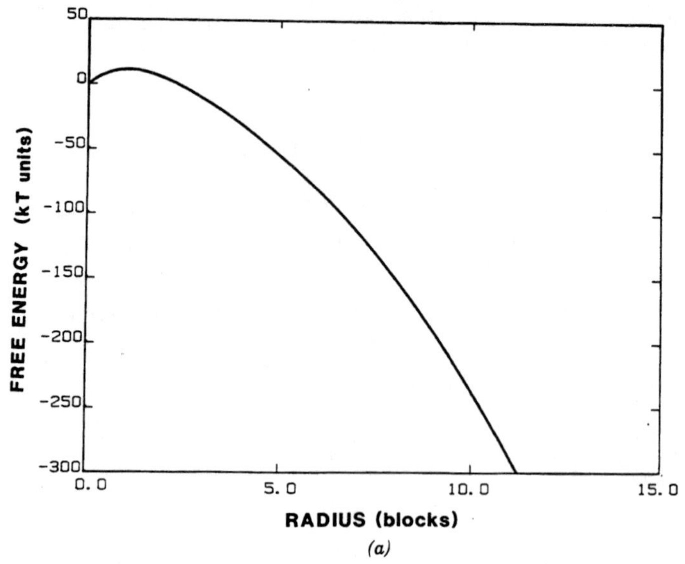

(a)

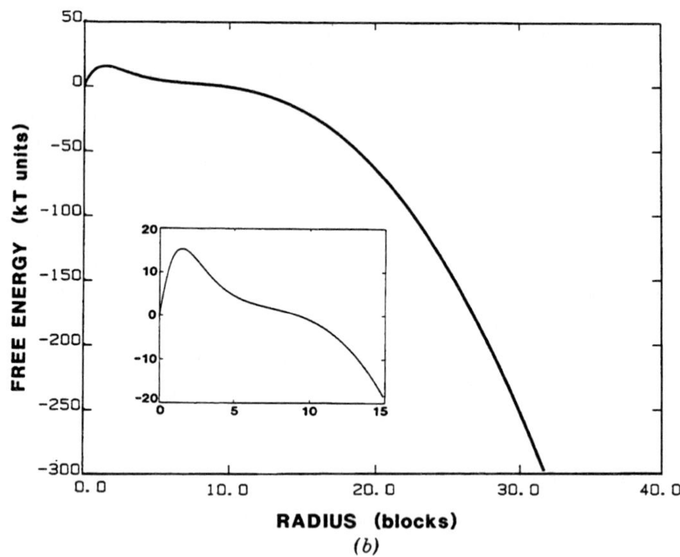

(b)

Figure 10.15. Free energy of formation of a pit as a function of pit radius r for various surface conditions; $\Phi/kT = 4.0$ (all). (*a*) $\Delta\mu/kT = -1.0$; (*b*) $\Delta\mu/kT = -0.25$; (*c*) $\Delta\mu/kT = -0.20$; (*d*) $\Delta\mu/kT = -0.10$.

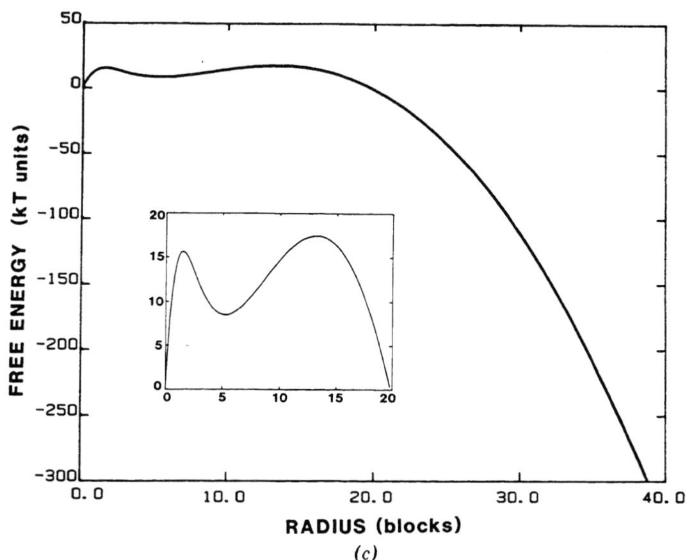

(c)

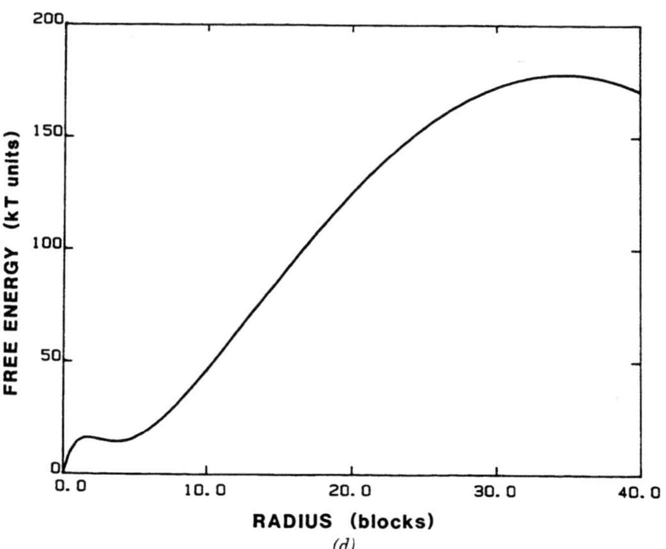

(d)

Figure 10.15. (*Continued*)

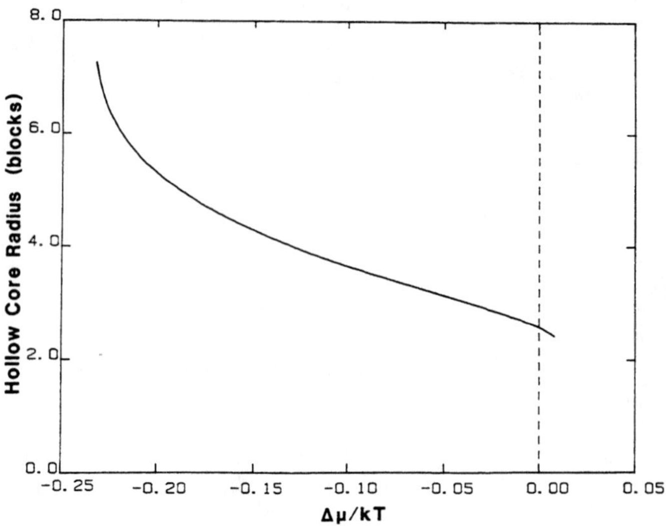

Figure 10.16. Hollow-core radius (minimum in ΔG_{pit} vs r curve) as a function of $\Delta\mu/kT$. Assumes $\Phi/kT = 4.0$ and a screw dislocation strain field.

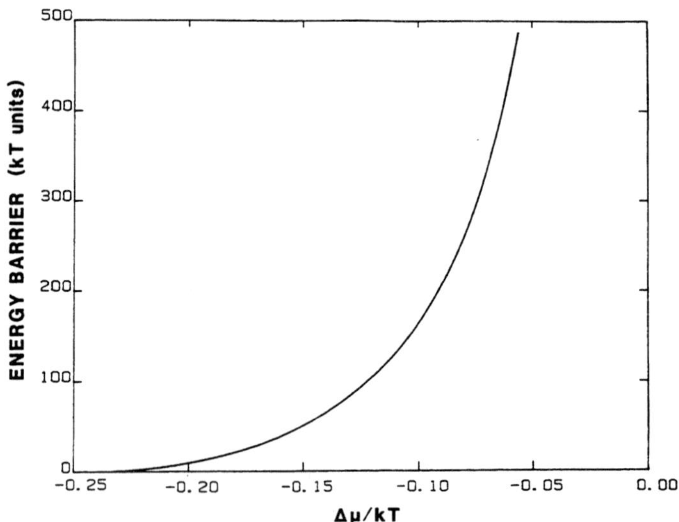

Figure 10.17. ΔG energy barrier to etch pit opening as a function of $\Delta\mu/kT$. Assumes $\Phi/kT = 4.0$ and a screw dislocation strain field.

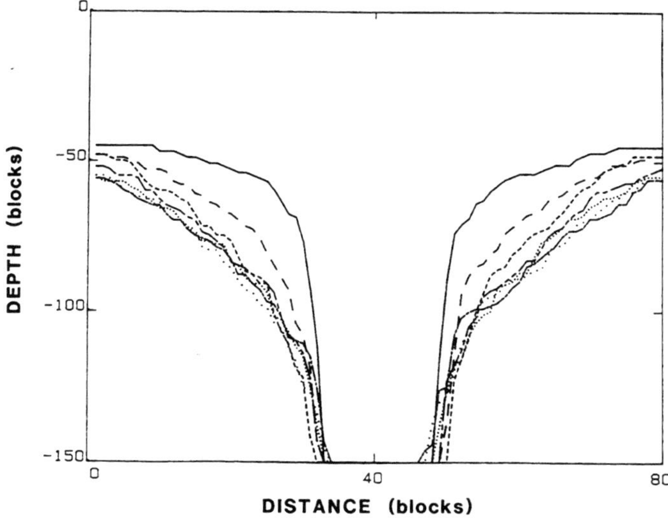

Figure 10.18. "Equal-volume" cross section of dislocation etch pits for a surface with $\Phi/kT = 4.0$ and a screw dislocation strain field. (———) $\Delta\mu/kT = -0.05$; (— —) $\Delta\mu/kT = -0.1$; (---) $\Delta\mu/kT = -0.2$; (—--) $\Delta\mu/kT = -0.25$; (···) $\Delta\mu/kT = -0.5$; (——-) $\Delta\mu/kT = -1.0$.

barrier to the expansion of the hollow core in kT energy units per unit depth as a function of saturation state. The energy barrier increases rapidly above $\Delta\mu/kT = -0.15$ and is much larger than the energy barrier at the core, which varies from 1 to 15 kT units.

Figure 10.18 shows the cross sections of etch pits generated by Monte Carlo simulations at several values of $\Delta\mu/kT$ for a surface with the stress field surrounding a screw dislocation but without inclusion of the step. All the profiles have a volume of 500,000 blocks dissolved. The etch pits are up to 1000 blocks deep, so the integrated area under the profiles on Figure 10.18 are not exactly equal. At undersaturations of $\Delta\mu/kT = -0.25$ to -5.0, the shape of the etch pit remains nearly identical, even though the dissolution rate increases by a factor of 3.5 (Fig. 10.14). However, there is narrowing of the etch pit shape at $\Delta\mu/kT = -0.2$, which becomes more pronounced as the solution saturation state decreases to -0.1 and -0.05.

At saturations below -0.25, dissolution becomes increasingly confined to the core region. This is in excellent agreement with analytic theory, which predicts the limit of hollow core stability at an undersaturation of -0.23, with the energy barrier confining the hollow core rising rapidly above -0.15. The simulations also show that the energy barrier at the dislocation core is easily surmounted and does not limit the dissolution rate. Even at $\Delta\mu/kT = -1.0$, where there is no activation energy to further outward movement of steps, the core region still dissolves more rapidly than the surrounding material (Fig. 10.12*b*). This suggests that above the critical undersaturation of -0.23, it is the horizontal velocity of step movement and not nucleation which is limiting the dissolution rate. The similarity of the etch pit profiles above the critical saturation reflects the absence of any activation energy barrier and the linear dependence of the dissolution rate on the horizontal step velocity, which must then yield an etch pit with a constant slope.

REFERENCES

Bennema, P. (1984), Spiral growth and surface roughening: Developments since Burton, Cabrera and Frank, *J. Cryst. Growth* **69,** 182–197.

Berner, R. A. (1978), Rate control of mineral dissolution under earth surface conditions, *Am. J. Sci.* **278,** 1235–1252.

Berner, R. A. (1980), *Early Diagenesis,* Princeton University Press, Princeton, NJ, 241 pp.

Berner, R. A. (1981), "Kinetics of Weathering and Diagenesis," in A. C. Lasaga and R. J. Kirkpatrick, Eds., *Kinetics of Geochemical Processes* (Rev. Mineral, Vol. 8), Mineralogical Society of America, Washington, DC, p. 111.

Berner, R. A., and Holdren, G. R., Jr. (1979), Mechanism of feldspar weathering. II. Observations of feldspars from soils, *Geochim. Cosmochim. Acta* **43,** 1173–1186.

Berner, R. A., and Morse, J. W. (1974), Dissolution kinetics of calcium carbonate in seawater: IV. Theory of calcite dissolution, *Am. J. Sci.* **274,** 108–134.

Berner, R. A., and Schott, J. (1982), Mechanism of pyroxene and amphibole weathering. II. Observations of soil grains, *Am. J. Sci.* **282,** 1214–1231.

Bjorklund, R. B., Lester, J. E., and Spears, K. G. (1977), Alkali halide surfaces: Measurements and modeling of adsorbate motions and reactions at surface defects, *J. Chem. Phys.* **66,** 3426–3436.

Bjorklund, R. B., and Spears, K. G. (1977), Alkali halide surfaces: Potential energy surfaces for diffusion and dimer dissociations, *J. Chem. Phys.* **66,** 3448–3454.

Brantley, S. L., Crane, S. R., Crerar, D. A., Hellmann, R., and Stallard, R. (1986), Dissolution at dislocation etch pits in quartz, *Geochim. Cosmochim. Acta,* **50,** 2349–2378.

Burton, W. K., Cabrera, N., and Frank, F. C. (1951), The growth of crystals and the equilibrium structure of their surfaces, *Phil. Trans. Roy. Soc. London, A* **243,** 299–358.

Cabrera, N., and Levine, M. M. (1956), On the dislocation theory of evaporation of crystals, *Phil. Mag.* **1,** 450–458.

Christian, J. W. (1975), *The Theory of Transformation in Metals and Alloys,* Part I, 2nd ed. Pergamon Press, Oxford.

Fung, P. C., Bird, G. W., McIntyre, N. S., Sapipelli, G. G., and Lopata, V. J. (1980), Aspects of feldspar dissolution, *Nucl. Tech.* **51,** 188–196.

Gilmer, G. H. (1976), Growth on imperfect crystal faces. I. Monte Carlo growth rates, *J. Cryst. Growth* **35,** 15–28.

Gilmer, G. H. (1977), Computer simulation of crystal growth, *J. Cryst. Growth* **42,** 3–10.

Gilmer, G. H. (1980), Computer models of crystal growth, *Science* **208,** 355–363.

Heimann, R. B. (1982), "Principles of Chemical Etching—The Art And Science of Etching Crystals," in J. Grabmaier, Ed., *Silicon/Chemical Etching,* Vol. 8, *Crystals,* Springer-Verlag, Berlin, pp. 173–224.

Helgeson, H. C. (1971), Kinetics of mass transfer among silicates and aqueous solutions, *Geochim. Cosmochim. Acta* **35,** 421–469.

Hirth, J. P., and Lothe, J. (1968), *Theory of Dislocations,* McGraw-Hill, New York, 780 pp.

Holdren, G. R., Jr., and Berner, R. A., (1979), Mechanism of feldspar weathering. I. Experimental studies, *Geochim. Cosmochim. Acta* **43,** 1161–1171.

Lasaga, A. C. (1981), "Rate Laws of Chemical Reactions," in A. C. Lasaga and R. J. Kirkpatrick, Eds., *Kinetics of Geochemical Processes* (Rev. Mineral., Vol. 8), Mineralogical Society of America, Washington, DC, pp. 1–67.

Lasaga, A. C. (1984), Chemical kinetics of water–rock interactions, *J. Geophys. Res.* **89,** 4009–4025.

Lasaga, A. C. (1986), Metamorphic reaction rate laws and developments of isograds, *Min. Mag.,* **50,** 359–373.

Lasaga, A. C., and Blum, A. E. (1986), Surface chemistry, etch pits and mineral-water reactions, *Geochim. Cosmochim. Acta* **50,** 2363–2379.

Lasaga, A. C., and Gibbs, G. V. (1986), Applications of quantum mechanical potential surfaces to mineral physics calculations, *Phys. Chem. Minerals,* in press.

Nielson, J. W., and Foster, F. G. (1960), Unusual etch pits in quartz crystals, *Am. Mineral.,* **45,** 299–310.

Ohara, M., and Reed, R. C. (1973), *Modeling Crystal Growth Rates from Solution,* Prentice-Hall, Englewood Cliffs, NJ, 272 pp.

Paces, T. (1973), Steady-state kinetics and equilibrium between ground water and granitic rock, *Geochim. Cosmochim Acta* **37,** 2641–2663.

Petrovic, R., Berner, R. A., and Goldhaber, M. B. (1976), Rate control in dissolution of alkali feldspars. I. Study of residual feldspar grains by X-ray photo-electron spectroscopy, *Geochim. Cosmochim. Acta* **40,** 537–548.

Schott, J., and Berner, R. A. (1983), X-ray photoelectron studies of the mechanism of iron silicate dissolution during weathering, *Geochim. Cosmochim. Acta* **47,** 2233–2240.

Schramke, J. A., Kerrick, D. M., and Lasaga, A. C. (1986), The kinetics of metamorphic hydration–dehydration reactions, *Am. J. Sci.* in press.

Sears, G. W. (1960), Dislocation etching, *J. Chem. Phys.* **32,** 1317–1322.

Tole, M. P., Lasaga, A. C., Pantano, C., and White, W. B. (1986), Kinetics of nepheline dissolution, *Geochim. Cosmochim. Acta* **50,** 379–392.

Tsukamoto, K., and Giling, L. J. (1982), On the origin of deep pit formation for gas phase HCl etched (111) silicon wafers, *J. Cryst. Growth* **50,** 338–342.

Van der Eerden, J. P., Bennema, P., and Cheiepanova, T. A. (1978), Survey of Monte Carlo simulations of crystal surfaces and crystal growth, *Progr. Crystal Growth Character.* **1,** 219–251.

Van der Hoek, B., Van der Eerden, J. P., and Bennema, P. (1982a), Thermodynamical stability conditions for the occurrence of hollow cores caused by stress of line and planar defects, *J. Crystal Growth* **56,** 621–632.

Van der Hoek, B., Van der Eerden, J. P., and Bennema, P. (1982b), The influence of stress on spiral growth, *J. Crystal Growth* **58,** 365–380.

Van der Hoek, B., Van Enckevort, W. J. P., and Van der Linden, W. H. (1983), Dissolution kinetics and etch pit studies of potassium aluminum sulphate, *J. Cryst. Growth* **61,** 181–193.

Wilson, M. J., and McHardy, W. J. (1980), Experimental etching of a microcline perthite and implications regarding natural weathering, *J. Microscopy* **120,** 291–302.

Wintsch, R. P., and Dunning, J. (1985), The effect of dislocation density on the aqueous solubility of quartz and some geologic implications: A theoretical approach, *J. Geophys. Res.* **90,** 3649–3657.

Wollast, R. (1967), Kinetics of the alteration of K-feldspar in buffered solutions at low temperature, *Geochim. Cosmochim. Acta* **35,** 635–648.

11

NEW EVIDENCE FOR THE MECHANISMS OF DISSOLUTION OF SILICATE MINERALS

Jacques Schott

Laboratory of Mineralogy and Crystallography, University Paul Sabatier, Toulouse, France

and

Jean-Claude Petit

DRDD/SESD, CEN-FAR, Fontenay-aux-Roses, France

Abstract

The basic processes involved at the reaction interface during the weathering of silicate minerals are still poorly understood and are a matter of controversy. In particular, the existence of a hydrated surficial layer, which has been invoked to explain the dissolution kinetics as well as the apparent overall incongruency, has not been demonstrated and has been disproved in recent studies. By using a resonant nuclear reaction (RNR), which allows hydrogen depth profiling, we present, in the case of diopside and albite, the first direct evidence that dissolution proceeds via a marked surfical hydration over thicknesses ≤1000 Å coupled with a decrease in the signal of all constituent elements observed with SIMS which indicates an increase in the porosity of the surface. The observed RNR and SIMS profiles cannot be entirely explained by simple models like surface reaction or diffusion exchange of H^+ *with cations. They rather suggest that the migration of molecular water in the crystal, which is facilitated by the presence of line defects intersecting the surface, could be a key step in the dissolution of silicate minerals.*

1. INTRODUCTION

The weathering of mineral assemblages by the chemical action of aqueous solutions is one of the major transformations occurring at the surface of the earth. For example, the weathering of silicate minerals such as feldspars, pyroxenes, amphiboles, or garnets constitutes a major process in the geochemical cycles of silicon, aluminum, magnesium, and iron. Thus, the modeling of the dissolution of silicate minerals is a necessary step in a quantitative approach to the geochemical cycles which control the evolution of the earth.

Although the aqueous corrosion of silicate minerals has received considerable attention since Daubree's pioneering work (1857, 1867), the basic processes involved at the reaction interface are still poorly understood and are a matter of controversy. Three main hypotheses have been advanced to account for the numerous available experimental observations.

1. *The surface reaction hypothesis* (Lagache et al., 1961; Lagache, 1976; Berner, 1978; Aagaard and Helgeson, 1982; Schott et al., 1981; Dibble, 1981; Stumm et al., 1985), where the rate of dissolution is controlled by reactions at the solid–aqueous solution interface.
2. *The armoring precipitate hypothesis* (Correns and Von Engelhardt, 1938; Correns, 1940; Wollast, 1967; Helgeson, 1971), where diffusion of species through a reprecipitated layer limits the rate of hydrolysis.
3. *The leached layer hypothesis* (Luce et al., 1972; Pačes, 1973; Busenberg and Clemency, 1976), where the preferential leaching of mobile elements such as alkalis occurs at the mineral surface and leads to the formation of a residual hydrated layer. Diffusion of reactants through this nonstoichiometric residuum controls the release of exchangeable cations, while the leached silicate or aluminosilicate framework dismantles at a slower rate. As the layer builds up, the rate of dissolution decreases until a steady state is reached when the rate of removal of silica from the surface keeps pace with the rate of removal of cations from deeper within the solid.

One can note that the last two models should result in a so-called parabolic rate of dissolution, while surface reaction control is consistent with a constant rate of dissolution. It is also worth noting that it is misleading to consider the armoring precipitate hypothesis as a control of initial dissolution. In fact, this hypothesis is only verified in the initial stages of the dissolution of Al or Fe silicates as the reacting solutions become oversaturated with respect to iron and/or aluminum hydroxides which precipitate at the surface of the dissolving silicate. Thus, interference from secondary precipitations should be avoided when studying exclusively the basic mechanisms which control the initial dissolution of a solid.

If the leached layer hypothesis is known to be verified for the dissolution of most silicate glasses [see, for example, Doremus (1975) and Lanford et al. (1979)], the first model—surface reaction—is now generally agreed to be the most likely for the aqueous corrosion of silicate minerals (Lasaga, 1984; Helgeson et al., 1984; Schott and Berner, 1985; Furrer and Stumm, 1983; Murphy, 1985).

The present preference for surface reaction models is based both on the analyses of reacting solutions and on mineral surface chemical analyses using modern spectroscopic methods. The aim of this chapter is to review the recent arguments which support surface reaction, with particular emphasis on surface analytical data, and also to present recent data obtained with new tools and techniques which suggest that the recent tendency to consider that the rate-limiting step during silicate dissolution is related only to a surface process may not be entirely justified.

2. EVIDENCE OF SURFACE REACTION CONTROL DURING DISSOLUTION OF SILICATE MINERALS

2.1. Analyses of the Reacting Solutions

If they are properly treated to remove ultrafine particles produced during sample preparation (i.e., grinding), most silicate minerals exhibit a constant rate of release of silica with time which is consistent with a reaction at the mineral surface. This is shown in Figure 11.1, where the amount of silica

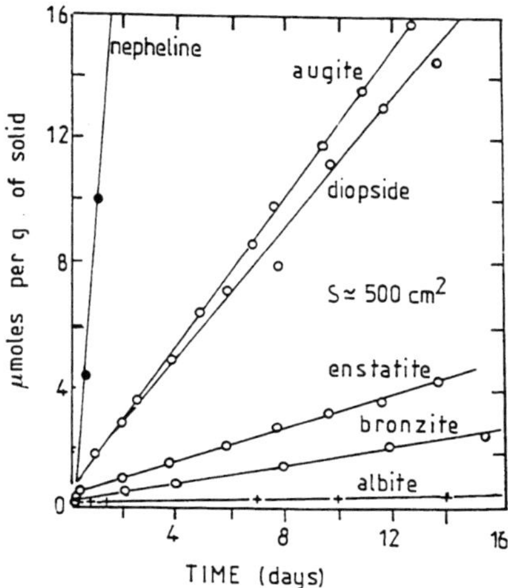

Figure 11.1. Plot of amount of silica release versus time for the dissolution of etched enstatite, bronzite (pO_2 = 0), diopside, augite, and albite at pH 6 and for nepheline at pH 7. T = 50°C for enstatite, diopside, and augite; 25°C for bronzite, albite, and nepheline. (○) Data of Schott and Berner (1985); (+) data of Holdren and Berner (1979); (●) data of Tole (1982).

released is plotted versus time for the dissolution of various silicates which were first pretreated with HF—H_2SO_4 to remove ultrafine particles and surface defects. In Figure 11.2 the behavior of untreated, sonicated, and etched (with HF—H_2SO_4) enstatite at pH 6 and 20°C are shown. It can be seen that the rate of silica release of untreated material and of sonicated material follows a parabolic law. This behavior is due to the initial rapid dissolution of ultrafine, supersoluble particles adhering to larger grains, which are produced during grinding (Holdren and Berner, 1979; Schott et al., 1981). From these results it is commonly inferred (Schott et al., 1981; Holdren and Berner, 1979; Helgeson et al., 1984) that the parabolic rate law found by previous investigators for the dissolution of silicate minerals is an artifact due to grinding, since preetching of samples was not performed in these earlier studies [e.g., note in Figure 11.2 that the curve based on the data of Luce et al. (1972) for enstatite has the same shape as that for sonicated and untreated enstatite].

Another important parameter affecting the rate of dissolution is temperature. For most silicate minerals, the temperature dependence of the rate of dissolution follows the Arrhenius equation:

$$r = Ae^{-\Delta E/RT} \tag{11.1}$$

where ΔE is the activation energy. Values of ΔE at steady state are given in Table 11.1 for a variety of silicate dissolution reactions. The values of ΔE

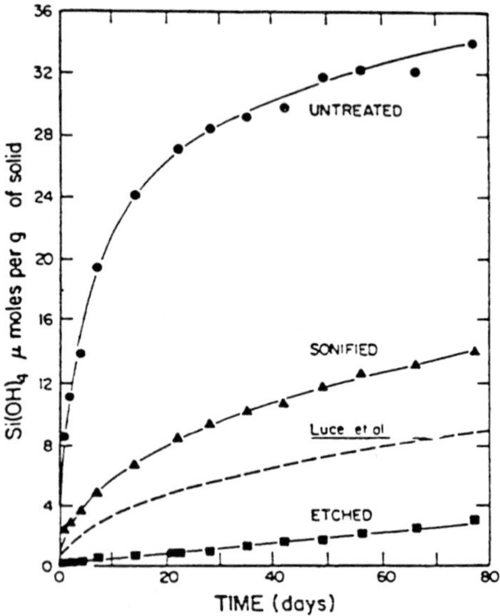

Figure 11.2. Silica release during the dissolution of untreated, sonicated, and etched enstatite at pH 6 and T = 20°C. The curve labeled Luce et al. (1972) refers to washed, but otherwise untreated, material from the same location. (After Schott et al., 1981.)

Table 11.1. Activation Energy for Silicate Dissolution Reactions

Mineral	ΔE (kJ mol^{-1})	Reference
Diopside	38 ($4 \leq pH \leq 6$) 81 (pH ~ 2)	Schott et al. (1981)
Enstatite	50	Schott et al. (1981)
Bronzite	45	Grandstaff (1977)
Olivine (Fo_{83})	38	Grandstaff (1981)
Augite	78	Schott and Berner (1985)
Nepheline	54–71	Tole (1982)
Anortite	35	Fleer (1982)
K-feldspar	38 ($3 \leq pH \leq 8$) 82 ($pH \leq 3$)	Helgeson et al. (1984)
Albite	38 ($3 \leq pH \leq 8$) 88.5 ($pH \leq 3$)	Helgeson et al. (1984)
Wollastonite	74	Murphy (1985)
Quartz	77	Rimstidt and Barnes (1980)

are notably higher than that measured for transport in solution [~20 kJ/mol, Tsao and Pask (1982)] and are consistent with reaction at the surface of the mineral. However, note that these activation energies are distinctly lower than those expected from breaking bonds in crystals (80–400 kJ/mol). The catalytic effects of adsorption on surfaces (see Stumm et al., 1985) probably reduce the activation energies to this intermediate range.

2.2. Analyses of the Surfaces of Dissolving Minerals

We have seen above that if dissolution is controlled by diffusion of reactants through the solid, this should result in the formation of a nonstoichiometric residual layer. X-ray photoelectron spectroscopy (XPS, also referred to as ESCA) is well suited to testing for the existence of such an altered surface layer. Indeed, XPS is a technique that allows the determination of binding energies of electrons, and it can be used for chemical analysis of the surfaces and very shallow (tens of angstroms) subsurfaces of solids and to deduce important data on oxidation states, bonding environment, and so on. [For an extended discussion of XPS and its application to the analysis of solid surfaces, see Carlson (1975), Petrović et al. (1976), Koppleman and Dillard (1979), and Siegbahn (1982).]

Until now, it has not been possible to identify such a residual layer either with the aid of XPS or with scanning electron microscopy. For example, using XPS, Petrović et al. (1976) and Holdren and Berner (1979) were unable to detect the presence of a leached layer thicker than 10 Å at the surface of weathered feldspars. With the same technique, but in the case of pyroxenes

Table 11.2. Calculated Thickness of Leached Layer Using XPS of Naturally (Soil Grains) and Artificially (Laboratory Dissolution) Weathered Pyroxenes and Amphiboles

		Thickness of Layer (Å)			
		Mg		Ca	
Mineral	pH	Total Depletion	Linear Increase with Depth	Total Depletion	Linear Increase with Depth
		Soil Grains			
Bronzite	—	5	10	—	—
Hypersthene	—	4	8	—	—
Diopside	—	7	15	5	11
Augite	—	2	4	0	0
Hornblende	—	7	15	10	21
		Laboratory Dissolution (T = 25°C, 26 days dissolution)			
Enstatite	6	3	5	—	—
	1	12	25	—	—
Diopside	6	0	0	3	6
	1	6	11	8	16
Tremolite	6	1.5	3	3	6
	1	4	8	4	8

Source: After Berner and Schott (1982) and Schott et al. (1981).

and amphiboles, Schott et al. (1981) and Berner and Schott (1982) showed that the cation-depleted layer was only a few angstroms thick and often less than one cell unit (see Table 11.2). Thus it appears that only the surface reaction hypothesis is consistent with X-ray photoelectron spectrographs of the surfaces of reacted or weathered grains of feldspars, enstatite, diopside, tremolite, and so on.

2.3. Rate Equations for Silicate Hydrolysis

Surface control of reaction rates is also consistent with transition state theory (Eyring, 1935; Lasaga, 1981; Aagard and Helgeson, 1982), in which the basic assumption is that the reactants should pass through a free energy maximum before they are converted to products (see an example in Fig. 11.3 for feldspars). The state defined by this energy maximum has been labeled for convenience the activated complex ($c^{\ddagger}$), although it may not be a discrete species.

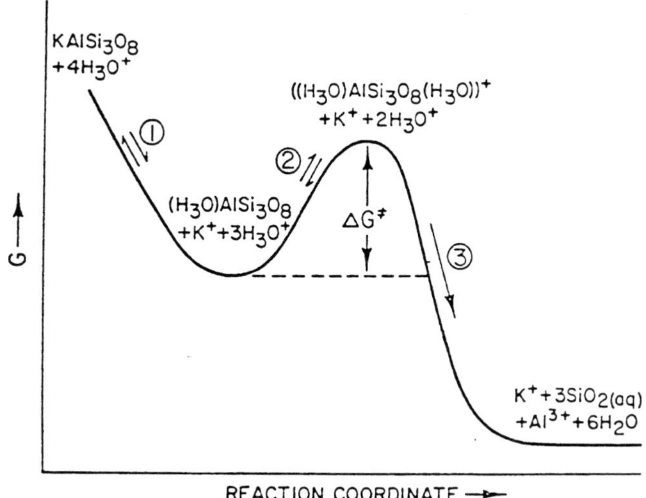

Figure 11.3. A schematic illustration of the free energy maximum through which reactants must pass to become products. Example of the hydrolysis of K-feldspar in acidic aqueous solution. (After Aagaard and Helgeson, 1982.)

The rate of reaction is assumed to be controlled by the decomposition of the activated complex and proportional to its concentration:

$$r = kc^{\ddagger} \tag{11.2}$$

where k is a rate constant.

For example, in the case of K-feldspar (Fig. 11.3), the formation of the activated complex may be written as

$$(H_3O)AlSi_3O_8 + H_3O^+ \rightleftharpoons (H_3O)AlSi_3O_8(H_3O)^+ \tag{11.3}$$

with the equilibrium constant

$$K = \frac{a_{(H_3O)AlSi_3O_8(H_3O)^+}}{a_{(H_3O)AlSi_3O_8}a_{H_3O^+}} \tag{11.4}$$

Combining Eqs. 11.2 and 11.4 and the definition of the activity coefficient leads to

$$r = kK \frac{a_{(H_3O)AlSi_3O_8}a_{H_3O^+}}{\gamma_{(H_3O)AlSi_3O_8(H_3O)^+}} \tag{11.5}$$

Table 11.3. Dependence of the Rate of Silicate Dissolution on H^+ Activity (T = 20°C)

Mineral	n^a	pH range	Reference
K-feldspar	1.0	<7	Helgeson et al. (1984)
	1.0	<5	Tole (1982)
	0.45	<5	Wollast and Chou (1985)
	−0.27	>7	
Sr-feldspar	1.0	<4	Fleer (1982)
	−0.28	>6	
Anorthite	0.54	2 < pH < 5.6	Fleer (1982)
Nepheline	1.0	<7	Lasaga (1984)
	−0.20	>7	
Olivine (Fo_{83})	1.0	3 ≤ pH ≤ 5	Grandstaff (1981)
Augite	0.7	1 ≤ pH ≤ 6	Schott and Berner (1985)
Enstatite	0.6	1 ≤ pH ≤ 6	Schott et al. (1981)
Diopside	0.5	2 ≤ pH ≤ 6	Schott et al. (1981)
Anorthite	0.54	2 < pH < 5.6	Fleer (1982)
Bronzite	0.5	<6	Grandstaff (1977)
Almandine	0.22	2 < pH < 6	Berner and Schott (1985), unpublished results; Nickel (1973)
	−0.2	>7	Nickel (1973)
Quartz	0.0	7	Rimstidt and Barnes (1980)

[a]Exponent in the expression r = constant $\cdot\ a^n_{H_3O^+}$.

It follows from Eq. 11.5, that the reaction rate is proportional to $a_{H_3O^+}$, if $a_{(H_3O)AlSi_3O_8}$ and $\gamma_{(H_3O)AlSi_3O_8(H_3O)^+}$ are constant, which is verified only when the concentration of the activated complex is small compared to the total number of $(H_3O)AlSi_3O_8$ sites. Otherwise the rate dependence on pH becomes more complex and the reaction order is no longer an integer (Wollast and Chou, 1985) as evidenced by data of Table 11.3 (r = constant $\cdot\ a^n_{H_3O^+}$).

3. NEW DATA ON THE DISSOLUTION OF SILICATES AND THE SURFACE REACTION HYPOTHESIS

If there is now general agreement that *steady-state dissolution* is controlled by surface reaction (with or without a leached layer), there is still a debate in the literature on the nature of the processes which control the initial stages of dissolution. Indeed, recent data suggest that the mechanisms of dissolution may be more complex than expected.

3.1. New Results on the Dissolution of Aluminosilicates

Until recently most experiments on the dissolution of silicates were performed in batch reactors. With this experimental approach, the composition of the reacting solution changes continuously with time, and it is difficult to identify and quantify the influence of the individual dissolved compounds on the reaction rate while the precipitation of secondary phases—particularly in the case of Al and Fe silicates—cannot be easily avoided. In order to work under better-controlled conditions, Chou and Wollast (1984) studied the weathering of albite with a continuous flow reactor based on the fluidized bed technique. This approach allowed Chou and Wollast to avoid the precipitation of aluminum hydroxides and permitted precise measurements of the rate of release of Na, Al, and Si during different stages of albite weathering. From the analysis of the reacting solutions they presented the evidence for the development of a layer of altered composition. In addition to being depleted in Na owing to rapid exchange with H^+, this layer is also depleted in Al under acidic conditions. It can be seen in Figure 11.4 that the thickness of these Na- and Al-free layers, which are assumed to be homogeneous, is 35 and 20 Å, respectively. Using the same technique, Holdren and Speyer (1985) confirmed most of Chou's and Wollast's results, namely, that (1) the dissolution is nonstoichiometric and (2) that under mildly acidic conditions a layer enriched in silica develops at the surface of alkali feldspars. It should be noted

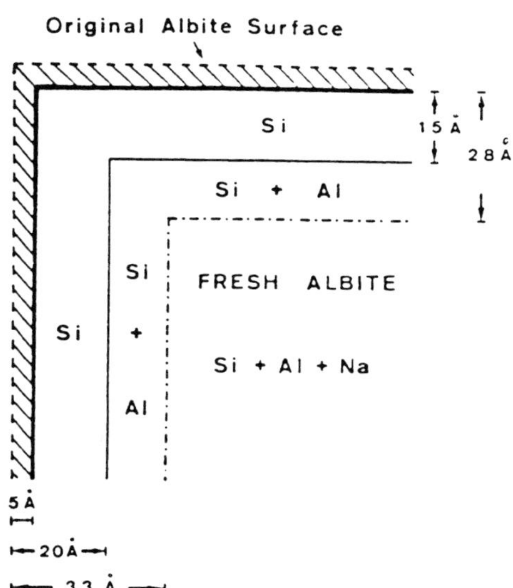

Figure 11.4. Schematic representation of an albite grain after 20 days of reaction in mildly acidic aqueous solutions. (After Chou and Wollast, 1984.)

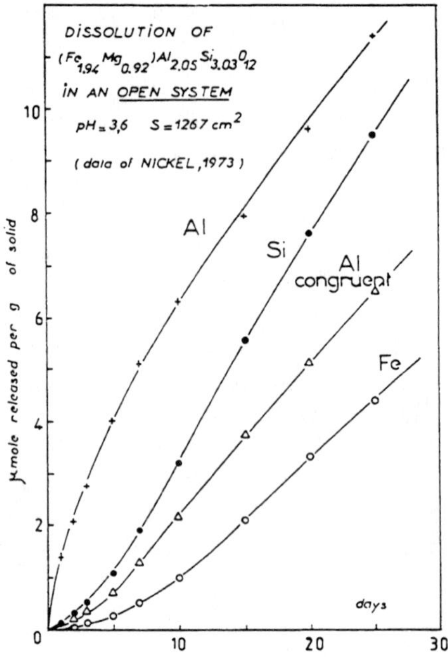

Figure 11.5. Plot of released Al, Fe, and Si for the dissolution of almandine at pH 3.6, $T = 20°C$. (Data of Nickel, 1973.)

that these results are in disagreement with those of Holdren and Berner (1979) and Berner and Holdren (1979), which were based on XPS surface analysis of dissolving feldspars.

Nonstoichiometric dissolution was also found for garnets by Nickel (1973), who performed experiments in an open system (i.e., the solvent was continuously added and removed) and by Berner and Schott (1985, unpublished results) with a batch reactor. It can be seen in Figure 11.5 that the dissolution of almandine garnet in mildly acidic conditions proceeds with preferential release of aluminum. The same behavior is observed in a closed system (Fig. 11.6) before precipitation of secondary phases takes place. The thickness of the Al-leached layer deduced from the amount of each element released and from the surface area of the solid is about 30 Å.

3.2. Formation of Etch Pits and XPS Analysis

It is commonly agreed that dissolution of a crystal surface is initiated at sites of high surface energy: edges, corners, cracks, scratches, holes, and, at a smaller scale, point defects, twin boundaries and dislocations, all these sites are favorable sites for rapid dissolution. Indeed, the omnipresence of etch

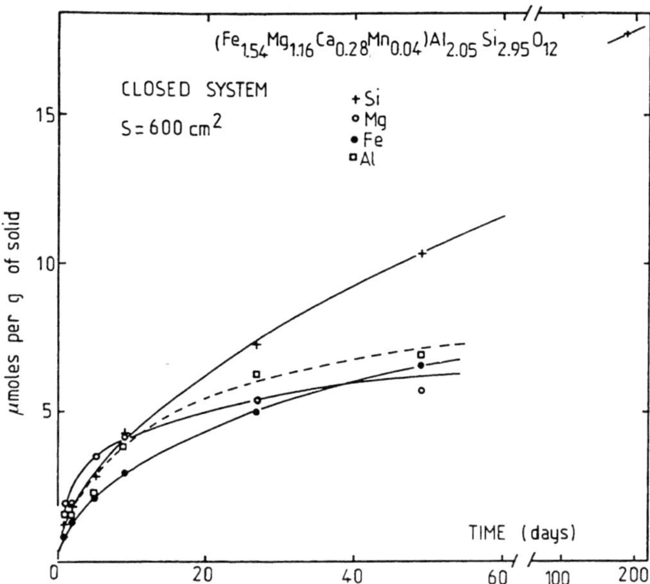

Figure 11.6. Plot of released Al, Fe, Mg, and Si for the dissolution of garnet at pH 2, T = 20°C. (Unpublished data of Berner and Schott, 1985.)

pits on dissolving minerals—particularly in the case of silicates (see micrographs, Fig. 11.7*a–e*) indicates that overall surface dissolution is nonuniform, with preferential attack at points of excess surface energy. The formation of etch pits is discussed by Blum and Lasaga in Chapter 10; we will simply note here that etch pits are likely to develop in most laboratory experiments where the affinity of the dissolution reaction is important (Brantley et al., 1986).

As pointed out by Berner et al. (1985), if dissolution occurs along deep cracks, tubes, holes, and so on, that intersect only a small portion of the mineral surface, X-ray photoelectron spectroscopy, which samples a large surface area to a small depth, is likely to miss cation depletions on the walls of these etch pits (see Fig. 11.7*f*). Hence, the failure to detect cation depletion does not prove that discontinuous altered layers are not formed, and the apparent disagreement between XPS (data of Holdren and Berner) and solution chemical analyses (Chou and Wollast) disappears. However, a problem remains because such a nonuniform and discontinuous layer should be much thicker than expected from solution chemical analyses. For example, the preferential Al loss during albite dissolution, which Chou and Wollast interpret in terms of a 20-Å thick uniform Al-free surface layer, can just as well be interpreted as representing Al removal along a number of holes, microcracks, and so on, extending 200 Å into the crystal and intersecting only 10% of the outermost crystal surface. The same type of calculations with garnet will lead to thicknesses of the order of 300 Å for the Al depletion.

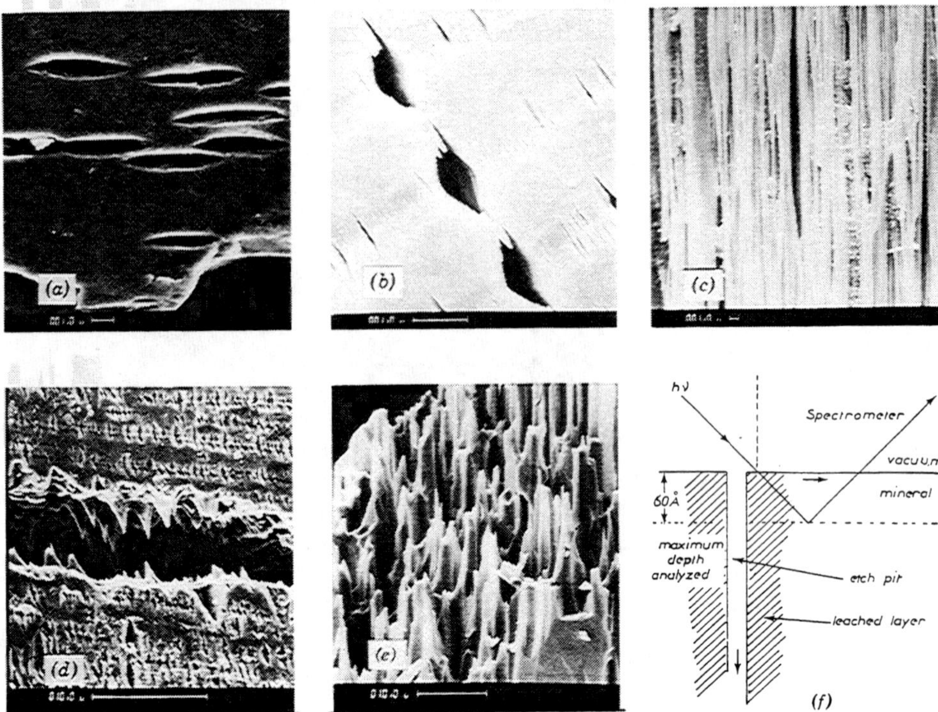

Figure 11.7. (*a*) Lens-shaped etch pits in a soil grain of hypersthene. (*b*) Side-by-side alignments of etch pits along lamellae boundaries in a soil grain of diopside (coalescence of etch pits can be observed). (*c*) End-to-end alignments of etch pits in a soil grain of diopside. (*d*) Formation of a microcrack in a soil grain of augite as a result of the coalescence of side-by-side aligned etch pits. (*e*) Teeth in hornblende produced in laboratory by etching with a HF aqueous solution. (*f*) Sample geometry during X-ray photoelectron spectroscopy (XPS) analysis. Micrographs *a–e* from Berner et al. (1980).

To check on the existence of discontinuously altered layers, we need a reliable spectroscopic technique which measures surface composition on a submicron scale laterally (and not a "centimeter scale" like XPS). This calls to mind Auger spectroscopy, but so far attempts to use Auger spectroscopy for silicate dissolution have been very disappointing because of its sensitivity to surface roughness. Finally it should be added that the analyses of reacting solutions have two limitations:

1. They do not tell where the ions come from.
2. The deduced thicknesses of leached layers are very questionable because the effective surface area affected by corrosion is not known with accuracy.

An alternative approach to studying the mechanism of dissolution and bringing to light altered layers may be to measure the hydrogen profiles of weathered silicates. Indeed, most mineral and silicate dissolution reactions can be viewed as an attack by H^+ or H_3O^+:

$$\text{Silicate} + n\text{H}^+ \longrightarrow \text{silica(aq)} + \text{alumina(aq)} + \text{cations}$$

which is demonstrated by the dependence of the rates of reaction on H^+ activity (see Table 11.3).

Then, depending on the mechanism of reaction, one can expect different distributions of hydrogen in the mineral:

Within the framework of surface reaction and transition state theory (in which dissolution is ruled by the formation and dissociation of a protonated activated complex formed at the very surface), hydrogen should not extend into the bulk of the mineral.

On the contrary, if dissolution is diffusion-controlled, the exchange reaction between H^+ (or H_3O^+) and cations should result in the formation of a residual protonated layer and thus in the presence of hydrogen deep in the solid.

However, until recently, there was no direct observation of a protonated layer in silicate minerals, and the possibility of the existence of such a layer and evaluations of its thickness were based only on chemical analyses of elements other than hydrogen.

3.3. Evidence for the Mechanisms of Dissolution of Silicates from Hydrogen Depth Profiling

Petit et al. (1987) presented the first direct evidence of surficial hydration of silicate minerals using a resonant nuclear reaction (RNR), which allows direct hydrogen profiling.

Hydrogen concentration versus depth was measured by means of the resonant α-γ (helium and γ-rays) reaction between hydrogen and ^{15}N which has been already applied with success to silicate glasses (Lanford et al., 1979; Della Mea et al., 1983):

$$^{15}\text{N} + {}^{1}\text{H} \longrightarrow {}^{12}\text{C} + {}^{4}\text{He} + 4.43 \text{ MeV } (\gamma\text{-rays}) \qquad (11.6)$$

In this reaction there is a narrow isolated resonance at 6.4 MeV (see Fig. 11.8)—in other words, a large probability for this reaction—whereas a few keV away from the resonance the probability is several orders of magnitude smaller.

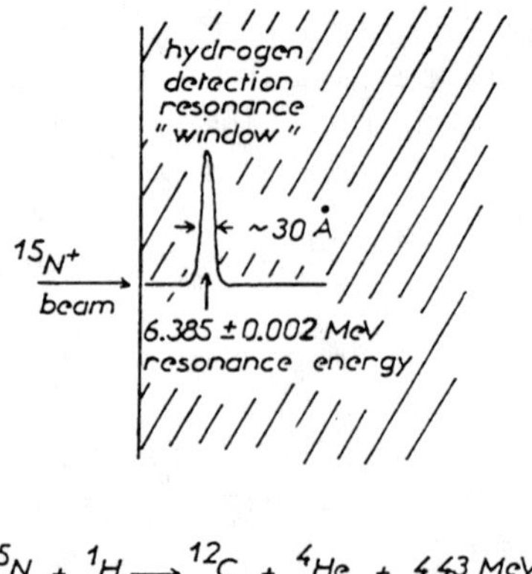

$$^{15}N + {}^{1}H \longrightarrow {}^{12}C + {}^{4}He + 4.43\ \text{MeV}$$

Figure 11.8. A schematic representation of the ^{15}N hydrogen profiling method.

To use the reaction as a probe for hydrogen, the sample is bombarded with ^{15}N of precisely controlled and variable energy produced by a Van de Graaff accelerator and the 4.4-MeV γ-rays produced are analyzed with a NaI scintillation detector. If the ^{15}N is at the resonance energy and if there is hydrogen at the surface of the sample, the yield of characteristic γ-rays is proportional to the amount of hydrogen on the surface. If the sample is bombarded with ^{15}N above the resonance energy, there are negligible γ-rays from H at the surface, and as the ^{15}N slows down in passing through the sample, it reaches the resonance energy at some depth; the γ-ray yield is then proportional to the H concentration at this particular depth. The relation depth of resonance versus ^{15}N energy is classically deduced from the theory of ion matter interaction (see, for example, Lanford et al., 1979) and is checked with ion-implanted samples. In the experiments of Petit et al., the incident ^{15}N was supplied by a 7-MeV accelerator. The beam size was ~3 mm^2, and its intensity ~1.5 A cm^{-2}. The H profile was obtained by increasing the energy in steps of ~15 keV (80 Å). The depth resolution was of the order of 30 Å at the surface and 100 Å at a depth of 0.1 μm, and the maximum depth where H can be measured ranged between 2 and 3 μm. Calibration was obtained with a Si sample implanted with 30-keV H^+ ions at a dose of 2×10^{17} ions cm^{-2} and with a hornblende of H^+ concentration of 4.6×10^{21} ions cm^{-3}.

It should be noted that a similar method can be used to profile sodium concentrations in minerals (e.g., albite). Here a narrow resonance at 0.308 MeV in the reaction $^{1}H + {}^{23}Na \rightarrow {}^{24}Mg + 1.32$ MeV (γ-ray) is used.

Dissolution experiments were conducted on centimeter-sized polished sections of single crystals of diopside from Rothenkopf and Val d'Alla of respective formula $Ca_{0.98}Mg_{0.93}Fe_{0.07}Si_2O_6$ and $Ca_{0.98}Mg_{0.97}Fe_{0.045}Si_2O_6$. The sections were polished carefully in ethanol and glycol.

The samples were subjected to dissolution in buffered aqueous solutions (pH 2 and 6) at 25, 50, and 100°C during periods of time ranging from 8 to 75 days. The ratio of surface area to solution volume was of the order of 2×10^{-3} cm^{-1}.

The main results were the following:

1. Before aqueous corrosion, both samples of diopside exhibited a very limited hydration as can be seen in Figure 11.9. The two profiles are identical and exhibit a maximum of less than 1% (atom percent) hydrogen at the surface.
2. By contrast, after reaction one can observe a marked hydration of the samples. For example, at pH 2 (Fig. 11.9) the two profiles are similar and present a maximum (~8% H) at about 250 Å, while the profiles extend to 800 Å from the surface of the sample. It should be emphasized that these profiles were perfectly reproducible after sample storage of up to one year and stable under the nitrogen beam, even after ½ hour of bombardment on the same spot.

These results are very preliminary and may still show some problems; however, it is worth noting that similar hydrogen penetrations have been observed in silicate glasses (Bunker et al., 1983) and in albite (Della Mea et al., 1984). In the case of albite, despite the very few available data (Fig. 11.10), it can be seen that the hydrogen penetration is combined, as expected, with superficial sodium depletion.

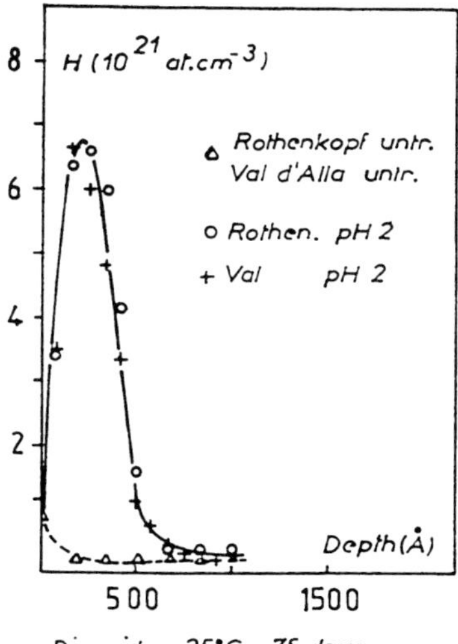

Figure 11.9. Hydrogen depth profiles by resonant nuclear reaction (RNR) for diopsides from Rothenkopf and Val d'Alla (pH 2, T = 25°C, t = 75 days).

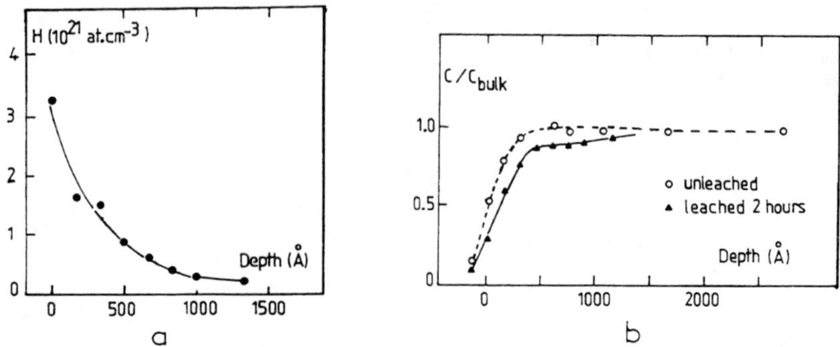

Figure 11.10. Ion depth profiles by RNR for albite (data of Della Mea et al., 1984). (*a*) Hydrogen profile of albite leached in water (T = 100°C, t = 24 h). (*b*) Sodium profile of albite leached in water (T = 100°C, t = 2 h).

A preliminary study has been carried out with diopside on the effect of pH on hydrogen penetration. Surprisingly, there was no marked effect of the pH of the reacting solution on the hydrogenation intensity (Fig. 11.11). This tends to support the assumption of penetration of H into the mineral in the form of molecular water. Similar conclusion was reached for silicate glasses by Smets and coworkers (Smets et al., 1984; Smets and Tholen, 1985) and Bunker et al. (1983).

A few samples were dried for 2 h at 150°C before RNR analysis. This heating only partially reduced the H concentration (Fig. 11.12), thus con-

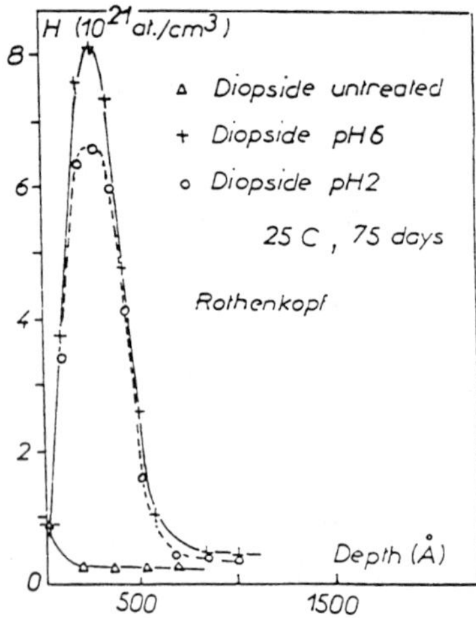

Figure 11.11. Effect of pH on hydrogen depth profiles for diopside from Rothenkopf (T = 25°C, t = 75 days).

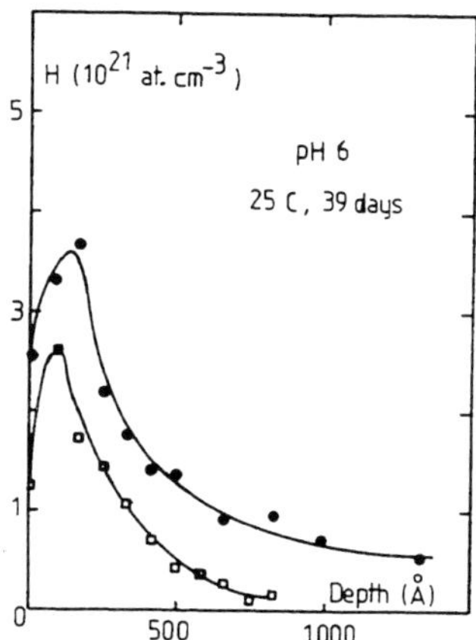

Figure 11.12. Effect of drying on hydrogen depth profiles for diopside ($T = 25°C$, pH 6, $t = 39$ days). The solid circles are for the sample dried in air at room temperature. The open squares are for a sample dried 2 h at 150°C.

firming the presence of molecular water and of OH-hydrogen bonded to the silicate network. This also suggests that the puzzling subsurface maximum of the hydrogen profile (which was also observed in silicate glasses by Bunker et al., 1983) may represent water loss to the atmosphere from the zones nearest the surface.

It is unlikely that this important hydration is an artifact or, in other words, that it takes place only in limited regions such as dislocations and microfractures, produced by sample preparation. Simple calculations show that if it is present as molecular water, the quantities of H measured represent a volume of 0.12 $cm^3\ cm^{-3}$. Thus an initial porosity of 12% would be necessary to allow a penetration of water to the depths shown. Similarly, in terms of dislocations produced by the preparation technique, it can be easily demonstrated that a density of dislocations of 10^{12}–$10^{13}\ cm^{-2}$ (compared to 10^4–$10^6\ cm^{-2}$ in normal crystals) would be required to account for the observed hydration. It is unlikely that sample preparation can produce such highly damaged surfaces. *However, it remains a fact that numerous line defects* intersecting the surface are likely to facilitate the diffusion of H_2O into the crystal as demonstrated by Burman and Lanford (1983) in the case of silicate glasses as a result of ion implantation with noble gas ions. Their results indicate that the creation of defects greatly enhances the reaction between water and the silicate network.

Secondary ion mass spectrometry (SIMS quadrupole-type instrument VG 3000), by detecting the mass 17 in the negative secondary ions, confirmed an increase in surficial hydrogen by a factor ~20 extending to an estimated depth of ~500 Å (Fig. 11.13). SIMS profiles performed on the same samples (Fig. 11.14) gave also complementary informations on other elements (Si,

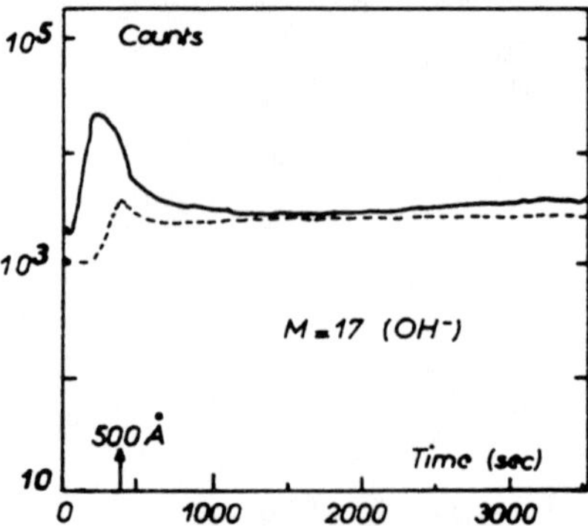

Figure 11.13. Secondary ion mass spectrometry (SIMS) depth profiles of OH^- for diopside from Rothenkopf before (dashed line) and after (solid line) 5 days leaching in deionized water (pH 5.5) at 100°C.

Ca, Mg, Fe). If almost no alteration of surfaces is detectable before dissolution, after dissolution marked depletions of all constituent elements (except Fe at the very surface) are observed over thicknesses ~1500–2000 Å for a 5-day leaching at 100°C (pH = 5.5). Quantitatively similar changes of the near-surface chemical composition of diopside have been also observed at pH ~2–6, temperatures ~25–100°C, and leaching time ~5–75 days. The decrease in

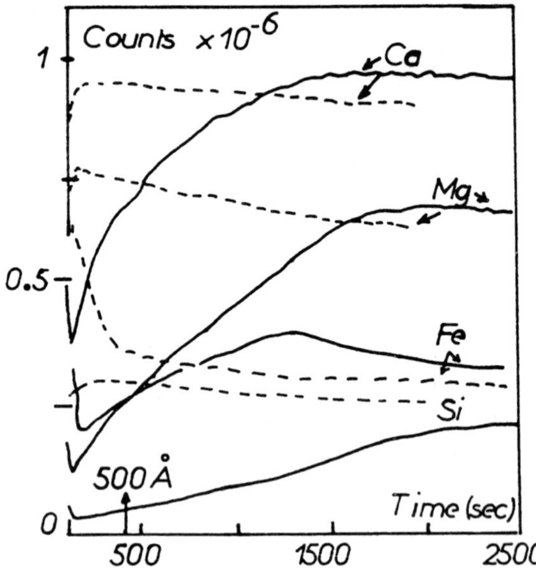

Figure 11.14. SIMS depth profiles of major positive secondary ions for diopside from Rothenkopf. The dashed lines refer to the unleached sample. The solid lines show the profiles after 5 days leaching in deionized water (pH 5.5) at 100°C.

all elements probably reflects an increase in the porosity of the surface, the pore space of which, on the basis of RNR results, is filled with water.

From hydrogen and SIMS profiles, therefore, there is clear evidence for the presence—during aqueous corrosion of diopside—of important quantities of hydrogen up to depths of ~1000 Å coupled with notable depletions in all elements marking an increase in porosity over this thickness. But there is no simple correlation between the penetration of hydrogen and the remobilization of cations.

These complex RNR and SIMS profiles cannot be entirely explained by simple models like surface reaction or diffusion exchange of H^+ with cations. In particular, the hydration of diopside to a depth as great as 1000 Å seems to rule out the dissociation of a surficial protonated complex as the only important step for dissolution.

In agreement with recent experimental results of Bunker et al. (1983) and Smets and coworkers (Smets et al., 1984; Smets and Tholen, 1985) on silicate glasses, these data rather suggest that an important step in the dissolution of silicates may be the diffusion of molecular water into the crystal and its reaction with the silicate network to produce a hydrated silicate. It should be emphasized that dislocations should play a major role in water diffusion (they permit water molecules to jump from one interstice to another) and that the breakdown of the silicate network via hydrolysis should allow more free molecular water in the structure.

The RNR profiles together with the SIMS data are consistent with the formation of a hydrated silicate depleted in Mg and Ca and enriched in Fe^{3+} and with precipitation of ferric oxide at the surface. This is in agreement with the experimental results of Schott and Berner (1983) on bronzite. These results are also consistent with the transmission electron microscope (TEM) observations by Eggleton and Boland (1982) of the weathering of enstatite to talc and iron oxides through topotactic transformations in the solid state which require diffusion of water but not of cations over long distances.

4. CONCLUDING REMARKS

This review of recent data on the dissolution of silicate minerals demonstrates the complexity of the kinetic mechanisms involved in heterogeneous systems where a solid is corroded by an aqueous solution. If steady-state dissolution is well described within the framework of transition state theory and coordination chemistry (see Stumm and Furrer, this volume, Chapter 8) by congruent reactions occurring at the mineral–solution interface, the initial steps of dissolution are diffusion-controlled and result in the formation of hydrated altered layers.

Numerous recent observations suggest that dissolution takes place preferentially at sites of high surface energy (edges, corners, cracks, dislocations) which intersect only a small fraction of the mineral surface. These lines of defects are favorable paths for the penetration of molecular water into the crystal and clarify recent hydrogen profiles which suggest that the diffusion of molecular water and its reaction with the silicate network can be an important step in the weathering of silicates. It should be added that diffusion of molecular water had already been detected during the dissolution of silicate glasses and was invoked by Veblen and Busek (1980) and Eggleton and Boland (1982) as the controlling step for the weathering of pyroxenes.

This complexity of the mechanisms of dissolution of silicate minerals calls for more experimental studies in which the parameters affecting the reaction (composition of the reacting solution, effective surface area involved in dissolution, density of dislocations, precipitation of secondary phases, etc.) are carefully controlled. For such a systematic approach it would be essential to combine chemical analysis of reacting solutions with the analysis of mineral surfaces with the aid of reliable surface techniques.

Acknowledgments

Some of the RNR experiments described in this paper were carried out with the help of J. C. Dran and G. Della Mea. The authors also thank R. A. Berner and J. C. Dran for helpful discussions.

REFERENCES

Aagaard, P., and Helgeson, H. C. (1982), Thermodynamic and kinetic constraints on reaction rates among minerals and aqueous solutions. I. Theoretical considerations, *Amer. J. Sci.* **282,** 237–285.

Berner, R. A. (1978), Rate control of mineral dissolution under earth surface conditions. *Amer. J. Sci.* **278,** 1235–1252.

Berner, R. A., and Holdren, G. R., Jr. (1979), Mechanisms of feldspar weathering—II. Observations of feldspars from soils, *Geochim. Cosmochim. Acta* **43,** 1173–1186.

Berner, R. A., Holdren, G. R., Jr., and Schott, J. (1985), Protective surface layers on dissolving silicates. Comments on the paper "Study of the weathering of albite at room temperature and pressure with a fluidized bed reactor" by L. Chou and R. Wollast [*Geochim. Cosmochim. Acta* **48,** 2205–2217 (1984)], *Geochim. Cosmochim. Acta* **49,** 1657–1658.

Berner, R. A., and Schott, J. (1985), Unpublished results.

Berner, R. A., and Schott, J. (1982), Mechanism of pyroxene and amphibole weathering II. Observations of soil grains, *Amer. J. Sci.* **282,** 1214–1231.

Berner, R. A., Sjöberg, E. L., Velbel, M. A., and Krom, M. D. (1980), Dissolution of pyroxenes and amphiboles during weathering, *Science* **207,** 1205–1206.

Blum, A., and Lasaga, A. C. (1987), "Monte Carlo Simulations of Surface Reaction Rate Laws", this volume, Chapter 10.

Brantley, S. L., Crane, S. R., Crerar, D. A., Hellmann, R., and Stallard, R. (1986), Dissolution at dislocation etch pits in quarts, *Geochim. Cosmochim. Acta* **50,** 2349–2361.

Bunker, B. C., Arnold, G. W., Beauchamp, E. K., and Day, D. E. (1983), Mechanisms for alkali leaching in mixed Na-K silicate glasses, *J. Non-Crystalline Solids* **58,** 295–322.

Burman, C., and Lanford, W. A. (1983), Radiation damage enhancement of the penetration of water in silica glass, *J. Appl. Phys.* **54,** 2312–2315.

Busenberg, E., and Clemency, C. V. (1976), The dissolution kinetics of feldspars at 25°C and 1 atm CO_2 partial pressure, *Geochim. Cosmochim. Acta* **40,** 41–49.

Carlson, T. H. (1975), *Photoelectron and Auger Spectroscopy,* Plenum, New York.

Chou, L., and Wollast, R. (1984), Study of the weathering of albite at room temperature and pressure with a fluidized bed reactor, *Geochim. Cosmochim. Acta* **48,** 2205–2218.

Correns, C. W. (1940), Die chemische Verwitterung der Silikate, *Naturwissenschaften* **28,** 369–376.

Correns, C. W., and Von Engelhardt, W. (1938), Neue Untersuchungen über die Verwitterung des Kalifeldspates, *Chem. Erde* **12,** 1–22.

Daubree, A. (1857), Observations sur le métamorphisme, *Ann. Mines* **12,** 289–326.

Daubree, A. (1867), Expériences sur les décompositions chimiques provoquées par les actions mécaniques dans divers minéraux tels que le feldspath, *C. R. Acad. Sci. Paris* **64,** 339–346.

Della Mea, G., Dran, J. C., Petit, J. C., Bezzon, G., and Rossi-Alvarez, C. (1983), Use of ion beam techniques for studying the leaching properties of lead implanted silicates, *Nucl. Instr. Method.* **218,** 493–499.

Della Mea, G., Dran, J. C., Petit, J. C., Bezzon, G., and Rossi-Alvarez, C. (1984), New data on ion-induced modifications of aqueous dissolution of silicates, *Mat. Res. Symp. Proc.* **26,** 747–754.

Dibble, W. E., Jr. (1981), Non-equilibrium water/rock interaction. I. Model for interface-controlled reactions, *Geochim. Cosmochim. Acta* **45,** 79–92.

Doremus, R. H. (1975), Interdiffusion of hydrogen and alkali ions in a glass surface, *J. Non-Crystalline Solids* **19,** 137–144.

Eggleton, R. A., and Bolland, J. N. (1982), Weathering of enstatite to talc through a sequence of transitional phases, *Clays Clay Minerals* **30,** 11–20.

Eyring, H. (1935), The activated complex in chemical reactions, *J. Chem. Phys.* **3,** 107–115.

Fleer, V. N. (1982), "The Dissolution Kinetics of Anorthite ($CaAl_2Si_2O_8$) and Synthetic Strontium Feldspar ($SrAl_2Si_2O_8$) in Aqueous Solutions at Temperatures below 100°C: With Applications to the Geological Disposal of Radioactive Nuclear Wastes," Ph.D. thesis, Pennsylvania State Univ., University Park.

Furrer, G., and Stumm, W. (1983), The role of surface coordination in the dissolution of δ-Al_2O_3 in dilute acids, *Chimia* **37,** 338–341.

Grandstaff, D. E. (1977), Some kinetics of bronzite orthopyroxene dissolution, *Geochim. Cosmochim. Acta,* **41,** 1097–1103.

Grandstaff, D. E. (1981), The dissolution rate of forsteritic olivine from hawaiian beach sand, *Third Intern. Symp. Water-Rock Interaction, Proc.,* pp. 72–74, Alberta Research Council, Edmonton.

Helgeson, H. C. (1971), Kinetics of mass transfer among silicates and aqueous solutions, *Geochim. Cosmochim. Acta* **35,** 421–469.

Helgeson, H. C., Murphy, W. M., and Aagaard, P. (1984), Thermodynamic and kinetic constraints on reaction rates among minerals and aqueous solutions. II. Rate constants, effective surface area, and the hydrolysis of feldspar, *Geochim. Cosmochim. Acta* **48,** 2405–2432.

Holdren, G. R., Jr., and Berner, R. A. (1979), Mechanism of feldspar weathering. I. Experimental studies, *Geochim. Cosmochim. Acta* **43,** 1161–1171.

Holdren, G. R., Jr., and Speyer, P. M. (1985), pH dependent changes in the rates and stoichiometry of dissolution of an alkali feldspar at room temperature, *Amer. J. Sci.* **285,** 994–1026.

Koppleman, M. H., and Dillard, J. G. (1979), "The Application of X-Ray Photoelectron Spectroscopy (XPS or ESCA) to the Study of Mineral Surface Chemistry," in M. M. Mortland and V. C. Farmer, Eds., *International Clay Conference 1978,* Elsevier, New York, pp. 153–166.

Lagache, M. (1976), New data on the kinetics of the dissolution of alkali feldspars at 200°C in CO_2 charged water, *Geochim. Cosmochim. Acta* **40,** 157–161.

Lagache, M., Wyart, J., and Sabatier, G. (1961), Mécanisme de la dissolution des feldspaths alcalins dans l'eau pure ou chargée de CO_2 à 200°C, *C. R. Acad. Sci. Paris* **253,** 2296–2299.

Lanford, W. A., Davis, K., Lamarche, P., Laursen, T., Groleau, R., and Doremus, R. H. (1979), Hydration of soda-lime glass, *J. Non-Crystalline Solids* **33,** 249–266.

Lasaga, A. C. (1981), "Transition State Theory," in A. C. Lasaga and R. J. Kirkpatrick, Eds., *Kinetics of Geochemical Processes* (Reviews in Mineralogy, Vol. 8), Mineral. Soc. Amer., Washington, DC, pp. 261–319.

Lasaga, A. C. (1984), Chemical kinetics of water-rock interactions, *J. Geophys. Res.* **89,** (B6), 4009–4025.

Luce, R. W., Bartlett, R. W., and Parks, G. A. (1972), Dissolution kinetics of magnesium silicates, *Geochim. Cosmochim. Acta* **36,** 35–50.

Murphy, W. (1985), "Thermodynamic and Kinetic Constraints on Reaction Rates among Minerals and Aqueous Solutions," Ph.D. thesis, University of California, Berkeley.

Nickel, E. (1973), Experimental dissolution of light and heavy minerals in comparison with weathering and intrastratal solution, *Contributions Sedimentol.* **1,** 1–68.

Pačes, T. (1973), Steady-state kinetics and equilibrium between ground water and granitic rock, *Geochim. Cosmochim. Acta* **37,** 2641–2643.

Petit, J. C., Della Mea, G., Dran, J. C., Schott, J., and Berner, R. A. (1987), Diopside dissolution: new evidence from H-depth profiling with a resonant nuclear reaction, *Nature*, in press.

Petrović, R., Berner, R. A., and Goldhaber, M. B. (1976), Rate control in dissolution of alkali feldspar. I. Study of residual grains by X-ray photoelectron spectroscopy, *Geochim. Cosmochim. Acta* **40,** 537–548.

Rimstidt, J. D., and Barnes, H. L. (1980), The kinetics of silica-water reactions, *Geochim. Cosmochim. Acta* **44,** 1683–1699.

Schott, J., and Berner, R. A. (1983), X-ray photoelectron studies of the mechanism of iron silicate dissolution during weathering, *Geochim. Cosmochim. Acta* **47,** 2333–2340.

Schott, J., and Berner, R. A. (1985), "Dissolution Mechanisms of Pyroxenes and

Olivinеś during Weathering," in J. I. Drever, Ed., *The Chemistry of Weathering,* D. Reidel Publ. Co., Dordrecht, pp. 35–53.

Schott, J., Berner, R. A., and Sjöberg, E. L. (1981), Mechanism of pyroxene and amphibole weathering. I. Experimental studies of iron-free minerals, *Geochim. Cosmochim. Acta* **45,** 2123–2135.

Siegbahn, K. (1982), Electron spectroscopy for atoms, molecules, and condensed matter, *Science* **217,** 111–121.

Smets, B. M. J., and Tholen, M. G. W. (1985), The pH dependence of the aqueous corrosion of glass, *Phys. Chem. Glasses* **26,** 60–63.

Smets, B. M. J., Tholen, M. G. W., and Lommen, T. P. A. (1984), The effect of divalent cations on the leaching kinetics of glass, *J. Non-Crystalline Solids* **65,** 319–332.

Stumm, W., and Furrer, G. (1987), "The Dissolution of Oxides and Aluminum-Silicates; Examples of Surface-Coordination Controlled Kinetics," this volume, Chapter 8.

Stumm, W., Furrer, G., Wieland, E., and Zinder, B. (1985), "The Effects of Complex-Forming Ligands on the Dissolution of Oxides and Alumino-Silicates," in J. I. Drever, Ed., *The Chemistry of Weathering,* D. Reidel Publ. Co., Dordrecht, pp. 55–74.

Tole, M. P. (1982), "Factors Controlling the Kinetics of Silicate–Water Interactions," Ph.D. thesis, Pennsylvania State Univ., University Park.

Tsao, S. T., and Pask, J. A. (1982), Reaction of glasses with hydrofluoric acid solution, *J. Am. Ceram. Soc.* **65,** 360–362.

Veblen, D. R., and Busek, P. R. (1980), Microstructures and reaction mechanisms in biopyriboles, *Amer. Mineral.* **65,** 599–623.

Wollast, R. (1967), Kinetics of the alteration of K-feldspar in buffered solutions at low temperature, *Geochim. Cosmochim. Acta* **31,** 635–648.

PART THREE

REGULATING THE COMPOSITION OF NATURAL WATERS

12

SURFACE CHEMICAL ASPECTS OF THE DISTRIBUTION AND FATE OF METAL IONS IN LAKES

Laura Sigg

Institute for Water Resources and Water Pollution Control (EAWAG), Swiss Federal Institute of Technology, Dübendorf, Switzerland

Abstract

For an understanding of the fate of metal ions in a lake, their interactions with settling particles are of primary importance. Surface chemical principles are an aid in the interpretation of field data on the binding and transport of metal ions in lakes. In freshwater lakes, the settling particles consist mostly of biological material, calcium carbonate, iron and manganese oxides, and silicate minerals. The role of these different particles in the transport of metal ions depends on their respective affinity for the binding of metal ions and on the number of available surface ligand sites. The field data on settling particles indicate that the biological particles play an important role in removing trace metals (especially Cu and Zn) from the water column and in regulating their concentrations. Iron and manganese oxides represent additional scavenger phases with large surface areas, while calcium carbonate is inefficient as a carrier phase.

1. INTRODUCTION

1.1. Fate of Metal Ions in Lakes

Trace metal inputs to a lake are ultimately removed either by transport to the sediments, where they are buried, or by transport out of the lake by the outflow. Metal contents in earlier lake sediments provide a historical record of the metal inputs in the past; in many examples it was found that the trace metal contents of the most recent sediments are much higher than the background natural level. The fraction of the metal inputs retained in lake sediments depends on the affinity of the metal ions for the settling material in the lake. The understanding of the fate of metal ions in a lake thus requires knowledge of the interactions of metal ions with the natural particles; since the binding of metal ions to these particles is mostly dependent on adsorption reactions, an application of the surface chemical principles derived from model systems is needed.

The objectives of this chapter are (1) to describe the distribution of metal ions between particulate and dissolved form in natural waters and attempt to apply surface chemical principles to natural particles; (2) to evaluate which are the most important particles for the removal of metal ions to lake sediments; and (3) to consider the interactions of metal ions with biological particles (e.g., algae, algal debris). Examples from studies in Lake Zurich and Lake Constance (Sigg, Sturm, Stumm, Mart, Nuernberg, 1982; Sturm et al., 1982; Sigg, Sturm, Kistler 1987) will be used to illustrate general principles.

1.2. A Summary of Some Relevant Chemical Properties of Metals in Freshwater Lakes, as Reflected in Representative Equilibrium Constants

The interactions of metal ions with dissolved and particulate ligands in lake water are dependent on their coordination chemistry properties. By considering the principles of coordination chemistry of metal ions in natural water systems, we can improve our ability to understand the factors that play a major controlling role in the biogeochemical cycles and transformations of metals.

Some data on the coordination chemical properties of a few selected elements are summarized in Table 12.1. The inorganic speciation is based on equilibrium calculations. The tendency to form organic complexes in natural waters is difficult to define. The complexation with a well-defined ligand such as salicylate may be taken as a crude measure for the interactions with natural organic ligands, although the polyelectrolytic properties of humic and fulvic acids are not accounted for with this model ligand. On the other hand, few studies on the interactions of metals with natural organic matter provide comparable results for different metal ions. The extent of organic complexation in natural waters has been studied by numerous authors and is a controversial issue (Buffle, 1984; Florence and Batley, 1980).

The solubility products of the oxides and sulfides illustrate the difference between *A* and *B* metals. Redox intensities are listed according to equilibrium predictions. It is uncertain whether the oxidation of Co(II) to Co(III) occurs in natural waters. Some of the oxidation reactions may have slow kinetics, so that equilibrium may not always be attained [for example, Cr(III)/Cr(VI)]. Many other interactions of metal ions with different surfaces are possible. Only some well-investigated systems are listed here. Unfortunately, the constants for the reactions with surfaces cannot always be compared, since they are based on different models. As a general rule, the constants for the complexation of cations with surface OH groups are related to the first hydrolysis constant of the cations. This has been found for different surfaces and is independent of the model used.

The data given in Table 12.1 provide a general framework that may help to predict the fate of metals under different environmental conditions.

2. INTERACTIONS OF METAL IONS WITH SURFACES OF NATURAL PARTICLES

The interaction of trace elements with the surfaces of solid particles is important in regulating the solute concentrations of many of these elements in natural waters. Solid particles in natural waters include, for example, oxides of aluminum, iron, and silicon, silicate minerals, and biological debris. The interactions of cations and anions with the particle surfaces cannot be explained by electrostatic interactions only; specific interactions have been interpreted in recent years in terms of surface coordination with surface functional groups, often OH groups, whose acid–base and other coordinative properties are similar to those of their counterparts in soluble compounds.

The interactions of metal ions with hydrous oxide surfaces have been extensively studied so they may serve as models for the particles encountered in natural waters and to allow quantitative predictions of the extent of adsorption (Stumm et al., 1976, 1980; Schindler et al., 1976, 1981; Davis and Leckie, 1978; Benjamin and Leckie, 1981, 1982; Bourg and Schindler, 1978, 1979; Schindler and Stumm, Chapter 4, and Westall, Chapter 1, in this book).

It is observed that metal ions are reversibly bound to these surfaces and that these reactions are strongly pH-dependent. These reactions can in principle be understood as an exchange of protons at the surface with metal ions. Several models describing the reactions of metal ions with hydrous oxide surfaces have been put forward. These models differ mainly in their way of describing (1) the location of the metal ions at the surface and the surface

Table 12.1. Summary of Some Relevant Chemical Properties of Selected Metals in Freshwater Lakes

Element	Main Soluble Inorganic Species[a]	Organic Complexes: log K of Salicylate Complex[b]	Solubility Products,[c] log K_{s0}: Oxide/ Hydroxide	Carbonate	Sulfide	Redox Processes,[a] $p\epsilon^0$	Formation of Volatile Methylated Compounds[d]	Tendency to Bind to Surfaces: SiO_2, log K_1^s	$FeOOH/Fe(OH)_3$, log K^s	Al_2O_3, log K_1
Al(III)	$Al(OH)_3(aq)$, $Al(OH)_4^-$	14.2	−33.5	—	—	—	—	—	—	—
Cr(VI)	CrO_4^{2-}	—	—	—	—	3.0	—	—	−10.6[e]	—
Cr(III)	$Cr(OH)_2^+$	—	−29.8	—	—		—	—	—	—
Mn(IV)	Insoluble	—	Ins. oxide	—	—	4.8	—	—	—	—
Mn(II)	Mn^{2+}	6.8	−12.8	−9.3	−13.5		—	—	—[f]	—
Fe(III)	$Fe(OH)_2^+$, $Fe(OH)_4^-$	17.6	−38.5	—	—	−2.5	—	−1.77[g]	—	—
Fe(II)	Fe^{2+}	7.4	−15.1	−10.7	−18.1		—	—	—	—
Co(III)	Not known	—	−44.5	—	—	5.5	—	—	—	—
Co(II)	Co^{2+}	7.5	−14.9	−9.98	−21.3		—	—	—[f]	—
Ni(II)	Ni^{2+}	7.8	−15.2	−6.87	−19.4/	—	—	—	—[f]	—
					−24.9/		—	—	—	—
					−26.6[c]					
Cu(II)	$CuCO_3^0$	11.5		−33.8,[h] −45.96[i]	−36.1	—	—	−5.52[g]	−3.0[f]	−2.1[g]
Zn(II)	Zn^{2+}, $ZnCO_3^0(aq)$	7.5	−15.5/[c] −16.8	−10.0	−24.7	—	—	—	−9.15[f]	—
As(V)	$H\ AsO_4^{2-}$	—	—	—	—	−1.95	$(CH_3)_3As$, $(CH_3)_2AsH$ CH_3AsH_2	—	—[e]	—[e]
As(III)	$As(OH)_3$	—	—	—	12.6			—	—	—

Se(IV)	SeO_4^{2-}	—	—	—	—	6.2 (Se(IV) and Se(VI))	$(CH_3)_2Se$, $(CH_3)_2Se_2$	—	-9.9^e	—
Se(VI)	H SeO_3^-	—	—	—	—		$(CH_3)_2SeO_2$	—	—	—
Ag(I)	Ag^+, $AgCl^0$	—	-7.7	-11.1	-50.1^c	—	—	—	-5.7^f	—
Cd(II)	Cd^{2+}, $CdCO_3^0(aq)$	6.4	-14.3	-13.7	-27.0^c	—	—	-6.09^g	-1.3^f	—
Hg(II)	$Hg(OH)_2^0$, $HgOHCl^0$	—	-25.4	-16.05	-53^c	6.2 (Hg(II) and Hg(0))	$(CH_3)_2Hg$	—	—	—
Hg(0)	Hg(0)	—	—	—	—		CH_3HgOH	—	—	—
Pb(II)	$PbCO_3^0(aq)$	—	-15.3	-13.3	-27.5	—	$Pb(CH_3)_4$	-5.09^g	-1.8^f	-2.2^g

[a]Speciation calculated for freshwater (pH8; [Alk] = 2.5×10^{-3} M, $[Cl^-] = 1.6 \times 10^{-4}$ M, $[SO_4^{2-}] = 3.6 \times 10^{-4}$ M, oxic conditions). Stability constants used are from Smith and Martell (1976) and Dyrssen and Wedborg (1980). In many natural waters, Cu(II) is often present to a large extent as an organic species. $p\epsilon^0$ for the redox reactions is calculated for the same conditions ($p\epsilon^0 = \frac{F}{2.3RT} \times E_H^0$, E_H^0 = standard redox potential, F = Faraday constant, R = gas constant. T = temperature (Stumm and Morgan, 1984)).

[b]Log K of 1:1 complex with salicylate.

[c]Solubility products are from Smith and Martell (1976) and from Morel (1983). (I = 0, 25°C). Different crystalline forms have different solubility products (Ni,Zn). Ag(I), Cd(II), and Hg(II) also form strong soluble sulfide complexes, so that the high solubility products do not mean insolubility. Arsenic is precipitated as sulfide according to the reaction: $As(OH)_3 + (3/2)H_2S \rightleftharpoons (1/2)\ As_2S_3 + 3H_2O$, log K = 12.6.

[d]Brinckman et al. (1982).

[e]The adsorption of AsO_4^{2-} on amorphous aluminum oxide was studied by Andersson et al. (1976) and on amorphous iron hydroxide by Pierce and Moore (1982). The adsorption of SeO_4^{2-} and of CrO_4^{2-} on amorphous iron oxide was investigated by Davis and Leckie (1980); the constants given are for the surface reactions $\equiv S{-}OH + H^+ + CrO_4^{2-} \rightleftharpoons \equiv SOH_2^+{-}CrO_4^{2-}$ and $\equiv S{-}OH + H^+ + SeO_4^{2-} \rightleftharpoons \equiv SOH_2^+{-}SeO_4^{2-}$, and are based on a triple-layer model.

[f]The constants for the adsorption of Cu, Pb, Zn, and Cd on α-FeOOH are from Balistrieri and Murray (1982) for seawater conditions. They are based on a triple-layer model for the reactions $\equiv SOH + Me^{2+} \rightleftharpoons SO{-}Me^+ + H^+$ for Cu, Pb, Cd and $\equiv SOH + Me^{2+} + H_2O \rightleftharpoons \equiv S{-}O{-}MeOH + 2H^+$ for Zn. For Mn^{2+}, Co^{2+}, and Ni^{2+}, the complexation by a FeOOH surface was shown to correlate with the first hydrolysis constant of these ions (Balistrieri and Murray, 1982). The constant for the adsorption of Ag^+ on amorphous $Fe(OH)_3$ is from Davis and Leckie (1978) with the same model.

[g]The constants for the adsorption of Fe(III), Cu, Cd, Pb on SiO_2 are from Schindler et al. (1976); those for the adsorption of Cu and Pb on Al_2O_3 are from Hohl and Stumm (1976). They are based on the surface complexation model for a reaction of the type: $\equiv SOH + Me^{2+} \rightleftharpoons \equiv S{-}O{-}Me^+ + H^+$.,

[h]$Cu_2(OH)_2CO_3$. [i]$Cu_3(OH)_2(CO_3)_2$.

species formed and (2) the electrostatic interactions. The surface complexation model (Stumm and Morgan, 1981; Schindler et al., 1976, 1984; Stumm et al., 1976) treats the binding of metal ions to oxide surfaces in analogy to the complexation by ligands in solution. Hydroxyl groups on a surface have coordinative properties similar to those of an oxygen-donor in a soluble compound (for example, carboxylate or phosphate). Protons and metal ions compete with each other for the available coordinating sites on the surface. The pH dependence of the adsorption of metal ions to an oxide surface is similar to that of the complexation by a weak acid.

The effects of soluble complexing ligands on the adsorption of metal ions on a hydrous oxide surface can be of two kinds; on the one hand, the complex formation may decrease the adsorbed concentrations on the surface if a simple competitive situation exists; on the other hand, the formation of ternary surface complexes occurs with certain ligands (Benjamin and Leckie, 1982; Bourg and Schindler, 1978; Bourg et al., 1979), so that in such cases adsorption may be increased in the presence of a ligand. Ternary surface complexes including a ligand L and a metal M may be of the type $\equiv$S—O—M—L or $\equiv$S—L—M, depending on the adsorption and complexation properties of M and L. Surface complexes may thus be included in speciation models of metal ions. Computer programs used for the calculation of the different species may include surface complexation equilibria (Westall et al., 1976; Westall and Hohl, 1980; Morel, 1983).

A high degree of sophistication can be attained in the modeling of laboratory data on specific adsorption to surfaces, as long as well-defined systems are considered. In applying the surface complex formation equilibrium constants to real systems or to field data, the following general principles of the surface complexation model are useful:

1. The complexation of metal ions by oxide surfaces is strongly pH-dependent; the extent of adsorption is a function of pH with an abrupt change within 1–2 pH units. For anions, an inverse pH dependence as for metal ions is observed.
2. A set of surface equilibrium constants permits estimation of the surface speciation of an oxide in natural water of a given composition; in ideal cases the surface charge of an oxide and its dependence on pH and solution variables can be predicted.
3. The tendency of metal ions to be bound to oxide surfaces is related to their tendency to interact with oxygen donor atoms or to form hydroxo complexes.
4. The complexation of metal ions by surface ligands is in competition with the complexation by soluble ligands.

In model studies of adsorption, one deals with simple, well-defined systems, where usually a single well-characterized solid phase is used and the composition of the ionic medium is known, so that reactions competing with the adsorption may be predicted. It is not a trivial problem to compare the results from such model studies with those from field studies, or to use model results for the interpretation of field data. In field studies, a complex mixture of solid phases and dissolved components, whose composition is only poorly known, has to be considered; competitive reactions of major ions and trace metal ions for adsorption may take place, and the speciation of the trace metal ions is often poorly understood. In order to relate field studies to model studies, distribution coefficients of elements between the dissolved and solid phases are useful. These distribution coefficients are of the following form:

$$K_D = \frac{c_s}{c_w} \ (\mathrm{m^3\ kg^{-1}}) \qquad (12.1)$$

where c_s is the concentration in the solid particles (mg kg^{-1}) and c_w is the concentration in water (mg m^{-3}).

The distribution coefficients are independent of the concentration of suspended solids in water, which can vary over a wide range; they thus give a better picture than the fraction of metal ions in solution. Such distribution coefficients can be predicted on the basis of the equilibrium constants defining the complexation of metals by surfaces and their complexation by solutes (Table 12.2).

Distribution coefficients based on adsorption equilibria are independent of the total concentrations of metal ions and suspended solids, as long as the

Table 12.2. Determination of the Distribution Coefficient K_D from Surface Complex Formation

Species at surface: $\equiv S{-}O{-}M^+$, $(\equiv SO)_2M$

Species in solution: M^{2+}, MOH^+, $M(OH)_m^{(2-m)+}$, ML_1, ML_2, where L_1 and L_2 are known soluble ligands.

$$K_D = \frac{\{\equiv S{-}O{-}M^+\} + \{(\equiv S{-}O)_2M\}}{[M^{2+}] + [MOH^+] + [M(OH)_m] + [ML_1] + [ML_2]} \left(\frac{\mathrm{mol/kg}}{\mathrm{mol/m^3}}\right)$$

$$K_D = \frac{K_{a_1}^s \{\equiv SOH\}/[H^+] + \beta_2^s \{\equiv SOH\}^2/[H^+]^2}{1 + K_{OH}[OH^-] + \beta_{OH_m}[OH^-]^m + K_1[L_1] + K_2[L_2]} \ (\mathrm{m^3/kg})$$

where { } denotes concentration in mol/kg of solid phase.

K_D depends on:
- pH
- kind of surface; number of OH groups per surface
- complexation in solution

Source: After Schindler (1984).

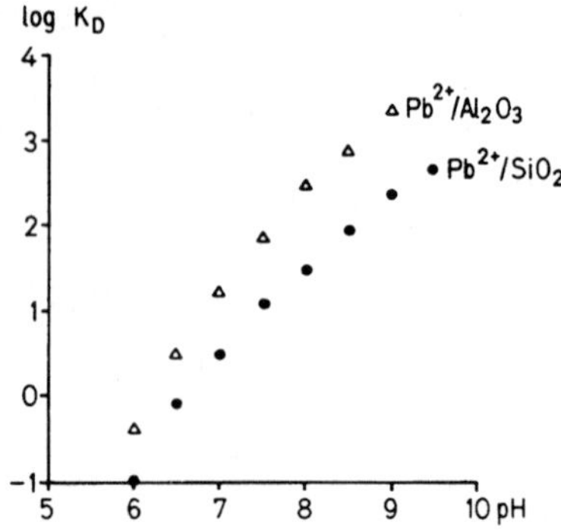

Figure 12.1. Distribution coefficients (m^3 kg^{-1}) calculated for surface complexation with ≡AlOH and ≡SiOH surface groups. The following species were taken into account for Pb: Pb^{2+}, $PbOH^+$, $PbCO_3^0$, $Pb(CO_3)_2^{2-}$, ≡Al—O—Pb^+, $(\equiv AlO)_2Pb^0$, and with SiO_2 $(\equiv SiO)_2Pb^0$ and ≡SiO—Pb^+. {≡AlOH} = 0.25 mol kg^{-1}; {≡SiOH} = 1.5 mol kg^{-1}

metal concentrations are small compared with the concentration of surface groups. Examples of the K_D obtained from calculations for model surfaces are presented in Figure 12.1. A strong pH dependence of these K_D values is observed. The pH range of natural lake and river waters (7–8.5) is in a favorable range for the adsorption of metal ions on hydrous oxides.

The influence of the complexation in solution is illustrated by Figure 12.2. Increasing concentrations of a soluble ligand cause a decrease in K_D for the simple case in which the complexation in solution and at the surface are competing with each other.

To relate the distribution between solution and particulate phases in natural waters with the surface complexation models, several approaches can be taken:

1. Distribution coefficients determined for the natural systems can be compared with the coefficients obtained in laboratory experiments for various materials; such comparisons may allow some conclusions about the nature of the solid material responsible for adsorption in the natural system.
2. The composition of the particles may be determined, and theoretical distribution coefficients may be calculated. The speciation of the metal ions in the solution phase should also be known.
3. Natural particulate material can be used in adsorption experiments, and conditional stability constants can be calculated (Mouvet and Bourg, 1983).

Distribution coefficients can be derived from field data:

1. By determining the dissolved and particulate concentrations of an element in individual water samples. This may be difficult to realize in waters

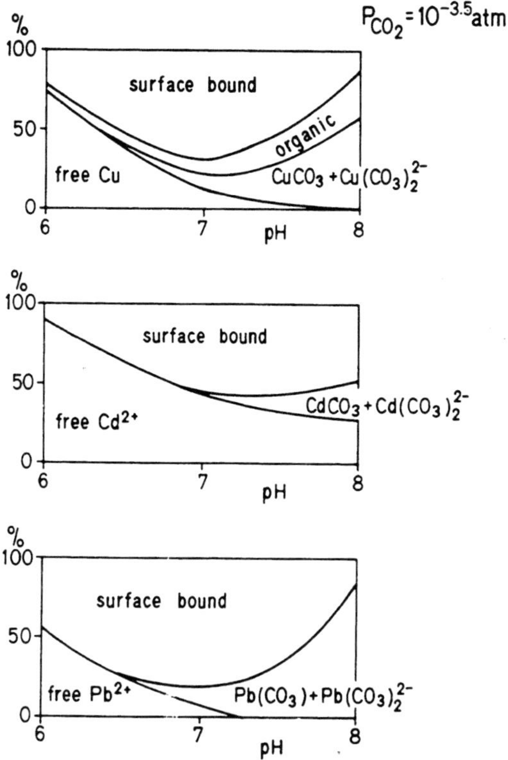

Figure 12.2. Competition between surface sequestration and complex formation in solution. Fresh water: pCa = 4.0, pMg = 4.3, pSO_4 = 4.3, and pCl = 3.0 plus 10^{-5} *M* salicylate (as organic complex-forming model substance) and silica (with 10^{-5} *M* ≡SiOH groups).

which contain only small concentrations of particulate matter, such as lake water, and small concentrations of metal ions.

2. By determining metals in the soluble phase and in the settling material (Sigg et al., 1982).
3. From residence time of trace elements in oceanic systems. This approach has been used to derive equilibrium constants describing the interaction of these elements with particulate matter in oceans (Balistrieri et al., 1981; Schindler, 1975).

Examples of distribution coefficients observed in Lake Constance and Lake Zurich are given in Table 12.3; the distribution coefficients were calculated from the metal concentrations measured in the settling particles and in the water column. They reflect the tendency of the different elements to be bound to the particulate phase [Fe(III) > Pb(II) > Zn(II) ≈ Cd(II) > Cu(II)]; this includes the formation of insoluble oxides [Fe(III)], adsorption on the settling particles, and the formation of dissolved complexes (Cu). The distribution

Table 12.3. Distribution Coefficients $K_D = \frac{c_{\text{particles}}}{c_{\text{water}}}$ (m³ kg⁻¹)[a]

	Fe	Zn	Cu	Cd	Pb
Lake Constance[b]	~1 × 10^4	100	35	50	1000
Lake Zurich[c]	10^3–10^4	40–140	10–100	100–500	500–3000

[a]Calculated from measurements on the settling particles and in the water column.
[b]Mean values.
[c]Range.

coefficients observed at different times of the year are found to be dependent on the composition of the particles (relative abundance of biological material, calcium carbonate, iron and manganese oxides, silicates), since these compounds have different affinities for the metal ions.

Distribution coefficients derived from field data are useful in rationalizing these data, but one must be aware that in most cases they reflect the result of a number of different reactions of an element: adsorption on the different available surfaces, precipitation of solid phases, incorporation in certain minerals such as clay minerals, uptake by biota, and complexation in solution.

Knowledge about the nature of the particles in a particular system allows us to evaluate the most important interactions. Conditional stability constants for surface complexation in natural systems may be calculated. For a river system with a variety of pH conditions, Johnson (1986) shows that the data can be interpreted on the basis of an adsorption of metal ions on iron oxides, which are the predominant solid phase in this case. Conditional stability constants are calculated; the stability of the surface complexes of different metal ions corresponds to the sequence observed with defined oxide systems in the laboratory. Balistrieri et al. (1981) estimate surface stability constants for different elements with marine particulate matter by using their scavenging residence time. A comparison of the stability constants obtained in this manner with the stability constants for hydroxo species of these elements reveals a correlation of the two sets of constants similar to the correlations obtained for pure oxides. The interactions between marine particulate matter and metals are found to be much stronger than the interactions of metals with model oxides (SiO_2, γ-Al_2O_3, α-FeOOH, amorphous Fe_2O_3). Balistrieri et al. concluded that organic material (especially surfaces of organisms and biological debris) must be responsible for the binding of metal ions to marine particulate matter.

Both the settling material and the suspended matter in lakes such as Lake Zurich and Lake Constance consist mainly of calcium carbonate, organic material, iron oxides, manganese oxides, and silicate minerals. In order to evaluate the relative contribution of these different phases to the adsorption of trace elements, their respective surface areas are compared.

For iron oxide, the surface area can be evaluated by assuming that iron exists entirely as amorphous iron oxide. Freshly precipitated iron oxide has surface areas of 200–500 $m^2\ g^{-1}$; with the typical Fe content in the settling particles of 5–10 mg g^{-1} this gives surface areas ranging from about 1 to 10 $m^2\ g^{-1}$ if the iron oxide is really very finely dispersed.

In order to evaluate the surface area of algal material, the number of cells and the corresponding surface areas are evaluated from the organic material content of the settling particles. Diatoms, as typically found in Lake Zurich, have the following dimensions (H. R. Bürgi, EAWAG, personal communication):

Volume 2000 μm^3 Dry weight $\approx$560 pg $cell^{-1}$
Surface 6000 μm^2

With organic material contents of 200–400 mg g^{-1}, surface areas of 2.3–4.5 $m^2\ g^{-1}$ are obtained. The results would, of course, vary if different algal species were taken into account, but this rough evaluation shows that the surface areas provided by algal material are large and come to the same order of magnitude as the surface areas of iron oxides. For calcium carbonate, the surface area can be evaluated by assuming cubic particles with an edge dimension of 5–30 μm.

Manganese oxides that are freshly precipitated at the redox boundary (oxic–anoxic) in the lake probably also have high surface areas; a specific surface area of 200 $m^2\ g^{-1}$ is assumed here. Figure 12.3 shows the calculated surface areas for a sample of settling particles from Lake Zurich. While the calcium carbonate represents an important fraction of the dry weight, the biological material and iron oxide are much more important in terms of surface area. In addition to the surface areas, the affinities of the different surfaces for the metal ions (surface complexation constants) should be known in order

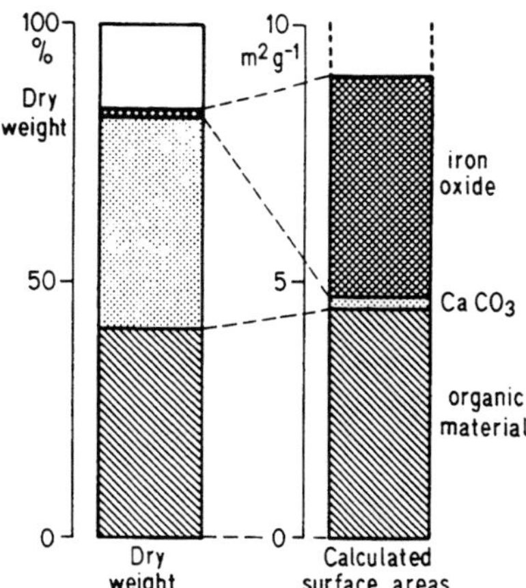

Figure 12.3. Composition of settling particles in Lake Zurich (collected June 6–26, 1984, depth 50 m) in terms of dry weight percentages and calculated surface areas. The surface areas were calculated from the specific surface area of amorphous iron hydroxide for the iron oxide and from the size and surface of typical algae for the organic (biological) material.

to establish their relative role. It can be assumed, for example, that surfaces of biological particles, which include amino groups, bind certain metal ions, like Cu, more specifically than oxide surfaces. It is thus necessary to examine more closely the interactions of metal ions with the biota.

3. INTERACTION OF METAL IONS WITH BIOTA: SIMPLE CHEMICAL MODELS

Metal ions are closely linked to biological processes. Some metals are essential to biological activity; others, although not required, may be taken up by cells. Above certain concentration thresholds, all elements can be toxic. The objective of the following discussion is to briefly review the interaction of metals with biota, specifically, to make use of some of the principles governing complex formation of metals with biological surfaces and biological ligands to elaborate on the role of plankton present in settling material as a carrier of metal ions.

It is, however, clearly beyond the scope of this treatment to discuss the biochemistry or the molecular mechanisms of the effects of trace elements on physiological and toxicological processes. Only some simplified chemical models are presented, which may give some preliminary insight into the reciprocal interactions between organisms and trace elements. The discussion is restricted to the consideration of monocellular algae.

3.1. Bioavailability and Toxicity of Metals

Organisms require for their growth and life a number of different elements; in addition to the major elements building up biological compounds (C, N, O, H, S, P), a number of other elements are needed in trace amounts. These essential trace elements must be present in a certain concentration range if they are to meet the requirements of the biota. If they are available in too small a concentration, they may become growth-limiting; on the other hand, excessive concentrations of these elements may have toxic effects (Williams, 1981).

For other, nonessential elements, no beneficial effects on organisms can be observed, and the adverse effects increase with increasing metal concentrations. Both the concentration range required for essential elements and the toxic concentration range differ for different metals and different algal species. Various organisms have adapted to diverse natural conditions with different environmental metal concentrations. Specialized organisms that are resistant to extreme natural conditions (for example, to naturally acidic environments) have been found (Wood and Wang, 1985). The anthropogenic mobilization of a number of toxic metals is rapidly changing local natural conditions, so that organisms can be adversely affected by these changes. Toxic concentration ranges of metals can be reached due to anthropogenic inputs into natural waters. Toxicity sequences of different metals have been established for various kinds of organisms (Nieboer and Richardson, 1980). The toxicity effects of different metal ions are closely related to their coordination chemical properties, since the mechanisms often include the binding of metals to sensitive cellular compounds, often enzymes and proteins. In general, B-type metal ions are more toxic than borderline (between A and B types) and A-type metal ions (Stumm and Morgan, 1981). This can be explained by their higher affinity toward S- and N-binding groups. For example, the very high affinity of Hg for S ligands may partly explain its strong toxicity, since Hg can interact with –SH and –S–S– groups of proteins.

The mechanisms of metal ion toxicity can include the replacement of an essential metal ion by a nonessential element with similar chemical properties, the blocking of essential biological functional groups of biomolecules, and the modification of the active conformation of biomolecules by complexation of metal ions.

Algae, through their cell membranes, are in direct contact with the aquatic environment. What is the influence of the natural water chemistry on the uptake and toxicity of metal ions? A great variety of chemical species of metal ions occur in natural waters; it is thus of primary importance to assess the effect of different metal species on algae. There is much experimental evidence that the responses of algae (e.g., growth rates) are related to changes in the *free* aquo metal ion activities and not directly to the *total* metal concentrations.

The responses of algae to free aquo metal ion activities have been demonstrated in ionic media containing different concentrations of metals and strong chelators (Sunda and Guillard, 1976; Jackson and Morgan, 1978; Anderson and Morel, 1978, 1982; Sunda and Huntsman, 1983); such media act as metal ion buffers.

3.2. Surface Chemical Models for the Binding of Metal Ions to Phytoplankton

In order to understand the role of algae in the limnological cycles of metals and of phytoplankton cell material that has become part of the settling particles in the vertical transport of metals into the deeper layers of the lake and its sediments, the binding of metal ions to the surface of algal cells may be tentatively described by simple surface chemical models. The nonspecific term "binding" used in this case means interaction with the surface as well as uptake into the cell. The primary objective of the models is to aid in answering questions such as

1. Which parameters are affecting the extent of metal binding to algal cells? Are biological surfaces better scavengers for heavy metals than the mineral surfaces also present in the settling particles?
2. What is the effect of an increased metal burden in an aquatic system on the phytoplankton?

3. To what extent and how do phytoplankton and its productivity regulate the residual dissolved metal ion concentration in lakes?

An important corollary to these questions is the inquiry as to whether phytoplankton contains in its mean elemental composition certain trace metals similar to the macronutrients N, P, and C in constant (stoichiometric) proportions or whether the trace metal composition of biological cells is influenced by the composition of the water. The mean composition of the settling particles of Lake Zurich and Lake Constance with respect to the main nutrients has been found (Sigg, Sturm, Davis and Stumm, 1982) to be in good agreement with the marine algal composition originally given by Redfield et al. (1963):

$$106CO_2 + 16NO_3^- + HPO_4^{2-} + 18\,H^+ + 122H_2O$$

$$\underset{\text{respiration}}{\overset{\text{photosynthesis}}{\rightleftharpoons}} \{(CH_2O)_{106}(NH_3)_{16}(H_3PO_4)_1 \cdots\} + 138O_2 \qquad (12.2)$$

For these nutrients the mean elemental stoichiometry of the composition of the phytoplankton is independent of that in the solution. Can the idea of a constant elemental algal composition be extended to certain trace metals, and could such a constancy be plausibly accounted for by a simple chemical model? Two models are compared: (1) an adsorption model considering only interactions of metals with a cell surface and (2) an uptake model that takes into account some mechanisms of transport into the cell.

3.2.1. Surface Coordination (Adsorption) Model. The metal ion becomes coordinated (adsorbed) to the algal surface, that is, to complex-forming groups on this surface, $-NH_2$ ligands, or carboxylic or hydroxo groups. Surface complex formation or adsorption equilibria, in principle similar to those discussed for oxide surfaces, can describe the extent of metal binding to cell surfaces:

$$>R-C(=O)OH + Me^{2+} \rightleftharpoons >R-C(=O)OMe^+ + H^+$$

$$>R-C(=O)OH \text{ (with } NH_2 \text{ on R)} + Me^{2+} \rightleftharpoons >R(-C(=O)-O-Me-NH-) + 2H^+$$

$$>R(OH)-C(=O)OH + Me^{2+} \rightleftharpoons >R(-C(=O)-O-Me-O-) + 2H^+ \qquad (12.3)$$

The transport of the surface-bound metal ions into the cell's interior is not considered in this simple model.

The number of ligand groups per cell should be constant for one cell species, as it depends only on the cell wall structure; it may depend on the life stage of the cell (changes in the cell wall structure). Since the chemical composition and the cell size are rather constant for a given algal species, the number of ligand groups in an algal suspension should be proportional to the organic carbon or phosphorus content and to the surface area of the suspended algae. Different ligand types are possible on one cell; the selectivity for different metal ions is different for –OH, carboxyl, or $-NH_2$ ligands. For example, Cu^{2+} will tend to bind first to amino acid surface groups, then to carboxylic or hydroxy-carboxylic groups, and eventually to hydroxo groups. The selectivity of NH_2 groups (amino acid groups) for metal ions is expected to vary with the Irvine–Williams order of complexation constants; that is Cu^{2+} would be especially strongly bound by such ligands (Williams, 1981, 1983). That such complexes are characterized by high stability constants has been confirmed by EPR measurements on surfaces of *Klebsiella pneumoniae* bacteria (Motschi, 1985). These results are in agreement with results obtained in a voltammetric study (Gonçalves et al., 1986) on the surface complex formation equilibria of Cu(II), Zn(II), and Pb(II) with the same bacterial cells.

A plausible number of high-affinity surface binding sites is about 10^7–10^8 per cell for a representative phytoplankton (Morel and Hudson, 1985). The equilibrium constant for the types of reactions described by (12.3) may be generalized as

$$K_1^s = \frac{\{\equiv S_1\text{—}M^+\}\,[H^+]}{\{\equiv S_1\text{—}H\}\,[M^{2+}]} \tag{12.4}$$

where $\{\equiv S_1H\}$ and $\{\equiv S_1M\}$ are the concentrations of surface groups bound by protons or metal ions (moles per gram dry weight), respectively, and where [ ] refers to aqueous phase concentrations (moles per liter).

The surface complex formation reactions of cell surfaces (similar to that already described for hydrous oxides) are defined as any processes that result in a close association, on the molecular level, between the metal and one or more functional groups on the cell surface. Conditional equilibrium constants can alternatively be defined for the reaction

$$a(\equiv SH_n) + M^{2+}(aq) \rightleftharpoons (\equiv S_aM^{2-an}) + an\ H^+\ (aq) \tag{12.5}$$

where a is restricted to the values 1 and 2 and n takes noninteger values. According to this model the cell surface is represented by noninteracting ligands which can form 1:1 and 1:2 complexes with a metal cation (Sposito, 1985).

Although no rigorous thermodynamic significance can be attributed to such constants, conditionally they can describe the observed equilibria for the conditions maintained in the sorption experiments they characterize. Because various functional groups have different affinities for the binding of metal ions and thus different stability constants, a spectrum of "microscopic" K_1^s values with changing metal/surface ligand ratio (Gamble et al., 1983) may give a better description of the metal-binding tendency.

Despite these shortcomings, some generalizations can be derived from Eq. 12.4. As with dissolved ligands and coordinating surfaces (e.g., hydrous oxides), complexation reactions of metal ions with ligands on cell surfaces are pH-dependent. For a fixed number of surface groups and a given pH, the concentration of metals bound to the surface is proportional to the free metal ions in solution; a competitive equilibrium with different metal ions is established.

The speciation of the metal in solution regulates the free aquo metal ion concentration and influences the resulting $\equiv S_1–M^+$ concentration. When pH and the aquo metal ion concentration are constant, the concentration of metal ions bound per gram of solid material becomes proportional to the number of sites per gram of material; the concentration $\{\equiv S_1–M^+\}$ is then proportional to the organic carbon or phosphorus content of a cell suspension. If surface sites with a very high affinity for certain metal ions exist, that is, if K is very large, equilibrium 12.3 can be shifted to the right and these surface sites may become saturated with metal ions. This means that the surface of the cells could carry a constant concentration of metal ions with respect to the cell composition if the metal ions were in excess of the concentration of high-affinity sites.

The competition of various metal ions can be described in this model by the different affinities (stability constants, Eq. 12.4) of the metal ions for the surface sites. If equilibrium is assumed, the reactions are reversible; bound metal ions are thus released upon changes in the chemical conditions, such as addition of dissolved ligands to the water or pH changes. The kinetics of such reactions are then governed by the exchange reactions (12.3) and by the kinetics of exchange reactions with dissolved ligands, as long as the cell surface area and thus the number of surface sites are constant. If the cells are growing, the number of available surface sites is increasing, and the rate at which metal ions disappear from solution will vary according to the growth rate.

Living and dead cells behave in a similar way with respect to adsorption of metal ions if the cell wall structure remains unchanged. A similar behavior of living and dead cells was described by Fisher et al. (1983) for the adsorption of transuranic elements. The adsorption of metal ions even on cell debris upon decay of an algal cell suspension is possible. This means that partial destruction of algal cells does not necessarily imply the total release of metal ions to the solution.

3.2.2. Uptake Model. This model implies a complexation of metals on the outside surface of the cell either by biologically released ligands or by surface functional ligand groups. Subsequent to this surface complex formation, metals are carried through a membrane to the inside of the cell.

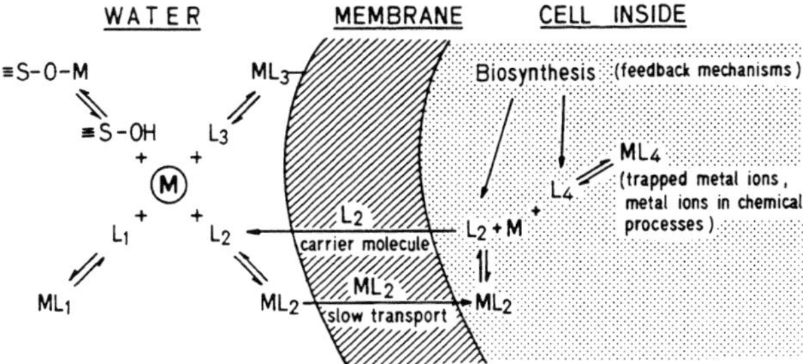

Figure 12.4. Uptake model. In a simplified model, the metal ions equilibrate on the outside of the cell with biologically produced and excreted ligands L_2 or ligands on the cell surface L_3; these reactions are followed by a slow transport step to the inside of the cell. In the cell, the metal ions may be used in biochemical processes or become trapped in inactive forms as a detoxification mechanism. After Williams (1981, 1983) and Wood and Wang (1985).

There the metals participate in various biochemical reactions or may be trapped by high-affinity ligands for either storage or detoxification (metal ion buffering) (Fig. 12.4). This model follows the concepts developed by Williams (1981, 1983) and Wood and Wang (1985).

Both thermodynamic and kinetic aspects are needed in this model. The thermodynamic aspects include the equilibria between the various ligands (L_1, L_2, L_3, L_4), the competition of different cations for these ligands, and the acid–base properties of the ligands. The kinetic aspects include the kinetics of the ligand exchange reactions, the rate of production and release of L_2, and the rate of transport of ML_2 across the membrane (by diffusion or by active processes), which is often slow in comparison to the other processes; the production and release of L_2, L_3, L_4 are related to the growth rate of the algae.

The selectivity of the uptake of certain metal ions is given by the selectivity of the ligand L_2. Carrier molecules are often proteins, and the stability of their complexes with different metal ions corresponds to the Irvine–Williams stability order (Williams 1983). Steric factors are also involved in the selectivity. Another selectivity mechanism is exerted by the size of the channels through which certain metal ions are allowed to diffuse into the cells; these channels must have the same radius as the hydrated metal ion for which they are intended (Williams, 1981). The concentration of M inside the cell depends on the availability of L_2, on the rate of transport across the membrane, and, possibly, on the rate of transport out of the cell. Algal cells may have different mechanisms regulating the concentration of a metal ion M inside the cell; the transport by L_2 over membranes may be regulated by a feedback mechanism, ML_2 acting on the biosynthesis of L_2. Such a mechanism allows a cell to maintain a constant concentration of an essential element, thus leading to a constant (stoichiometric) chemical composition of the cell material.

This means that the stoichiometric composition of algal cell material, as given by Redfield (1963) for the major elements, may be extended to trace elements:

$$(CH_2O)_{106}(NH_3)_{16}H_3PO_4Cu_{0.006}Zn_{0.03} \cdots$$

(The coefficients for Cu and Zn are tentative coefficients based on the study of settling material in Lake Zurich.) The mean composition of algal material for the major elements is usually close to this ratio, maintained independently of varying C:P:N ratios in water (see, e.g., Schindler, 1985).

It is unclear whether a constant proportion of trace elements may be maintained independently of the concentrations outside the cell. One further problem is whether such mechanisms act also on biologically nonessential elements. Nonessential elements, like Cd^{2+}, may be mistaken as micronutrients by the cell. Other mechanisms allow the rejection of unwanted elements by transport out of the cell or by trapping in an inactive form inside or at the surface of the cell (Wood and Wang, 1985). The metals are thus not bound in a reversible way inside the cell; changes in the external chemical conditions may not simply lead to release or to increased uptake of metal ions, since the cells have control mechanisms that can accommodate for changes over a certain range.

Since such uptake mechanisms require metabolic activity for the production of ligands and for active transport across membranes, differences between living and dead cells should be noted in their behavior toward metal ions. Upon decay of cells, no further binding should occur, but metals may remain trapped as insoluble complexes.

Different algal species have different requirements for essential elements; the mechanisms for dealing with nonessential and toxic elements differ and lead to varying metal accumulation. The two models for the interactions of metal ions with algal cells, namely, those based on (1) adsorption and surface complexation or (2) active uptake, have been treated separately to clearly emphasize their respective implications and consequences. It must be recognized, however, that both models are strongly simplified pictures and that some features of both may be implied in the interactions of metals with algal cells. Adsorption of metal ions on cellular surface ligands may be a first step in the uptake of metals, so that pseudo-equilibrium conditions are attained. It is important to note that both models can account for the response of cells to free aquo metal ion concentrations, since the binding or uptake of metal ions depends on equilibria between different ligands and metal ions (Turner and Whitfield, 1980). However, this does not imply that the free aquo metal ion is the species actually transported into the cell, but only that the equilibria implied are dependent on its concentration.

On the basis of these two models, various mechanisms acting as a control on the concentration of metals in algal cells may be put forward that result in a constant concentration of metals in the algal material (e.g., constant metal/phosphorus ratios):

1. The saturation of a fixed number of surface sites with metal ions gives a constant concentration on the basis of the high affinity of these sites.
2. Due to enzymatic controls, algal mechanisms that control the transport of metal ions into the cell result in constant metal concentrations; such mechanisms are especially important for essential elements.
3. The concentrations of free aquo metal ions outside the cell may be regulated by the excretion of exudates which complex the metal ions, indirectly exerting control on the concentrations of metals bound to the cell.
4. The mechanism of colimitation of algal growth in oceans by several elements, which would result in the uptake of different trace elements in fixed required amounts, has been postulated by Morel et al. (1985).

4. THE ROLE OF SETTLING PARTICLES IN THE VERTICAL TRANSPORT OF METAL IONS IN LAKES

In order to be transported to the sediments, the metals have to be bound in or on the particulate phase. Several mechanisms may be responsible for the association with the solid phase. Interactions with the biota have been discussed; the binding of metals to biological particles will lead to their transport to the sediments together with the biological material. Further mechanisms include the precipitation of solid phases (hydroxides, carbonates, sulfides); for the conditions encountered in Lake Zurich or Lake Constance, precipitation of solid phases of Cu, Zn, Pb, Cd, and Cr is not likely to occur, since the concentrations of these metals in the water column are much lower than the calculated equilibrium concentrations with respect to the solid hydroxides and carbonates; sulfide formation does not occur in these lakes, but in anaerobic environments sulfide precipitation can be an important mechanism (Jacobs and Emerson, 1982). Adsorptive interactions (surface complexation) with different kinds of surfaces are dependent on the affinity of different metal ions for the surface groups, on chemical parameters (pH, speciation of metal in solution, etc.), and on the concentrations of both metal ions and surface sites. The settling particles in lakes contain, in addition to the biological particles, iron and manganese oxides and hydroxides and, to a minor extent, silicate minerals, which all offer suitable surface sites for the adsorption of metal ions. Furthermore, coagulation effects with biological particles may induce the sedimentation of small colloidal particles (see, e.g., O'Melia, this volume, Chapter 14) which would otherwise remain suspended (e.g., iron hydroxide colloids). The major elemental composition of the settling particles, as well as their heavy metal content, have to be known in order to gain insight into these various interactions.

The composition of the settling particles, as well as the total fluxes of material to the sediments, depend strongly, of course, on the biological cycles in the lake and vary over the course of the year. The following considerations are based on the study of settling particles collected in Lake Constance (summer 1981 and summer/fall 1982) and in Lake Zurich (July 1983 to December 1984) (Sigg, Sturm, Stumm, Mart, Nuernberg, 1982; Sigg, Sturm, Davis, Stumm, 1982; Sturm et al., 1982; Sigg, Sturm and Kistler 1987). The settling particles were collected in traps suspended at various depths and exposed for 3 weeks.

Figure 12.5 gives some representative examples of the major composition of settling particles from Lake Zurich and Lake Constance. The main components of these particles are as follows (all percentages given are relative to the dry weight of the particles):

1. Calcium carbonate varies between about 20 and 80%. Its precipitation depends on the degree of supersaturation in the upper layers of the lake, where the pH is higher as a consequence of the seasonally varying photosynthetic activity.
2. The biological material represents about 10–50% of the settling particles. Although it consists of different algal species and also of products of the zooplankton activity (fecal pellets), the biological material can in a simplified way be thought of as having a constant stoichiometric composition.

According to the Redfield stoichiometry (Redfield et al., 1963), the mean composition of algal material is $(CH_2O)_{106}(NH_3)_{16}H_3PO_4$. In the settling particles from Lake Zurich, a mean C:P ratio = 97:1 was found, while in Lake Constance the mean ratio was C:N:P = 113:15:1. In both cases, the ratios are very close to the Redfield ratios; the difference is related to the kinds of samples collected. In Lake Zurich the samples were collected over a whole year, while in Lake Constance samples were obtained only in the summer and fall. The C:P ratios in samples from Lake Zurich at different times of the year vary around this mean ratio (Table 12.4).

A constant composition of the settling particles, independently of time and depth, presupposes the following conditions:

Constant elemental ratios are maintained in the biological material.

The degradation of biological material by respiration occurs congruently, that is, photosynthesis and respiration can be represented by the stoichiometric equation 12.2.

No other processes influence the composition of the settling particles (e.g., adsorption of phosphate on nonbiological phases). The *deviations from ideal stoichiometry* in the particles can be explained by assuming that the different conditions are not always valid.

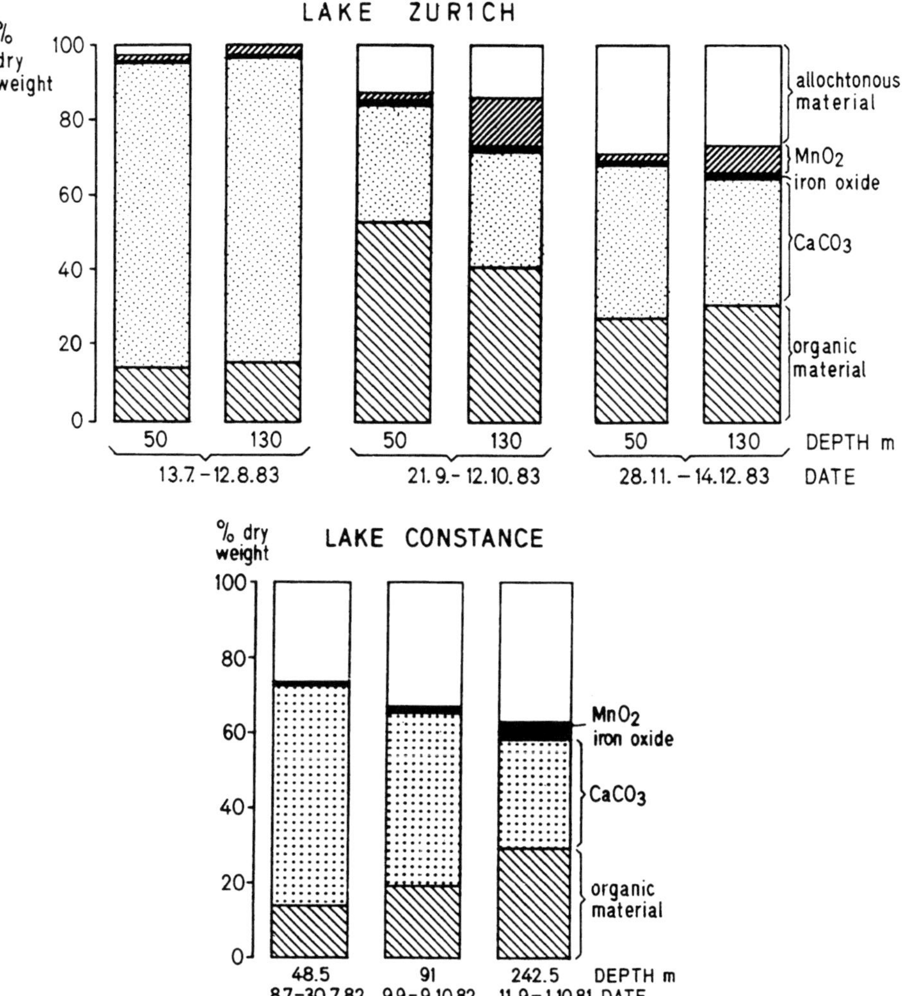

Figure 12.5. Some examples of the composition of settling particles from Lake Zurich and Lake Constance. The percentage of organic material was calculated from the measured content of organic C in the particles. The remainder (white area) may represent allochthonous material (material brought into the lake from outside) such as Al silicates.

3. Iron oxides represent about 1–3% of the settling particles in Lake Zurich and Lake Constance. A major part of the iron enters the lake in the form of small iron oxide or hydroxide particles; the removal of these small particles (partly of colloidal size) from the water column is favored by the settling of biological particles due to coagulation effects. A covariance of iron with the biological particles is found in the settling material. An additional source of iron could be the diffusion of reduced Fe(II) from

Table 12.4. Mean Elementary Composition of Settling Material, Relative to Phosphorus[a]

	C	N	P	Fe	Zn	Cu
Lake Zurich[b]						
50 m	120 ± 28	—	1	2.3 ± 1.2	$3.7 \times 10^{-2} \pm 1.2 \times 10^{-2}$	$7.5 \times 10^{-3} \pm 2.7 \times 10^{-3}$
130 m	84 ± 17	—	1	2.0 ± 0.8	$3.5 \times 10^{-2} \pm 1.4 \times 10^{-2}$	$7.0 \times 10^{-3} \pm 1.5 \times 10^{-3}$
Lake Constance[c]	113 ± 22	15 ± 2.5	1	4.0 ± 1.8	$6.0 \times 10^{-2} \pm 3.4 \times 10^{-2}$	$8.0 \times 10^{-3} \pm 2.5 \times 10^{-3}$

	Pb	Cr	Cd
Lake Zurich[b]			
50 m	$6.4 \times 10^{-3} \pm 2.6 \times 10^{-3}$	$5.4 \times 10^{-3} \pm 3.5 \times 10^{-3}$	$2.3 \times 10^{-4} \pm 1.3 \times 10^{-4}$
130 m	$4.0 \times 10^{-3} \pm 2.0 \times 10^{-3}$	$3.9 \times 10^{-3} \pm 2.1 \times 10^{-3}$	$1.6 \times 10^{-4} \pm 0.7 \times 10^{-4}$
Lake Constance[c]	$3.7 \times 10^{-3} \pm 1.5 \times 10^{-3}$	—	—

[a]Moles per gram relative to moles P per gram.
[b]Samples were collected in sediment traps for three weeks each (20 collection periods 1983/84). (Sigg et al. 1987.)
[c]Samples were collected at various depths in interval sediment traps (maximum eight samples in 3 weeks, 13 collection periods May–October 1981 and 1982, (Sigg, 1985).

the sediments. Fe(II) diffusing from deeper (and more reducing) sediment layers is probably oxidized rapidly in the water column and thus appears as iron oxide.

4. Manganese oxides represent less than 1% of the settling material in the upper layers of the lake, but they can make up as much as 3% in the deeper part of Lake Constance and 2–15% in the deeper part of Lake Zurich.
5. The fraction of allochthonous material (consisting mainly of silicates such as quartz, feldspars, and micas) was not measured directly, but it can be calculated on the basis of a summation of the other phases. Aluminum was measured in some samples and can be taken as a measure of the allochthonous material. The allochthonous material is more important in Lake Constance than in Lake Zurich as a consequence of the far more important tributaries (Rhine, Schussen). The fraction of allochthonous material in Lake Constance is about 20–40%. In Lake Zurich, allochthonous material represents only a few percent of the settling material in summer samples and 10–30% in winter samples.

No direct method is available for determining the metal contents associated with each of the carrier phases. We attempted to gain evidence for the association of trace elements with the different carrier phases by correlating the trace element contents in the settling particles with the proportion of the various carrier phases and by considering seasonal variations in the trace element contents of the settling particles and the fluxes to the sediments.

Examination of the data leads to the following conclusions:

1. *Calcium carbonate.* Although $CaCO_3$ often makes up a large part of the settling material, it seems that it is unimportant as a carrier phase. Negative correlations are obtained for the concentrations of Zn, Cu, Pb, and Cd versus Ca in the settling particles; the same is also true for P and Fe versus Ca. Extrapolation of these correlations to 100% $CaCO_3$ leads to concentrations of Zn, Cu, Pb ≈ 0 in pure $CaCO_3$. The effect of $CaCO_3$ on the sediment composition can then be thought of as the dilution of the other phases by an inert phase. Although $CaCO_3$ is for a large part formed in the upper layer of the water column, it does not take up other elements; crystallization of $CaCO_3$ and the subsequent settling occur at fast rates (Kunz, 1983; Sturm et al., 1982), and the crystals formed are rather large (length d = 5–30 μm) and thus offer only small surfaces for adsorption. Coprecipitation of other elements seems unfavorable, possibly because of the low dissolved concentrations and the competition of other phases for adsorption and uptake of the metal ions. Baccini and Joller (1981) found no precipitation of Cu and Zn with $CaCO_3$ in laboratory experiments. Collier and Edmond (1984) reported very low metal concentrations in $CaCO_3$ particles picked out of marine phytoplankton samples and reached similar conclusions about the inefficiency of $CaCO_3$ as a carrier phase in oceanic systems.

2. *Binding to biological material.* In order to evaluate the affinity of trace elements for the biological material (phytoplankton and its debris), correlations between trace elements and phosphorus in the settling material were investigated. Phosphorus is used as a measure of the biological material, since its stoichiometric composition is nearly constant.

The binding of trace elements to the biological material may be expected to result in stoichiometric proportions of these elements. Rather good correlations of Cu and Zn with P are found, indicating that these elements are mainly associated with the biological material (Fig. 12.6). Alternatively, the mean elementary composition of the settling material, relative to phosphorus, is given in Table 12.4. The relationship between Cu, Zn, and P can be interpreted as a stoichiometric relationship. From the data of Lake Zurich, it is calculated that 6–7 × 10^{-3} mole of Cu and 3–4 × 10^{-2} mole of Zn, respectively, are taken up together with 1 mole of P (calculated from the correlation of Cu and Zn, respectively, with P or from the mean elementary composition). The mean composition of the samples from Lake Constance gives 8 × 10^{-3} mole Cu and 6 × 10^{-2} mole Zn for 1 mole P. The Cu:P ratio is thus comparable in both sets of samples, and the results for Cu, Zn, and P from Lake Constance are found to scatter around the correlation line obtained with data from Lake Zurich. For Zn, the higher Zn:P ratio found in samples from Lake Constance may be a result of higher inputs; it is un-

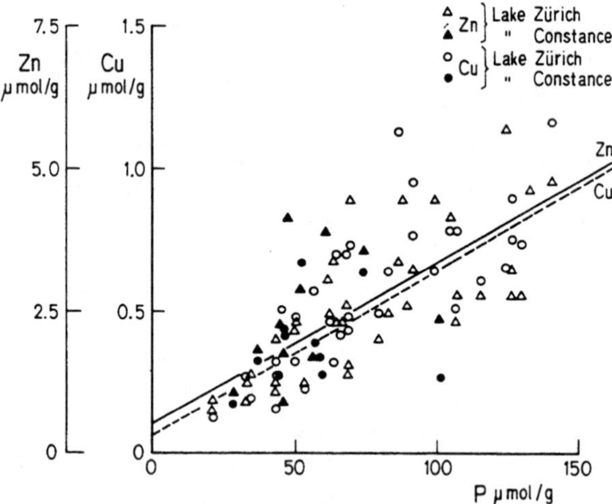

Figure 12.6. Concentrations of Cu, Zn as a function of P in the settling material from Lake Zurich and Lake Constance. P is used as a measure of the concentration of biological material in these settling particles. Each point is for settling material collected for 3 weeks. The regression lines are calculated from the data of Lake Zurich only; the regressions for Zn and Cu fall nearly together, indicating that rather constant ratios Cu:Zn:P are found in this material. These correlations indicate an association of Cu and Zn with P in the settling material and thus a binding of these metals to the biological material.

certain, however, whether the limited number of samples from Lake Constance biased the results; the contamination risks are also higher for Zn. The similar ratios obtained for Cu and Zn relative to P in both sets of samples indicate the occurrence of similar processes.

As an extension of the Redfield stoichiometry, a tentative composition of the biological material may be derived from the data for Lake Zurich:

$$(CH_2O)_{97}(NH_3)_{16}H_3PO_4Cu_{0.006}Zn_{0.03}$$

The molar ratio of Zn to Cu in the settling material is approximately 4–5:1, being comparable with the molar concentration ratios in the water column (Zn:Cu $\approx$ 3–4). Furthermore, fluxes of Cu and Zn to the sediments in Lake Zurich, as calculated from sedimentation rates and from the content of these elements in the sedimenting material, are higher at times of important sedimentation of biological material (June–September) than in the fall and winter (Fig. 12.7).

Ratios of Pb, Cr, and Cd to P in the settling material are given in Table 12.4; the results are less systematic than for Cu and Zn. Data on the fluxes to the sediments in Lake Zurich (Sigg et al., 1987) indicate, however, that the fluxes of Pb and Cd are also highest at times when the sedimentation of biological material is highest (Fig. 12.7). It can be assumed that for these elements the mechanisms of binding to different kinds of surfaces may play a role in addition to the binding to biological material. Pb, and perhaps also Cd, enters the lake in part already bound to particles (from atmospheric depositions and rivers). Furthermore, the interactions of Pb and Cd with biological surfaces may be different from that of Zn and Cu, since Cd and Pb are not essential elements; adsorption to biological surfaces as well as coagulative interactions are possible. Especially for Cd, we have to take into account that the measured values are in a low range and thus subject to lower analytical reproducibility than those obtained for the other elements. Cd has been shown to be strongly correlated with P in the oceanic water column (Bruland, 1980). In the ocean, Cd is present primarily as soluble chloro complexes; it can be taken up by phytoplankton cells that may mistake Cd for a micronutrient.

3. *Iron oxide.* The molar ratio of iron to phosphorus, Fe:P, of 1–3:1 (Lake Zurich) and up to 8:1 (Lake Constance) is too large to be accounted for by biological binding alone. Morel and Hudson (1985) give the ratio Fe:P = 0.01:1 for marine phytoplankton, while Bowen (1979) indicates Fe:P ratios ranging from 0.03 to 0.6 for marine algae. Fe probably enters the lake mainly as small iron oxide particles or as iron oxide coatings on other particles. It is possible that the removal of the small iron oxide particles by settling is more favored at times of high sedimentation of biological material due to sorption to biological surfaces or coagulation, resulting in additional removal of Fe together with the biological material.

Iron oxides have large surfaces to which many cations and anions may

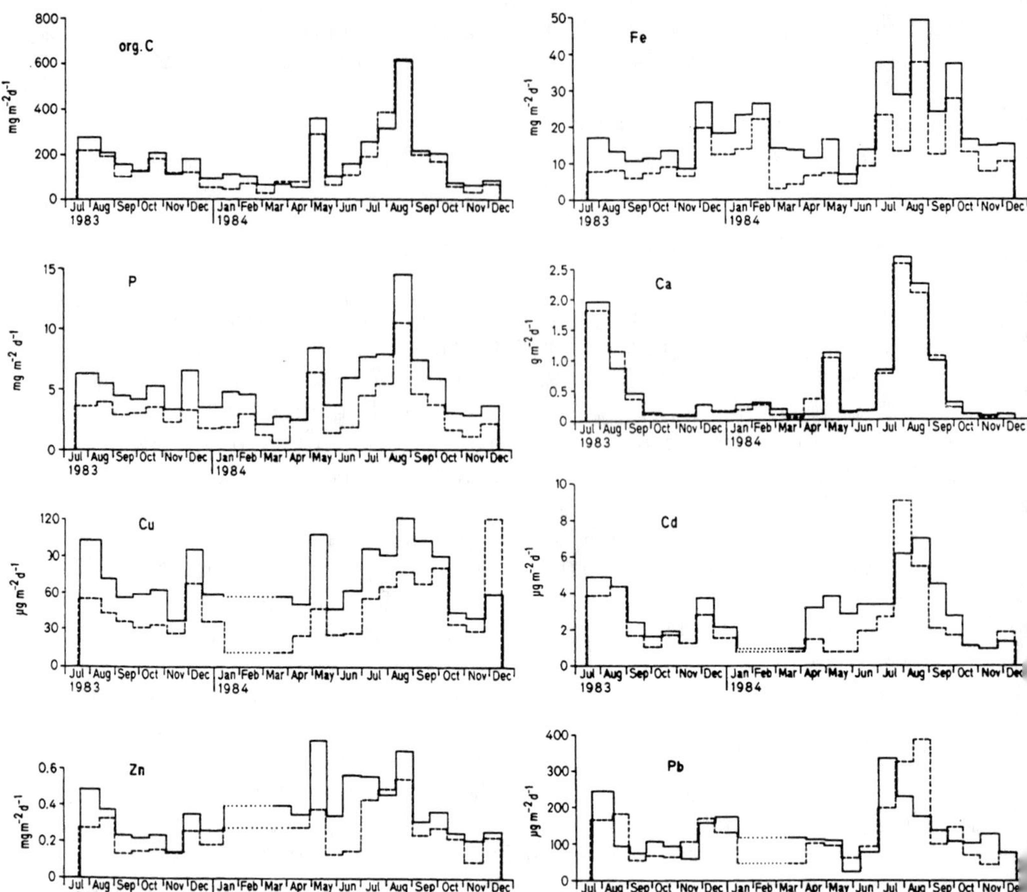

Figure 12.7. Fluxes of C (organic), P, Cu, Zn, Fe, Ca, Cd, and Pb to the sediments as a function of time (mg m^{-2} day^{-1} for C, P, Zn, Fe; µg m^{-2} day^{-1} for Cu, Cd, Pb; g m^{-2} day^{-1} for Ca). High fluxes of Cu, Zn, Fe and also of Cd and Pb in summer show the importance of biological material in the removal of these elements. The fluxes were calculated from the metal concentrations in the sediment trap material and the total sediment fluxes in the sediment traps at 50 m (dashed line) and at 130 m (solid line). Each bar represents a 3-week sampling period. Some sampling periods in January–February 1984 are omitted for Cu, Zn, Cd, and Pb, since contamination of these samples was suspected. From Sigg, Sturm, and Kistler (1987).

adsorb (Balistieri and Murray, 1982; Sigg and Stumm, 1981). They represent another scavenging phase for other metal ions. For example, Pb is found to correlate reasonably well with iron in the settling particles of Lake Zurich. This indicates that Pb may be preferentially bound to such iron oxide surfaces. Cu and Zn also correlate quite well with iron; since these elements are all related to P, the role of iron oxide in their removal cannot be decided on the basis of these correlations.

4. *Manganese cycling.* In the deeper part of lakes, a very active Mn cycling may take place. This occurs in Lake Zurich; Mn^{2+} released from sediments is reoxidized back to MnO_2 in contact with O_2-containing water and sediments. This manganese oxide is, in principle, a very good scavenger for heavy metals. The comparison of fluxes in the 50-m and 130-m traps in Lake Zurich shows that higher fluxes of Cu and Zn are often observed at a depth of 130 m (Fig. 12.7). These differences are assumed to be due to binding to freshly precipitated MnO_x in the deep part of the lake, which represents an additional scavenging mechanism.

5. *Allochthonous material.* The metal-scavenging role of allochthonous material, consisting for the most part of silicates (quartz, feldspars, and micas) is probably a minor one. Indirect evidence is based on the very similar concentrations of different metals in samples of Lake Constance and Lake Zurich, despite the different contributions of allochthonous material to these samples. The surface areas of the allochthonous material are probably not very large and not very specific for adsorption of metals. Baccini (1976) also found that allochthonous materials are rather inefficient scavengers of metals.

Thus we tend to infer that the biota comprise the major carrying phase which, in addition to binding Cu and Zn and possibly other elements, also scavenges iron oxide colloids, which in turn act as solid substrate for other elements such as Pb and Cr.

Collier and Edmond (1984) concluded from a study on marine phytoplankton that the biological organic material was the major carrier of Cu, Zn, Ni, Cd, Mn, Ba, and Fe in the oceans. They found very small percentages of the trace elements associated with calcium carbonate and with opal compared to that associated with plankton. The findings in the lakes are thus in agreement with these observations in the marine environment.

5. CONCLUSIONS

Surface chemical principles are an aid in the interpretation of field data on the binding and transport of metal ions in lakes, although many open questions remain about the surface chemical properties of natural particles. In freshwater lakes, the settling particles consist mostly of biological material, calcium carbonate, iron and manganese oxides, and silicate minerals. In order to evaluate the role of these different particles in the transport of metal ions, their affinity for the binding of metal ions and the number of available surface ligand sites, which is dependent on the specific surface area, should be known. The interactions of metal ions with the biological material may tentatively be described in terms of simplified chemical models, which reveal a high affinity of some metal ions (for example, Cu) for these surfaces and pH-dependent binding reactions. Ratios of metals to phosphorus in the settling material from Lake Zurich and Lake Constance were taken as a measure of the binding to biological material. Rather constant Cu:P and Zn:P ratios are observed in settling material from different depths at various times, indicating a preferential binding of these metals to the biological material.

Constant metals/phosphorus ratios can be expected from an adsorption (surface complexation) model for algal cells, since constant proportions of surface sites to phosphorus would be expected. Binding to surfaces may be a preliminary step of the uptake of metal ions. Recent work on the adsorption of metal ions on bacteria cell surfaces has shown the high affinity of these surfaces for copper (Gonçalves et al., 1986). In this adsorption study on dead bacterial cells, very large surface stability constants were obtained for the binding of copper to the biological surface sites. Algal cells are expected to show similar properties. This means that if algal material and iron oxide are present in the settling particles of a lake with similar surface areas (and number of surface sites), copper will preferentially bind to the algal cell surfaces, since their affinity is higher.

In aquatic systems with high biological production and thus a high sedimentation rate, trace metals may be eliminated efficiently from the water column; this is especially the case for trace metals that are preferentially bound to biological surfaces. This means that lakes with high nutrient (i.e., phosphorus) supply and thus biological production of phytoplankton particles may be more efficient in counterbalancing the effects of increased metal loading. High sedimentation rates also mean high metal fluxes to the sediments.

The resulting metal concentrations in the water column which finally act on the biota may be very low in spite of high metal inputs into the lake owing to the efficient removal. If increased metal loadings caused the concentrations to increase up to growth-inhibitory levels for algae, the algal production and thus the sedimentation rate would decrease; the metal concentrations would in this case increase because of smaller transfers to the sediments.

The field data on settling particles thus indicate that biological particles play an important role in removing trace metals (especially Cu and Zn) from the water column and in regulating their concentrations. Iron and manganese oxides represent additional scavenger phases with large surface areas.

Acknowledgment

I would like to thank Michael Sturm for carrying out the sedimentological part of the lake projects, and Werner Stumm for many stimulating discussions.

REFERENCES

Anderson, D. M., and Morel, F. M. M. (1978), Copper sensitivity of *Gonyaulax tamarensis, Limnol. Oceanogr.* **23,** 283.

Anderson, M. A., Ferguson, J. F., and Gavis, J. (1976), Arsenate adsorption on amorphous aluminum oxide, *J. Colloid Interface Sci.* **54,** 391.

Anderson, M. A., and Morel, F. M. M. (1982), The influence of aqueous iron chemistry on the uptake of iron by the coastal diatom *Thalensiosira weissflogii, Limnol. Oceanogr.* **27,** 789.

Baccini, P. (1976), Untersuchungen über den Schwermetallhaushalt in Seen, *Schweiz. Z. Hydrol.* **38,** 121.

Baccini, P., and Joller, T. (1981), Transport processes of copper and zinc in a highly eutrophic and meromictic lake, *Schweiz. Z. Hydrol.* **43,** 176.

Balistrieri, L. S., Brewer, P. G., and Murray, J. W. (1981), Scavenging residence times of trace metals and surface chemistry of sinking particles in the deep ocean, *Deep-Sea Res.* **28A,** 101.

Balistrieri, L. S., and Murray, J. W. (1982), The adsorption of Cu, Pb, Zn and Cd on goethite from major ion seawater, *Geochim. Cosmochim. Acta* **46,** 1253.

Benjamin, M. M., and Leckie, J. O. (1981), Conceptual model for metal–ligand–surface interactions during adsorption, *Environ. Sci. Technol.* **15,** 1050.

Benjamin, M. M., and Leckie, J. O. (1982), Effects of complexation by Cl, SO_4 and S_2O_3 on adsorption behavior of Cd on oxide surfaces, *Environ. Sci. Technol.* **16,** 162.

Bourg, A. C. M., Joss, S., and Schindler, P. W. (1979), Ternary surface complexes. 2. Complex formation in the system silica–Cu(II)-2,2′-bipyridyl, *Chimia* **33,** 19.

Bourg, A. C. M., and Schindler, P. W. (1978), Ternary surface complexes. 1. Complex formation in the system silica–Cu(II)-ethylenediamine, *Chimia* **32,** 166.

Bowen, H. J. M. (1979), *Environmental Chemistry of the Elements,* Academic Press, London.

Brinckman, F. F., Olson, G. J., and Inverson, W. P. (1982), "The Production and Fate of Volatile Molecular Species in the Environment: Metals and Metalloids," in E. D. Goldberg, Ed., *Atmospheric Chemistry,* Springer-Verlag, Berlin, pp. 231–249.

Bruland, K. W. (1980), Oceanographic distributions of cadmium, zinc, nickel and copper in the North Pacific, *Earth Planet. Sci. Lett.* **47,** 176.

Buffle, J. (1984), "Natural Organic Matter and Metal–Organic Interactions in Aquatic Systems," in H. Sigel, Ed. *Metal Ions in Biological Systems,* Vol. 18, Marcel Dekker, New York.

Collier, R., and Edmond, J. (1984), The trace element geochemistry of marine biogenic particulate matter, *Prog. Oceanogr,* **13,** 113.

Davis, J. A., and Leckie, J. O. (1978), Surface ionization and complexation at the oxide/water interface. II. Surface properties of amorphous iron oxyhydroxide and adsorption of metal ions, *J. Colloid Interface Sci.* **67,** 90.

Davis, J. A., and Leckie, J. O. (1980), Surface ionization and complexation at the oxide/water interface. 3. Adsorption of anions. *J. Colloid Interf. Sci.* **76,** 32.

Dyrssen, D., and Wedborg, M. (1980), "Major and Minor Elements, Chemical Speciation in Estuarine Waters", in E. Olausson and I. Cato, Eds., *Chemistry and Biochemistry of Estuaries,* Wiley, New York.

Fisher, N. S., Bjerregaard, P., and Fowler, S. W. (1983), Interactions of marine plankton with transuranic elements. 1. Biokinetics of neptunium, plutonium, americium and californium in phytoplankton, *Limnol. Oceanogr.* **28,** 432.

Florence, T. M., and Batley, G. E. (1980), Chemical speciation in natural waters, *Crit. Rev. Anal. Chem.* **9,** 219.

Gamble, D. S., Schnitzer, M., Kerndorff, H., and Langford, C. H. (1983), Multiple metal ion exchange equilibria with humic acid, *Geochim. Cosmochim. Acta* **47,** 1311.

Gonçalves, M. d. L. S., Sigg, L., Reutlinger, M., and Stumm, W. (1986), Metal ion binding by biological surfaces; voltammetric assessment in presence of bacteria. *Sci. Tot. Environ.* in press.

Hohl, H., and Stumm, W. (1976), Interaction of Pb^{2+} with hydrous γ-Al_2O_3, *J. Colloid Interface Sci.* **55,** 281.

Jackson, G. A., and Morgan, J. J. (1978), Trace metal–chelator interactions and phytoplankton growth in seawater media: Theoretical analysis and comparison with reported observations, *Limnol. Oceanogr.* **23,** 268.

Jacobs, L., and Emerson, S. (1982), Trace metal solubility in an anoxic fjord, *Earth Planet Sci. Lett.* **60,** 237.

Johnson, C. A. (1986), The regulation of trace element concentrations in river and estuarine waters contaminated with acid mine drainage: The adsorption of Cu, Zn, P and As on amorphous Fe hydroxides, *Geochim. Cosmochim. Acta,* **50,** 2433.

Kunz, B. (1983), "Heterogene Nukleierung und Kristallwachstum von $CaCO_3$ (Calcit) in natürlichen Gewässern," Ph.D. thesis, ETH Zürich, No., 7355.

Morel, F. M. M. (1983), *Principles of Aquatic Chemistry,* Wiley, New York.

Morel, F. M. M., and Hudson, R. J. M. (1985), "The Geobiological Cycle of Trace Elements in Aquatic Systems: Redfield Revisited," in W. Stumm, Ed., *Chemical Processes in Lakes,* Wiley, New York, pp. 251–281.

Motschi, H. (1985), Cu(II) EPR: a complementary method for the thermodynamic description of surface complexation, *Adsorption Sci. Technol.* **2,** 39.

Mouvet, C., and Bourg, A. C. M. (1983), Speciation (including adsorbed species) of copper, lead, nickel and zinc in the Meuse river. Observed results compared to values calculated with a chemical equilibrium program, *Water Res.* **17,** 641.

Nieboer, E., and Richardson, D. H. S. (1980), The replacement of the nondescript term "heavy metals" by a biologically and chemically significant classification of metal ions, *Environ. Pollution* (*Ser. B*) **1,** 3.

Pierce, M. L., and Moore, C. B: (1982), Adsorption of arsenite and arsenate on amorphous iron hydroxide, *Water Res.* **16,** 1247.

Redfield, A.C., Ketchum, B. H., and Richards, F. A. (1963), The influence of organisms on the composition of sea water, in M. N. Hill, Ed. *The Sea,* Vol. 2, Wiley Interscience, New York, pp. 27–77.

Schindler, D. W. (1985), "The Coupling of Elemental Cycles by Organisms: Evidence from Whole-Lake Chemical Pertubations," in W. Stumm, Ed., *Chemical Processes in Lakes,* Wiley, New York.

Schindler, P. W. (1975), Removal of trace metals from the oceans: A zero order model, *Thal. Jugosl.* **11,** 101.

Schindler, P. W. (1981), "Surface Complexes at Oxide–Water Interfaces," in M. A. Anderson and A. Rubin, Ed., *Adsorption of Inorganics at Solid-Liquid Interfaces,* Ann Arbor Science Publ, Ann Arbor, MI.

Schindler, P. W. (1984), "Surface Complexation," in H. Sigel, Ed., *Metal Ions in Biological Systems,* Vol. 18, Marcel Dekker, New York.

Schindler, P. W., Fuerst, B., Dick, R., and Wolf, P. U. (1976), Ligand properties of surface silanol groups. 1. Surface complex formation with Fe^{3+}, Cu^{2+}, Cd^{2+} and Pb^{2+}, *J. Colloid Interface Sci.* **55,** 469.

Sigg, L. (1985), "Metal Transfer Mechanisms in Lakes; The Role of Settling Particles," in W. Stumm, Ed., *Chemical Processes in Lakes,* Wiley, New York.

Sigg, L., and Stumm, W. (1981), The interaction of anions and weak acids with the hydrous goethite (α-FeOOH) surface, *Colloids Surf.* **2,** 101.

Sigg, L., Sturm, M., Davis, J., and Stumm, W. (1982), Metal transfer mechanisms in lakes, *Thal. Jugosl.* **18,** 293.

Sigg, L., Sturm, M., and Kistler, D. (1987), Vertical transport of heavy metals by settling particles in Lake Zürich, *Limnol. Oceanogr.* **32,** 112.

Sigg, L., Sturm, M., Stumm, W., Mart, L., and Nuernberg, H. W. (1982), Schwermetalle im Bodensee; Mechanismen der Konzentrationsregulierung, *Naturwissenschaften* **69,** 546.

Smith, R. M., and Martell, A. E. (1976), *Critical Stability Constants,* Vol. 4, *Inorganic Complexes,* Plenum, New York.

Sposito, G. (1985), "The Distribution of Potentially Hazardous Trace Metals," in H. Sigel, Ed. *Metal Ions in Biological Systems,* Vol. 20, Marcel Dekker, New York.

Stumm, W., Hohl, H., and Dalang, F. (1976), Interactions of metal ions with hydrous oxide surfaces, *Croat. Chem. Acta* **48,** 491.

Stumm, W., Kummert, R., and Sigg, L. (1980), A ligand exchange model for the adsorption of inorganic and organic ligands at hydrous oxide interfaces, *Croat. Chem. Acta.* **53,** 291.

Stumm, W., and Morgan, J. J. (1981), *Aquatic Chemistry,* 2nd ed., Wiley, New York.

Sturm, M., Zeh, U., Mueller, J., Sigg, L., and Stabel, H. H. (1982). Schwebestoffuntersuchungen im Bodensee mit Intervall-Sedimentationsfallen, *Eclogae Geol. Helv.* **75,** 579.

Sunda, W. G., and Guillard, R. R. L. (1976), The relationship between cupric ion activity and the toxicity of copper to phytoplankton, *J. Marine Res.* **34,** 511.

Sunda, W. G., and Huntsman, S. A. (1983), Effect of competitive interactions between manganese and copper on cellular manganese and growth in estuarine and oceanic species of the diatom *Thalassiosira, Limnol. Oceanogr.* **28,** 924.

Turner, D. R., and Whitfield, M. (1980), Chemical definition of the biologically available fraction of trace metals in natural waters, *Thal. Jugosl.* **16,** 231.

Westall, J., and Hohl, H. (1980), A comparison of electrostatic models for the oxide solution interface, *Adv. Colloid Interface Sci.* **12,** 265.

Westall, J., Zachary, J., and Morel, F. (1976), "MINEQL": A computer program for the calculation of chemical equilibrium composition of aqueous systems, *Tech. Note* No. 18, Ralph M. Parson Laboratory, MIT, Cambridge, MA.

Williams, R. J. P. (1981), Physico-chemical aspects of inorganic element transfer through membranes, *Phil. Trans. Roy. Soc. London B* **294,** 57.

Williams, R. J. P. (1983), The symbiosis of metal ion and protein chemistry, *Pure Appl. Chem.* **55,** 35.

Wood, J. M., and Wang, H. K. (1985), "Strategies for Microbial Resistance to Heavy Metals," in W. Stumm, Ed., *Chemical Processes in Lakes,* Wiley, New York.

13

INTERPRETATION OF METAL COMPLEXATION BY HETEROGENEOUS COMPLEXANTS

Jacques Buffle and R. Scott Altmann

Department of Inorganic, Analytical, and Applied Chemistry, University of Geneva, Geneva, Switzerland

Abstract

The chemical form of many trace metal cations present in aqueous environments is known to greatly influence their effects on organisms, whether toxic or beneficial. For many important ecological systems, cation reactions with such heterogeneous complexing agents as fulvic acids and hydrous metal oxides have been strongly implicated as playing a decisive role in trace cation chemistry. Our ability to measure and interpret the cation complexation equilibria of these substances is, however, severely limited by their being inherently ill-defined and by their common possession of certain complex qualities. These common characteristics impose constraints on the type of information which can be obtained experimentally and, ultimately, determine the most effective means for representation and interpretation of their complexation properties. The purpose of this chapter is to address these latter difficulties.

1. INTRODUCTION

The healthy functioning of our biosphere is dependent on the continued maintenance of an infinite array of aqueous chemical reactions within limits consistent with the well-being of all living organisms. Man's ability to transgress these limits by perturbing both the rate at which elements move through the biosphere (Nriagu, 1979) and their chemical forms has been clearly demonstrated by such events as Hg pollution in Minimata Bay or, more recently, the many inferred consequences of acid rain. Our ability to minimize such adverse consequences is, to a great extent, dependent on our knowledge of how aqueous environments respond to changes in their chemical composition (e.g., an increase in total Hg^{2+} concentration, a lowering of pH).

Although dynamic considerations are very important in environmental processes, our present knowledge of the biogeochemistry of aquatic systems is based predominantly on the concepts of equilibrium thermodynamics (classical or statistical), coordination chemistry, and nonrelativistic physics. The purpose of this chapter is to discuss how one can represent and ultimately interpret, within an equilibrium thermodynamic framework, the observable characteristics of reactions occurring between metal cations and such natural heterogeneous compounds as organic macromolecules (e.g., humic substances), mineral surfaces (e.g., clays, oxides), and bacterial exopolymers. These types of reactions have been shown to be particularly important in determining the behavior of many transition and heavy metal cations (Cu^{2+}, Pb^{2+}, Cd^{2+}, etc.) in aqueous environments ranging from river, lake, and sea waters to soil and sediment solutions.

2. BACKGROUND: METALS AND LIGANDS

2.1. Metals

The vast majority of chemical reactions that occur between metal cations and ligands in aquatic environments can be classified as *complexation* reactions. Metal cations, as a consequence of their different outer electron configurations and the resulting variation in the stability of bonds formed with potential ligand atoms, can be grouped according to similarities in their chemical behavior (i.e., bond strengths, ligand preference) as exemplified in Table 13.1. Not surprisingly, such differences in chemical behavior lead to differences in the distribution of the total amounts of each of the metals present in a given environmental system among the often enormous variety of chemical structures containing available ligand atoms. As a result, any given metal may potentially be found in many diverse forms (Fig. 13.1), including:

1. The hydrated "free" ion (H_2O ligand).
2. Complexes with "simple" inorganic and organic ligands.
3. Complexes with ligand atoms (sites) which are part of the structure of naturally occurring organic macromolecules or colloids or exposed at the surface of mineral particles.

We shall be particularly interested here in the differences in metal binding behavior exhibited by the third category of compounds vis-à-vis "simple"

Table 13.1. Classification of Metals as a Function of Their Preference for Ligand Atoms[a]

Metal Groups	Preference for Ligand Atoms
A Cations	
(H^+), Li^+, Na^+, K^+	N ≫ P
Be^{2+}, Mg^{2+}, Ca^{2+}	O ≫ S
Sr^{3+}, Al^{3+}, Sc^{3+}	F ≫ Cl
Transition Cations	
Cr^{2+}, Mn^{2+}, Fe^{2+} Co^{2+}, Ni^{2+}, Cu^{2+} V^{3+}, Cr^{3+}, Mn^{3+}, Fe^{3+}	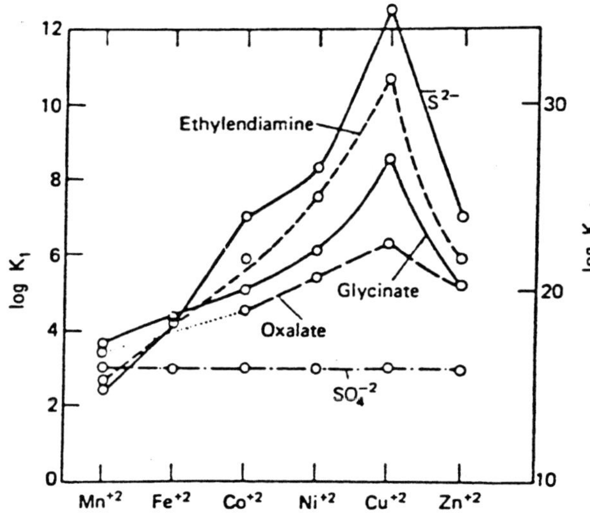
B Cations	
Ag^+, Au^+, Tl^+, Cu^+	P ≫ N
Zn^{2+}, Cd^{2+}, Hg^{2+}, Pb^{2+}, Sn^{2+}	S ≫ O
Tl^{3+}, In^{3+}, Bi^{3+}	I ≫ F

[a] The graph (from Stumm and Morgan, 1981, p. 344) represents the so-called Irving–Williams series for the stability of 1:1 complexes of transition elements. K_{so} = solubility product with S^{2-}; K_1 = complexation constant with other ligands.

ligands. Knowledge concerning metal distribution among the different chemical forms is critical since virtually all of the experimental and theoretical research of the last two or three decades has led to the conclusion that the nature of a metal-containing compound plays a central role in determining the metal's biogeochemical circulation and biological effects. This importance has been demonstrated, in particular, by Whitfield and Turner (1979, 1983) for the global distribution and residence times of both metals and nonmetals in seawater and freshwater and by Luoma (Luoma and Jenne, 1977; Luoma

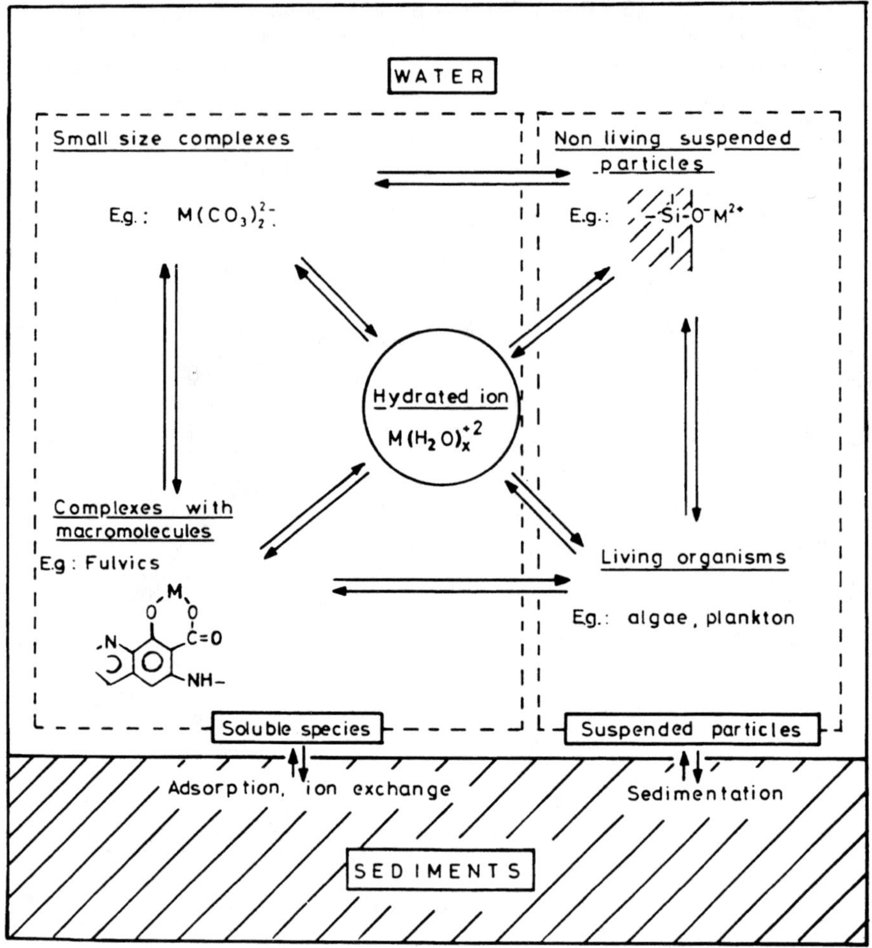

Figure 13.1. Schematic representation of the reaction of a metal ion M with the different types of aquatic system constituents. From Buffle, (1987).

and Bryan, 1981) for the accumulation of toxic metals by estuarine bivalves. Metals in aqueous environments are commonly classified in two ways, both being directly dependent on metal chemical reactivity:

Environmental abundance. The combination of differences in the crustal abundance of metallic elements and their chemical reactivity with environmental ligands together control their total concentrations, those at high concentrations (i.e., $>10^{-4}$ *M*), largely A cations (Table 13.1) being called the *major* cations, while others (mostly transition and B cations) are present at much lower levels (Fig. 13.2) and are therefore termed *minor* or *trace* metals.

Biological role. It is well known that complexation reactions determine, to a large extent, the role played by metal cations in organism functioning. Many metal

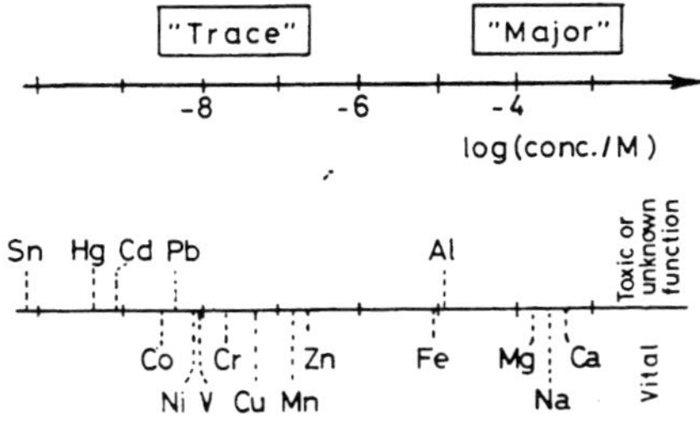

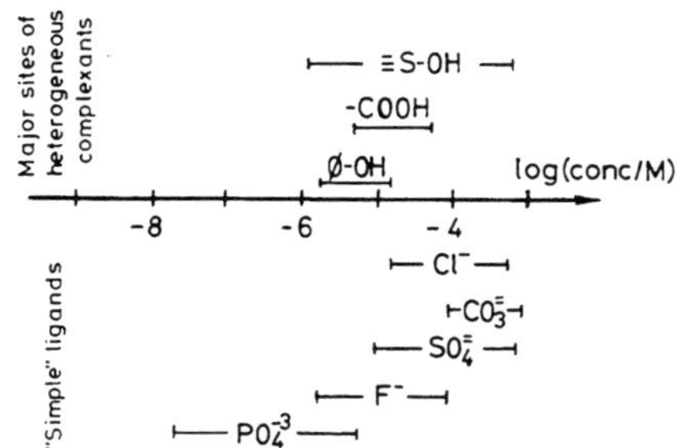

Figure 13.2. Average values of total concentration of metals and ranges of concentrations of ligands or complexing sites in natural fresh waters. ≡S—OH, inorganic solid surface sites. —COOH and Ø–OH, total concentration of —COOH and phenolic sites, respectively, of natural organic matter. Adapted from Buffle (1987).

ions (called *vital*) are required for the proper functioning of living processes, for example, as electrolyte and counterions to maintain the three-dimensional structure of cellular proteins (mostly A cations) or in enzyme reactions (mostly transition cations) (Fig. 13.2). Other metals are *toxic* (predominantly B cations) at very low doses or have no known biochemical function.

Of particular importance regarding the toxic or vital role of cations is the large body of experimental evidence demonstrating that, for many cations, observed uptake rates are correlated with the concentration of the *free aquo cation* in the external solution and not with the total metal concentration (e.g., Sunda and Guillard, 1976). Our ability to interpret and foresee the ecological impact of metal ions under differing environmental conditions is, then, directly related to our ability to measure and, ultimately, to predict the partitioning between free cation and complexed species in very heterogeneous natural systems.

2.2. Ligands

As mentioned above and suggested schematically in Figure 13.3, there exist great differences in character among the ligands that are present in aqueous

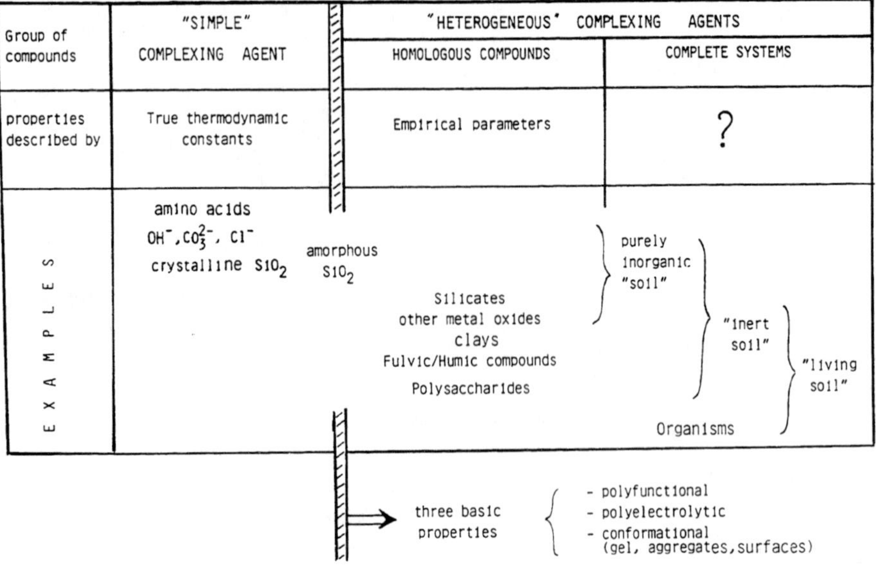

Figure 13.3. Classification of natural complexing agents according to their degree of heterogeneity. A complete complexing system such as a soil is composed of several subsystems, for example, organisms or nonliving components. The latter include "simple" compounds and groups of homologous compounds (clays, humics, polysaccharides), which are themselves operationally defined (see text). Heterogeneity increases from right to left and must be considered a basic property of homologous compounds and complete systems.

environments. They can be roughly divided, as a useful first step, into two groups, which we shall call "simple" and "heterogeneous" complexing agents, respectively, based both on differences in structural complexity and on our current ability (or limitations thereof) to describe both the compounds themselves and their metal complexation chemistry.

"*Simple*" *compounds* (e.g., see Fig. 13.3) are those for which

The actual molecular structure (composition and geometry) of the compound is well-defined for all solution conditions of interest.

The amounts of the compound and its complexes present can be expressed in molar concentration (ultimately, activity) units.

The stoichiometry of each complex formed with all metal ions to be considered is known.

The free energy of formation (ΔG^0) of each metal–ligand complex has a unique value and is known for the range of system conditions being studied.

There exist specific quantitative analytical methods.

For ligands meeting the above criteria, the equilibrium concentrations of all species are uniquely determined once the thermodynamic formation constant K for each metal–ligand complex is known and the total concentration of each of the metals and ligands is specified.

Heterogeneous complexing agents, on the other hand, include those ligand-site-containing structures found in a given aqueous environment that do not satisfy all the "simple" compound criteria given above. Heterogeneous complexing agents (e.g., Fig. 13.4), almost by definition, cannot be isolated from environmental samples in a pure state (i.e., consisting of a single molecular structure). They can, however, often be separated from such samples and fractionated into groups of so-called *homologous compounds* having similar operationally defined physicochemical characteristics. Examples of such groups of homologous compounds are the humic and fulvic acids (Fig. 13.4*c*), hydrous metal oxides, and clays. Homologous groups are always made up of a collection of highly complex and poorly (but hopefully, ever better) characterized molecular structures. Collectively they are able to form metal cation complexes having a nearly infinite number of different molecular scale bonding characteristics. All homologous compounds exhibit, to varying degrees, three major properties (Fig. 13.3):

1. Their *polyfunctionality* (i.e., having many coordinating sites of differing nature present on the same molecule).
2. Their *polyelectrolytic* character (i.e., having, per molecule, a high electric charge density due to the presence of a large number of dissociated functional groups).
3. The importance of *conformational* factors (e.g., reactions on surfaces or within "gels," formation of aggregates).

As a consequence of the foregoing, the parameters that can be used to globally represent the experimentally measured properties of homologous groups tend to become more empirical and therefore less useful as the complexity of the system increases. We will be specifically concerned in what follows with the proper representation and interpretation of experimental data that have been, or can be, acquired for such systems so as to extract the maximum amount of information in a rigorous fashion.

3. MEASUREMENT OF METAL–LIGAND COMPLEXATION EQUILIBRIUM DATA

3.1. Titration of Simple and Heterogeneous Complexing Agents

Before discussing various means of representing and interpreting experimental complexation data, it is necessary to briefly summarize the nature of directly accessible experimental variables and the way in which they can be obtained. This

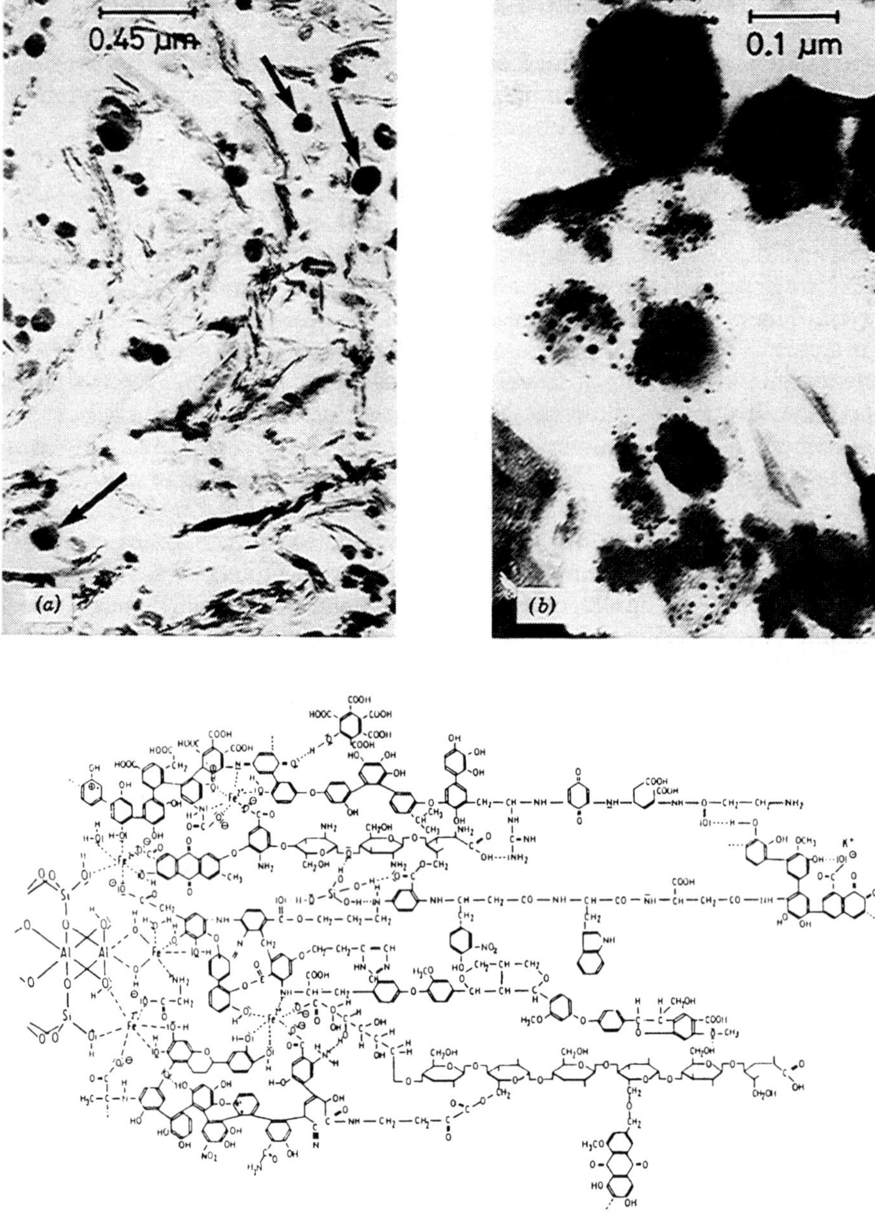

Figure 13.4. Examples of heterogeneity at various levels: "complete system" (*a*) and homologous compounds (*b* and *c*). (*a*) Mixed iron phosphate and iron oxide particles (dense "balls," see arrows) embedded in organic gel (most of the light gray matter). Transmission electron microscopy on sample taken from a eutrophic lake, after "purifying" fractionation for maximum enrichment in iron oxide particles. After Leppard, De Vitre, and Buffle (unpublished results). (*b*) Detail of the iron oxide "balls" of (*a*), showing that each ball is itself composed of much smaller globules (a few nm) and porous material. (*c*) Model chemical structure of soil humic acid. After Kleinhempel (1970).

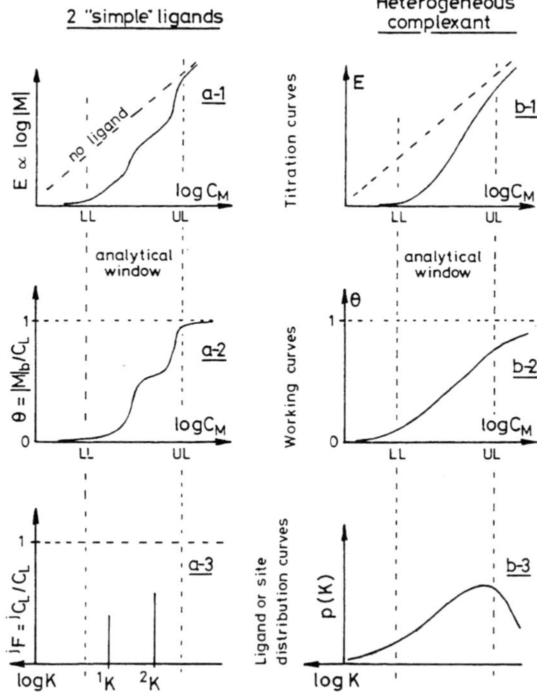

Figure 13.5. Schematic representation of titration curves, working curves and distribution curves for a mixture of two simple ligands (left) and heterogeneous complexant (right). Curves *a*-1 and *b*-1: Experimental titration curve of complexant system by metal M, and the corresponding useful analytical window: LL (lower limit) to UL. *E* is the potential of the ion-selective electrode used for measuring |M| ($E \propto \log|M|$). Curves *a*-2 and *b*-2: The corresponding working curves $\theta = f(C_M)$; θ = metal loading = $(C_M - |M|)/C_L$. C_M, C_L are total metal and ligand concentrations, respectively. Curves *a*-3 and *b*-3: The corresponding distribution curves. Curve *a*-3: fraction ${}^jF$ of a given site *j* as a function of its *K* value (${}^jK$). Curve *b*-3: Affinity spectrum, $p(K) = f(\log K)$; the surface under $p(K)$ in a given interval $\Delta\log K$ is the probability of finding sites with log *K* values located in this interval. For the relation between *F* and $p(K)$, see Section 4.3.

will be done below, by comparing (Fig. 13.5) the case of simple ligands with that of heterogeneous complexing agents and by showing, in particular, how limitations that are not important in the former may become essential in the latter.

Let us begin with a simple ligand system, taking, for example, a solution containing as its sole reactants (other than H_2O), one metal cation M and one simple ligand L at known total concentrations C_M and C_L. As stated in Section 2.1, this system is fully characterized thermodynamically once the equilibrium constant *K* of the complex ML is known, given C_M and C_L:

$$K = \frac{|ML|}{|M| \cdot |L|} \tag{13.1}$$

K is, then, the sought-after parameter. (Note that, in reality, L can often be protonated, but for simplifying purposes this will not be considered here.) At equilibrium, if $|M|$ is measurable, $|ML|$ and $|L|$ (and therefore K) can be obtained from the mass balance equations:

$$C_M = |M| + |ML| \tag{13.2}$$

$$C_L = |L| + |ML| \tag{13.3}$$

For this particular case, a single measurement of $|M|$ for one (C_M, C_L) couple is sufficient, in principle, to determine K. Once K is known, one can compute the M and L distribution for any other (C_M, C_L) condition.

We can now consider a mixture of two simple ligands, ^1L and ^2L, at total known concentrations ^1C_L and ^2C_L, able to form two 1 : 1 complexes, M^1L and M^2L. The system is now fully characterized by three known concentrations (C_M, ^1C_L, and ^2C_L) and two sought-after constants ^1K and ^2K. If $|M|$ is the only measurable parameter, as is often the case in natural mixtures, the K values must be calculated using the set of five equations:

$$C_M = |M| + |M^1L| + |M^2L| \tag{13.4}$$

$$^1C_L = |^1L| + |M^1L| \tag{13.5}$$

$$^2C_L = |^2L| + |M^2L| \tag{13.6}$$

$$^1K = \frac{|M^1L|}{|M| \cdot |^1L|} \tag{13.7}$$

$$^2K = \frac{|M^2L|}{|M| \cdot |^2L|} \tag{13.8}$$

These equations contain six unknowns, and therefore at least two $|M|$ values must be measured, with two different data sets (^1C_L, ^2C_L, C_M), in order to be able to find the K's. Obviously, the larger the number of ligands in the mixture, the larger the minimum number of points required to find the corresponding K's. In practice, even for two ligands, precise K values are obtainable only by measuring a large number of $|M|$ values. This may be accomplished, for instance, by employing an ion-selective electrode whose potential E is proportional to log $|M|$. $|M|$ is caused to vary by progressively changing one of the total concentrations, usually C_M, in small increments. This procedure is called a *titration* (Fig. 13.5, curve *a*-1).

Titration procedures are also required for characterization of heterogeneous complexing agents for at least three reasons:

1. The number of different complexing sites is very large (polyfunctional character; Section 2.2).
2. The stability of complexes formed at each type of complexing site is influenced by several secondary factors, in particular polyelectrolytic and conformational effects (Section 2.2), whose intensity may vary over the range of a titration. As a result, each site is no longer characterizable by a single K value and, since more physicochemical parameters must be considered, more experimental points are necessary.

3. The individual ${}^jC_L$ values for each site ${}^jL$ are generally unknown. Only the total concentration of all sites, C_L ($= \Sigma_j\, {}^jC_L$) is known (even obtaining an unambiguous value for C_L is often difficult). As a consequence, the number of equations which can be employed to find the K's is decreased.

Note that as a result of the first two reasons above, the titration curves of heterogeneous complexants (Fig. 13.5, curve *b*-1) never exhibit the distinct inflections that are observed when titrating a mixture of a few simple ligands (Fig. 13.5, curve *a*-1). It is also important to note that ligands (in a mixture of simple compounds) or complexing sites (for heterogeneous complexants) will be titrated *sequentially*, as a function of their K values, as C_M (and, necessarily, $|M|$) is increased, the ligand having the largest K value (i.e., forming the most stable complex) being titrated first.

3.2. Titration "Working Curves"

Given the above limitations, the maximum information which can be obtained from a titration consists of C_L *plus a set of* (C_M, $|M|$) *data pairs*. These data are generally transformed into "working curves" which relate the bound metal concentration to the corresponding $|M|$ or C_M (Fig. 13.5, curves *a*-2, *b*-2). Since, for heterogeneous complexants, only C_L is known, we are left only with the possibility of computing the *total* collective concentration of all ML forms (i.e., bound M : $|M|_b$) from a mass balance on the metal:

$$C_M = |M| + \sum_j |M\,{}^jL| \tag{13.9}$$

$$C_M = |M| + |M|_b \tag{13.10}$$

$$|M|_b = C_M - |M| \tag{13.11}$$

Titration data then are represented as a set of (θ, $|M|$) or (θ, C_M) data pairs, where

$$\theta = \frac{|M|_b}{C_L} \tag{13.12}$$

θ is the fraction of all sites occupied by M and will vary from 0 to 1 as C_M is increased during a titration. The mass action and mass balance equations for all complexes and ligands can be combined to give

$$\theta = \frac{1}{C_L} \sum_j \frac{{}^jK|M|}{1 + {}^jK|M|}\, {}^jC_L \tag{13.13}$$

A full characterization of a heterogeneous complexant consists, therefore, in attempting to extract all the (${}^jC_L$, ${}^jK$) couples from a $\theta = f(|M|)$ or $\theta = f(C_M)$ titration curve using Eq. 13.13. Determination of one of these curves is, then, the critical first step in representing and interpreting equilibrium complexation behavior (Fig. 13.5, curves *a*-2 and *b*-2). Note that when C_L is unknown, only $|M|_b = f(C_M$ or $|M|)$ functions can be obtained.

3.3. Titration "Window"

There are limits to the range of solution compositions (e.g., minimum and maximum C_M values) from which usable experimental data can be obtained. The lower limit (LL) of this experimental titration "window" is determined by the ability to accurately measure $|M|$ and is therefore a function of the sensitivity of the analytical method used to detect this species. The upper limit (UL) derives from the fact that we are able to compute $|M|_b$ only by using Eq. 13.11. Inspection of Figure 13.5, curves *a*-1 and *b*-1, will show that when C_M becomes sufficiently large, such that $|M|_b/C_L \rightarrow 1$, $|M|_b$ is computed from an ever smaller difference between two increasingly large numbers. Usable data cannot be obtained when this difference approaches the statistical precision of the experiment. Any given method, therefore, can only provide experimental data corresponding to its own *titration window*. For highly heterogeneous complexants it must be realized that this window may be much smaller than that required to "cover" all the complexing sites. All ligands having K values that are larger than that corresponding to the lower $|M|$ detection limit will not be "seen" by the titration (there will be a consumption of M added but little or no detectable change in $|M|$), nor will those whose K is too small to be titrated before the upper limit condition is reached. The existence of the upper limit on the window also explains why C_L is not easily accessible. With presently available measurement methods, window width rarely exceeds 5 log K units and is often smaller.

3.4. Ligand or Site Differentiation

The ability to distinguish titrimetrically the various ligands or sites from one another by the presence of clear inflections (e.g., Fig. 13.5 *a*-2) is dependent on their K values being sufficiently different ($\Delta \log K \geq \pm 1.3$). The "resolution" of the method can be estimated by considering the case where a given C_L is divided equally among a large number (n) of ligand types (${}^jC_L = C_L/n$) having log K values equally spaced over the analytical window (say, 5 log K units). In such a case, it is no longer possible to distinguish the effects of more than $n = 5/(\Delta \log K) \approx 4$ individual ligands (or sites). This number is even smaller when secondary effects (i.e., polyelectrolytic or conformational) are superimposed, as is the case for heterogeneous complexants.

3.5. Summary

We can summarize the foregoing by recalling that characterization of a ligand mixture consists of extracting an accurate thermodynamic description of the ligands or complexing sites present, in terms of the (${}^jC_L$, ${}^jK$) couples, from the experimental $\theta = f(|M|)$ titration data. Suffice it to say that this is possible only for systems containing a *small* number of ligand types which have *well-separated* K values falling *within* the measurement window. From the previous discussion, it is obvious that heterogeneous homologous complexants cannot be resolved sufficiently to allow accurate assignment of discrete $({}^jK, {}^jC_L)$ values for all the sites. A new approach must therefore be found for interpreting these systems.

4. REPRESENTATION OF COMPLEXATION EQUILIBRIA

4.1. The Need for a Normalized Representation

In order to be practically useful for environmental purposes (Section 2), a description of metal complexation by homologous complexants should be able, in particular, to:

1. Predict the partitioning of some total metal concentration C_M between M and all its M^jL forms. In particular, it must enable the *computation* of $|M|$, the master variable for biogeochemical processes involving M.
2. Predict the change in $|M|$ that would be observed for a specified change in C_M. This is quantified by the so-called *buffer capacity* of the system for the metal M, defined as

$$\tau = \frac{d \log |M|}{dC_M} \tag{13.14}$$

   This property is also very important for understanding the role of M in natural systems (Section 5).
3. Allow *comparison of the complexation properties* of widely different complexants or of the same complexant under widely different conditions (e.g., C_M, pH, ionic strength), on a rationally normalized basis. Such comparisons are necessary if analogies are to be made between heterogeneous complexants and well-defined model compounds (a valuable aid for interpreting the properties of the former at the molecular level). The ability to compare complexants having very different properties (such as Cl^-, organic macromolecules, or metal oxide particles) is also a necessary requirement to obtain a rigorous overall picture of the complexing properties of complete natural aquatic systems (Fig. 13.3).

This necessity of comparing very different and complicated systems implies seeking a method that enables us to *normalize* experimental data in a rational manner that is independent of the complexant nature. Continuing with the comparison between simple and heterogeneous complexants, one can state that:

—For the former, interpretation at the molecular level is done by assuming, a priori, the nature of the possible complexes formed between M and L (ML, ML_2, MHL, . . .) and by testing whether the corresponding equilibrium constants enable us to explain the experimental titration curves. The values of these constants can then be subsequently employed for comparing various ligands on a rational basis.

—For heterogeneous complexants, no a priori assumption can be made concerning the nature of the complexation reaction at each site since, owing to their very great structural complexity, any realistic guess would require measurement of too large a number of physicochemical parameters vis-à-vis the number observable in the experimental analytical window (i.e., a maximum of 4, Section 3.4). As regards heterogeneous complexants, then, not only must the rational normalization be done without any a priori chemical assumption, but, furthermore, it must serve as a tool for determining the nature of the chemical reactions. Since the principal difference between heterogeneous and simple ligands is the very large number of possible reactivity modes of the former, a rational normalization of their properties can be based on statistical representations, by applying the classical concept of a probability distribution to complexation reactions.

4.2. Chemical Basis for Application of Statistical Concepts

In order for normalization procedures based on probability functions to be applicable to a particular complexant, the complexant must incorporate a very large (almost infinite) number of different types of complexing sites (L) *regardless of the physical or chemical reasons for the differences.* This is the case for most natural (generally polymeric) heterogeneous complexants (P). For these complexants, differences between sites may be due to many factors, including, for example, the particular coordinating atoms involved, and the molecular conformation or variations in electric field strength. In this context, site-type distinction can be based only on differences in the *free energy* of the complexation reaction, ΔG^0. Although we shall continue to use the K parameter to represent this quality (since $\Delta G^0 = 2.3\, RT \log K$), it no longer has the molecular significance of the K for simple ligands.

The infinite-site-number conceptualization [see Buffle (1987) for a detailed discussion] necessarily implies that

1. Only one-to-one metal–ligand complexes must be considered (since an ML_2 complex, for instance, is simply treated as a complex with a second site, L_2, different from L).

2. The total concentration of any given site j, ${}^jC_L$, as well as the corresponding values of $|{}^jL|$ and $|M{}^jL|$, become infinitely small and can therefore be represented as differentials (dC_L, $d|L|$, and $d|ML|$) of the respective total concentrations. By considering the definition of the metal loading (Eq. 13.12), one has

$$|M{}^jL| \approx d|ML| = C_L\, d(\theta) \qquad \text{and} \tag{13.15a}$$

$$|{}^jL| \approx d|L| = C_L\, d(1 - \theta) \tag{13.15b}$$

3. A given value of the equilibrium constant ${}^jK$,

$${}^jK = \frac{|M{}^jL|}{|M| \cdot |{}^jL|} \tag{13.16}$$

is representative of the complexation properties of P only at a given point in the titration. It will vary continuously with the metal loading θ, since the nature of ${}^jL$ will vary accordingly. This variation will take the form of a continuous decrease for the case of P titration by M, since the stronger binding sites will be saturated before weaker ones. The ensemble of ${}^jK$ can therefore be replaced by a continuous function

$$K = \frac{d|ML|}{|M| \cdot d|L|} = f(\theta) \tag{13.17}$$

The K function is termed the *differential equilibrium function* (see section 4.4), and it represents the complexation properties of a heterogeneous system, under given experimental conditions in a manner analogous to the unique complexation constant existing for a simple ligand. (Remember that, for the sake of simplicity, protonation of L is not considered here.)

The question before us now is: how can we derive probability-distribution-based normalized representations for homologous compounds from their M titration data? Two approaches are currently available, the so-called *affinity spectrum* and *differential equilibrium functions*. The details of their mathematical derivations are described elsewhere [e.g., Ferry (1970); Thakur et al. (1980); Shuman et al. (1983); Olson and Shuman (1985) for the former, Gamble and coworkers (1973, 1980, 1983) and Buffle (1984) for the latter, and Buffle (1987) for a general comparative discussion]. We shall give only the essential principles below and discuss their relative advantages and limitations.

4.3. Affinity Spectrum

The affinity spectrum approach is based on representing complexation properties using a classical *probability density function* [denoted below by $p(K)$]. Such a function (Fig. 13.6) gives the probability of finding a type L site which forms a complex ML having a particular value of K. The best known density function is the Gaussian (Fig. 13.6*a*). When applying a density function to complexation properties, K is used as the independent variable of the distribution (i.e., K is considered a continuously variable parameter and is plotted along the abscissa). The mathematical form of the $p(K)$ function is such that any surface area under the curve corresponding to a given interval of K (e.g., ${}^{\mathrm{A}}K$ to ${}^{\mathrm{B}}K$) represents the probability of finding sites whose K values are in this interval (${}^{\mathrm{A}}K > K > {}^{\mathrm{B}}K$). This probability can be seen, more simply, as being equivalent to the fraction ${}^{\mathrm{AB}}F$ of the total sites present that have K values in the A–B defined interval. In concentration terms, ${}^{\mathrm{AB}}F$ is simply given by the concentration of sites in the A–B interval, ${}^{\mathrm{AB}}C_{\mathrm{L}}$, divided by the total site concentration C_{L}:

$$ {}^{\mathrm{AB}}F = \frac{{}^{\mathrm{AB}}C_{\mathrm{L}}}{C_{\mathrm{L}}} \tag{13.18} $$

In chemical parlance, the probability density function $p(K)$ is called the *affinity spectrum*. In general, for heterogeneous complexants, the $p(K)$ function is unknown and certainly not Gaussian (those that are known have been found to be highly asymmetrical as schematically illustrated in Fig. 13.6*c*). In fact, when studying heterogeneous complexant, the $p(K)$ function is the information sought since it represents the entirety of its complexation properties. In this respect it is qualitatively equivalent to the single values for the (${}^{j}C_{\mathrm{L}}$, ${}^{j}K$) couple that represents the properties of a simple ligand ${}^{j}\mathrm{L}$.

4.3.1. Relationship between Site Concentrations and p(K).

The mathematical relationship between F or C_{L} and $p(K)$ is illustrated in Figure 13.6*a*. In a fashion similar to that followed for defining ${}^{\mathrm{AB}}F$, Figure 13.6*a* shows that for a very small log K interval, d log K, the fraction of sites [corresponding to the surface area under $p(K)$] is given by

$$ \frac{dC_{\mathrm{L}}}{C_{\mathrm{L}}} = dF = p(K) \cdot d \log K \tag{13.19} $$

and since C_{L} is constant for a given heterogeneous complexant,

$$ dC_{\mathrm{L}} = C_{\mathrm{L}}\, p(K) \cdot d \log K \tag{13.20} $$

Therefore, for any A–B interval (Fig. 13.6*a*),

$$ {}^{\mathrm{AB}}C_{\mathrm{L}} = C_{\mathrm{L}} \int_{{}^{\mathrm{A}}K}^{{}^{\mathrm{B}}K} p(K) \cdot d \log K \tag{13.21a} $$

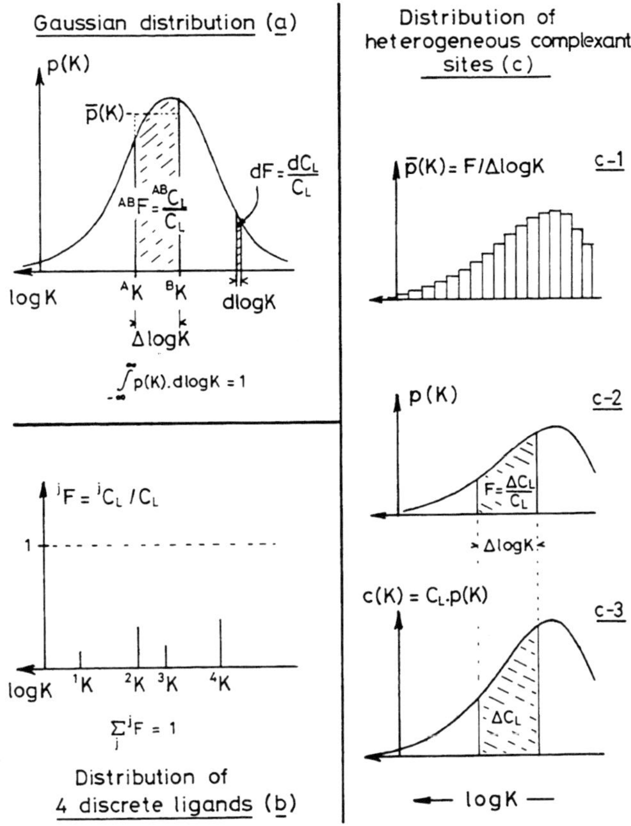

Figure 13.6. Distribution functions of complexing sites (or ligands) as a function of the variable K. Note that in this figure, as in all subsequent ones, K decreases from left to right (i.e., the corresponding free energy, ΔG^0, increases). This convention also reflects what happens during a titration, when the sites react in order of decreasing K values. Progression toward smaller K values (from left to right) therefore corresponds to increasing C_M, as in Figure 13.5. (*a*) Gaussian probability density function and definition of the corresponding parameters. The surface under the curve in any ${}^{A}K$–${}^{B}K$ interval is the probability of finding a site with ${}^{A}K > K > {}^{B}K$. $\bar{p}(K)$ is the average value of $p(K)$ in this interval. (*b*) Distribution of a mixture of four simple ligands, each having a single K value. Note that $\Sigma_j {}^{j}\mathrm{F} = 1$. (*c*) Mathematically nondescribable distribution function for a heterogeneous complexant for which K can take continuous values. (*c*-1) Estimation $\bar{p}(K)$ of the probability density function by a histogram where sites are arbitrarily divided into classes of various small $\Delta\log K$. The surface of each block is the probability (F) of finding sites with K values within the corresponding interval. Note that $\bar{p}(K) = F$ only if $\Delta\log K = 1$. (*c*-2) $p(K)$ function corresponding to the histogram in (*c*-1). (*c*-3) Corresponding distribution curve, $c(K)$, obtainable from $|\mathrm{M}|_b = f(|\mathrm{M}|)$ data when C_L is unknown. The surface area under the curve in an interval gives a concentration of sites, ΔC_L, equal to $\Delta C_L = \int_{\Delta\log K} c(K) \cdot d\log K$. The average K value, K_{av}, in this interval, is

$$K_{av} = \frac{1}{C_L} \cdot \int_{\Delta\log K} \log K \cdot c(K) \cdot d\log K = \int_{\Delta\log K} \log K \cdot p(K) \cdot d\log K$$

If the $\Delta \log K$ interval is not too large, this can also be expressed in an approximate form that will be useful later for comparative purposes:

$$^{AB}C_L \sim C_L \bar{p}(K) \cdot \Delta \log K \qquad (13.21b)$$

where $\bar{p}(K)$ is the average value of $p(K)$ in the interval $\Delta \log K$ (see Fig. 13.6*a*).

The analogy between a mixture of a few discrete ligands and a heterogeneous complexant is shown in Figure 13.6*b* and *c* (see also Figure 13.5 curves a3 and b3). The equivalent of a distribution function for a mixture of simple ligands (Figure 13.6*b*) is obtained by plotting the fraction $^iF = {}^iC_L/C_L$ of each ligand ^{i}L as a function of ^iK. This gives a series of vertical lines, since each ^iK has a unique value. For complexants possessing a very large number of sites, these sites can be classified into a number of arbitrary groups in such a way that the log K values of all sites in each group are in the same $\Delta \log K$ interval. By calling ΔC_L the global concentration of sites in a given group (e.g., $\Delta C_L = {}^{AB}C_L$ in Figure 13.6*a*) the fraction of the total sites in this group is $F = \Delta C_L/C_L$, and if the interval is not too large an estimation of the value of the density function for the group is given by (eq. 13.21b)

$$\bar{p}(K) = \frac{\Delta C_L/C_L}{\Delta \log K} \qquad (13.22)$$

Therefore, on plotting $\bar{p}(K) = f(\log K)$, one obtains a histogram (Figure 13.6*c*-1). This histogram is an estimation of the actual density function $p(K)$ which is obtained when $\Delta \log K$ becomes infinitely small ($\Delta \log K \rightarrow d \log K$: Figure 13.6*c*-2). The surface area for each site class in Figure 13.6*c*-1, is the corresponding fraction F, that is, an estimation of the probability of finding sites in the corresponding interval $\Delta \log K$. In Figure 13.6*c*-2 ($\Delta \log K \rightarrow d \log K$), this probability is defined as in Figure 13.6*a*.

4.3.2. Affinity Spectrum Determination. In practice, $p(K)$ is obtained from Eq. 13.13, where the ^iKs are replaced by the continuous variable K, the ^iC_L ($\equiv dC_L$) is replaced by $C_L \cdot p(K) \cdot d \log (K)$ (Eq. 13.20), and summation is replaced by integration:

$$\theta = \int_{-\infty}^{+\infty} \frac{K|\mathrm{M}|}{1 + K|\mathrm{M}|} p(K) \cdot d \log K \qquad (13.23a)$$

$|\mathrm{M}|$ and θ being obtained experimentally from the titration data, $p(K)$ can be obtained after mathematical transformation (Ferry, 1970; Shuman et al., 1983; Thakur et al., 1980). Recall that for heterogeneous ligands, C_L is often unknown and only $|\mathrm{M}|_b$ and $|\mathrm{M}|$ are measurable. In this case Eq. 13.23a must be used in the form

$$|\mathrm{M}|_b = \theta C_L = \int_{-\infty}^{+\infty} \frac{K|\mathrm{M}|}{1 + K|\mathrm{M}|} c(K) \cdot d \log K \qquad (13.23b)$$

As a result, only the function $c(K)$ is obtainable from the titration data (Fig. 13.6*c*-3). Comparison of Eqs. 13.23a and 13.23b shows that $c(K) = C_L p(K)$. The surface area under $c(K)$ within an interval $\Delta\log K$ directly yields the corresponding value of ΔC_L. If the experimentally observable range of log K is large enough, the value of C_L can be found from this graph, since

$$C_L = \int_{-\infty}^{+\infty} c(K) \cdot d \log K \tag{13.24}$$

The principal limitation on the use of $p(K)$ or $c(K)$ functions for practical interpretation of the complexation properties of heterogeneous complexants results from the great complexity of the required mathematical transformations (Thakur et al., 1980; Buffle 1987). As a result, it is only possible to obtain approximate solutions, and there are great risks of mathematical artifacts.

4.3.3. Usefulness of an Affinity Spectrum. Despite the above-mentioned practical limitations, the affinity spectrum concept is potentially useful for understanding

1. The qualitative effects of certain physicochemical factors on the overall complexation properties of P, since any change in chemical conditions which influences its properties will be reflected by a change in the shape of the $p(K)$ function.
2. The exact nature of any K and ΔC_L "averaging" inherent in the analytical method employed or resulting from interpreting continuously distributed K complexants in terms of discrete site models.

The latter point is extremely important if data are to be properly interpreted. In particular, the affinity spectrum concept enables us to understand that (Fig. 13.5*b*; 13.6*c*):

The window of the analytical method used for the titration (Fig. 13.5) can be seen as a particular case of $\Delta\log K$, as defined in Figure 13.6. Therefore, if the experimental data are transformed to compute an average equilibrium constant and a corresponding "global" site concentration (as is often the case in the literature), the resulting parameters correspond, in effect, to an experimental "integration" across the window of that particular technique. It is evident, then, that such parameters are *not comparable when they are obtained by different methods.* Utilization of affinity spectra offers the possibility of specifying their degree of similarity.

The thermodynamic significance of the average K, K_{av}, corresponding to a particular $\Delta\log K$ decreases as $\Delta\log K$ increases, since the interval comes to incorporate sites having ever greater differences in character. One should, therefore, try to determine only *differential K values,* that is, K values in very small $d \log K$ intervals.

It should be apparent from Figure 13.6*c* that the average K (K_{av}) and the corresponding site concentration (ΔC_L) measured in a given window are not independent (in contrast to the case for the complexation constant and site concentration of a simple ligand). For a heterogeneous complexant, some significance can be attached only to a (K_{av}, ΔC_L) couple where K_{av} and ΔC_L are measured in the *same interval.*

4.4. Differential Equilibrium Functions

Compared with the computation difficulties of affinity spectra, calculation of a differential equilibrium function as defined in Eq. 13.17 is much simpler and presents fewer risks of introducing mathematical artifacts. The means by which it can be obtained for metal complexation or proton binding have been demonstrated and discussed in detail by Gamble and coworkers (1973, 1980, 1983). Only its basic principles will be presented below, for which the simplest case, $|M|_b \ll C_L$ (large excess of complexing sites) and pH constant, will be taken as an example.

4.4.1. Determination of Differential K Values. For any point during a titration of P by M ($\theta = f(C_M)$; see Figure 13.5*b*), it is possible to compute

$$\overline{K}' = \frac{|M|_b}{|M| \cdot |L'|} \tag{13.25}$$

where $\overline{K}'$ is an average equilibrium quotient valid for the pH considered and $|M|_b$ is the total bound metal ($|M|_b = C_M - |M| = \theta C_L$; Eqs. 13.11 and 13.12). $|L'|$ represents all the sites not combined with M and is computed as $|L'| = \Sigma_j |^jL| = C_L - |M|_b$. $|L'|$ is used now, instead of $|L|$, to include the protonated forms that were considered nonexistent in the previous section. For the imposed conditions (Buffle, 1984; Buffle et al., 1984) $|L'|/|L| =$ constant and therefore $\overline{K}'$ is a conditional parameter valid only at the pH considered. As can be seen from Eq. 13.26, it is not a thermodynamic constant but, rather, only an average equilibrium parameter that varies with C_M (or θ) due to the corresponding changes in $\Sigma_j |M^jL|$ and $\Sigma_j |^jL'|$:

$$\overline{K}' = \frac{\sum_j |M^jL|}{|M| \sum_j |^jL'|} \tag{13.26}$$

Combining Eqs. 13.26 and 13.16 (where ${}^{j}K'$ and $|{}^{j}L'|$ replaces ${}^{j}K$ and $|{}^{j}L|$, respectively) one obtains

$$\overline{K}' = \frac{\sum_j {}^{j}K' \, |{}^{j}L'|}{\sum_j |{}^{j}L'|} \tag{13.27}$$

This equation shows that $\overline{K}'$ calculated for a given point (θ, C_M) during the titration is an average of the ensemble of constants for all sites, weighted by the concentration of sites not yet occupied by M at that point. Since, as a result of the infinite number of sites, the ensemble of ${}^{j}K'$ constitutes a continuous function, $K' = f(\theta)$ (= differential equilibrium function; Eq. 13.17), Eq. 13.27 shows that $\overline{K}'$ is also a continuous function of θ. This latter is called the *average equilibrium function.* $\overline{K}'$ must be considered only as an *experimental parameter* that can be used in succeeding calculations in a manner similar, for instance, to θ.

The differential and average functions can be shown (Gamble et al., 1980; Buffle, 1984) to be related by

$$K' = -\frac{d(\overline{K}'(1 - \theta))]}{d\theta} \tag{13.28}$$

In practice, $\overline{K}'$ and θ are obtained very simply from Eqs. 13.12 and 13.25 after which $\overline{K}'(1 - \theta)$ is plotted as function of θ with $K' = f(\theta)$ then being obtained by measuring the slope of the curve for each value of θ. It is also possible to transform Eq. 13.28 in such a way (Buffle 1984) as to permit calculation of K' directly from C_M, $\alpha = C_M/|M|$ and $\{L\}_t$, where { } indicates the complexant concentration expressed in any units (e.g., g L^{-1}). This latter transformation indicates that the actual molar total concentration of sites, C_L, is not required for computing K' (only an invariant proportionality between C_L and $\{L\}_t$ must be assured). This feature is important for heterogeneous complexants for which C_L is often unknown.

4.4.2. The G(K') Function: a Cumulative Distribution. The function $K' = f(\theta)$ must, in fact, be interpreted in terms of a probability distribution. Indeed, its inverse, $\theta = f(K')$, which we shall call the *Gamble function* $G(K')$, is simply the *cumulative distribution* corresponding to $p(K')$ and is illustrated in Figure 13.7 [which also shows $G'(K') = C_L G(K')$ and $c(K') = C_L p(K')$ so as to permit discussion of the distributions in terms of $|M|_b$ instead of $\theta(|M|_b = \theta C_L)$]. Let us consider the case of a titration of P by M stopped at some first point (subscript 1) (in Figure 13.7) where $K' = {}^{p}K'$ and $|M|_b = |M|_{b,1}$. As a first approximation it can be considered that all sites stronger than p ($j < p$; $K' > {}^{p}K'$) are saturated and that all weaker sites ($j > p$) are metal-free. $|M|_{b,1}$ corresponds then to the area under the $c(K')$ curve, integrated as

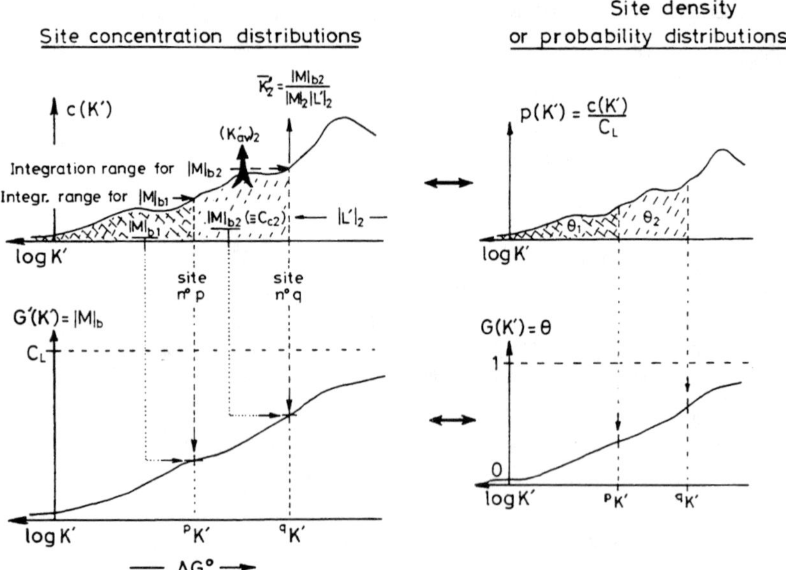

Figure 13.7. Relationship between the affinity spectrum $p(K')$ and the Gamble function $G(K')$, and between $c(K') = C_L \cdot p(K')$ and $G'(K') = C_L \cdot G(K')$. $(K_{av})_2$ is the average constant corresponding to $|M|_{b,2}$, as defined in Figure 13.6, that is, the weighted average of all the K' between $K' = \infty$ and $K' = {}^pK'$. It can be obtained by integration of $p(K')$. The curve at top left shows that $(K'_{av})_2 > {}^qK'$. C_c, the complexation capacity, is defined as the point of a titration where all sites are "seen," by the method used, as saturated by M. C_c therefore depends on the UL of the method (Fig. 13.5*b*-1). It may correspond to a first point (subscript 1) for one method and to a later point (subscript 2) for another. In the latter case, one finds $C_{c,2} = |M|_{b,2}$. The value of the average equilibrium function, $\overline{K}'$, at that point is $\overline{K}_2' = |M|_{b,2}/|M|_2\,|L'|_2$. Note that $|M|_{b,2}$ are complexes with sites most of which are such that $j < q$, whereas $|L'|_2$ are all free sites for most of which $j > q$ (i.e., these two site types are completely different). Furthermore, $|M|_2$ is mostly fixed by sites with $j \sim q$. Note that $\overline{K}_2'$ is obtained by an averaging process different from that used to get $(K'_{av})_2$. The average equilibrium quotient, $\tilde{K}_2'$, corresponding to $C_{c,2}$ (most often found in the literature) is an average of $\overline{K}'$ values computed with various $|M|_b$ values in the vicinity of stage 2.

in Eq. 13.21a but between $\log K' = \log {}^pK'$ and $\log K' = \infty$. The same reasoning is applicable when the titration is stopped at a later stage (subscript 2) corresponding to ${}^qK'$ and $|M|_{b,2}$. The $G'(K')$ function is directly obtained, therefore, by plotting all $|M|_b$ as a function of the corresponding K' (computed as explained above). $G(K')$, which is simply $\theta = |M|_b/C_L = f(K')$, can be obtained similarly when C_L is known.

The above approximation obviously neglects the fact that sites in the neighborhood of p (at point 1) are neither totally occupied nor completely free. It does, however, enable us to see that at any point (e.g., θ_1 corresponding to ${}^pK'$), θ is equivalent to a *cumulative probability;* θ_1, for instance, is the probability of forming a complex with any K' value such that $K' \geq {}^pK'$. The interpretation of the $G(K')$ function in cumulative probabilistic terms helps us to understand the actual meaning of two operational complexation parameters frequently used in the literature in attempting to represent, in a global fashion, the complexation properties of heterogeneous complexants. They are known as

1. *The complexation capacity* (C_c), obtained from a titration curve and defined as the value of C_M for which all sites are saturated (i.e., no additional observable increase in $|M|_b$ with increasing C_M).
2. *The average equilibrium quotient* ($\tilde{K}'$), which is the average of a number of $\overline{K}'$ values in a given window situated in the vicinity of C_c and within which $\overline{K}'$ is considered to be constant within experimental error.

These parameters, which are used below, are discussed in detail elsewhere (Buffle, 1984; Buffle et al., 1984) and illustrated in Figure 13.7. Suffice it to say that any C_c rarely represents a true saturation point (see Section 3.3) but that the couple $(C_c, \tilde{K}')$ can be considered an estimation of $(|M|_b, K')$ at $|M|_b = C_c$, provided C_c and $\tilde{K}'$ are computed for the same window.

4.4.3. Potential Usefulness of the G(K′) Function. In addition to its much greater mathematical simplicity relative to a $p(K')$ spectrum, the $G(K')$ function or a similar representation mode (see below) offers very important advantages. This is particularly the case for plots of $\log \theta = f(\log K')$ (Figure 13.8):

In order to interpret the properties of a given heterogeneous complexant *at the molecular level,* its complexation response must be compared under varying conditions (pH, ionic strength, T, etc.). For such comparisons to be valid, however, K' values (representative of the complexation free energy ΔG^0) must be compared at equivalent θ (or $|M|_b/\{L\}_t$), since this is the best and simplest way to ensure that at least similar (if not exactly the same) types of sites are concerned. This condition is easily met using the $G(K')$ function and has been found to be a highly useful approach for interpreting the properties of fulvic compounds (Section 5; Buffle, 1984, 1987; Buffle et al., 1984).

As was mentioned in Section 4.1, a very important property of natural systems is their *metal buffering capacity* τ, that is, their ability to "absorb" increases in C_M while maintaining $|M|$ as constant as possible. It may be shown (Altmann and Buffle, in preparation) that τ can be readily computed for any metal loading, from the slope Γ of the curve $\log(|M|_b/\{L\}_t) = f(\log K')$.

In order to understand the role and circulation of metals in natural systems (e.g., soil), it is necessary to compare the properties of the individual homologous complexants which are present (e.g., humics, clays, oxides). In such cases, comparisons are most valuable when made at constant free metal ion concentration $|M|$ (or, preferably, activity), since this is the only common variable (i.e., master variable, Stumm and Morgan,

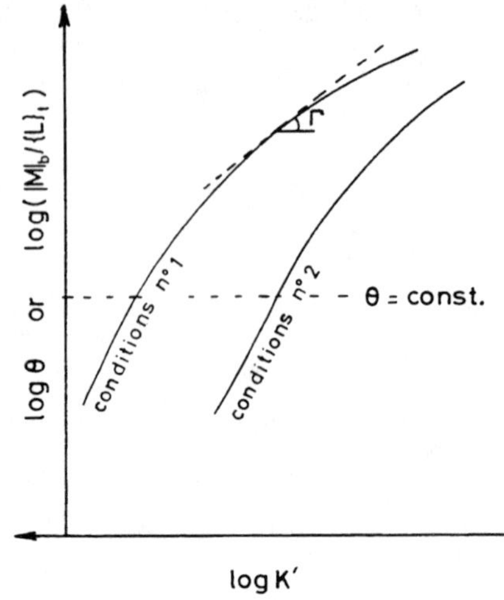

Figure 13.8. Comparison of cumulative distribution functions for interpretation of complexation properties of heterogeneous complexants.

1981) for all complexants in the same system. A plot of $\log(|M|_b/\{L\}_t) = f(\log |M|)$ is easily obtained from experimental data since $|M|$ is measured and $|M|_b = |M|_t - |M|$ ($\{L\}_t$ must then be expressed in the same units, e.g., g L^{-1}, for all complexants). It may, in fact, be shown that such a plot is equivalent to $\log(|M|_b/\{L\}_t) = f(\log K')$ and that at any metal loading $|M|$ is linked to K' and Γ by a simple mathematical expression (Buffle and Altmann, in preparation). In any case, the set of $\log(|M|_b/\{L\}_t) = f(\log |M|)$ curves for all complexants can be used to compare both the degree of saturation and the buffering contribution of each complexant at any given value of $|M|$.

5. APPLICATION OF THE $G(K')$ FUNCTION TO INTERPRETATION OF HETEROGENEOUS COMPLEXANT PROPERTIES (using metal–fulvic interactions)

In actuality, little of the data in the literature for metal titrations of homologous compounds have been represented or interpreted from the perspective of continuous distributions, the exceptions being the large work by Gamble and Langford (Gamble and Schnitzer, 1973; Gamble et al., 1980, 1983) and the application of affinity spectra by Shuman et al., 1983; and Olson and Shuman, 1985 both for interpretation of fulvic compound properties. For these latter compounds (structure: see Fig. 13.4) a particularly rich database is available but, in nearly all instances, what is reported are average equilibrium quotients $\tilde{K}'$ and the associated measured complexation capacity C_c. As discussed in Section 4.3.3 and Figure 13.7, these parameters are highly con-

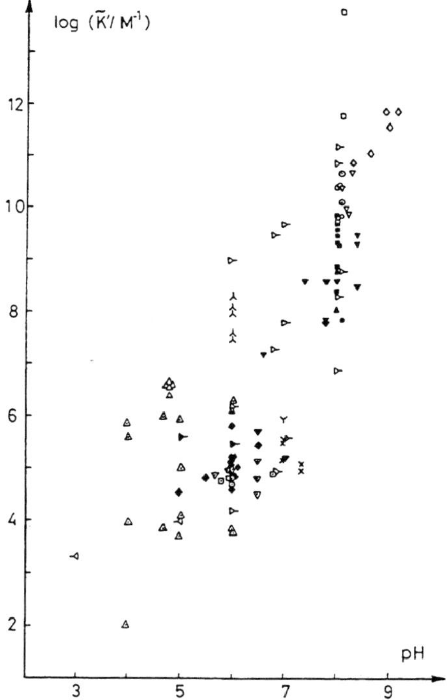

Figure 13.9. Compilation of literature data for equilibrium quotients, $\tilde{K}'$, for complexation of Cu(II) by pedogenic fulvic compounds. Each symbol represents a different reference. In all cases, $\tilde{K}'$ is computed by assuming the formation of "average" 1:1 complexes between Cu(II) and fulvics. After Buffle, et al. (1984).

ditional and are therefore of little use when treated as independent parameters. This is demonstrated by the log $\tilde{K}'$ values plotted in Figure 13.9, which compiles all values reported for Cu(II) complexation by fulvic compounds found in a recent survey (Buffle, 1984b). It can be seen that at a given pH, the dispersion of $\tilde{K}'$ values extends over a range of about 6–7 logarithmic units.

G(K) to the Rescue: The fact that $\tilde{K}'$ and C_c are meaningful when treated as a couple (C_c, $\tilde{K}'$) has been discussed above, and it was pointed out that, together, they can be used as an approximation of the ($|M|_b$, K') couple at $|M|_b = C_c$. This being so, the $G(K')$ function can then be applied to the large number of ($\tilde{K}'$, C_c) values cited in the literature for fulvic compounds. This has been done by Buffle (1984a,b) and the resulting $G(\tilde{K}')$ functions (e.g., Fig. 13.10) prove very useful despite the inevitable dispersion arising when attempting to compare data obtained by different workers using varying methods applied to diverse samples. First, the good correlation evident in Figure 13.10 between log C_c and log $\tilde{K}'$ for any given restricted pH range shows that

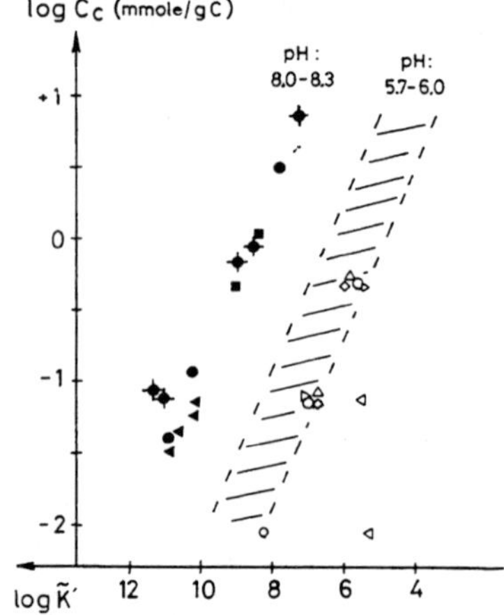

Figure 13.10. Changes in the complexation capacity (C_c) with the corresponding equilibrium quotient ($\tilde{K}'$) for the complexation of Cu(II) by pedogenic fulvic compounds (from Buffle, 1987). Solid symbols: All literature data taken from Figure 13.9, for pH 8.0–8.3. Open symbols: Results obtained with a single sample at pH 6.25 by the same authors using five different methods. Hatched zone: Ensemble of literature data for pH 5.7–6.0. A given symbol (solid or open) refers to the same method as follows: (◁) ASV, (○) ISE, (-⟡-) bioassays, (□) gel chromatography, (▷) fluorescence, (△) ligand competition with UV absorption, (◇) ultrafiltration. This figure illustrates the influence of pH, nature of samples, workers (different for solid symbols), and analytical methods. For instance, open symbols show that all methods give coherent results except voltammetry. This conclusion is difficult to draw by simply comparing $\tilde{K}'$ values without taking C_c values into account.

the vertical dispersion of log $\tilde{K}'$ values in Figure 13.9 is clearly due to their differing C_c values. Since the sample origins and techniques are widely different, the fact that the points are well grouped around a single line indicates that, at constant pH, any variation in $\tilde{K}'$ is largely dependent on the metal loading (represented here by C_c). The $G(K')$ function also proves to be very useful for interpreting a number of other important aspects summarized below.

5.1. Application to Interpretation of Reactions at the Molecular Level

5.1.1. Effect and Comparison of Analytical Technique. The fact that experimental complexation results depend on the window of the technique used was discussed in Sections 3.3 and 4.3.3. Figure 13.10 shows that the $G(\tilde{K}')$ representation of data is useful for differentiating between such operational effects and the more fundamental influence of metal loading (see figure legend for details).

5.1.2. Role of pH. Figure 13.9 shows that unless the C_c effect is taken into account, no unambiguous trend can be ascribed to the influence of pH

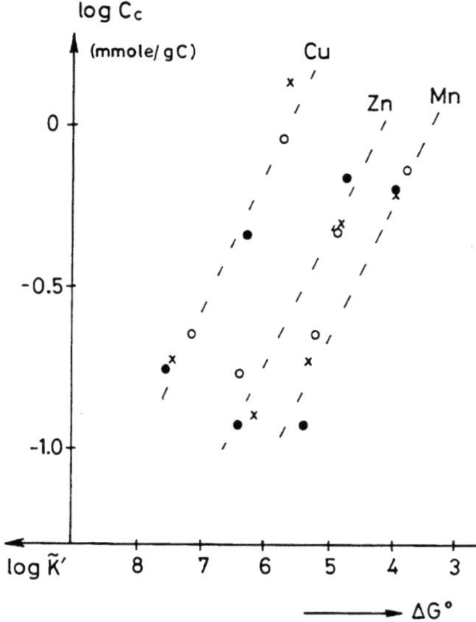

Figure 13.11. Influence of the nature of the metal ion on the change in log C_c with log $\tilde{K}'$. (From Buffle, 1984.) Each symbol refers to a different sample of sediment organic matter. pH 8.0. Comparison of $\tilde{K}'$ for the different metals can be only qualitative if C_c is not taken into account. Plotting log C_c as a function of log $\tilde{K}'$ permits unambiguous comparison of $\tilde{K}'$ at constant C_c.

on $\tilde{K}'$. A great reduction in data dispersion is obtained when log $\tilde{K}' = f(\text{pH})$ are plotted at constant C_c (corresponding to a horizontal line in Fig. 13.10). Values of (Δlog K')/ΔpH of 1.2, 0.5, and 0.24 were obtained for Cu(II), Pb(II), and Zn(II) at C_c = 2 mmol $(\text{gC})^{-1}$ (Buffle, 1987), a trend which is consistent with other known characteristics of these metals.

5.1.3. Influence of the Metal Ion Nature; Irving–Williams Series. Figure 13.11 shows that, as was true for pH, the complexation properties of various metal ions must also necessarily be compared at constant metal loadings. Application of this approach showed that the normal Irving–Williams series (Table 13.1) is observed for C_c = 2 mmole $(\text{gC})^{-1}$.

5.1.4. Kinetic Properties. Preliminary experimental work concerning the dissociation kinetics of Cu(II)–fulvic complexes also shows that $p(K')$-type normalizations can be used to characterize the kinetic behavior of these complexes (Shuman et al., 1983; Olson and Shuman, 1985). The $G(K')$ function also enables a quantitative evaluation as to why complex dissociation times increase over a very wide range (milliseconds to hours) as metal loading decreases (Buffle, 1987).

5.1.5. Polyfunctional versus Polyelectrolytic Effects; Major and Minor Sites. Figure 13.12 can be used to show how the $G(\tilde{K}')$ function can help

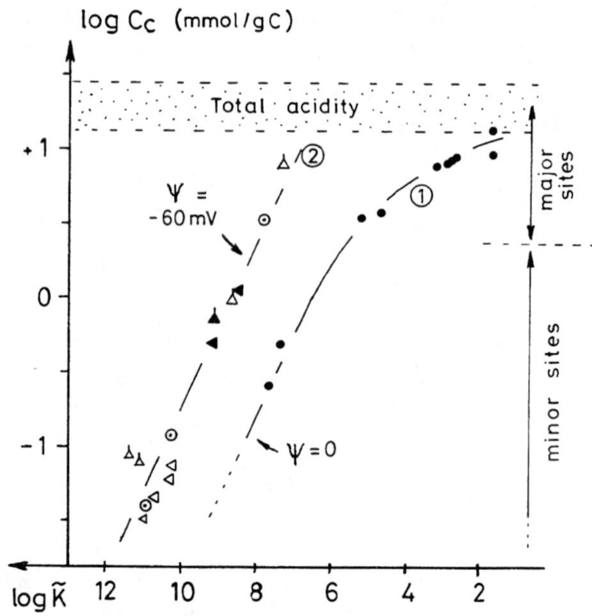

Figure 13.12. Comparison of the relative importance of polyelectrolytic and polyfunctional factors for the complexation of Cu(II) by pedogenic fulvic compounds at pH 8.0. The domain of total acidity is estimated from the average properties of pedogenic and soil fulvic compounds (Buffle, 1984, 1986). The points on curve 2 are literature data and are identical to the solid points in Figure 13.10. (In curve 2 the different symbols serve simply to identify the different references.) The solid circles (curve 1) are computed from the data by Murray and Linder (1983), who have given the structure and density of sites of fulvics "synthesized" by computer (see Fig. 13.13), and from the literature values for the complexation constants of these sites, for which charge, that is polyelectrolytic effect, is considered as zero ($\psi = 0$). For real fulvic compounds (line 2) the charge density is largely due to the dissociation of –COOH groups and is therefore close to the total acidity (10–15 mmole $(gC)^{-1}$. The influence of bound M on the charge density is then negligible when $|M|_b/\{L\}_t$ (or C_c) is much smaller than total acidity (say a few millimoles per gram of carbon. In that case, (1) change in $\tilde{K}'$ is not due to charge effect and only to polyfunctional effects, and (2) the $\Delta \log K'$ (~2) between lines 1 and 2 gives an estimation of the charge contribution to the stability of complexes. ($\Delta \log K = 2$ corresponds to an average potential $\psi = -60$ mV.)

in elucidating the molecular reasons for the change in K' with metal loading. In particular, it allows estimation of the relative roles played by the polyfunctional (binding by a large number of different coordinating atoms) and polyelectrolytic (electrostatic attraction due to a high electric charge density) characteristics of fulvic complexants (see figure legend). The latter derives from deprotonation of all dissociable groups and at pH 8 may be estimated as being close to the total acidity of the fulvics (see Fig. 13.12). By comparing C_c values to the total acidity, one can distinguish two types of sites on the $G(\tilde{K}')$ functions:

Minor (*strongest*) *sites* contribute only negligibly to the overall charge density. Since they are likely to be the predominant ones occupied by M at C_c (or $|M|_b/\{L\}_t$) less than a few mmole $(gC)^{-1}$, the charge density is independent of metal loading in this range. Therefore any *change* in $\tilde{K}'$ (i.e., not its absolute value) with C_c is overwhelmingly due to polyfunctionality.

Major sites determine the charge density. These become increasingly occupied by M as $|M|_b/\{L\}_t$ becomes greater than a few mmole $(gC)^{-1}$; any change in $\tilde{K}'$ with C_c is then attributable to changes in both net charge and the nature of the complexing sites.

This division into two site types is likely to be valid for other heterogeneous complexants and has important implications for their behavior vis-à-vis the various types of metal cations (Section 5.2).

Chemical Nature of the Coordinating Sites. Figure 13.12 can also be used to show that the $G(\tilde{K}')$ function is fruitful in considering probable fulvic compound composition. The solid circles in Figure 13.12 are based on a structure for fulvic acids (Fig. 13.13) computed by Murray and Linder (1983) using a theoretical mathematical model. They fall along a line (curve 1) closely parallelling that of the experimental data (curve 2), which could be taken as indicating some correspondence between the computed site types and those existing in real samples. As mentioned in the legend of Figure 13.12, the horizontal shift between lines 1 and 2 is due to and can be used to estimate the polyelectrolytic properties of the latter, since monomeric ligand mixtures are not subject to charge effects.

5.2. Application to Understanding Natural Systems

The polyfunctional character of heterogeneous compounds is of particular interest vis-à-vis their role in determining the chemistry of trace metal ions. A few aspects are discussed below, again taking the fulvic compounds as an example.

As discussed above, two groups of fulvic compound sites can be discriminated: the minor, strong sites and the major, weak ones. If their properties and concentrations in natural waters are compared with those of the three important metal groups (Table 13.1, Fig. 13.2), it can be seen (Buffle, 1984, 1987) that:

The transition (vital) metals and the B (toxic) cations are most likely to react with the minor sites.

The major cations (A cations, particularly H^+ and Ca^{2+}) will react principally with major sites.

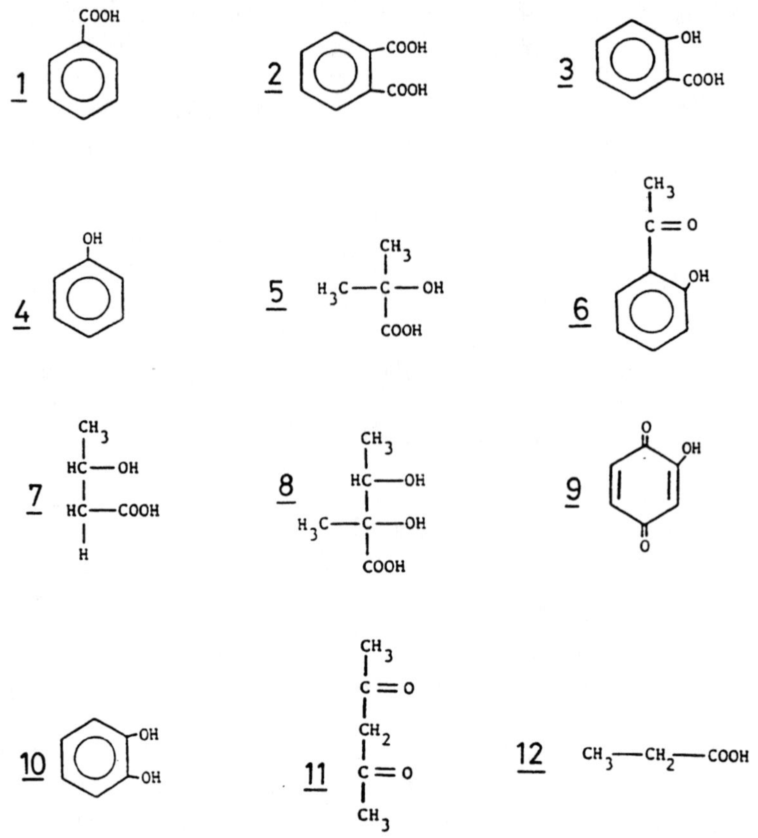

Figure 13.13. Nature of sites computed theoretically by Murray and Linder (1983) for fulvic compounds, based on elemental composition and 50% aromaticity [corresponding to soil-derived (pedogenic) fulvics]. The site numbering increases with decreasing order of site density (expressed in mmole $(gC)^{-1}$.

One can therefore anticipate that pH and Ca^{2+} will control the importance of the charge density effect and, as a consequence, will control the effects due to the conformation of the macromolecules. Stated another way, the major cations will control the structure and physical properties of these macromolecular complexants. Although this effect will contribute in the form of a constant term, as regards the net complexation energy of trace metals with minor sites, it is principally the polyfunctional quality of these sites that will be responsible for their regulation. Note that the same two differing roles for A and B (plus transition) metal ions, respectively, have also been noted for living biological systems (as mentioned in Section 2.1).

It is interesting to note that all of the log $(|M|_b/\{L\}_t) = f(\log \tilde{K}')$ functions reported for fulvic compounds, including the Murray and Linder data, present a striking feature in common. Regardless of the pH (Figure 13.10), the nature of the metal complexed (Figure 13.11), or the analytical method employed (Figure 13.10), all the plots are straight lines having a slope in the vicinity of -0.5. It may be shown using a rigorous mathematical derivation (Altmann and Buffle) that this behavior is that of an "ideal" heterogeneous system having (1) an infinite number of sites and (2) K' values that increase regularly as the corresponding site fraction $({}^jC_L/C_L)$ decreases. The mathematical characteristics of such "ideal" systems allow other important properties to be predicted with the *metal buffer capacity* τ being particularly interesting. It was mentioned earlier (Section 4.3.3), for example, that τ is easily computable from the slope of the $\log \theta = f(\log K')$ curve, and results show that for fulvic compounds these plots are observed to be straight lines (within experimental error and in the explored range of metal loading). In such a case, it may be shown theoretically (Buffle and Altmann, in preparation) that τ is proportional (and close) to the total metal concentration. As a consequence of this property, the complexant is able to "absorb" the metal over a large concentration range (at least several decades) with only a gradual increase in the biologically critical free metal ion concentration (i.e., no sharp concentration jumps are observed). This property, which is very different from that of classical buffer systems, is important ecologically since it helps to assure that living organisms have the greatest chance of adapting to chemical changes in their environment.

6. CONCLUSIONS

The foregoing discussion shows that the use of probabilistic concepts and, in particular, the cumulative frequency distribution for rationally normalizing experimental data on complexation is a powerful tool for the interpretation and comparison of heterogeneous complexant behavior. When it is applied to fulvic compounds, it shows that they behave as "ideal" heterogeneous complexants whose peculiar metal buffering action is particularly beneficial for maintenance of life. This important role is one of several indications that natural organic matter should not be considered simply as detritic metabolic waste but rather as a basic structure of nature that is essential for the maintenance of life. This interpretation is supported by the convergence of both statistical computations (i.e., Murray and Linder results) and the natural processes responsible for pedogenic fulvic compound formation in widely varying climatic and geologic regimes: all appear to result in mixtures possessing globally remarkably reproducible properties, despite the vast number of different mixtures that such processes potentially could produce.

REFERENCES

Buffle, J. (1984), "Natural Organic Matter and Metal-Organic Interactions in Aquatic Systems," H. Sigel, Ed., *Metal Ions in Biological Systems,* vol. 18, Marcel Dekker, New York.

Buffle, J. (1987), *Complexation Reactions in Aquatic Systems. An Analytical Approach,* Ellis Horwood, Chichester.

Buffle, J., Tessier, A., and Haerdi, W. (1984), "Interpretation of Trace Metal Complexation by Aquatic Organic Matter," in C. J. M. Kramer and J. C. Duinker, Eds., *Complexation of Trace Metals in Natural Waters,* Martinus Nijhoff/Dr W. Junk, Publishers. The Hague.

Ferry, J. D. (1970), *Viscoelastic Properties of Polymers,* 2nd ed., Wiley, New York.

Gamble, D. S., and Schnitzer, M. (1973), "The Chemistry of Fulvic Acid and Its Reactions with Metal Ions," Chapter 9 in P. C. Singer, Ed., *Trace Metals and Metal Organic Interactions in Natural Waters,* Ann Arbor Science, Ann Arbor, MI.

Gamble, D. S., Schnitzer, M., Kerndorff, H., and Langford, C. H. (1983), Multiple metal ion exchange equilibria with humic acids, *Geochim. Cosmochim. Acta* **47,** 1311–1323.

Gamble, D. S., Underdown, A. W., and Langford, C. H. (1980), Cu(II) titration of fulvic acid ligand sites with theoretical potentiometric and spectrophotometric analysis, *Anal. Chem.* **52,** 1901–1908.

Kleinhempel, (1970) Albrecht Thaer Archiv, **14,** 3

Luoma, S. N., and Bryan, G. W. (1981), A statistical assessment of the form of trace metals in oxidized estuarine sediments employing chemical extractants, *Sci. Total Environment* **17,** 165–196.

Luoma, S. N., and Jenne, E. A. (1977), "Forms of Trace Elements in Soils, Sediments, and Associated Waters: An Overview of Their Determination and Biological Availability, in R. E. Wildung and H. Drucker, Eds., *Biological Implications of Metals in the Environment,* Conf. 750929, NTIS, Springfield, VA.

Murray, K., and Linder, P. W. (1983), Fulvic acids: structure and metal binding. I. A random molecular model, *J. Soil Sci.* **34,** 511–523.

Nriagu, J. O. (1979), Global inventory of natural and anthropogenic emissions of trace metals to the atmosphere, *Nature* **279,** 409–411.

Olson, D. L., and Shuman, M. S. (1985), Copper dissociation from estuarine humic materials, *Geochim. Cosmochim. Acta* **49,** 1371–1375.

Shuman, M. S., Collins, B. J., Fitzgerald, P. J., and Olson, D. L., (1983), "Distribution of Stability and Dissociation Rate Constants of Cu-Organic Complexes," in R. F. Christman and E. T. Gjessing, Eds., *Aquatic and Terrestrial Humic Materials,* Ann Arbor Science, Ann Arbor, MI, Ch. 17.

Stumm, W., and Morgan, J. J. (1981), *Aquatic Chemistry,* Wiley, New York.

Sunda, G. W., and Guillard, R. R. N. (1976), The relationship between cupric ion activity and the toxicity of copper to phytoplankton, *J. Marine Res.* **34,** 511–529.

Thakur, A. K., Munson, P. J., Hunston, D. L., and Rodbard, D. (1980), Continuous affinity distribution of arbitrary shape, *Anal. Biochem.* **103,** 240–254.

Whitfield, M., and Turner, D. R. (1979), Water rock partition coefficients and the composition of seawater and river water, *Nature* **278**(5700), 132–137.

Whitfield, M., and Turner, D. R. (1983), "Chemical Periodicity and the Speciation and Cycling of the Elements," in C. S. Wong, E. Boyle, K. W. Bruland, J. D. Burton, and E. Goldberg, Eds., *Trace metals in seawater,* NATO Conf. Ser. #4, Vol. 9, Plenum, New York.

14

PARTICLE–PARTICLE INTERACTIONS

Charles R. O'Melia

Department of Geography and Environmental Engineering, The Johns Hopkins University, Baltimore, Maryland

Abstract

The study of particle–particle interactions in aquatic systems has both theoretical and practical stimuli. Hydrodynamics, solution chemistry, and surface science are involved. Applications encompass diverse physical, chemical, and biological phenomena. In this chapter, present theories of particle–particle interactions are reviewed; are assessed, using observed kinetics of coagulation, deposition, and filtration processes; and are used to interpret phenomena in natural waters and in water and wastewater treatment systems.

1. INTRODUCTION

Particulate problems, processes, and people.

Transport and fate and particles and particle-reactive pollutants in natural waters. Cadmium in constance. Ferrum in vierwaldstättersee. Synthetic organic substances in superior ontario sees. Asbestos deposition in lungs. Acid fog collection on trees. Oil-water separation. Packedbedfiltration. Coagulationinwaterandwastewatertreatment. Biofilms. Heatransfrictionalresistance. Corrosion. Etc.

Brownian diffusion, fluid shear, and gravity. Hydrodynamic retardation and van der Waals forces. Electrostatic and specific chemical interactions. Surface complex formation. GouySternIHOHslippingplanes. Constant charge, constant potential, and regulated interactions. Patches and bumps.

Smoluchowskilevichfriedlander. Werner VODL Stummorgan. Fuchspielmanruckenstein.

The subject of colloidal particle–particle interactions has a diverse conceptual base, a long history, extensive current activity, and wide application in environmental science and engineering. The objectives of this chapter are (1) to summarize our present understanding of these interactions, (2) to point out strengths and deficiencies in our ability to describe these phenomena quantitatively, and (3) to indicate some applications of this knowledge to environmental problems.

Interactions between solid particles have been conveniently segregated into two sequential steps: transport and attachment. The results of these interactions are given terms such as coagulation, deposition, and filtration. Particle transport depends upon hydrodynamics and external forces such as gravity; it is primarily a physical process. The attachment of two particles can be dominated by the surface properties of the solid particles and by solution chemistry; it can be primarily a chemical process. This distinction between transport and attachment, or physics and chemistry, is not perfectly sharp. Some colloidal forces must be present to allow transport of one particle to the surface of another. Similarly, surface chemistry does not always influence attachment; physical forces are sufficient to describe "favorable" interactions.

This separation of physics from chemistry introduces significant limitations. A synthesis rather than a segregation is needed to describe particle–particle interactions in a holistic manner and to provide a quantitative basis for solving many particulate problems. Such a synthesis is also useful for identifying the strong points and shortcomings in each step.

These two sequential steps are summarized separately in Sections 2 and 3, and in Section 4 they are synthesized in an overview of particle–particle interactions. Applications in treatment processes and natural waters are considered in Section 5, and in Section 6, conclusions are presented.

2. PARTICLE TRANSPORT

Three physical processes are emphasized here. The first is Brownian diffusion, in which random motion of small particles is brought about by thermal effects. The driving force for this transport is a function of kT, the product of Boltzmann's constant k and absolute temperature T. Kinetic energy is transferred from fluid molecules to small particles during the continuous bombardment of the particles by the surrounding water molecules.

A second process affecting particle transport in aquatic systems is fluid shear, either laminar or turbulent. Velocity differences or gradients occur within all real flowing fluids. Hence, particles that follow the motion of the suspending fluid will travel at different velocities. These spatial fluid and particle velocity differences or gradients can produce interparticle contacts. In this case, particle transport depends upon the fluid velocity gradient G.

The third force considered here is gravity, which produces vertical transport of particles and depends on the buoyant weight of these particles, represented by $(\pi/6)\ (\rho_p - \rho)gd_p^3$ in which ρ_p and ρ are the densities of the particles and the fluid, respectively, g is the gravity acceleration, and d_p is the particle diameter.

These physical processes are few in number and simple in concept. They are also ubiquitous in occurrence and complex in effect. They will be included here in examinations of two environmental problems: coagulation in a lake and packed bed filtration in water and wastewater treatment.

The effects of Brownian motion (perikinetic flocculation) and velocity gradients (orthokinetic flocculation) were quantitatively addressed by Smoluchowski in 1917. Gravitational coagulation was considered by Findheisen (1939). Levich (1962) coupled fluid convection with particle diffusion and presented analytical expressions for the transport of small particles to plates, spheres, and rotating disks. Friedlander (1977) has considered convective diffusion with external force fields and has emphasized the effects of the polydisperse size distribution of natural particles on coagulation and deposition. When the diffusion coefficient is constant and the flow is incompressible, the convective diffusion equation can be written in the following form:

$$\underset{(a)}{\frac{\partial n}{\partial t}} + \underset{(b)}{\vec{v}\cdot\nabla n} = \underset{(c)}{D\nabla^2 n} - \underset{(d)}{\nabla\cdot\vec{w}n} \tag{14.1}$$

Here n is the particle concentration, $\vec{v}$ is the fluid velocity vector, D is the Brownian diffusion coefficient of the particles, and $\vec{w}$ is the particle migration velocity vector resulting from an external force field. The term $\vec{v}\cdot\nabla n$ describes particle transport by bulk fluid flow, $D\nabla^2 n$ denotes transport by Brownian diffusion, and $\nabla\cdot\vec{w}n$ represents transport by external forces. Yao et al. (1971) developed and tested analytical and numerical solutions for the convective transport of particles by diffusion, fluid flow, and gravity to spherical collectors in packed beds (packed bed filtration). In this case, $-\nabla\cdot\vec{w}n = w_s\partial n/\partial z$, where w_s is the Stokes settling velocity and z is the vertical dimension.

Simplifying Eq. 14.1 so that term (a) = term (c) results in Smoluchowski's formulation for perikinetic flocculation (1917). Levich (1962) added fluid flow [term (b)], and Friedlander has emphasized external force fields [term (d)], yielding the complete convective diffusion equation. Smoluchowskilevich-friendlander.

As two particles approach each other to within a distance of a few radii, $\vec{v}$, $\vec{w}$, and D in Eq. 14.1 are altered by the hydrodynamic interactions that occur. The motion of a particle in close proximity to another particle or to a collector must necessarily deviate from that of an undisturbed streamline, because the continuum description of fluid motion at both solid surfaces produces infinitesimally slow drainage of fluid from the gap between them as they approach. Diffusion coefficients, for example, are no longer constant and, in fact, approach zero as the separating distance approaches zero. Considering only these hydrodynamic interactions, particle–particle contacts cannot occur. This anomaly is resolved, however, when attractive van der Waals forces are considered. These increase rapidly as the particles approach each other, tending toward infinity as the particles come into contact, and overcome the otherwise slow fluid drainage.

The collective action of these hydrodynamic and van der Waals forces is termed *hydrodynamic retardation.* For most cases it results in a decrease in the contact rate compared to Smoluchowski's analysis. Neglect of hydrodynamic and van der Waals forces, or considering that they are exactly equal, has been termed the Smoluchowski–Levich assumption (Adamczyk et al., 1983). Useful summaries of hydrodynamic retardation and its effects are given by these investigators, by Spielman (1977), and by Rajagopalan and Tien (1979).

An important practical difficulty is that numerical solutions are required for Eq. 14.1 when hydrodynamic retardation is considered. Some graphical and approximate analytical results obtained in this way are available.

3. CHEMICAL ASPECTS

The viewpoint presented here can be considered a combination of coordination chemistry proposed by Alfred Werner in 1893 with the electric double-layer model for the solid–solution interface and for particle–particle interaction developed in the late 1930s and early 1940s by Verwey, Overbeek, Derjaguin, and Landau (Verwey and Overbeek, 1948). The need for a combination of these approaches has been indicated by Stumm and Morgan (1962). Werner VODL Stummorgan.

Three sequential components of the chemical aspects of particle–particle interactions are considered here:

1. The origins of particle stability.
2. The structure of the solid–liquid interface.
3. Chemical interactions between two such interfacial regions.

Two distinct mechanisms are important in colloidal stability; these are termed *electrostatic* and *steric stabilization.* The latter is briefly addressed first, after which a somewhat more extensive review of electrostatic stabilization follows.

Steric stabilization can result from the adsorption of polymers at solid–water interfaces. Large polymers can form adsorbed segments on a solid surface with loops and tails extending into solution (Lyklema, 1978). A stabilizing polymer may contain two types of groups, one with a high affinity for the particle surface and a second, hydrophilic group that is left "dangling" in the water (Gregory, 1978). The configuration of such an interfacial region is difficult to characterize theoretically or experimentally. Statistical mechanical and semiempirical approaches to describe the conformational properties of polymers at interfaces have met with only limited success. This, in turn, limits quantitative formulation of the interaction between two such regions during a particle–particle encounter. Some useful qualitative descriptions can, however, be made.

Gregory (1978) summarized two processes that can produce a repulsion when two polymer-coated surfaces interact at close distances. First, the adsorbed layers can each be compressed by the collision, reducing the volume available for the adsorbed molecules and thereby reducing the number of possible configurations available to the molecules (a reduction in entropy), increasing the free energy, and causing repulsion between the particles. Second, and more frequently, the two adsorbed layers may interpenetrate on collision, increasing the concentration of polymer segments in the mixed region. If the polymer segments are strongly hydrophilic, they can prefer the solvent to other polymer segments. An overlap or mixing of the adsorbed layers then leads to an increase in the free energy and to repulsion. These two processes are separate from and in addition to the effects of polymer adsorption on the charge on the particles and the van der Waals interaction between the particles.

Most particles in water are charged. This (primary) charge produces a volumetric (opposite) charge density and an electric field (diffuse double layer) in the aqueous solution near the particle surface. This, in turn, generally produces a repulsive force when two similarly charged particles approach each other close enough for their diffuse double layers to interact. Like hydrodynamic interactions, these electrostatic "chemical" forces can be offset by attractive van der Waals forces which, as stated previously, increase rapidly as the interparticle separation approaches zero. When such electrostatic repulsion dominates during particle–particle interactions, the particles are said to be electrostatically stabilized, and the results are termed "slow" coagulation and "unfavorable" deposition and filtration. Origins of surface charge, interfacial structure, and electrostatic stabilization were presented by Verwey and Overbeek (1948). A good didactic summary has been given by Lyklema (1978).

The Verwey–Overbeek–Derjaguin–Landau (VODL) theory for colloidal stability has successfully predicted some characteristics of particle–particle interactions and has been unsuccessful in predicting others. Early validations of the theory were based on coagulation studies. Predictions of the effects of counterion valence on the occurrence of coagulation (as evidenced, for example, by observations summarized in the Schulze–Hardy rule) are accurate and persuasive. The effects of salt concentration on coagulation kinetics, however, have not been modeled well by the theory (Ottewill and Shaw, 1966). Some of the many modifications of the VODL theory and some other experimental evaluations are discussed subsequently.

Proton transfer, metal complexation, and ligand exchange have been proposed by Stumm, Schindler, and coworkers (e.g., Schindler, Wälti, and Furst, 1976; Stumm, Kummert, and Sigg, 1980; Hohl, Sigg, and Stumm, 1980) to describe the surface charge, acid–base, and adsorption behavior of metal oxides. Several models have been developed that can accurately describe the results of potentiometric and adsorption experiments. Surface charge can be determined, but these models do not distinguish unambiguously among proposed surface reactions (Westall and Hohl, 1980) and are not sensitive to the detailed structure of the interfacial region (Sposito, 1984). For a quantitative formulation of the chemical aspects of particle–particle interactions, accurate descriptions of *both* surface charge and interfacial structure are probably required. This difficulty is addressed in more detail subsequently.

Calculation of the repulsive force between two charged particles is often made using one of two extreme cases, constant potential or constant charge. Lyklema (1980) has proposed that diffuse layers can relax or equilibrate during a Brownian encounter, while a layer of more fixed charge (Stern layer) may not be able to do so. The result is that the kinetics of charge (ion) transport can affect the interaction between two particles. Prieve and Ruckenstein (1976), considering biological surfaces, argued that acid–base equilibria at the surface will be shifted as two particles approach each other, altering both surface charge and surface potential. Healy et al. (1980) have extended this analysis to latex particles and some oxides and consider the "regulation" of surface charge and potential as equilibrium is maintained during particle–particle interactions.

Many assumptions, models, and calculations have been made for chemical particle–particle interactions. There are few, if any, definitive experiments. For two identical surfaces, all predictions reside between the limits described by the constant-potential and constant-charge cases (Chan and Mitchell, 1983). Later discussions will suggest that all are inadequate to describe the kinetics of unfavorable coagulation, deposition, or filtration.

4. A PHYSICOCHEMICAL SYNTHESIS

Smoluchowski addressed slow coagulation and defined α as the fraction of interparticle contacts that result in aggregation but did not relate α to particle–particle interactions or to variables such as electrolyte concentration. In 1934, Fuchs first combined transport and chemistry. He considered Brownian diffusion in a force field and defined the stability ratio W as the ratio of the collision rate to the aggregation rate ($W = \alpha^{-1}$). W has been expressed in terms of the maximum or peak net repulsive interaction between two colliding charged particles and, in turn, to certain solution and surface characteristics. Fuchs's approach has been extended by several investigators to include advection, other particle transport processes, hydrodynamic retardation, and various forms of diffuse layer interactions. Representative examples are particle deposition by convective diffusion on spheres and cylinders (Spielman and Friedlander, 1974), non-Brownian deposition on spheres and rotating disks (Spielman and Cukor, 1973), and consideration of the effects of the charging mechanism on deposition (Ruckenstein and Prieve, 1980). Fuch-spielmanruckenstein.

Some excellent experimental studies of the kinetics of particle–particle interactions have been made of the coagulation of colloidal suspensions, deposition on rotating disks, and filtration by packed beds. Before summarizing some of these, it is convenient to classify particle–particle interactions as rapid or favorable and slow or unfavorable. In rapid or favorable coagulation, deposition, or filtration, there are no repulsive interactions caused by particle charge. Mass transport establishes the contact and attachment rate; $W = 1 = \alpha$. In slow or unfavorable interactions, a net repulsive interaction as small as a few kT substantially affects particle attachment. The interparticle contact rate is substantially greater than the successful attachment rate; W approaches infinity and α approaches zero.

Hull and Kitchener (1969) studied the deposition of monodisperse latex particles on rotating disks. When the latex particles were negatively charged and the rotating disk was positively charged, the observed deposition rates were in agreement with Levich's theory for convective diffusion to a rotating disk (1962); deposition was favorable or rapid. When both the suspended particles and the rotating disk were negatively charged, deposition was retarded but the kinetics could not be characterized by incorporation of VODL theory into the physical transport model. These authors wrote, "When W values were evaluated from [theory], theoretical values were orders of magnitude higher than the experimental values. . . . The evidence strongly suggests that . . . anomalous deposition occurred preferentially onto areas of locally favorable potential or geometry (or both)."

Gregory and Wishart (1980) studied the packed bed filtration of submicron polystyrene latex particles by alumina fibers and applied the convective diffusion equation with surface reaction as developed by Spielman and Friedlander (1974) to their results. "Unfavorable" deposition of negatively charged latex particles on the negatively charged alumina fibers was much less than for the favorable case but was also substantially greater than predicted by theory; α was <1 but also much greater than expected. These authors wrote, "The observed deposition [of negatively charged latex particles] on the negative fibers is much greater than predicted from theory and may possibly be explained by a heterogeneity of fiber surface charge."

Bowen and Epstein (1979) studied the deposition of uniform spherical submicrometer silica particles from an aqueous suspension in laminar flow onto a smooth parallel plate channel. Their results and conclusions are similar to those obtained in different systems by Hull and Kitchener (1969) and Gregory and Wishart (1980). For the deposition of negatively charged silica particles onto a positively charged plastic substrate, good quantitative agreement was found between mass transport theory and experiment. For negative particles and negative channel walls, the measured deposition rates were always much greater than those predicted theoretically considering hydrodynamic interactions, van der Waals forces, and electric double layer interaction forces. The authors suggested that "the primary cause of this discrepancy is the failure of the model to account for surface heterogeneity, which could result in preferential deposition onto areas of locally favorable potential or geometry."

These three examples are representative of kinetic studies of coagulation, deposition, and filtration in the recent literature. For favorable cases, the kinetics of particle–particle attachment can be described by mass transport models that include hydrodynamic retardation. For unfavorable cases, all applications of chemical models, whether constant charge, constant potential, regulated, or otherwise mixed, substantially underestimate attachment.

Many reasons can be proposed to explain these observations. Five are noted here. Three deal with the structure of the interfacial region of a single particle, and two are based on consideration of the interactions between two charged particles.

The interfacial region is usually considered a uniform continuum in two dimensions, with a structure that varies only with distance from the interface. The proposals by Hull, Kitchener, Gregory, Wishart, Bowen, Epstein, and others that significant variations in chemical and physical characteristics occur *at the surface* are plausible. Quantitative predictions and experimental testing of the effects of such variations on particle–particle interactions are lacking, but the real possibility exists that chemical patches and physical bumps invalidate the two-dimensional uniform continuum assumed in the VODL model and its many modifications.

Many planes have been assumed to exist at many locations parallel to solid surfaces. Stern, inner Helmholtz, and outer Helmholtz planes are examples. All would be altered by patches and bumps. All are proposed structures invoked to model certain observed phenomena, but none has been proposed for the primary purpose of understanding the kinetics of coagulation, deposition, or filtration. When observations of these phenomena do not agree with theories using one or more of the model planes, it is plausible that the theories and their associated uniform planes are inadequate.

A diffuse or Gouy–Chapman layer is assumed to exist and to extend from the outermost plane to the bulk solution. Ions in this layer are treated as mobile point charges, electrostatically attracted or repelled by the charged solid and its assembly of planes, and diffusing from or to the bulk solution. These ions are also assumed to lose all of their specific chemical reactivity and some of their electrostatic reactivity in the diffuse layer. When residing in the bulk solution, an ion may form inner-sphere complexes and ion pairs with other solutes and participate in electrostatic clustering as described by the Debye–Hückel theory. Similar reactivity is permitted for these ions at the solid surface and in some of its planes. Only in the region between the surface and the bulk solution, that is, in the diffuse layer, are chemistry and physics among these ions restricted, although it is predominantly from the interaction between two such restricted diffuse layers that unfavorable particle–particle interactions (electrostatic stabilizations) arise. Sposito (1984) has used Monte Carlo techniques while allowing only coulombic effects among ions in the diffuse layer and has demonstrated that such interactions can produce ion concentrations in the diffuse layer that are substantially different from those given by the Gouy–Chapman model.

Several types of double-layer interactions are possible during an encounter between two charged particles. Lyklema (1980) has illustrated important possibilities during a Brownian encounter. For a particle with a diameter of 1 μm in a 0.1 *M* solution of 1:1 electrolyte, the time for an unretarded Brownian encounter is about 10^{-5} s. Considering hydrodynamic retardation, this estimate is raised to 10^{-4} s. This time is considerably longer than the equilibration or relaxation time of ions in the diffuse layer, about 10^{-8} s. Rates of adjustment of surface layers to the encounter can be substantially longer. For an AgI sol, Lyklema estimates a relaxation time for the Stern layer of 10^{-3} s; this may be the time to transfer a potential-determining ion across the interface. It is considerably longer than the Brownian encounter.

If the surface potential of a solid is determined by the adsorption equilibrium of potential-determining ions, then a fully relaxed interaction between two particles (during which equilibrium is maintained at the two surfaces) will result in the surface potential remaining constant while a reduction in the surface charge occurs as potential-determining ions are desorbed. If the interacting double layers do not relax or equilibrate at all, then all potential-determining ions remain at the surface and the surface potential increases during the interaction. The results reported by Lyklema (1980) suggest that this latter interaction mode can occur with AgI sols and probably some others. In general, the type of electric double-layer interaction will depend on the kinetics of ion and particle transport. Constant-charge interactions can produce the largest repulsive interaction energies; constant-potential interactions yield the lowest calculated repulsive interaction energies. Intermediate interaction modes would produce intermediate repulsive interaction energies.

Where does the water go? The slow drainage of fluid from the space between two interacting particles gives rise to hydrodynamic retardation. In the VODL model, however, it does not appear to carry any solute with it. Stated another way, the concentration of solutes in the small volume occupied by two interacting electric double layers could be affected by the slowly draining flow of fluid in this region, but this is not considered in charge or ion balances used in calculating interaction energies.

These five factors (bumps, patches, ion interactions within a diffuse layer, relaxation kinetics, and fluid flow) are only a few of the many proposals for improvements in the modeling of solid–solution interfaces and particle–particle interactions. They are selected to be illustrative, potentially important, and informative. The path from present knowledge of interfacial structure and the chemical aspects of particle–particle interactions to successful quantitative description of the kinetics of unfavorable coagulation, deposition, and filtration will be complex and difficult. An excellent comprehensive review and assessment of the present knowledge of such colloidal phenomena has been written by Hirtzel and Rajagopalan (1985).

5. ENVIRONMENTAL APPLICATIONS

Two systems are selected to illustrate the extent and importance of particle–particle interactions in environmental engineering and chemistry. The first is a conceptual and laboratory study of the chemical aspects of packed bed filtration (Wang, 1986). In the second, (1) the role of natural organic matter in establishing the surface properties of particles in natural waters, (2) the significance of this role on coagulation in lakes, and (3) the possible consequences of humic substances and coagulation on particle transport and deposition in lakes are reviewed (Ali et al., 1984; Ali, 1985).

5.1. Packed Bed Filtration

Wang (1986) has related solution chemistry, the surface characteristics of solids, and hydrodynamics to particle deposition in the packed bed filtration of aqueous suspensions. This was accomplished by combining the surface complex formation model of Schindler, Stumm, and coworkers with present particle transport and interaction models. The surface charges of suspended particles and stationary collectors were calculated from the acid–base and adsorption properties of the solids [intrinsic constants, zero proton conditions (pH_{zpc}), isoelectric points (pH_{iep})]. Two types of suspension were used: (1) carboxyl latex particles in solutions of constant ionic strength and varying pH and calcium concentration, and (2) submicrometer hematite particles in solutions of constant ionic strength and varying pH and phosphate concentration. Glass beads comprised the filter media.

Among the conclusions of this work are the following. First, for favorable filtration, transport by convective diffusion with hydrodynamic retardation can account for observed deposition. This is illustrated by theoretical and experimental results of hematite filtrations (Fig. 14.1). The pH_{zpc} of the hematite used in this work was observed to be pH 7 by alkalimetric titration and by microscopic electrophoresis. For the glass media, pH_{zpc} was estimated to be less than 3. At $pH < 7$, the hematite particles are positively charged, the glass is negatively charged, and filtration is favorable. Under these solution conditions, theoretical and experimental values of α are near 1. Physical models and experimental measurements agree.

Second, for unfavorable filtration, particle deposition is substantially greater than predicted using present double-layer structures and interaction mechanisms. This is illustrated for hematite filtration in Figure 14.1 and for

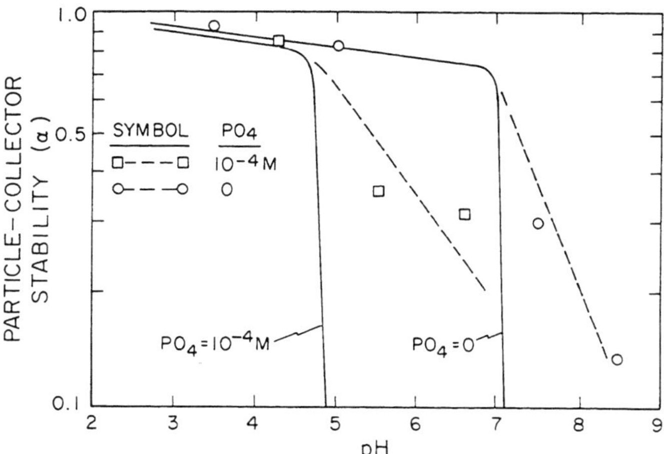

Figure 14.1. Comparison of theoretical and experimental values of particle-collector stability (α) as functions of pH. Solid lines denote predicted α values; dashed lines and points denote experimental results. Based on filtrations of hematite particles in 10^{-3} M $NaNO_3$, with and without added NaH_2PO_4. From Wang (1986).

latex filtration in Figure 14.2. At all pH's tested, the latex particles are negatively charged. All models predict a sharp decrease in α toward zero as the pH increases slightly above pH_{zpc} (or pH_{iep} with specific adsorption), due to the formation of a repulsive energy barrier of a few kT. Most experimental measurements indicate values of $\alpha > 0.1$ ($W < 10$); the lowest α measured was 0.03 ($W = 33$). Experimental values of α, while much less than 1, are much greater than zero. This suggests that present electrostatic and complex formation models for the solid–solution interface do not adequately describe the three-dimensional structure of that interface and also that particle–particle interaction models do not adequately describe the chemical aspects of these interactions. These two conclusions are similar to those of recent studies of other systems discussed previously; they extend these conclusions to filtration with spherical media and to models that include surface complex formation.

Finally, the use of surface complex formation modeling in this work has permitted good predictions of pH_{iep} for suspended particles in solutions containing adsorbable species. This, in turn, has provided good predictions of the solution conditions under which filtration is favorable ($\alpha = 1$) and unfavorable ($\alpha \ll 1$). This is illustrated in Figure 14.1. The surface charge on hematite is positive at pH 5 in the absence of phosphate; the surface on the glass media is negative at this pH. There is no barrier to the attachment of hematite on glass; filtration is calculated and observed to be favorable. Stated another way, theoretical and experimental values of α were near 1. In the presence of 10^{-4} M orthophosphate, pH_{iep} for hematite is predicted to be 4.8 with surface complex formation (ligand-exchange) modeling. This is confirmed with microscopic electrophoresis measurements. At pH 5.5, both the

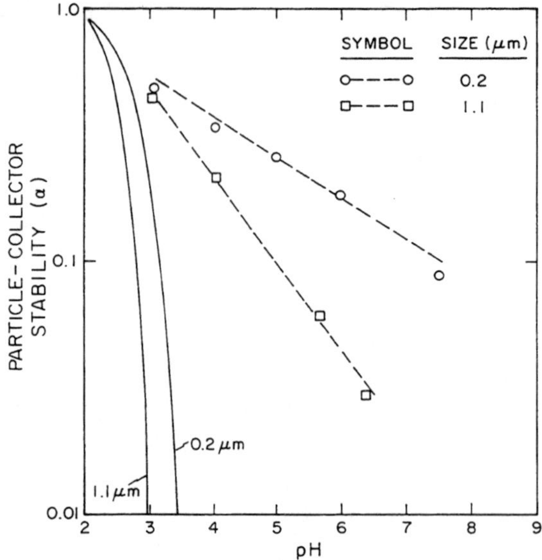

Figure 14.2. Comparison of theoretical and experimental values of particle-collector stability (α) as functions of pH. Solid lines denote predicted α values; dashed lines and points denote experimental results. Based on filtration of two sizes of latex particles in 10^{-3} M $NaNO_3$. Adapted from Wang (1986).

hematite and the glass beads are negatively charged, an electrostatic energy barrier exists, and filtration is unfavorable. Both experimental and theoretical values of α are less than 1, although the dramatic drop in α predicted by theory is again not observed.

5.2. Particle–Particle Interactions in Natural Waters

There is extensive evidence from fresh waters, estuaries, and the oceans that the surface properties and colloidal stability of particles in natural waters are affected by naturally occurring organic substances dissolved in these waters. The majority of such compounds are classified operationally as humic substances that are derived from soil or produced within natural waters and sediments by biological and chemical processes. Humic substances are anionic polyelectrolytes of low to moderate molecular weight; their charge is due primarily to carboxyl and phenolic groups; they have both aromatic and aliphatic components and can be surface-active; they are refractive and can persist for centuries or longer.

Some investigators have noted that major divalent metals in natural waters (in particular, Ca^{2+}) can exert destabilizing effects on natural and synthetic particles at low metal ion concentrations. A brief review of some results relating natural organic matter (NOM) and divalent metal ions with the surface properties and colloidal stability of particles in natural waters follows.

Niehof and Loeb (1972) determined the electrophoretic mobility of several inorganic solids after exposure to natural seawater and to seawater pretreated with ultraviolet radiation to photooxidize organic matter. All solids became moderately electronegative in natural seawater and all showed charges consistent with their chemical composition when immersed in organic-free seawater. The authors concluded that the surface charge exhibited by solids in seawater is due to adsorbed organic matter. Edzwald et al. (1974) studied the kinetics of particle–particle interactions in the coagulation of clays and estuarine sediments by solutions of NaCl and of synthetic (inorganic) seawater. All suspensions coagulated more rapidly in dilutions of artificial seawater than in NaCl solutions of equal ionic strength. The authors concluded that calcium and magnesium ions in seawater exert specific destabilizing effects on suspended particles. Sholkovitz and coworkers (Eckert and Sholkovitz, 1976; Boyle, Edmond, and Sholkovitz, 1977) studied the extent of flocculation of iron, aluminum, and humic substances in estuaries. They observed that iron colloids in rivers can be stabilized by dissolved organic matter; these colloids are destabilized in estuaries by seawater cations (magnesium and especially calcium ions) through specific chemical and electrostatic interactions with humic materials. Hunter and Liss (1979) determined the electrophoretic mobility of particles collected from four estuaries and evaluated them in terms of the salinity in the estuary at the site of collection. All particles at all locations were negatively charged. The authors concluded that dissolved organic matter formed surface coatings that gave the particles a consistent electronegative charge. They also showed that in calcareous river waters containing high calcium concentrations, the electrophoretic mobilities of the particles were smaller (less negative) than in rivers low in Ca^{2+}; they concluded that some of the charge on the particles can be neutralized by incorporation of Ca^{2+} "into the fixed part of the double layer." Tipping and coworkers (Tipping, 1981; Tipping and Cooke, 1982; Tipping and Higgins, 1982; Tipping and Heaton, 1983) studied the effects of humic substances and divalent metal ions on the surface properties of hematite and two manganese oxides. Effects on the colloidal stability of hematite were also determined. Measurements included electrophoretic mobility, adsorption of humic substances, and coagulation kinetics. Values of α were estimated to range from 0.01 to 1 for hematite coagulated in the presence of varying concentrations of Ca^{2+} and humic substances (Tipping and Higgins, 1982). Tipping and coworkers concluded that adsorbed humic substances produced negative charges on the particles studied. This charge was reduced by the presence of divalent metal ions, especially Ca^{2+}. Humic substances slowed coagulation rates and calcium ions enhanced them, possibly by specific chemical interactions with the humic substances. In addition to enhancing coagulation rates, calcium also substantially increased the adsorption of humic substances on hematite. Davis (1982) studied the adsorption on alumina and kaolinite of natural organic matter extracted from lake sediments. Measurements included electrophoretic mobility, charge titration, and adsorption of NOM. The author concluded that NOM was readily adsorbed on these surfaces, giving them a negative charge. Calcium enhanced the adsorption of organic matter on alumina at high pH and reduced its adsorption at pH less than 7–8. Davis (1982) proposed that adsorption of NOM occurs by complex formation between surface hydroxyl groups and acidic functional groups on the organic matter. Gibbs (1983) studied the effects of natural organic coatings on the kinetics of coagulation of four natural samples. Results for natural particles were compared with results using particles treated with sodium hypochlorite to remove their organic coatings. Gibbs observed that particles with natural organic coatings coagulated less rapidly than particles treated with hypochlorite and concluded that the natural organic coatings stabilized the particles.

Ali et al. (1984) reported the results of experimental measurements of α for the coagulation by fluid shear of natural particles obtained from the sediments of the Loch Raven Reservoir (Baltimore County, Maryland, USA), in solutions buffered at pH 8.1 and containing various combinations of calcium and of fulvic acids isolated from the Great Dismal Swamp (Virginia, USA). Some results are presented in Figures 14.3 and 14.4. In the absence of calcium ions, α decreased from 0.035 to 0.006 (W increased from 30 to 170) as the concentration of fulvic acid added to the suspensions was increased from 0 to 20 mg L^{-1} expressed as dissolved organic carbon (DOC, Fig. 14.3). Most of this increase in colloidal stability (decrease in α) occurred by a fulvic acid addition of 5 mg L^{-1} DOC; higher dosages had little additional effect. Calcium acted as a destabilizing agent; α increased from 0.035 to 0.2 (W decreased from 30 to 5) as calcium was increased from 0 to 10^{-3} M (Fig. 14.4).

Concurrent addition of fulvic acid and calcium produced intermediate effects. For example, in the presence of 5 mg L^{-1} fulvic acid as DOC, calcium

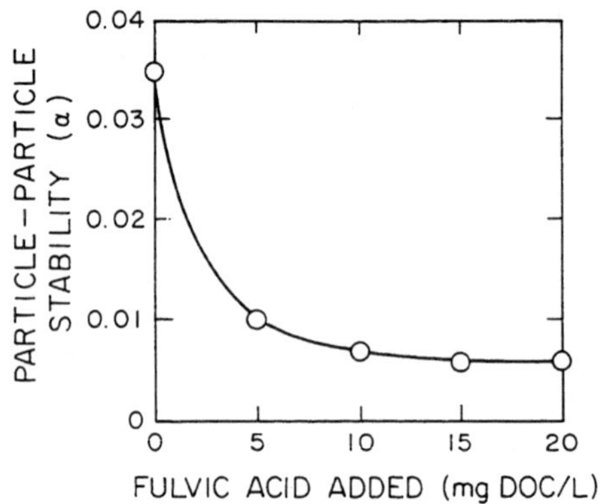

Figure 14.3. Effects of fulvic acid on the colloidal stability (α) of natural particles. Adopted from Ali et al. (1984).

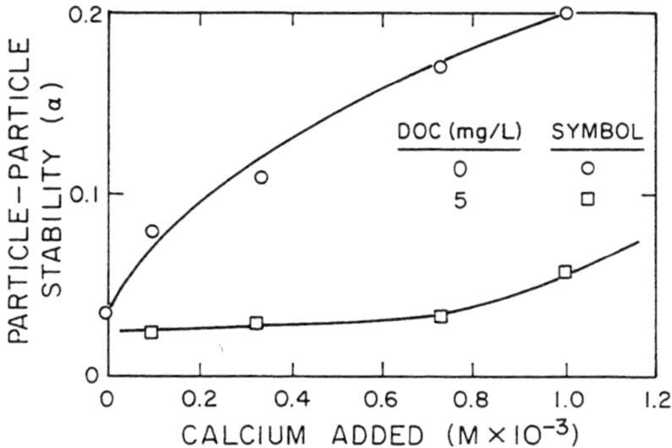

Figure 14.4. Effects of calcium alone and of calcium and fulvic acid (FA) in combination on the colloidal stability (α) of natural particles. Adopted from Ali et al. (1984).

additions of 7×10^{-4} *M* and lower had little effect on particle stability. Higher Ca^{2+} concentrations destabilized the particles, and α reached a value of 0.23 ($W = 4$) in the presence of 5 mg L^{-1} (DOC) fulvic acid and 3.3×10^{-3} *M* Ca^{2+} (not shown). These ranges of Ca^{2+} (0 to 3×10^{-3} *M* and fulvic acid (0 to 5 mg L^{-1} DOC) are representative of most aquatic systems and indicate that wide variation in α can be expected among natural waters. All measured values of α (0.006–0.23), however, are greater than expected from theoretical predictions, for which $\alpha \rightarrow 0$ and $W \rightarrow \infty$.

Ali (1985) reported measurements of α for the surface waters of the Loch Raven Reservoir. Samples were obtained over an 18-month period; α ranged from 0.01 to 0.09 with an average of 0.032. The DOC in the water ranged from 1.4 to 2.7 mg L^{-1}, and the hardness ($Ca^{2+} + Mg^{2+}$) ranged from 6.1×10^{-4} to 7.8×10^{-4} eq L^{-1} with an average of 6.8×10^{-4} eq L^{-1}. Statistical analysis of the data suggests that variations in α were related to variations in DOC. Measurements of α were also made for other lakes. In agreement with the laboratory studies reported in Figures 14.3 and 14.4, lakes low in DOC and high in calcium have high values of α.

The possible effects of coagulation on the transport and fate of particles and particle-reactive pollutants in lakes have been discussed by O'Melia and coworkers (O'Melia, 1980; O'Melia and Bowman, 1984; O'Melia et al., 1985). Recent experimental and modeling studies are reported by Ali et al. (1984), Ali (1985), and Weilenmann (1986). Ali (1985) reported that a coagulation and sedimentation model based on Smoluchowski kinetics (O'Melia and Bowman, 1984) successfully predicted measured total particle concentrations and trends with depth in the Loch Raven Reservoir, indicating that coagulation significantly affects particle concentrations in the reservoir. However, the model had limited success in predicting particle size distributions, generally overpredicting the concentrations of large (>25 μm) particles in the water column.

Weilenmann (1986) has applied coagulation modeling to Swiss hardwater lakes. The effect of coagulation in lakes is to increase settling velocities of particles by aggregation, thereby reducing their residence time in the water column and increasing the particulate flux to the sediments. Measured stability factors (values of α) indicate substantial variation in particle stability among Swiss hardwater lakes. These differences can be interpreted qualitatively with existing concepts of surface and solution chemistry. Divalent cations decrease particle stability, probably by specific adsorption. Humic substances increase particle stability due to electrostatic stabilization and perhaps also to steric effects. Their DOC concentration can be used to classify the Swiss lakes investigated into two groups. Those with DOC concentrations of less than 2 mg L^{-1} (Lakes Zurich and Luzern) have particle stabilities roughly an order of magnitude lower than lakes with DOC concentrations in the 3–5 mg L^{-1} range (Lakes Sempach and Greifen). For example, α for Lake Zurich is about 0.1, while α for Lake Sempach is less than 0.01. Modeling indicates that coagulation has substantial effects on the sedimentation of particles in Lake Zurich, owing to the unstable particles in this lake. Modeling of particle concentrations and fluxes in Lake Sempach was less successful, perhaps due to more extensive mineralization and dissolution arising from the longer hydraulic and particle detention times in that lake.

6. CONCLUSIONS

Chemistry is crucial to an understanding of particle–particle interactions. Much has been learned, but much more remains to be discovered.

Simultaneous consideration of chemical and physical aspects of particle–particle interactions indicates that the physical phenomena involved have been well described for several cases. When chemical knowledge indicates that chemical effects can be neglected (i.e., for favorable interactions when $\alpha = W = 1$), present physical transport models can often predict the kinetics of particle coagulation/deposition/filtration. In these cases, knowledge of physical characteristics of the particles (e.g., concentration, size, density) and of the system (e.g., size, flow conditions) is required.

This understanding of physical aspects permits an evaluation of our chemical understanding when chemistry is important (i.e., for unfavorable interactions when $\alpha \ll 1$, $W \gg 1$). The results demonstrate that the chemical aspects of particle–particle interactions are poorly understood. Specific adsorption is one key. It is the origin of most colloidal stability and a requirement for most particle destabilization. Second, better descriptions of the three-dimensional structure of solid–water interfaces are needed. These are a requirement for quantitative assessment of particle–particle interactions.

Chemical considerations allow successful prediction of favorable and/or unfavorable regions of interaction in some cases. Specific adsorption is again a key; surface complex formation models have been successful in describing the influence of several inorganic cations and anions on particle–particle interaction regions. The effects of organic molecules have not been well characterized.

For most "real" particle problems in natural waters and in water and wastewater treatment systems, the chemical characteristics of the primary particles do not affect the kinetics of their coagulation, deposition, and/or filtration. Natural organic matter (e.g., fulvic acid) acts as a natural dispersing agent and can make particle–particle interactions in these systems unfavorable. This effect is ubiquitous; it can affect the fate of particles in natural waters and determine the design and operation of water and wastewater treatment units.

Nevertheless, coagulation, deposition, and/or filtration occur more rapidly and more extensively than can be anticipated by present theoretical models of colloidal stability and particle–particle interactions. This knowledge is important for our understanding of the functioning of many natural systems, ranging from the transport of organisms in groundwater aquifers to the fate of particle-reactive pollutants in lakes. Yes.

Particulate problems, proce . . . .

Acknowledgments

It is a pleasure and an honor to acknowledge Stumms, students, spouse, and sponsors.

Aliclarkdempseyedzwalderalphundtlawlerolsonsnodgrasswangweilenmannwiesner-yao.

Research reported in this paper has been supported by the National Science Foundation (US) under Grant No. CEE81-21501 and the Environmental Protection Agency (US) under Grant No. 810094.

REFERENCES

Adamczyk, Z., Dabros, T., Czarnecki, J., and van de Ven, T. G. M. (1983), Particle transfer to solid surfaces, *Adv. Colloid Interface Sci.* **19,** 183–252.

Ali, W. (1985), "Chemical Aspects of Coagulation in Lakes," unpublished doctoral dissertation, The Johns Hopkins University, Baltimore, Maryland.

Ali, W., O'Melia, C. R., and Edzwald, J. K. (1984), Colloidal stability of particles in lakes: Measurement and significance, *Water Sci. Technol.* **17,** 701–712.

Bowen, B. D., and Epstein, N. (1979), Fine particle deposition in smooth parallel-plate channels, *J. Colloid Interface Sci.* **72,** 81–97.

Boyle, E. A., Edmond, J. M., and Sholkovitz, E. R. (1977), The mechanism of iron removal in estuaries, *Geochim. Cosmochim. Acta* **41,** 1313–1324.

Chan, D. Y. C., and Mitchell, D. J. (1983), The free energy of an electrical double layer, *J. Colloid Interface Sci.* **95,** 193–197.

Davis, J. A. (1982), Adsorption of natural dissolved organic matter at the oxide/water interface, *Geochim. Cosmochim. Acta* **46,** 2381–2393.

Eckert, J. M., and Sholkovitz, E. R. (1976), The flocculation of iron, aluminum, and humates in river waters by electrolytes, *Geochim. Cosmochim. Acta* **40,** 847–848.

Edzwald, J. K., Upchurch, J. B., and O'Melia, C. R. (1974), Coagulation in estuaries, *Environ. Sci. Technol.* **8,** 58–63.

Findeisen, W. (1939), Zur Frage der Regentropfenbildung in reinen Wasserwolken, *Meteorol. Z.* **56,** 365–368.

Friedlander, S. K. (1977), *Smoke, Dust, and Haze,* Wiley-Interscience, New York.

Fuchs, M. (1934), Über die Stabilität und Aufladung der Aerosole, *Z. Physik* **89,** 736–743.

Gibbs, R. J. (1983), Effect of natural organic coatings on the coagulation of particles, *Environ. Sci. Technol.* **17,** 237–240.

Gregory, J. (1978), "Effects of Polymers on Colloid Stability," in K. J. Ives, Ed., *The Scientific Basis of Flocculation,* Sijthoff and Noordhoff, The Netherlands, pp. 101–130.

Gregory, J., and Wishart, A. J. (1980), Deposition of latex particles on alumina fibers, *Colloids Surfaces* **1,** 313–334.

Healy, T. W., Chan, D., and White, L. R. (1980), Colloidal behavior of materials with ionizable group surfaces, *Pure Appl. Chem.* **52,** 1207–1219.

Hirtzel, C. J., and Rajagopalan, R. (1985), *Colloidal Phenomena,* Noyes Publications, Park Ridge, NJ.

Hohl, H., Sigg, L., and Stumm, W. (1980), "Characterization of Surface Chemical Properties of Oxides in Natural Waters: the Role of Specific Adsorption in Determining Surface Charge," in M. C. Kavanaugh and J. O. Leckie, Eds., *Particulates in Water* (Adv. Chem. Ser., No. 189), American Chemical Society, Washington, DC, pp. 1–31.

Hull, M., and Kitchener, J. A. (1969), Interaction of spherical colloidal particles with planar surfaces, *Trans. Faraday Soc.* **65,** 3093–3104.

Hunter, K. A., and Liss, P. S. (1979), The surface charge of suspended particles in estuarine and coastal waters, *Nature* **282,** 823–825.

Levich, V. G. (1962), *Physicochemical Hydrodynamics,* Prentice-Hall, Englewood Cliffs, NJ.

Lyklema, J. (1978), "Surface Chemistry of Colloids in Connection with Stability," in K. J. Ives, Ed., *The Scientific Basis of Flocculation,* Sijthoff and Noordhoff, The Netherlands, pp. 3–36.

Lyklema, J. (1980), Colloid stability as a dynamic phenomena, *Pure Appl. Chem.* **52,** 1221–1227.

Niehof, R. A., and Loeb, G. I. (1972), The surface charge of particulate matter in seawater, *Limnol. Oceanogr.* **17,** 7–16.

O'Melia, C. R. (1980), Aquasols: The behavior of small particles in aquatic systems, *Environ. Sci. Technol.* **14,** 1052–1060.

O'Melia, C. R., and Bowman, K. S. (1984), Origins and effects of coagulation in lakes, *Schweiz. Z. Hydrol.* **46,** 64–85.

O'Melia, C. R., Wiesner, M., Weilenmann, U., and Ali, W. (1985), "The Influence of Coagulation and Sedimentation on the Fate of Particles, Associated Pollutants, and Nutrients in Lakes," in W. Stumm, Ed., *Chemical Processes in Lakes,* Wiley-Interscience, New York, pp. 207–224.

Ottewill, R. H., and Shaw, J. N. (1966), Stability of monodisperse polystyrene latex dispersions of various sizes, *Disc. Faraday Soc.* **42,** 154–163.

Prieve, D. C., and Ruckenstein, E. (1976), The surface potential of and double-layer interaction force between surfaces characterized by multiple ionizable groups, *J. Theoret. Biol.* **56,** 205–228.

Rajagopalan, R., and Tien, C. (1979), "The Theory of Deep Bed Filtration," in R. J. Wakeman, Ed., *Progress in Filtration and Separation,* Volume 1, Elsevier, New ork, pp. 179–269.

Ruckenstein, E., and Prieve, D. C. (1980), "Role of Physicochemical Properties in the Deposition of Hydrosols," in J. K. Beddow and T. P. Meloy, Eds., *Testing and Characterization of Powders and Fine Particles,* Heyden, Philadelphia, pp. 107–137.

Schindler, P. W., Wälti, E., and Furst, B. (1976), The role of surface hydroxyl groups in the surface chemistry of metal oxides, *Chimia* **30,** 107–109.

Smoluchowski, M. (1917), Versuch einer mathematischen Theorie der Koagulationskinetik kolloider Lösungen, *Z. Physikal. Chem.* **92,** 129–168.

Spielman, L. A. (1977), Particle capture from low-speed laminar flows, *Ann. Rev. Fluid Mech.* **9,** 297–319.

Spielman, L. A., and Cukor, P. M. (1973), Deposition of non-Brownian particles under colloidal forces, *J. Colloid Interface Sci.* **43,** 51–65.

Spielman, L. A., and Friedlander, S. K. (1974), Role of the electrical double layer in particle deposition by convective diffusion, *J. Colloid Interface Sci.* **46,** 22–31.

Sposito, G. (1984), *The Surface Chemistry of Soils,* Oxford University Press, New York.

Stumm, W., and Morgan, J. J. (1962), Chemical aspects of coagulation, *J. Amer. Water Works Assoc.* **54,** 971–992.

Stumm, W., Kummert, R., and Sigg, L. (1980), A ligand exchange model for the adsorption of inorganic and organic ligands at hydrous oxide interfaces, *Croat. Chem. Acta* **53,** 291–312.

Tipping, E. (1981), The adsorption of aquatic humic substances by iron oxides, *Geochim. Cosmochim. Acta* **45,** 191–199.

Tipping, E., and Cooke, D. (1982), The effects of adsorbed humic substances on the surface charge of goethite (α-FeOOH) in freshwaters, *Geochim. Cosmochim. Acta* **46,** 75–80.

Tipping, E., and Higgins, D. C. (1982), The effects of adsorbed humic substances on the colloid stability of haematite particles, *Colloids Surfaces* **5,** 85–92.

Tipping, E., and Heaton, M. J. (1983), The adsorption of aquatic humic substances by two oxides of manganese, *Geochim. Cosmochim. Acta* **47,** 1393–1397.

Verwey, E. J. W., and Overbeek, J. Th. G. (1948), *Theory of the Stability of Lyophobic Colloids,* Elsevier, Amsterdam.

Wang, Z. (1986), "Chemical Aspects of Deep Bed Filtration," unpublished doctoral dissertation, The Johns Hopkins University, Baltimore.

Weilenmann, U. (1986), "The Role of Coagulation for the Removal of Particles by Sedimentation in Lakes," unpublished doctoral dissertation, Eidgenössischen Technischen Hochschulen, Zürich.

Westall, J., and Hohl, H. (1980), A comparison of electrostatic models for the oxide/solution interface, *Adv. Colloid Interface Sci.* **12,** 265–294.

Yao, K-mu, Habibian, M. T., and O'Melia, C. R. (1971), Water and wastewater filtration: Concepts and applications, *Environ. Sci. Technol.* **5,** 1105–1112.

15

THE ROLE OF COLLOIDS IN THE PARTITIONING OF SOLUTES IN NATURAL WATERS

François M. M. Morel and Philip M. Gschwend
Department of Civil Engineering, Massachusetts Institute of Technology, Cambridge, Massachusetts

Abstract

Thermodynamic models for the sorption of ionic and hydrophobic solutes on natural particles are now well established. However, these models do not predict widely observed trends in the experimental data; in particular, the effect of particle concentration on sorption is the object of controversy between empiricists and theorists. We suggest that the presence of colloids (solutes for some, solids for others) is the reason for these discrepancies. Such colloids are not experimentally separated from the aqueous phase and possess sorbent properties similar to those of separable particles. Colloids coagulate to form larger particles; they also appear to be continuously formed from larger particles. The result is a quasi-steady-state distribution between colloidal and particulate fractions. The general applicability of existing sorption models to natural waters awaits the development of a theoretical or empirical description of this relationship.

1. INTRODUCTION

From the point of view of the geochemist or of the environmental chemist, there is probably no more important issue than that of the partitioning of solutes between the solid and aqueous phases. This partitioning controls the transport and influences to a large extent the transformations affecting natural and anthropogenic chemicals. As a result, over the past several years, much effort has been directed at developing a physicochemical understanding of solute interactions with the solid phases present in natural waters (Stumm et al., 1976; Schindler et al., 1976; Bowden et al., 1977; Davis et al., 1978; Karickhoff et al., 1979; Chiou et al., 1979; Means et al., 1980; Li et al., 1984; Farley et al., 1985). The eventual goal is to obtain a mathematical description of solid–solution partitioning that permits a quantitative prediction on the basis of general chemical data concerning the nature of the solute, the properties of the natural sorbent particles, and the chemistry of the solution.

Although the solubility of mineral phases controls the solution concentrations of some solutes (mostly the major ions), it is a process of sorption on the suspended phases that most often governs the dissolved levels of many trace metals and organic compounds. Extensive research has led to widely accepted thermodynamic models of sorption for both inorganic and organic sorbates. However, these models, which elegantly describe experimental laboratory systems, do not appear to predict accurately the situation encountered in natural waters. In fact, the results of statistical studies of solid–solution partitioning in aquatic systems appear at odds with our physicochemical wisdom, spurring a kind of revisionism among the more fickle.

On the basis of the available information, we believe that colloids are the reason for many of the apparent discrepancies between experimental facts and theory. By the term *colloid,* we mean a component that has the physicochemical properties of a solid but is not separated from the solution phase by whatever separation technique (e.g., filtration or centrifugation) is employed. In other words, we are referring to a suspended solid that is operationally considered a solute.

It is our purpose in this chapter to demonstrate how the presence of colloids does or should affect the solid–solution partitioning of aquatic solutes in a way that matches experimental observations. Due to the elusive nature of aquatic colloids, there is little experimental information concerning either their existence or their properties. In fact, even the theories concerning the dynamic properties of colloids in suspension appear inadequate. As a result, this chapter is partly speculative; we believe, however, that it focuses the question of apparent solid–solution partitioning in natural waters on the appropriate issue: the dynamic balance between the aggregation and the disaggregation of small particles.

2. SURFACE REACTIONS OF IONS AND POLAR COMPOUNDS

As discussed in Chapter 1 (Westall, 1987), the sorption of solutes on solids can be accounted for by ionic and/or hydrophobic interactions. In the first case, which applies to ions and polar compounds, sorption is viewed as a chemical reaction with atoms at the surface of the solid. The thermodynamic description of such adsorption is thus given by a law of mass action between solutes and reactive surface sites:

$$\frac{\text{moles sorbate}}{\text{free surface sites}} = K_{\text{ads}} \frac{\text{moles solute}}{\text{volume solution}} \tag{15.1}$$

The corresponding equilibrium constant must account for both chemical and electrostatic energies of interaction. Since the electrostatic energy of adsorption on one surface group is affected by long-range coulombic interactions with neighboring surface groups, the electrostatic energy term is usually separated from the chemical (or "intrinsic") energy term and formulated by some variation of the double-layer theory:

$$K_{\text{ads}} = K_{\text{chem}} K_{\text{coul}} \tag{15.2}$$

Various adsorption models for ionic compounds differ chiefly by the way in which the electrostatic interactions are quantified, that is, the way in which K_{coul} is calculated (Morel, 1983). In the end, the corresponding parameters are fitted to experimental acid–base titration data. The strength of the resulting surface complexation models lies in their ability to predict accurately (on the basis of one sorption isotherm and one acid–base titration) the effect of pH ("pH edges") on cation and anion adsorption and to rationalize relative sorption properties on the basis of the acidity or hydrolyzability of the ions (Kavanaugh and Leckie, 1980).

3. ORGANIC FILM SOLVATION OF HYDROPHOBIC COMPOUNDS

The sorption of nonpolar hydrophobic compounds is viewed as a dissolution of the compound into the bulk of a polymeric natural organic phase, which is usually adsorbed onto some mineral phase. Such a hydrophobic sorption process is accordingly described as the partitioning of the solute between two solvents—one aqueous, one organic. The partition coefficient K_p is simply

$$K_p = \frac{\text{moles sorbate/mass solid}}{\text{moles solute/volume solution}} \tag{15.3}$$

K_p is empirically found to be proportional to the organic content of the solid (f_{oc}) and to the "hydrophobicity" of the compound (K_{oc}):

$$K_p = f_{oc} K_{oc} \tag{15.4}$$

The parameter K_{oc}, which measures the tendency of the compound to dissolve in the organic phase, is determined principally by the aqueous activity coefficient of the sorbate of interest. This parameter can generally be estimated from the chemical's octanol–water partition coefficient. The partition coefficient of hydrophobic compounds is thus separated into a factor adjustable from soil to soil (f_{oc}) and a compound-specific partition constant that reflects binding with respect to unit organic carbon content (K_{oc}). The organic film solvation model allows prediction of hydrophobic compound sorption solely on the basis of the carbon content of the solid and the hydrophobicity of the compound.

4. STRANGE OBSERVATIONS

The two general models we have just sketched provide a satisfying physicochemical picture of sorption, and each results in a thermodynamic description that yields a coherent and parsimonious fit of the experimental data. However, more complicated models have been proposed recently by various authors (DiToro and Horzempa, 1982; DiToro, 1985; Voice and Weber, 1985). These models, whose physicochemical basis is at best unclear, are prompted by two sets of "strange" observations—observations that appear inconsistent with any simple physicochemical mechanism. In that sense, the new "models" are merely ad hoc mathematical descriptions of experimental data; they do not provide a thermodynamic interpretation of the data.

One set of observations that do not fit our physicochemical picture of the sorption mechanism concerns the "particle concentration effect" (O'Connor and Connolly, 1980). This experimental result involves the less than proportional increase in the sorbate/solute ratio with increasing solid concentration C. It is usually expressed as a *decrease* in partition coefficient (K_p or K_{oc}) with increasing C (Fig. 15.1*a*). The mass law for the reaction of ionic solutes with surface sites predicts that the partition coefficient should be independent of C in the linear region of the isotherm (Eq. 15.1: free surface site concentration proportional to C) and should *increase* with C near saturation (Eq. 15.1: free surface site concentration increasing more than proportionally with C). In the same way, the proportional dissolution of hydrophobic compounds in the solvating soil organic phase results in a partition coefficient that is independent of C (Eq. 15.4). Consequently, the observed particle concentration effect must either involve analytical artifacts or represent a fundamental error in our physicochemical conceptualization of sorption processes.

The second set of observations at variance with thermodynamic expectations comes from desorption studies. Upon resuspension into a fresh aqueous medium, solids previously exposed to sorbing contaminants desorb less solute than expected (DiToro and Horzempa, 1982; DiToro et al., 1986). The partition coefficients measured in desorption experiments are markedly larger than those measured in adsorption* experiments. In consecutive desorption

*The prefix *ad* is taken here to represent the forward sorption reaction rather than to indicate a surface process.

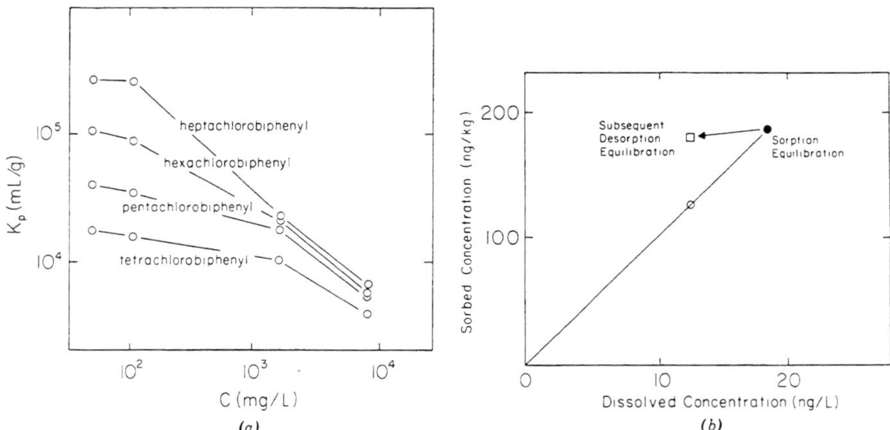

Figure 15.1. "Strange" sorption phenomena inconsistent with theoretical thermodynamic treatments. (*a*) Partition coefficients decrease with increasing suspended solids concentrations in batch experiments. Data from Wu (1986). (*b*) Sorbed concentrations observed from desorption equilibrations (square) greatly exceed expected value (open circle) based on proportional relation to dissolved concentrations derived from previous adsorption experiment (filled circle).

experiments, the sorbate concentration decreases much less than proportionally with the soluble concentration (Fig. 15.1*b*), leading to an apparent hysteresis in sorption isotherms. However, only one sorbate concentration can be at thermodynamic equilibrium with a given solute concentration regardless of the way that concentration is obtained. Again, we are forced to conclude that this apparent hysteretic behavior involves experimental difficulties or a theoretical failure!

Some type of particle–particle interaction is usually invoked to explain the effect of particle concentration on partition coefficients (DiToro, 1985). For example, an increase in the aggregation of particles at high concentrations could result in a decrease in surface area and available sorbing sites. Note, however, that such an explanation is inconsistent with a hydrophobic sorption mechanism where sorbent volume, rather than surface area, should be the operative parameter. Some irreversible binding with the solid phase is used to explain the apparent differences in adsorption and desorption isotherms (DiToro and Horzempa, 1982). The sorbate concentration is then separated into a (rapidly) desorbing fraction and an effectively irreversibly bound fraction. Such a kinetic explanation immediately precludes thermodynamic treatment.

While it remains possible that such mechanisms may explain some of the observed solid concentration effect and hysteresis, we have concluded that most of these observations are due to experimental artifacts leading to inaccurate measurements of equilibrium solute and/or sorbate concentrations. We base our conclusion on the facts that these "strange" observations all but disappear in carefully designed and executed experiments and that they can be qualitatively and quantitatively predicted by taking into account all sources of error in the experimental measurements.

5. EXPERIMENTAL ARTIFACTS

Most experimental artifacts in sorption experiments stem from the difficulty (impossibility?) in obtaining a separation of the solid and aqueous phases that corresponds to the thermodynamic definitions of these phases. Some water is always associated with the separated solid phase; some concentration of small (i.e., unfilterable or unsettleable) solids is invariably included with the aqueous phase. In addition, the concentration of material sorbed to the vessel walls or to the filter (when filtration is used) may sometimes be important. We can write an expression for *observed* K_p values that includes these experimental inadequacies and allows us to judge when such artifacts dominate truly thermodynamic effects. The observed partition constants may reflect the inclusion of dissolved materials in what we measure as sorbed, as well as sorbed loads in what we ascertain to be dissolved:

$$K_p^{\text{observed}}\ (\text{mL g}^{-1}) = \frac{\dfrac{\text{total moles of chemical with filter or pellet}}{\text{mass of solid}}}{\dfrac{\text{total moles of chemical with filtrate or supernatant}}{\text{volume of solution}}} \qquad (15.5)$$

$$K_p^{\text{observed}}\ (\text{mL g}^{-1}) = \frac{\text{particle-sorbed conc.} + (\text{dissolved conc.})(V_w)}{\text{dissolved conc.} + (\text{colloid-sorbed conc.})(M_c)} \qquad (15.6)$$

where V_w is the volume of solution retained with the filter or pellet per mass of filtered or sedimented particles (mL g^{-1}) and M_c is the mass of colloids retained with the filtrate or supernatant per volume of solution (g mL^{-1}). If we assume sorption equilibrium so that

$$K_p = \frac{\text{particle-sorbed conc.}}{\text{dissolved conc.}} \qquad (15.7)$$

and

$$K_c = \frac{\text{colloid-sorbed conc.}}{\text{dissolved conc.}} \qquad (15.8)$$

we may simplify our observed partition coefficient expression:

$$K_p^{\text{observed}} = \frac{K_p + V_w}{1 + K_c M_c} \tag{15.9}$$

Consider, for example, a series of centrifugation or filtration experiments involving various solid concentrations and compounds with a range of propensities to sorb (i.e., various K_p's). For the more water-soluble compounds, it is possible that significant loads of solutes in the water associated with the solid pellet or solid cake and the filter will lead to an overestimation of the sorbate concentration ($V_w > K_p$). Since the volume of water associated with the filter or in the pellet and on the walls of the centrifuge tube will be essentially constant in a series of batch experiments, this overestimation will be particularly important at low solid concentrations. The net result is an increase in the apparent partition coefficient at low solid concentrations (Fig. 15.2, region *a*).

For the sparingly soluble compounds, the loads of sorbed molecules associated with the small solids (colloids) that pass through the filter or that will not effectively settle will lead to an overestimation of the solute concentration. If, as might be expected, the unfilterable solid concentration covaries with the total solid concentration, then this effect will be maximized at high solid concentrations ($K_c M_c > 1$), resulting in a decrease in the apparent partition coefficient (Fig. 15.2, region *c*). Qualitatively, these experimental

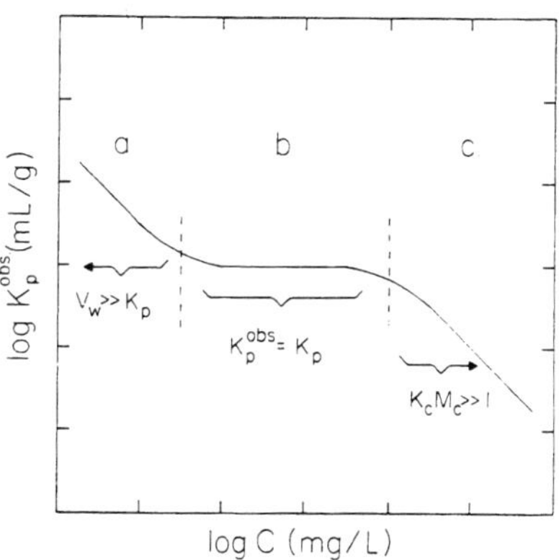

Figure 15.2. Schematized variation of experimentally determined partition coefficients when phase separations are used to isolate sorbed and dissolved states. Region *a*, reflecting the conditions of extremely low solids masses in batch equilibrations, yields thermodynamically nonmeaningful values of K_p as described by Eq. 15.17. Region *b* involves intermediate ranges of solids loads in which experimentally determined K_p equals the thermodynamically defined value. Region *c*, reflecting the conditions of extremely high solids masses in batch equilibrations and consequently corresponding to large quantities of colloids in the system, again illustrates thermodynamically nonmeaningful values of K_p as described by Eq. 15.11.

artifacts should thus lead to a decrease in apparent partition coefficients with increasing solid concentrations, as is commonly observed. Gschwend and Wu (1985) and Gschwend and coworkers (unpublished) have shown that such crossover solid and aqueous phases can, in fact, be observed and their abundances used to quantitatively explain thermodynamically inconsistent sorption results.

For the well-studied case of solutes that have a strong propensity to sorb, at high solid concentrations the apparent partition coefficient becomes independent of the nature of the solute and inversely proportional to the solid concentration:

$$K_p^{\text{observed}} = \frac{K_p}{K_c} \frac{1}{M_c} \tag{15.10}$$

If the large and small solid fractions sorb the chemicals of interest similarly, then we may expect K_p to be nearly the same as K_c. Therefore, Eq. 15.10 reduces to

$$K_p^{\text{observed}} = \frac{1}{M_c} \tag{15.11}$$

If M_c is proportional to C,

$$K_p^{\text{observed}} \propto \frac{1}{C} \tag{15.12}$$

In other words, the measured sorbate/solute ratio becomes independent of the nature of the compound and of the solid concentration:

$$\begin{aligned} \frac{\text{Moles separated with solid}}{\text{Moles in solution}} &= K_p^{\text{observed}} C \\ &= \frac{C}{M_c} \\ &\sim \text{constant} \end{aligned} \tag{15.13}$$

For hydrophobic organic compounds, the tendencies to sorb to the separable solid phase and to the colloids are very similar when normalized to the organic carbon content of each (Brownawell, 1986):

$$K_{oc}^{\text{colloids}} = K_{oc}^{\text{particles}} \tag{15.14}$$

and the concentration of macromolecular or colloidal organic carbon can be approximated by the total dissolved organic carbon. Then the measured partition coefficient reduces to the inverse of the dissolved organic carbon concentration (DOC) in units of g mL^{-1}:

$$K_{oc}^{\text{observed}} = \frac{1}{[\text{organic colloid}]} \simeq \frac{1}{\text{DOC}} \tag{15.15}$$

In this case, at high colloid concentrations the ratio of the moles of the compound retained by a filter or sedimenting in a pellet to moles remaining with the solution becomes simply the POC/DOC ratio:

$$\begin{aligned} \frac{\text{Moles separated with solid}}{\text{Moles with solution}} &= K_{oc}^{\text{observed}}\,\text{POC} \\ &= \text{POC/DOC} \end{aligned} \tag{15.16}$$

The assumptions that the measured dissolved organic carbon in natural waters may approximately all be considered as colloidal (or macromolecular) and that it posseses the same solvating properties as the separable particulate organic phase appear to be correct within a factor of 3 or so.

At low solid concentrations (but where V_w is still less than K_p), the measured partition coefficients of strongly sorbed solutes approach their true values. The converse is true for highly soluble compounds: the measured partition coefficients approach their true values at high solid concentrations when the influence of the water retained with the solid and the filter becomes small (but where $K_c M_c < 1$). At low solid concentrations, the partition coefficients of these relatively low hydrophobicity compounds measured by filtration all become equal to the ratio of the volume of water retained in the filter to the mass of solid used in the experiment:

$$K_p = V_w = \frac{\text{volume retained}}{C \cdot \text{batch volume}} \tag{15.17}$$

The measured sorbate/solute ratio then also becomes independent of the solid concentration and of the nature of the compound:

$$\begin{aligned} \frac{\text{Moles separated with solid}}{\text{Moles with solution}} &= K_p C \\ &= \frac{\text{vol. retained}}{\text{batch vol.}} \end{aligned} \tag{15.18}$$

6. REVERSIBILITY OF COLLOID AGGREGATION

At high solid concentrations, the measured partition constants of highly sorbed solutes are dominated by the presence of colloids. To understand what happens in various experimental systems or in natural waters, it is thus essential to understand what determines the concentration of colloids. It is clear from the experimental data that the colloid concentration is roughly proportional to the total solids concentration and that the separable solid phase releases "new" colloids upon resuspension in a fresh aqueous phase. In fact, experiments designed to disprove the "colloid explanation" show that colloids are released (become more concentrated) by simple concentration of the suspended phase (DiToro et al., 1986). Conversely, colloids are aggregated with the separable particulate phase (become less concentrated) upon dilution of the suspended phase. Some reversible equilibrium process thus seems to control the colloid concentration in proportion to the total solids concentration.

However, the results of the desorption experiments demonstrate that the colloid concentration does not remain perfectly proportional to the total solids. The increase in partition coefficient observed upon resuspension into a fresh aqueous phase (the desorption hysteresis, Fig. 15.1*b*) is due to a decrease in the colloid concentration while the total solids concentration remains practically unchanged. Gschwend and Wu (1985) have quantified this decrease as a function of the number of washes for a particular soil, and in this instance they observed a nearly exponential initial decrease (a factor of 2 at each wash) followed by an eventual plateauing at about 10% of the initial value (ca. 0.5% of the separable solid mass). These results can be used to predict the observed desorption hysteresis (Gschwend and Wu, 1985); the effects of an undetected colloidal fraction rather than slow kinetics thus account for this strange observation also.

7. ARTIFACTS AND REALITY

From the point of view of the physical chemist, the poor effective separation of small solids from the aqueous phase by filtration or centrifugation indeed leads to experimental artifacts. For the physical chemist, solids and solutes are defined by thermodynamic considerations. From the point of view of the geochemist interested in the cycle or fate of elements and compounds, however, transport considerations define what is considered a solid or a solute. For the geochemist, colloidal material, which possesses no appreciable settling velocity and which moves by advective and diffusive processes, is effectively in solution. One's artifact is the other's reality.

Even for a physical chemist, the distinction between particulate and dissolved fractions becomes difficult when dealing with a macromolecular component such as humic compounds or polymeric hydroxides. The notions of sorption and complexation may both cover the same reality of physical–chemical binding of small solutes. The aggregation of the macromolecular compounds with the particulate phase may be described alternatively as an adsorption or as a coagulation process.

If we are ever to incorporate our knowledge of physicochemical characteristics of chemicals into fundamentally predictive geochemical models, the important differences in the notions of particulate and dissolved phases applicable to speciation and transport issues must be explicitly recognized. The key is the colloidal fraction whose concentration is governed by the dynamics of aggregation (coagulation) and disaggregation (peptization) processes and which controls, in turn, the apparent (or effective) partition of highly sorbed solutes.

8. COAGULATION KINETICS AND SIMULATIONS OF PARTICLE-SIZE DISTRIBUTIONS

According to the classical theory of Smoluchowski (1916, 1917), the kinetics of coagulation may be described by the sum of three separate terms representing particle collisions due to Brownian motion, fluid shear, and differential settling (O'Melia, this volume, Chapter 14). The theory includes no peptization term corresponding to either thermal or shear processes; however, large particles may be eliminated from a control volume by Stokes settling.

Using numerical simulations of the Smoluchowski equations, Farley and Morel (1986) have studied the dynamics of coagulation and settling in a well-mixed water column. Regardless of initial conditions, they have observed that the particle-size distribution rapidly reaches a characteristic form corresponding to a quasi-steady state between the formation of large particles by coagulation and their elimination by settling. This characteristic size distribution is only a function of the solids concentration as is illustrated in Figure 15.3. The figure is presented as the differential of the total volume as a function of the logarithm of the particle volume to show how the total volume (and hence the particle mass) is distributed among size classes. On such a graph, a horizontal line corresponds to an equipartition of particle volume among logarithmically spaced size classes and to the commonly observed negative fourth power dependency of the differential of the particle number on the particle radius (Morel, 1983).

One observes in Figure 15.3 that the high end of the particle spectrum is gradually eliminated as the particle concentration decreases. This may be understood by considering that the upper right corner of the distribution corresponds to particles for which the loss by settling is roughly balanced by coagulation of smaller particles. As the total concentration of particles becomes smaller, the coagulation rate into that critical size class becomes slower, since it depends on the frequency of particle collisions. The balancing settling rate must thus also be slower and the corresponding particles smaller. The net result is a characteristic size distribution that shifts toward smaller particles as the solid concentration decreases.

A consequence of the decrease in size and in settling velocity of the particles that provide the bulk of the removal rate at any point in time is that the total

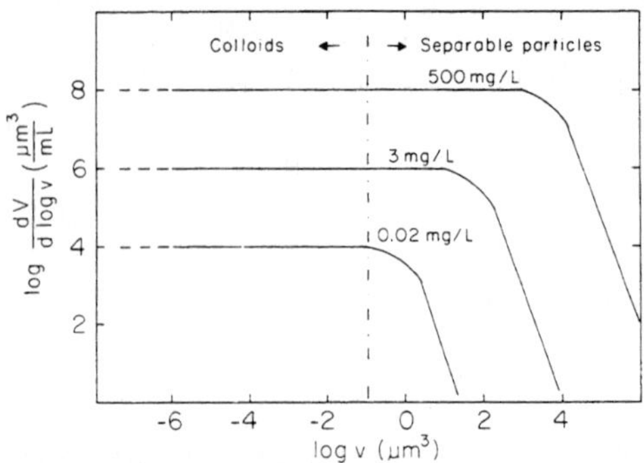

Figure 15.3. Expanded and idealized particle volume distribution curves reflecting materials ranging in size from macromolecular colloids to silts for three widely different total solids concentrations. The vertical dashed line indicates the typical operational size distribution using filtering between particulate and dissolved species.

removal kinetics are high order. (In these conditions, removal kinetics are found to be roughly second-order.) This result provides another means of observing that coagulation must be at least partly reversible. Consider a settling column containing a concentrated solid suspension and another column containing the same suspension diluted, say, by a factor of 10. If coagulation were not reversible, the fast initial settling rate observed in the first column would also be observed in the second column, since it would be controlled by the same large particles. The *experimental* observation that the initial settling rate in the second diluted column is much slower (roughly a hundred times slower) than in the first column (Fig. 15.4*a*) must then reflect the fact that the high end of the size distribution in that column corresponds to smaller particles (Fig. 15.4*b*). Upon dilution, the larger aggregates must have been broken up.

In a numerical study of coagulation kinetics, Mercier (1984) has noted that the particle size increase by coagulation is almost entirely made up of aggregations of particles of whatever size with particles of the smallest size class, never of the aggregation of two similarly sized particles. This is because of the much larger number of collisions with the smallest particles, which are the most abundant by far. The formation of large aggregates thus results from an incremental growth process in which tiny colloids are continuously added to a particle. This observation may be important as it underlines the fact that sorption of a macromolecule on a solid and particle coagulation may be a single process. By extension, the thermal reversibility of one may help explain the reversibility of the other.

9. SIZE DISTRIBUTIONS IN PEPTIZABLE SUSPENSIONS

Smoluchowski's theory of coagulation, which is the basis of the numerical studies just discussed, includes no explicit peptization mechanism correspond-

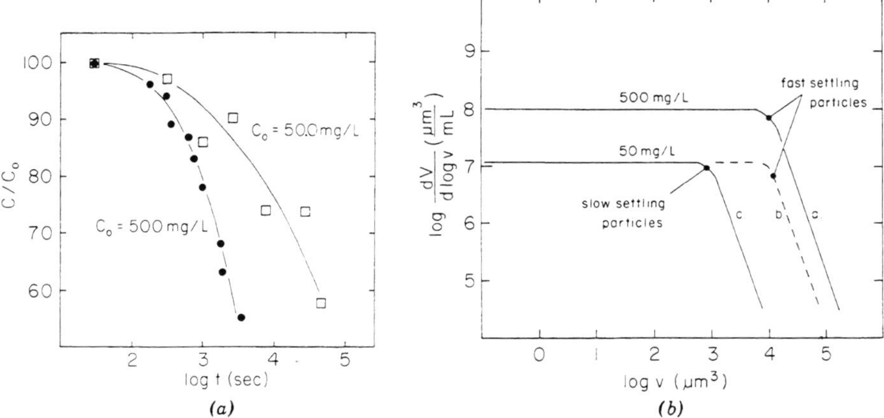

Figure 15.4. Evidence for the disaggregation or peptization process. (*a*) Data from Faisst (1976) showing the disproportionately diminished removal rate of diluted particles. (*b*) Particle size distribution for 500 mg L^{-1} suspension (curve *a*) after it is diluted 10× (curve *b*). Curve *c* indicates distribution corresponding to the low sedimentation rate experimentally observed after dilution yielding evidence that disaggregation must have occurred.

ing to either thermal or shear processes. If we take the indirect evidence from sorption and settling studies as convincing, however, such a reverse coagulation mechanism must be operative. The question then is what the particle size spectrum might be in a system where both coagulation and peptization are occurring. In the absence of known functionalities for the peptization terms, we cannot answer that question exactly; however, we can perhaps extend the results obtained for settling suspensions to describe *grosso modo* what particle size spectra might be like in peptizing systems. In so doing, it is worth remembering that the results are described in wide-scale log–log graphs representing quasi-steady-state conditions that are very insensitive to the exact functionalities of the dynamic equations.

It seems reasonable to expect that a quasi-steady state representing an approximate balance between coagulation and peptization will be achieved in any particle suspension. Coagulation is indeed rapid, and, according to kinetic theory, it is the faster of the forward or backward reaction that determines the rate of achievement of equilibrium or steady state (Morel, 1983). Furthermore, because the energy of eddies increases with their size, the rate of mass disaggregation by shear almost certainly increases with particle volume. Thus in a mixed system we expect the high end of the size spectrum to decrease rapidly as in a settling system. (Stokes settling eliminates mass as the $\frac{5}{3}$ power of the particle volume.) Thus we expect the same general shape of the particle spectra in both systems regardless of the exact functionality of the peptization process. The position of the upper right corner of the size distribution may be different, but the general features of the $dV/(d \log v)$ versus $\log v$ graphs must be similar (see Fig. 15.3), as must be the relative positions of the graphs corresponding to different total solid concentrations.

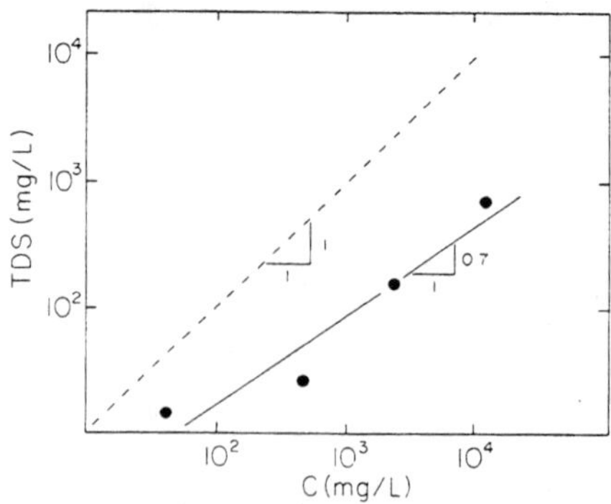

Figure 15.5. The observed covariation of colloid concentration (measured as total dissolved solids, TDS) with solids concentration C. Data from Gschwend and Wu (1986).

If we assume that the characteristic size spectrum is roughly flat on a graph of $dV/(d \log v)$ versus $\log v$, we can estimate the order of magnitude of the particle/colloid concentration ratio. For example, with an upper particle size of $\sim 10^5$ μm^3 for a 100-mg L^{-1} solid suspension, a lower limit of $\sim 10^{-7}$ μm^3 (i.e., a molecular weight of 10^5–10^6), and an effective separation of ~ 0.1 μm^3, one would expect equal masses (volumes) of filterable and nonfilterable solids. This result is very insensitive to the size limits but very sensitive to the slope of the size distribution on the logarithmic graph (Fig. 15.3). The fact that the experimental results with various suspensions of natural sediments cluster on a suspended solids to dissolved solids ratio of 10 (Fig. 15.5), only a factor of 10 off our most naive expectation, indicates that the actual size spectrum is indeed very nearly flat, at least near the colloid–solid separation line.

Due to the increase in the high end of the distribution with increasing solids concentration, we also expect that the colloid concentration should increase less than linearly with C. At higher total solids concentrations, the porportion of larger, separable solids should be larger and the proportion of nonseparable solids conversely smaller. In systems sedimenting under the influence of gravity, this effect is of the order of 20% (0.9 log unit increase in colloidal solids per log unit increase in total solids). Using centrifugation, Gschwend and Wu (Fig. 15.5) observed a somewhat weaker increase in nonsettling solids concentration with increasing C (0.7 log unit per log unit). More generally, the DOC versus POC correlation in natural waters appears to follow a similar relationship, with the colloidal abundance increase as a function of total particle concentration also less than 1:1. For example, in lakes, streams, and seawater, where the total particle loads are small, the DOC/POC ratio is nearly always around 10:1 (Wangersky, 1965; Wetzel, 1975). Conversely, recent evidence (Krom and Sholkovitz, 1977; Brownawell and Farrington, 1986; Brownawell, 1986) suggests that the very substantial colloidal organic matter load suspended in sediment beds (10–100 mg L^{-1}), where the total particle load is obviously extremely high, is only about 1/1000 of the sedimentary organic fraction. Interestingly, the laboratory experiments of Voice et al. (1983) and Gschwend and Wu (1985) produced intermediate DOC/TOC ratios for intermediate total sediment conditions.

10. APPARENT PARTITION COEFFICIENTS IN NATURAL WATERS

Much of the initial interest in the issue of the solids concentration effect came from regression studies pulling together all available data on the partitioning of hydrophobic compounds in natural waters (O'Connor and Connolly, 1980; DiToro, 1985). Newer studies have extended the results to metal ions (HydroQual, 1984). We can thus qualitatively and quantitatively compare these experimental results on apparent partition coefficients with our view of the expected effects of aquatic colloids. As seen in Figure 15.6, the most striking result of the experimental data taken in their totality is an almost 1 to 1 decrease in the partition coefficients with increasing solids concentration when we consider the high solids concentration ends of the plots. It does not seem to matter which metal or organic sorbate we consider (shown for cobalt and DDT in the figure). Just as striking is the absence of correlation of metal partitioning with other dominant chemical parameters such as pH or alkalinity. These are precisely the results one would expect if the solute partitioning

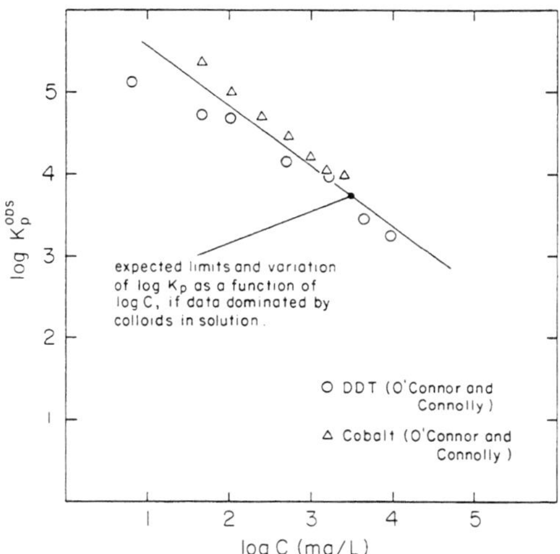

Figure 15.6. Comparison of experimentally obtained partition coefficients at high solids concentrations for metals and organic compounds with results predicted if such observations reduce to limiting K_p values given by Eq. 15.11 and colloids vary with total solids as indicated by Figure 15.5. Similar results have been obtained for a host of hydrophobic organic compounds and trace metals (see text for references).

were controlled by the solid/colloid ratio and had nothing to do with the physicochemical properties of the sorbates, sorbents, and solutions!

Consequently, in cases where the observed partition coefficient should be given directly by the inverse colloid concentration, we can use the empirical relationship shown in Figure 15.5 for colloid abundance to predict these experimental distribution constants. Indeed, we see that the apparent partition coefficient for metals is commonly $\sim 10^4$ mL g^{-1} at a total solids concentration ~ 1000 mg L^{-1}. If K_p is given by 1/(colloid) under these high solids concentration conditions, in light of Figure 15.5 we would expect colloids at ~ 100 mg L^{-1} and therefore $K_p = 10^{-2}$ L $mg^{-1} = 10^4$ mL g^{-1} as observed. A similar result is obtained for hydrophobic organic compounds.

11. CONCLUSION

Over the past several years, aquatic chemists have developed sound and practical models for describing quantitatively the equilibrium sorption of solutes on aquatic particles. Yet these models do not give us a prediction of the effective solid–solution partitioning in natural waters because they do not take into account the colloidal phase that is invariably present. Such a phase has the physicochemical properties of a solid but is analyzed and transported with the soluble fraction. In order to utilize our understanding of sorption processes for predicting the fate of natural and anthropogenic chemicals, we need an additional model, one that describes the concentration of colloids in the system.

Present models of the dynamics of particle suspensions consider only aggregation processes. There is mounting indirect evidence for the importance of disaggregation processes, but the corresponding theoretical kinetic expressions are difficult to establish. Experimentally, our ability to observe and measure directly the concentration of colloids and associated substances is limited by the practical size cutoffs associated with all separation techniques and by the artifacts introduced by the separation step itself.

New nonintrusive optical techniques for counting and sizing colloidal particles are being developed. They offer some hope that we may soon be able to account, at least empirically, for the role of small particles in the partitioning of chemicals, both in beakers and in natural waters. On the theoretical side, it may be that a complete kinetic description of particle dynamics is too difficult and unnecessary. Aggregation and disaggregation processes appear to be relatively fast. A quasi-equilibrium (steady-state) description of the relationship between the concentration of (small) colloids and that of (large) particles as a function of system parameters may then be possible. Such a relationship would provide the link that would allow us to translate our physicochemical models of sorption into an effective solid–solution partitioning and hence into a prediction of transport in natural waters.

REFERENCES

Bowden, J. W., Posner, A. M., and Quirk, J. P. (1977), Ionic adsorption on variable charge mineral surfaces. Theoretical-charge development and titration curves, *Aust. J. Soil Res.* **15,** 121.

Brownawell, B. J. (1986), "The Role of Colloidal Organic Matter in the Marine Geochemistry of PCBs," Ph.D. thesis, MIT-WHOI joint program.

Brownawell, B. J., and Farrington, J. W. (1986), Biogeochemistry of PCBs in interstitial waters of a coastal marine sediment, *Geochim. Cosmochim. Acta* **50,** 157.

Chiou, C. T., Peters, L. J., and Freed, V. H. (1979), A physical concept of soil–water equilibria for nonionic organic compounds, *Science* **206,** 831.

Davis, J. A., James, R. O., and Leckie, J. O. (1978), Surface ionization and complexation at the oxide/water interface. 1. Computation of electrical double layer properties in simple electrolytes, *J. Colloid Interface Sci.* **63,** 480.

DiToro, D. M., and Horzempa, L. M. (1982), Reversible and resistant components of PCB adsorption–desorption: Isotherms, *Environ. Sci.* **16,** 594.

DiToro, D. M. (1985), A particle interaction model of reversible organic chemical sorption. *Chemosphere,* **14,** 1503–1538.

DiToro, D. M., Mahony, J. D., Kirchgraber, P. R., O'Bryne, A. L., Pasquale, L. R., and Piccirilli, D. C. (1986), Effects of nonreversibility, particle concentration, and ionic strength on heavy metal sorption, *Environ. Sci. Technol.* **20,** 55.

Faisst, W. K. (1976), Digested sewage sludge: Characterization of a residual and modeling for its disposal in the ocean off Southern California. EQL Report No. 13, CALTECH, Pasadena, CA.

Farley, K. J., Dzombak, D. A., and Morel, F. M. M. (1985), A surface precipitation model for the sorption of cations on metal oxides, *J. Colloid Interface Sci.* **106,** 226.

Farley, K. J., and Morel, F. M. M. (1986), Role of coagulation in the kinetics of sedimentation, *Environ. Sci. Technol.* **20,** 187.

Gschwend, P. M., and Wu, S.-C. (1985), On the constancy of sediment–water partition coefficients of hydrophobic organic pollutants, *Environ. Sci. Technol.* **19,** 90.

HydroQual (1984), "Water–Sediment Partition Coefficients for Priority Metals," attachment I to *Technical Guidance Manual for Performing Waste Load Allocations* by R. C. G. Delos, W. L. Richardson, J. V. DePinto, R. B. Ambrose, P. W. Rodgers, K. Rygwelski, J. P. St. John, W. J. Shaughnessy, T. A. Faha, and W. M. Christie, Office of Water Regulations and Standards, U.S. Environmental Protection Agency, Washington, DC.

Karickhoff, S. W., Brown, D. S., and Scott, T. A. (1979), Sorption of hydrophobic pollutants on natural sediments, *Water Res.* **13,** 241.

Kavanaugh, M. C., and Leckie, J. O. (1980), *Particulates in Water: Characterization, Fate, Effects, and Removal,* ACS Adv. Chem. Ser. No. 189, American Chemical Society, Washington, DC.

Krom, M. D., and Sholkovitz, E. R. (1977), Nature and reactions of dissolved organic matter in the interstitial waters of marine sediments, *Geochim. Cosmochim. Acta* **41,** 1565.

Li, Y.-H., Burkhardt, L., Buchholtz, M., O'Hara, P., and Santschi, P. H. (1984), *Geochim. Cosmochim. Acta* **48,** 2011.

Means, J. C., Wood, S. G., Hassett, J. J., and Banwart, W. L. (1980), Sorption of polynuclear aromatic hydrocarbons by sediments and soils, *Environ. Sci. Technol.* **14,** 1524.

Mercier, R. S. (1984), "The Reactive Transport of Suspended Particles: Mechanisms and Modeling," Ph.D. thesis, MIT–Woods Hole Joint Program.

Morel, F. M. M. (1983), *Principles of Aquatic Chemistry,* Wiley-Interscience, New York.

O'Connor, D. J., and Conolly, J. P. (1980), The effect of concentration of adsorbing solids on the partition coefficient, *Water Res.* **14,** 1517.

O'Melia, C. R. (1987), "Particle–Particle Interactions," this book, Chapter 14.

Schindler, P., Walk, E., and Furst, B. (1976), The role of surface hydroxyl groups in the surface chemistry of metal oxides, *Chimia,* **30,** 107–109.

Smoluchowski, M. (1916), *Z. Phys. Chem. (Munich)* **17,** 537.

Smoluchowski, M. (1917), *Z. Phys. Chem. (Munich)* **92,** 129.

Stumm, W., Hohl, H., and Dalang, F. (1976), Interaction of metal ions with hydrous oxide surfaces, *Croat. Chem. Acta* **48,** 491.

Voice, T. C., Rice, C. P., and Weber, W. J., Jr. (1983), Effect of solids concentration in the sorptive partitioning of hydrophobic pollutants in aquatic systems, *Environ. Sci. Technol.* **17,** 513.

Voice, T. C., and Weber, W. J., Jr. (1985), Sorbent concentration effects in liquid/solid partitioning, *Environ. Sci. Technol.* **19,** 789.

Westall, J. C. (1987), "Adsorption Mechanisms in Aquatic Surface Chemistry," this book, Chapter 1.

Wetzel, R. G. (1975), *Limnology,* Saunders, Philadelphia, PA, p. 540.

Wangersky, P. J. (1965), The organic chemistry of sea water, *Amer. Sci.* **53,** 358.

Wu, S.-C. (1986), "The Transport of Hydrophobic Organic Compounds Between Water and Sediment," Ph.D. thesis. Department of Civil Engineering, M.I.T., Cambridge, MA.

16

ABIOTIC TRANSFORMATION OF ORGANIC CHEMICALS AT THE PARTICLE–WATER INTERFACE

Richard G. Zepp and N. Lee Wolfe

Environmental Research Laboratory, U.S. Environmental Protection Agency, Athens, Georgia

It is only after you have come to know the surface of things that you can venture to seek what is underneath. But the surface of things is inexhaustible.

ITALO CALVINO, *Palomar*, 1983

Abstract

Heterogeneous reactions are important contributors to the transformation of organic chemicals in the environment and are increasingly being used in the treatment of polluted water. Kinetic models are presented here that describe selected environmentally relevant thermal and photochemical reactions in particle–water systems, including reactions in approximately equilibrated heterogeneous systems and in systems involving reactive site saturation. The effects of slow intersite diffusion on heterogeneous reactions are discussed, as are recent models that describe the kinetics of such processes.

To illustrate the application of the kinetic models, examples are provided of kinetic treatments of thermal and photochemical reactions involving organic chemicals in aqueous suspensions of sediments, metal oxides, or algae. Thermal reactions that are discussed include the hydrolysis of sorbed organic chemicals and the reduction of organic compounds in anaerobic sediments. Kinetic relations that govern photoreactions in particle suspensions are discussed, and illustrative examples are provided of direct photoreactions of surface complexes in metal oxide suspensions, semiconductor mechanisms for photochemical redox processes, and photosensitized reactions on surfaces of algae, sediments, and soil surfaces.

1. INTRODUCTION

During the past decade, studies of heterogeneous chemistry involving various types of particles in water have burgeoned. This interest is reflected in the American Chemical Society's publication of the journal *Langmuir,* which is entirely devoted to surface chemistry. A major impetus for environmentally related research in this area has been the need for reliable estimates of the lifetimes of pollutants in surface water and groundwater. Interest in the kinetics of metal oxide dissolution as related to fields of aquatic geochemistry and chemical oceanography and limnology has also prompted a number of recent studies dealing with the chemistry of organic chemicals adsorbed on oxides in water. In atmospheric research, concern over toxic effects of precipitation has led to investigations of heterogeneous reactions in cloud droplets and aqueous aerosols. In addition, the energy crisis of the mid-1970s stimulated research on the development of heterogeneous photocatalytic processes that can be used to capture solar energy. Such photocatalytic processes also have been applied to oxidize pollutants in wastewaters.

In this chapter we discuss only a portion of the research in this active area. Primary emphasis is on the kinetics of abiotic transformations of organic chemicals in particle–water systems that are relevant to surface waters. Our considerations are further limited to conditions in which reactant concentrations are typical of those usually encountered in aquatic environments, that is, less than 10 μM and/or below the aqueous solubility limit.

The predominant sorbents of organic solutes in freshwater systems are sediments. Sediments are mixtures of inorganic minerals that are usually associated with organic and inorganic polyelectrolytes. A substantial part of the sediments in freshwaters is derived from soil runoff. Thus, the abiotic transformations of organic chemicals in sediment–water and soil–water mixtures receive a major emphasis. In addition, photoreactions involving algae and certain metal oxides are discussed, because such particles are important components of the suspended matter in lakes.

Heterogeneous chemical reactions in aqueous systems involve the partitioning of the substrate between water and one or more other phases. Estimation of the extent of partitioning of micropollutants to nonaqueous phases such as sediments, biota, colloidal organic substances, and surface films has been and continues to be the subject of much current research. Here we refer to the partitioning of a solute from water onto an aquatic particle as *sorption*. Sorption includes partitioning to particle interiors as well as to the surface itself.

Our discussion of heterogeneous reactions is organized under three topical areas. First, we consider the kinetic models and equations that describe thermal and photochemical reactions on aquatic particulates. Then we present a detailed examination of two types of organic heterogeneous reactions that are significant in aquatic environments: hydrolysis and reduction. Finally, we discuss the kinetics and types of organic heterogeneous photoreactions of environmental interest, covering both direct and indirect photoprocesses.

2. KINETIC MODELS AND EQUATIONS

Our discussion begins with the simplest case, a description of multiphase reaction kinetics in equilibrated systems. Then we examine more complex models that apply to heterogeneous reactions in nonequilibrated systems. For purpose of illustration, only the simplest models that apply to this vast research area are discussed here. Useful discussions of heterogeneous catalysis can be found in a monograph by Clark (1970). Laidler and Bunting's (1973) treatments of enzyme kinetics also provide a relevant extension to our discussions.

2.1. Heterogeneous Kinetics in Equilibrated Systems

A reactive organic chemical, or *reactant,* in an aqueous suspension of natural particles is likely to be distributed into several different phases each of which influences the reaction kinetics. Here we assume that the movement of reactant from one phase to another is much more rapid than chemical reaction and that the system is approximately at equilibrium. More detailed discussions of equilibrium sorption are provided in Westall's chapter on adsorption mechanisms (this volume, Chapter 1) and in a recent paper by Karickhoff (1984).

For illustration, consider a model system in which a chemical is distributed among three rapidly equilibrated phases, one of which is water and the other two of which are components of a natural particle, two particles with differing reactivities, etc. The chemical reacts in water and in one of the two particle phases. Reaction rates are assumed to be first-order with respect to reactant concentration.

$$\begin{array}{ccccc} W & \underset{k_{-1}}{\overset{k_1}{\rightleftharpoons}} & R & \underset{k_U}{\overset{k_U}{\rightleftharpoons}} & U \\ {\scriptstyle k_W}\downarrow & & {\scriptstyle k_S}\downarrow & & \\ \text{Products} & & \text{Products} & & \end{array} \tag{16.1}$$

In Eq. 16.1, W represents aqueous phase reactant, R denotes reactant sorbed in a phase where reaction occurs, and U denotes reactant sorbed in an unreactive phase. The first-order rate constants, k_W and k_S, correspond to reaction in aqueous and sorbed states, respectively. It is further assumed in this case that no interactions occur among sorbed reactant molecules and that an infinite number of reactive sites are available on the particle. In a subsequent section, the effect of site saturation will be considered.

The fraction of reactant residing in a given phase j at equilibrium is

$$F_j^e = \frac{\rho_j K_j}{\sum_i \rho_i K_i} \tag{16.2}$$

where K_i are partition coefficients that equal the ratio of concentrations in phase i to the concentration in some reference phase and ρ_i are the relative amounts of the two phases. If water is selected as the reference phase, then ρ_W and K_W for the reactant in aqueous phase W are unity and

$$\sum_i \rho_i K_i = 1 + \rho_R K_R + \rho_U K_U$$

The concentration C_j (expressed as amount per system volume) in a given phase j at equilibrium equals $F_j^e C_T$, where C_T is the total concentration. It follows that the rate of chemical reaction, $-dC_T/dt$, denoted by v, is expressed by

$$v = C_T(F_W^e k_W + F_R^e k_R) \tag{16.3}$$

Note that the term in the rate expression that is attributable to reaction in a given phase is "retarded" by the factor F_j^e. If the rate constant for reaction on the particle is less than that for reaction in aqueous solution, that is, if $k_R < k_W$, the net effect of sorption onto natural particles is a reduction in the chemical reaction rate compared to that in pure water.

Through generalization of this model to a system that includes a total of i equilibrated phases with varying reactivities, it can be shown that v is described by

$$v = C_T \sum_i F_i^e k_i \tag{16.4}$$

where k_i are first-order rate constants for reaction in phases i. Under the equilibrium assumption, the observed disappearance rate in any phase must equal the total disappearance rate.

The neutral hydrolysis of the organophosphorothioate insecticide chlorpyrifos provides a good example of an organic transformation in an equilibrated sediment–water system (Fig. 16.1) (Macalady and Wolfe, 1985). The kinetic data in both the aqueous phase and sediment phase are described by first-order rate expressions. Computer analysis of the concentration–time data in both phases indicated that sorption and desorption of chlorpyrifos are

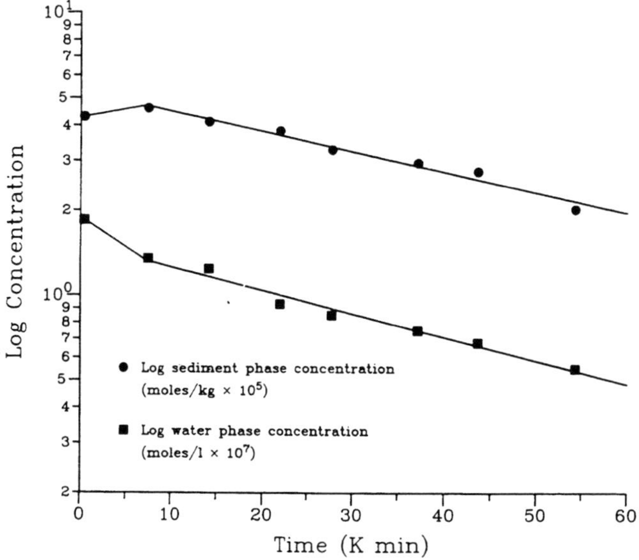

Figure 16.1. Chlorpyrifos hydrolysis in aqueous suspension of sediments (Macalady and Wolfe, 1985).

considerably more rapid than hydrolysis. The constancy of the ratio of sorbed to dissolved reactant in this system is another indication that the two phases are equilibrated during the reaction, although such a constant ratio could also result even if the system was not equilibrated. (See following discussion of steady-state model.)

2.2. Kinetic Models involving Site Saturation

The preceding section described equilibrium models that assumed unlimited reaction sites on particles. Many studies of heterogeneous reactions, however, have shown that site limitation must be taken into account in any general description of reaction kinetics. Langmuir's pioneering studies (Gaines et al., 1983) in the case of heterogeneous catalysis of gas-phase reactions were the first to develop the concept of site saturation.

A number of kinetic situations can develop that require consideration of site saturation. Here, for illustration, we consider the simplest case:

$$\mathrm{W} \underset{k_{-1}}{\overset{k_1}{\rightleftharpoons}} \mathrm{R} \xrightarrow{k_S} \text{products} \tag{16.5}$$

In this case phase R is assumed to have a limited number of reaction sites and the products are assumed to rapidly leave the surface. If θ denotes the fraction of reactive sites that are occupied, ρ is the particle concentration and R_t denotes the total concentration of sites, then the rate equations that describe the kinetic model are

$$v = k_S \rho C_R \tag{16.6}$$

and

$$\frac{dC_R}{dt} = k_1(1 - \theta)R_t C_W - k_{-1}C_R - k_S C_R \tag{16.7}$$

As reaction proceeds, the concentration of sorbed reactant reaches a steady state and dC_R/dt becomes nearly zero. Under these conditions,

$$C_R = \frac{k_1(1 - \theta)R_t C_W}{k_{-1} + k_S} \tag{16.8}$$

By substitution of $\theta = C_R/R_t$ and Eq. 16.8 into Eq. 16.6, it can be seen that

$$v = \frac{\rho k_1 k_S R_t C_W}{k_{-1} + k_S + k_1 C_W} \tag{16.9}$$

Both algal-induced photoreaction of aniline (Fig. 16.2) and the photodissolution of the iron oxide lepidocrocite (Fig. 16.3) are examples of het-

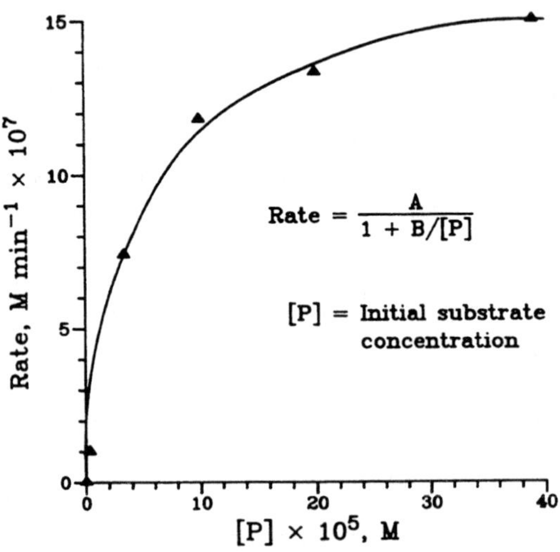

Figure 16.2. Rate versus substrate concentration in the sunlight-induced oxidation of aniline in suspension of *Chlamydomonas* sp. (Zepp and Schlotzhaüer, 1983).

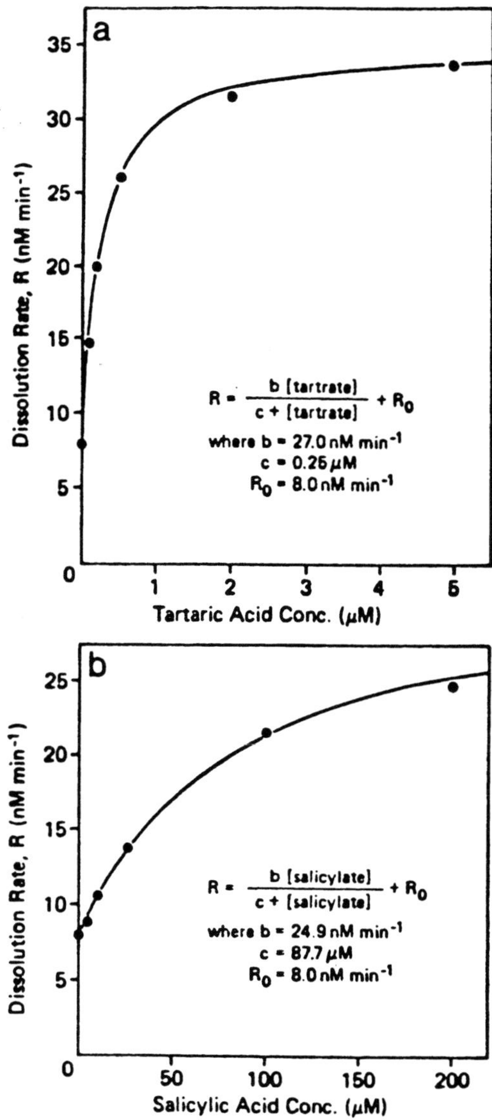

Figure 16.3. Dependence of the dissolution rate of 5 μM γ-FeOOH (pH 4.0, 0.01 M NaCl) on concentration of tartaric and salicylic acid. Used with permission of Waite (1986).

erogeneous reactions that exhibit the hyperbolic relationship described by Eq. 16.9. These reactions are discussed later in more detail.

Equation 16.9 simplifies under two conditions. When reactant concentration is sufficiently large that k_1C_W is much larger than $k_{-1} + k_S$, the rate becomes zero-order with respect to reactant concentration. Under these conditions, all the reactive sites are "saturated." The maximum rate is proportional to the rate constant for reaction of substrate sorbed on reactive sites as well as to the total concentration of such sites. The latter is proportional to the particle/water ratio ρ, as well as to the site density. Smaller particles, with their higher surface area to mass ratio, are expected to exhibit a higher R_t/ρ ratio than larger particles.

On the other hand, with very low reactant concentrations, such that $k_{-1} + k_S$ is much larger than k_1C_W and $1 - \theta \simeq 1$, the usual case in aquatic environments, the rate becomes first-order with respect to C_W.

$$v = \frac{k_1}{k_{-1} + k_S} \rho k_S R_t C_W \tag{16.10}$$

Also, under these assumed steady-state conditions, C_W, and thus the rate, is proportional to C_T, the total reactant concentration.

Under the special condition that equilibrium sorption exists, that is, when k_{-1} is much greater than k_S, the first-order rate constant becomes equal to Kk_SR_t, where $K = k_1/k_{-1}$. This K is sometimes referred to as the equilibrium constant of sorption, but it differs from the partition coefficients discussed in Section 2.1. It can be shown that when reactant is very dilute, the fraction sorbed is $\theta = KC_W$. Because $\theta R_t = C_R$, it follows that $KR_t = K_R$, the partition coefficient.

That the concentration dependence of a reaction rate in an aqueous particle suspension is described by a hyperbolic expression does not necessarily mean that the reaction occurs on a surface. Another kinetic model that does not involve a heterogeneous reaction can also be described by a hyperbolic rate expression that is equivalent to Eq. 16.9. Consider a particle–water system in which some reactive species X undergoes reversible exchange between the solution phase and the particle surface. This species reacts in the solution phase with some reactant M that is added to the system.

$$\text{S--X} \underset{k_a}{\overset{k_d}{\rightleftharpoons}} \text{X} \tag{16.11}$$

$$\text{X} + \text{M} \xrightarrow{k_X} \text{products} \tag{16.12}$$

If X is highly reactive, its concentration should reach a steady state shortly after reactant is added, a condition that is defined by

$$C_X^{ss} = \frac{k_d C_{S\text{--}X}}{k_a + k_X C_M} \tag{16.13}$$

where $C_{S\text{--}X}$ equals the concentration of sorbed reactant and C_M equals the concentration of M. For simplification we assume that $C_{S\text{--}X}$ is much larger than C_M and thus approximately constant throughout the reaction. The reaction rate is defined by

$$v = k_X C_M C_X^{ss} \tag{16.14}$$

where k_X is the bimolecular rate constant for the reaction. By substitution of Eq. 16.13 into Eq. 16.14, we obtain

$$v = \frac{k_d k_X C_{S\text{--}X} C_M}{k_a + k_X C_M} \tag{16.15}$$

which is of the same general form as Eq. 16.9. A possible example of such a reaction is discussed in Section 3.2.

In the above considerations, the steady-state approximation has been applied to define relationships that describe the reaction rate. This approximation is valid when concentrations of reaction intermediates, such as R in Eq. 16.5, remain approximately constant during reaction. In some systems, however, this may not be a valid assumption, and non-steady-state kinetics apply. The exact solutions for a non-steady-state treatment of a variety of kinetic models have been derived (Moore and Pearson, 1981), but the equations are cumbersome and generally of little use. A more profitable approach involves the use of computer programs to analyze the kinetic data and calculate values for the rate constants that appear in the differential equations.

We see from this discussion that the saturation effect must always be assumed to occur in heterogeneous reactions, especially in those cases in which site-specific sorption is expected, for example, partitioning that involves chemisorption or ion exchange. The concentration of reactant required for saturation depends on properties of both the reactant and the particle, and usually it is difficult to predict a priori.

The situation is complicated even further in systems involving reactions in natural sediment–water suspensions, because sediments are mixtures of competing sorbents. Saturation of reactive sites may occur at reactant concentrations that are much lower than those required to saturate sorption of the reactant on a sediment particle. Thus, an important first step in any study of chemical reactions in particle–water systems is the determination of the dependence of the initial rate on reactant concentration.

2.3. Heterogeneous Reactions involving Slow Interphase Transport

Two general diffusional processes are important in heterogeneous reactions. The first involves movement of the reactant molecules to and from the particle surface, and the second involves mass transport into and back out of particle interiors. If reaction is faster than these processes, then the equilibrium considerations discussed earlier, especially in Section 2.1, do not apply. In this section we mainly focus on kinetic models that have been used to analyze environmentally significant heterogeneous reactions in which intraparticle or interparticle transport is slow compared to reaction.

Diffusion to the particle surface becomes rate-limiting when reaction on the particle is much more rapid than transport to and from the particle surface. This type of transport limitation has been considered in detail in various engineering publications (Satterfield and Sherwood, 1963). Pasciak and Gavis' (1975) treatment of transport limitation to nutrient uptake in lake microbiota provides other useful discussions and illustrations of this process.

The great industrial importance of heterogeneous catalysis has prompted numerous studies on the effects of pore diffusion and particle geometry on rates of catalytic reactions. The previously mentioned text by Clark (1970) provides an introductory discussion and major references to this research area.

Considerably less attention has been paid to diffusion effects on processes occurring on natural particles. In the remainder of this section we discuss the application of two models, the *two-component* and *dispersed kinetics* models, to the analysis of kinetics of processes in heterogeneous systems.

Studies by Karickhoff and coworkers (1980, 1984, 1985) of sorption processes involving sediments have provided much useful information concerning sorption kinetics. Using the gas purging of hydrophobic chemicals from sediment suspensions to examine desorption kinetics, Karickhoff and Morris (1985) found that desorption can be factored into at least two components, a rapid release of "labile" sorbed chemical and a slower release of "nonlabile"

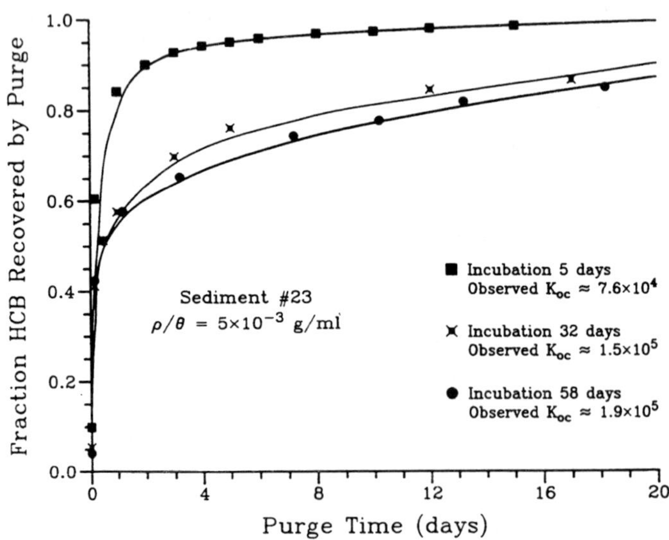

Figure 16.4. Purged volatilization of hexachlorobenzene from sediment suspension in water (Karickhoff, 1984).

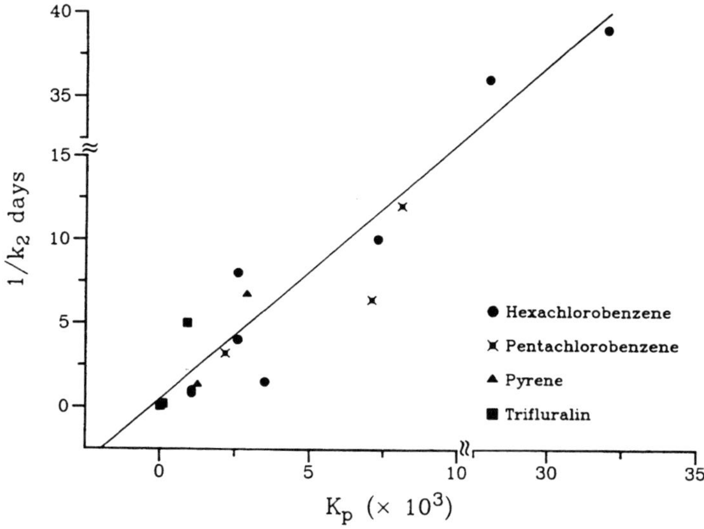

Figure 16.5. Dependence of nonlabile release constant k_2 on sorption equilibrium constant K_p (Karickhoff and Morris, 1985).

sorbed chemical. The purged release of hexachlorobenzene (HCB) provides a good example (Fig. 16.4). Karickhoff (1980) conceptualized the system using a model similar to Eq. 16.1, assuming that loss of chemical was due solely to purging from the aqueous phase and that one of the particle phases, the "nonlabile" phase, was accessed from water much more slowly than the other. He found that sorption to and desorption from the "nonlabile" phase of the particles could be described by a first-order rate expression with rate constant k_2. Based on detailed kinetic studies of four chemicals in equilibrated suspensions of several different sediments (Karickhoff and Morris, 1985), it was found that the following relation approximately described k_2 (Fig. 16.5):

$$\frac{1}{k_2}\,(\text{hr}) = 0.03\ K_p(\text{L kg}^{-1}) \tag{16.16}$$

where K_p is the equilibrium partition coefficient corresponding to the ratio of concentrations of sorbed to dissolved chemical in the suspensions. The process quantified by k_2 involves sorption to or desorption from sediment components, probably particle interiors, that do not have direct access to water. Indeed, Wu and Gschwend (1986) have found that sorption kinetics involving natural sediments can be interpreted assuming diffusion of sorbate into pores of spherical particles.

Photolysis of the hydrophobic aromatic hydrocarbon 1,1-bis(*p*-chlorophenyl)-2,2-dichloroethylene (DDE) in sediment suspensions (Zepp and Schlotzhauer, 1981) also involved diffusive rate limitations. Plots of log con-

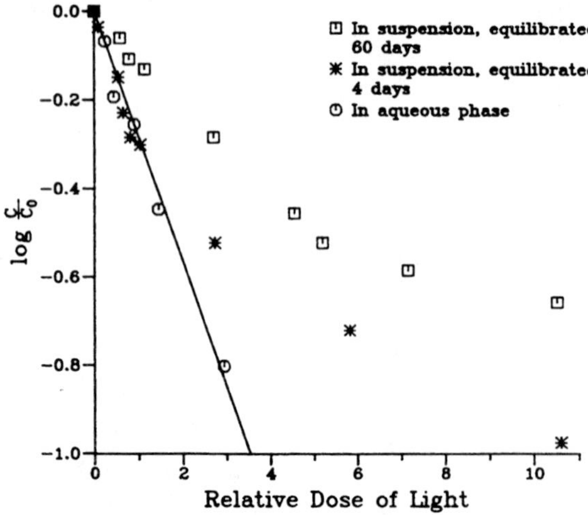

Figure 16.6. Photoreaction of 2,2-dichloro-1,1-bis(*p*-chlorophenyl)ethylene (DDE) in aqueous suspension of Ohio River sediments and in supernatant of centrifuged sediment suspension (Zepp and Schlotzhauer, 1981).

centration versus time were linear for DDE photolysis in water, but in the sediment suspensions the plots, like those for the purge of HCB (Fig. 16.4), were nonlinear (Fig. 16.6). The kinetic model shown in Eq. 16.1 was used to interpret the results, with W representing aqueous phase DDE, and R and U representing DDE sorbed in reactive and unreactive phases, respectively. The results indicated that U was accessed much more slowly than R, thus accounting for the dependence of the photolysis kinetics on equilibration time prior to irradiation of the suspension. The aqueous and reactive phases were assumed to be in a rapid equilibrium that is described by the constant $F_R K_p$, where F_R is the fraction in phase R.

Comparisons of the kinetic analysis of the HCB purge with that of the DDE photolysis strongly suggest that the "nonlabile" and "unreactive" phases were approximately the same. Values of k_U evaluated from the DDE experiments (Table 16.1) (Zepp and Schlotzhauer, 1981) were the same as values of k_2 for HCB (Karickhoff and Morris, 1985) within a factor of 2, as was predicted by assuming that k_U can be computed using Eq. 16.16. The lack of photoreactivity for DDE sorbed in the unreactive phase could simply reflect a lack of light penetration into particle interiors or, alternatively, highly efficient excited state quenching. That quenching could account for the results is supported by studies of DellaGuardia and Thomas (1983), who found that photoprocesses of a ruthenium complex were strongly quenched on various clay minerals. Kinetic analysis of the DDE data also indicated that the photolysis rate constant of DDE sorbed in phase R was 3 times as large as the rate constant in water (Table 16.1).

The nonlinear behavior exhibited in Figures 16.4 and 16.6 is by no means

Table 16.1. Kinetic Parameters for the Photolysis (313 nm) of DDE Sorbed on Clay-Sized Particles of Soil and Sediment

Parameter[a]	Ohio Soil	Ohio River Sediment	Mississippi River Sediment
F_R^e	0.56	0.62	0.43
k_U, hr^{-1}	5.8×10^{-3}	4.7×10^{-3}	3.7×10^{-3}
k_S/k_W	3.2	—[b]	3.1

[a]See Eqs. 16.1–16.4 for definitions.
[b]Not determined.
Source: Zepp and Schlotzhauer (1981).

limited to processes involving natural sediments. Such behavior has been noted by De Mayo and coworkers (1985) in the case of pyrene fluorescence decay on silica gel surfaces, by Thomas and coworkers (DellaGuardia and Thomas, 1983; Nakamura and Thomas, 1985) in the photoprocesses of tris(bipyridyl)ruthenium(II) cation sorbed on various clays, and by Brown and coworkers (1985) in the photoreduction of methylviologen cations sorbed on colloidal TiO_2. In all of the cases noted, the chemicals were tightly sorbed to the particle surface, and interparticle or intraparticle transfer was much slower than the photoprocess. Typically, these kinetic data have been fit by a double exponential function.

Very recently, Albery and coworkers (1985) have developed another general kinetic model for the analysis of chemical reactions in heterogeneous systems. This "dispersed kinetics" model starts with two basic assumptions: (1) that heterogeneous reactions involve parallel reactions on multiple sites exhibiting different reactivities that can be described by a Gaussian distribution function and (2) that transport of reactant from one site to another is slow compared to reaction. When reaction is initiated, reactant molecules sorbed to the most reactive sites decay first, leaving an altered distribution of less reactive sorbed molecules. As a result, even though the reaction may be first-order with respect to reactant concentration for a given site, a logarithmic plot of total concentration versus time exhibits a decreasing slope with increasing conversion. To describe this system, Albery and coworkers (1985) assumed that the dispersion in rate constants could be described by a Gaussian distribution in ln (k), where k are the rate constants. The resulting integrated rate expression is

$$\frac{C}{C_0} = \frac{1}{\pi^{1/2}} \int_{-\infty}^{+\infty} \exp(-x^2) \exp[-\tau \exp(\gamma x)]dx \qquad (16.17)$$

where C_0 and C represent reactant concentration initially and at dimensionless time τ $(= \bar{k}t)$; $\exp(-x^2)$ is the normal distribution; and $\exp(\gamma x)$ expresses the dispersion. ($\bar{k}$ is the mean rate constant, and γ is a fitting parameter that describes the spread of the distribution.) If $\gamma = 0$, then there is no dispersion and Eq. 16.17 reduces to the simple first-order rate expression

$$\frac{C}{C_0} = \exp(-\tau) \tag{16.18}$$

Examples of the use of this model are discussed by Albery and coworkers (1985) and by Brown and coworkers (1985), who analyzed kinetic data for photoinduced reductions on metal oxide surfaces.

2.4. Ecotoxicological Applications

The preceding sections have described models for reactions in heterogeneous systems, ranging from the most generalizable models that deal with equilibrated systems to the most empirical systems in which reaction is much more rapid than interphase transport. Here we use the models to make general assessments of heterogeneous reactions of organic micropollutants on suspended particles and on particles in bottom sediments or soil–water mixtures.

Most organic micropollutants are nonionic, hydrophobic chemicals. As discussed by Karickhoff (1984) and Westall (1987), studies of equilibrium sorption of nonionic organic chemicals to sediments have shown that at low concentrations of the chemical, the total sorbed concentration C_S is related to the aqueous phase concentration C_W by

$$C_S = K_p C_W \tag{16.19}$$

where K_p (expressed in this chapter in units of mL g^{-1} or L kg^{-1}) is the *sorption partition coefficient.*

Sorption of nonionic organic chemicals to sediments mainly occurs on particles that are clay- or silt-sized. Moreover, sorption is strongly controlled by the organic carbon content of the sediment, although the organic component is typically only a minor fraction of the sediment mass. That is, for a given chemical on a series of sediments, K_{oc}, the ratio of K_p to the fractional organic carbon content of each sediment, f_{oc}, is approximately constant (Westall, 1987). The K_{oc} for a given organic chemical is approximately proportional to its octanol–water partition coefficient K_{ow} (Karickhoff, 1984).

Veith (1985) has shown that the median value of K_{ow} for about 20,000 industrial chemicals regulated by the U.S. Environmental Protection Agency is about 1000. This value corresponds to a K_{oc} of about 440 and to K_p values in the 5–50 range for sediments having f_{oc} in the range of 1–10%. Although nonionic organic micropollutants with K_p values in excess of 10,000 are not usual, such hydrophobic chemicals do enter the aquatic environment, where, owing to their great tendency to bioconcentrate, they often cause serious problems. Examples include DDT, PCBs, dioxins, and other notorious pollutants.

Assuming that sorption is governed by equilibrium relations, the relative effects of reactions on particles can be assessed using the equation

$$\frac{k_{obs}}{k_W} = \frac{F_W^e k_W + F_S^e k_S'}{k_W} \tag{16.20}$$

where F_S^e is the total fraction sorbed, k_S' is the average rate constant in the sorbed state, k_{obs} is the rate constant for reaction in the particle suspension, and k_W is the rate constant in the absence of particles. The value of $\Sigma_i \rho_i K_i$, defined in terms of K_p values, is $1 + \rho K_p$, where ρ is the weight ratio of sediment to water. Combining Eqs. 6.2 and 6.20, we obtain

$$\frac{k_{obs}}{k_W} = \frac{1}{1 + \rho K_p}\left[1 + \rho K_p\left(\frac{k_S'}{k_W}\right)\right] \tag{16.21}$$

Table 16.2 summarizes computed relationships (Eq. 16.21) between k_{obs}/k_W and k_S/k_W. Examination of this table leads to three main conclusions concerning environmental reactions of nonionic micropollutants on particle surfaces:

1. For most nonionic micropollutants ($K_p \leq 50$), reaction on suspended particles is not likely to be significant. Suspended particle concentrations in lakes usually do not exceed 10 mg L^{-1} ($\rho = 10^{-5}$). Thus, the value of ρK_p for a typical nonionic organic chemical is expected to be no higher than about 0.001, that is, only 0.1% is sorbed. For such a ρK_p, the rate constant in the sorbed state would have to be over three orders of magnitude larger than k_W in order for the overall reaction rate constants, k_{obs}, to be a factor of 2 greater than the rate constant for reaction in particle-free water (Table 16.2). In addition, chemical reaction would

Table 16.2. Estimated Reaction Rate Constants in Equilibrated Particle–Water Systems (Eq. 16.21)

	k_{obs}/k_W[a]		
k_S'/k_W	$\rho K_p = 10^{-5}$	$\rho K_p = 10^{-3}$	$\rho K_p = 1$
0.1	1.00	1.00	0.60
1	1.00	1.00	1.50
10	1.00	1.0	10.5
100	1.00	1.10	100
1000	1.01	2.00	1000

[a]Estimated ratio of rate constant in particle suspension to rate constant in particle-free water.

have to be at least as rapid as other removal processes in order to be important. For example, the rate constant for epilimnetic water renewal in Lake Zurich is about 0.01 day^{-1} (Imboden and Schwarzenbach, 1985). Thus, for a chemical with $K_p = 50$ L kg^{-1}, k_S for a photoreaction on the suspended particles in the epilimnion would have to be at least 10 day^{-1} to be significant. Such reactivity is not likely to be generally observed, although photoreactions of anilines on algal surfaces occur with rate constants in this range (Zepp and Schlotzhauer, 1983).

2. For extremely hydrophobic chemicals with K_p values in excess of 10,000 L kg^{-1} (ρK_p in the range 0.1–1.0), the probability of reaction being significant on suspended particles is greatly enhanced. In addition, kinetic studies indicate that desorption lifetimes of such chemicals on sediments can be on the order of months (Fig. 16.5), further enhancing the likelihood of reactions in the sorbed state.
3. In bottom sediments and soils, the sorbent/water ratio ($\rho \geq 1$) is at least 10^6 times as great as in the water column of a lake. Under these conditions most nonionic chemicals are predominantly sorbed and k_{obs}/k_W becomes approximately equal to k_S/k_W (Table 16.2). Reaction in the sorbed state has a much higher probability of being significant in such environments.

The above considerations apply to nonionic organic chemicals, but some important micropollutants are ionic. Organic cations, like metal ions, usually sorb very strongly to natural particles. For example, the dication methyl viologen (also known as paraquat) sorbs very tightly to clay-sized particles; K_p typically exceeds 10^6 (Karickhoff and Brown, 1978). Methylviologen is known to be very susceptible to photoredox reactions on particles (Brown et al., 1985). Moreover, certain organic ligands, such as carboxylic acids, sorb strongly to metal oxides (Waite, 1985), thus enhancing the rate of metal dissolution (Stone and Morgan, this volume, Chapter 9; Waite, 1986). Heterogeneous reactions of such ionizable organic chemicals are likely to be significant in water columns as well as in bottom sediments or soils.

3. HETEROGENEOUS ORGANIC REACTIONS IN SEDIMENT SUSPENSIONS

This section emphasizes studies of hydrolysis and reduction, two major types of abiotic organic processes that are believed to be important in natural waters. The area of heterogeneous organic reactions involving metal oxides is discussed by Stone (1986) and Stone and Morgan (1987), in Chapter 9 of this volume. Useful discussions also are found among Theng's treatments of clay structure and interactions with organic polymers (Theng, 1979).

As used here, the term "abiotic" refers to processes that do not involve direct participation of metabolically active organisms. By this definition, abiotic processes include reactions involving chemicals of biological orgin, such as extracellular enzymes or iron porphyrins. The results discussed here were obtained under conditions that isolated the abiotic from the biological processes. The isolation procedures used and their possible effects on sediment reactivity are discussed by Macalady et al. (1986).

3.1. Hydrolysis Reactions

The types of hydrolysis reactions of organic chemicals that have been studied include base-mediated, neutral, and acid-mediated reactions (Mabey and Mill, 1978; Wolfe, 1980; Perdue and Wolfe, 1983). The general rate expression for hydrolysis is

$$v_{hy} = \left[\sum_i (k_{B,i} C_{B,i}) + k_N + \sum_i (k_{A,i} C_{A,i}) + \right] C \qquad (16.22)$$

where v_{hy} is the rate; $k_{B,i}$ and $C_{B,i}$ are rate constants and concentrations of bases, respectively, that mediate hydrolysis; k_N is the rate constant for neutral hydrolysis; and $k_{A,i}$ and $C_{A,i}$ are rate constants and concentrations, respectively, for acid hydrolysis. Perdue and Wolfe (1983) have shown that for solution-phase abiotic hydrolysis reactions in aquatic environments, specific acid or base catalysis (by H^+ and by OH^-) is far more important than reactions mediated by other general acids and bases. Thus, the observed rate constant k_{hy} is expressed as

$$k_{hy} = k_{ac}[H^+] + k_N + k_b[OH^-] \qquad (16.23)$$

At constant pH, the various terms and the overall rate become pseudo-first-order, with the contribution of a given hydrolysis process being dependent on the pH and the various rate constants. Changes in pH may result in changes in the observed hydrolysis rate constant that are reflected in a pH–rate profile. A variety of shapes describe pH–rate profiles for organic chemicals that enter aquatic environments. By judicious selection of pH and reactant, it is possible to conduct an experiment under conditions in which one of the three hydrolysis processes dominates. The remainder of this section provides examples of studies designed to determine the rates of these various types of hydrolysis processes for organic chemicals in sediment suspensions.

Neutral hydrolysis involves the pH-independent reaction of a chemical with water. Macalady and Wolfe (1984) have studied the neutral hydrolysis of six nonionic organic chemicals in suspensions of well-characterized sediments obtained from several U.S. water bodies. Under the short time domains used in the kinetic experiments, a simple one-site sorption model (Eq. 16.1 in which U is ignored) could be used to analyze the kinetic results. Typical kinetic data for the neutral hydrolysis of the organophosphorothioate ester chlorpyrifos were discussed earlier (Fig. 16.1). Computer analysis of these data indicated that k_S and k_W were very similar in magnitude. Similar results were obtained in suspensions of other sediments as well (Table 16.3). Studies such as these were also conducted for five other organic chemicals with the same general finding: neutral hydrolysis proceeded at about the same rate for nonionic organic chemicals dissolved in water or sorbed on sediments.

Studies also were conducted of the alkaline hydrolysis of chlorpyrifos at high pH values in sediment suspensions (Macalady and Wolfe, 1985). Changes in the pH had no effect on the sorption partition coefficient of chlorpyrifos. In sharp contrast to the results for neutral hydrolysis, however, the sorbed chlorpyrifos reacted much more slowly than aqueous phase chlorpyrifos. The amount of retardation was close to that predicted using Eq. 16.4, assuming no reaction in the sorbed state.

Alkaline hydrolysis of hydrophobic esters also was strongly retarded when the esters were sorbed to sediments. This retarding effect is graphically illustrated in Figure 16.7, which compares the strongly sorbed octyl ester with the weakly sorbed methyl ester of 2,4-dichlorophenoxyacetic acid. The experiments were conducted by first equilibrating the sediment–water system for 3 days at pH 7 prior to jumping to pH 10. As discussed in Section 2.3, the curvature in the plot for the octyl ester also shows that the aqueous phase reaction occurred more rapidly than desorption. Note the similarities between Figures 16.4, 16.6, and 16.7.

In summary, the studies of hydrolysis on sediment surfaces have shown that sorption of nonionic organic chemicals to sediments has little effect on neutral hydrolysis, which involves attack on reactant by water molecules. On the other hand, attack by hydroxide ions is strongly retarded. The retardation

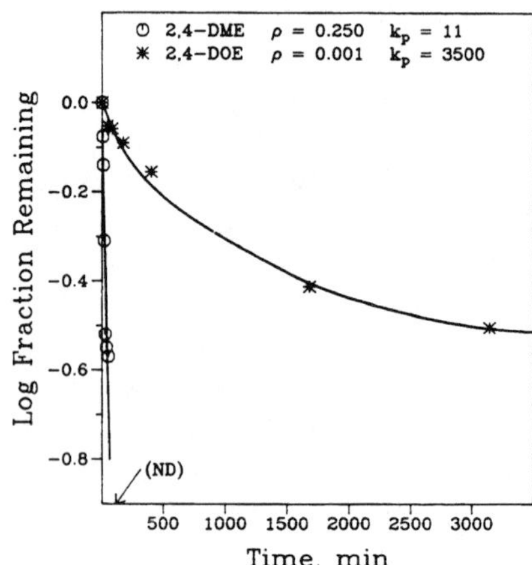

Figure 16.7. Comparison of alkaline hydrolysis (pH 10) of the methyl ester of 2,4-dichlorophenoxyacetic acid (2,4-DME) with its octyl ester (2,4-DOE) in sediment suspension (Macalady and Wolfe, 1984).

Table 16.3. Kinetic Data for Neutral Hydrolysis of Selected Nonionic Organic Chemicals in Sediment Suspensions (At 35°C unless indicated otherwise)

Chemical	Sediment	ρ	Fraction Sorbed	k_{hy}, min^{-1}	k_W, min^{-1}	k_S, min^{-1}
Chlorpyrifos	EPA-14	0.20	0.94	1.0×10^{-5}	1.0×10^{-5}	6.9×10^{-6}
	EPA-23	0.016	0.87	1.6×10^{-5}	1.0×10^{-5}	1.2×10^{-5}
Diazinon	EPA-26	0.040	0.64	3×10^{-5}	3.8×10^{-5}	2.9×10^{-5}
Ronnel	EPA-26	0.040	0.96	2.7×10^{-5}	3.8×10^{-5}	2.6×10^{-5}
Benzyl chloride[a]	EPA-13	0.025	0.05	1.4×10^{-3}	1.25×10^{-3}	—
	EPA-2	1.0	0.87	1.15×10^{-3}	1.21×10^{-3}	—
4-(*p*-Chlorophenoxy)butyl bromide	EPA-12	0.050	0.80	—	7.9×10^{-5}	5.1×10^{-5}
Hexachlorocyclopentadiene[b]	EPA-13	0.20	—	2.2×10^{-4}	—	—
		1.0	—	2.0×10^{-4}	—	—
		2.0	—	1.3×10^{-4}	—	—

[a] At 25°C.
[b] At 30°C.
Source: Macalady and Wolfe (1984).

of alkaline hydrolysis was attributed to the destabilizing effect of the negative surface charge of sediments on the negatively charged transition state that mediates alkaline hydrolysis of esters. The hydrolysis results in sediments are strikingly similar to those reported for hydrolysis reactions of nonionic organic chemicals in other negatively charged particles designed to mimic biomembranes such as micelles and vesicles (Fendler and Fendler, 1975).

3.2. Organic Reductions in Anaerobic Sediment–Water Systems

Hydrophobic organic chemicals are often found in bottom sediments of lakes and in soils. In such environments, microbial activity sometimes greatly reduces the concentration of dioxygen, and the systems are said to be anaerobic. From past studies, it is known that several types of organic compounds are generally susceptible to reduction in anaerobic sediments: chlorinated hydrocarbons, nitroaromatic compounds, N- and O-alkylated organic compounds, azo compounds, and quinones, among others. A recent paper by Macalady and coworkers (1986) thoroughly discusses this work.

Few quantitative data are available on reductions of organic chemicals sorbed on sediments. A recent study by Sanders and Wolfe (1986), however, illustrates some of the models we discussed earlier. The reaction studied involved a six-electron reduction of nitroaromatic compounds to the corresponding anilines in sterilized bottom sediments from local water bodies (Fig. 16.8). Under the reaction conditions, the rate was first-order with respect to nitroaromatic concentration and sediment concentration. For a series of alkyl-substituted nitrobenzenes, the rate constant decreased with increasing sorption K_p (Fig. 16.9).

Sanders and Wolfe (1986) analyzed the kinetic results for the nitroaromatic reductions using a kinetic model similar to Eq. 16.1 in which the reactant is partitioned between a reactive phase and an unreactive phase. The effect of partitioning into the unreactive phase was, as in the above-cited case of alkaline hydrolysis, to reduce the fraction of reactant in the reactive phase. Reaction was postulated to occur either at the particle–water interface or,

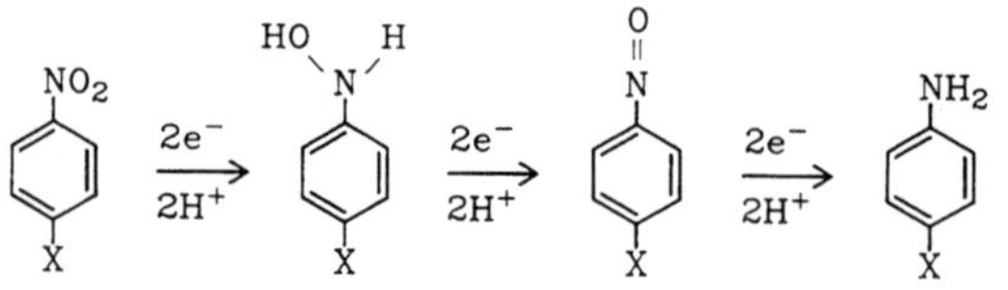

Figure 16.8. Reduction of nitroaromatic compounds in anaerobic sediments (Sanders and Wolfe, 1985).

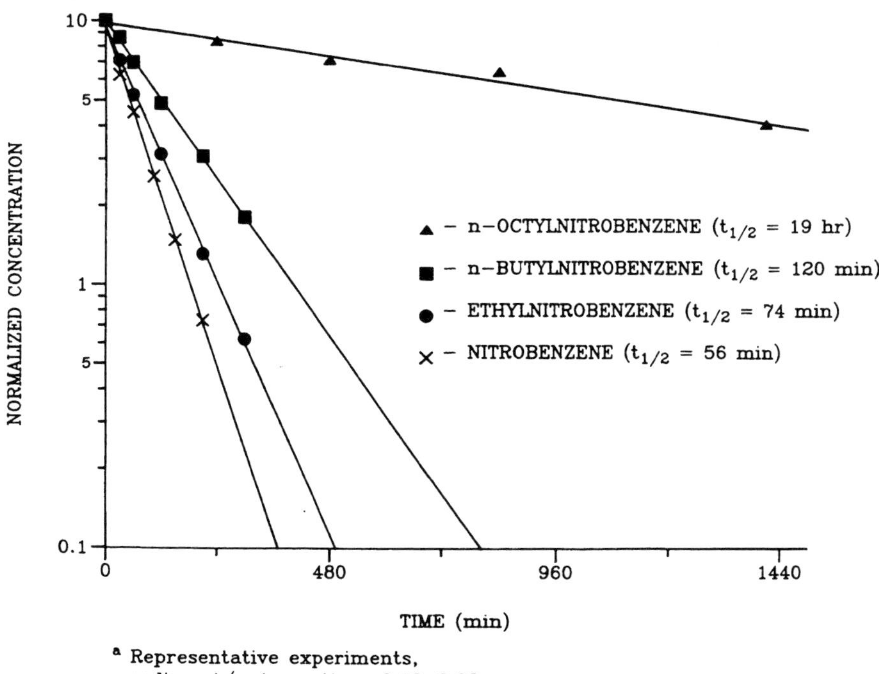

Figure 16.9. Kinetic data for reaction of nitroaromatic compounds in anaerobic sediments (Sanders and Wolfe, 1985).

alternatively, with a reductant released into the aqueous phase by the sediment. This latter mechanism was discussed in more detail in Section 2.2 (Eqs. 16.11 and 16.12). The rate of reduction was reduced by 80–90% on heating, but it was unaffected by *m*-cresol, a reductive dehydrogenase enzyme blocking agent. These results indicate that catalysis by bioorganics or by one or more extracellular enzymes in the sediment may have been involved in the reduction. Similar effects have been observed in kinetic studies of the reduction of methyl parathion, a nitroaromatic pesticide, in bottom sediments (Wolfe et al., 1985).

4. HETEROGENEOUS PHOTOREACTIONS

Particles in the upper layers of aquatic environments, on soil surfaces, and in water droplets in the atmosphere are exposed to sunlight. The particles in water bodies are among the most important sunlight absorbers, and in many natural waters, the biota, detritus, and/or suspended sediments are almost completely responsible for light attenuation. No doubt a large portion of the solar radiation absorbed by these natural particles is converted into thermal energy, but a fraction of the absorbed radiation is diverted into photochemical reactions.

The widespread occurrence of photosynthetic organisms in aquatic environments is the most obvious evidence of the importance of heterogeneous photoreactions in water bodies. Recent research, however, has resulted in the discovery of other environmentally significant abiotic photoreactions involving particle surfaces. Also, heterogeneous photoreactions on metal oxide surfaces are increasingly being used as a method for wastewater and groundwater treatment.

The main goals of this section are twofold. First, we supplement the general discussion of heterogeneous reactions presented above by a brief examination of some of the basic factors that govern heterogeneous photoreactivity. Second, we provide examples of various types of heterogeneous photoreactions that are believed to be of environmental significance.

4.1. Rate Equations and Types of Photoreactions

The rate of any type of photoreaction depends upon the rate of light absorption by the photoactive species, I_a, and the quantum efficiency of the reaction, ϕ. The rate of light absorption is affected, in turn, by the spectral overlap between the light source and the spectrum of the chemical that initiates the photoreaction (Zepp and Cline, 1977).

$$\text{Rate} = I_a\phi \qquad (16.24)$$

The light absorption rate in a water body depends upon the solar spectral irradiance as it is transmitted down into the water, and it also depends on the geometry of the light field. Particles in the photic zone influence the irradiance, and thus photolysis rates, through both light absorption and light scattering (Miller and Zepp, 1979a; Smith et al., 1983). The concepts and terminology used in modeling the transmission of sunlight in natural waters are described in detail by Baker and Smith (1982), Morel and Smith (1982), and Jerlov (1976). The complex effects of light attenuation and scattering by sediments are illustrated by the interesting dependence of the direct photolysis rate of the ketone γ-methoxy-*m*-trifluoromethylbutyrophenone on the concentrations of various types of clays suspended in water (Fig. 16.10). The photolysis of the ketone in these systems occurs completely in the aqueous phase. The enhancement of the photolysis rate in some of the suspensions was attributed to increases in the mean light path length that were caused by scattering (Miller and Zepp, 1979a).

Photochemical processes at the particle–water interface can be subdivided into two main categories—direct and indirect. Direct photolysis processes involve the direct absorption of light by the sorbed substrate:

$$\text{P} + \text{light} \longrightarrow \text{P}^* \longrightarrow \text{products} \qquad (16.25)$$

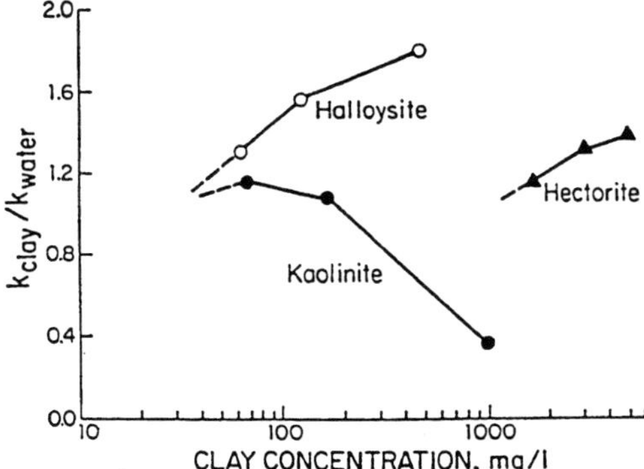

Figure 16.10. Direct photolysis of γ-methoxy-*m*-trifluoromethylbutyrophenone in aqueous suspensions of clay minerals. k_{clay} and k_{water} are first-order rate constants in clay suspension and distilled water, respectively. Substrate is over 95% dissolved in solution phase in all cases. (From Miller and Zepp, 1979a.)

where P is some photoreactive substance on the particle surface and P* is the excited state that reacts to form products. On the other hand, heterogeneous photoreactions can also involve indirect photoprocesses, which are initiated through light absorption by photoreactive components of the particle, symbolized by "sens" because such substances are sometimes referred to as photosensitizers.

$$\text{sens} + \text{light} \longrightarrow \text{sens}^* \tag{16.26}$$

$$\text{sens}^* + \text{P} \longrightarrow \text{products} \tag{16.27}$$

In some cases of indirect photolysis, no direct reaction occurs between the excited surface component and the reactant, but rather the reaction involves short-lived intermediates derived from photolysis of the surface component. Indirect photoreactions involving the semiconductor mechanism are a special case that is discussed in more detail in Section 4.3.

4.2. Direct Heterogeneous Photoreactions

Considerable research has focused on heterogeneous photoreactions of iron oxides that apparently involve direct photolysis of inner-sphere surface complexes of iron with various ligands (Faust and Hoffmann, 1986; Waite and Morel, 1984; Waite, 1986). Manganese oxides (Sunda et al., 1983) also undergo enhanced photoreductive dissolution in the presence of natural organic solutes. This research has been recently reviewed by Waite (1985). Waite and Morel (1984) have proposed a mechanism similar to Eq. 16.5 to account for the effects of various organic carboxylic acids on the photoreductive disso-

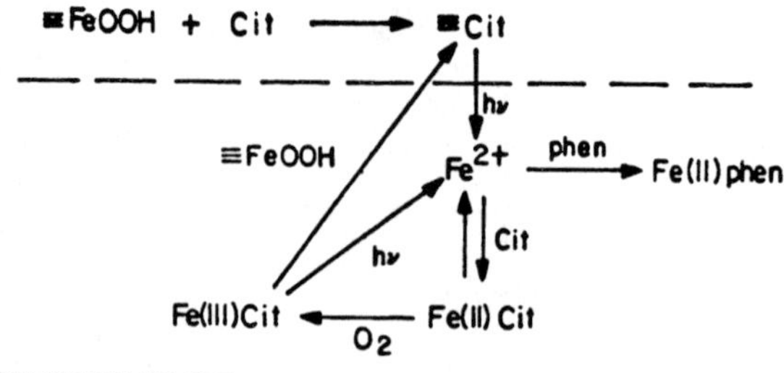

Figure 16.11. Model for major processes occurring in irradiated suspensions of iron oxides in aqueous citrate solutions. Used with permission of Waite and Morel (1984).

lution of the iron oxide lepidocrocite (Fig. 16.11). A similar model has been proposed by Faust and Hoffmann (1986) to describe the heterogeneous photooxidation of S(IV) species by the iron oxide hematite.

Kinetic results obtained by Waite (1985), such as those in Figure 16.3, indicated that the rate of dissolution of lepidocrocite upon site saturation was about the same for four organic ligands, despite the fact that the solution-phase photolysis rates of iron(III) complexes of these ligands were quite different (Table 16.4). The extremely inefficient photoreduction of the iron oxide goethite in the presence of the weakly sorbed ethylene glycol apparently involved an outer-sphere complex (Cunningham et al., 1985). Unlike carboxylic acids, ethylene glycol is incapable of forming inner-sphere complexes with the iron oxide surface.

Several studies have appeared on the photolysis of organic chemicals sorbed on clay minerals or on natural sediments. These studies indicate that sorption

Table 16.4. Kinetic Parameters That Describe Photodissolution (365 nm) of γ-FeOOH in Aqueous Solutions (pH 4, 0.01 *M* NaCl) of Selected Organic Carboxylic Acids (See Eqs. 16.5–16.9.)

Organic Acid	k_{soln}, min^{-1} [a]	k_S, min^{-1} [b]	$\frac{k_1}{k_{-1} + k_S}$, M^{-1}
Tartaric	0.49	0.032	$10^{6.6}$
Citric	0.27	0.028	$10^{5.7}$
Oxalic	0.63	0.034	$10^{5.3}$
Salicylic	0.67	0.030	$10^{4.1}$
Phthalic	0.14	—	—

[a]Photolysis rate constant for Fe(III) complex in solution.
[b]Rate constant for photoreduction step in dissolution of γ-FeOOH.
Source: Waite (1985).

of organic substrates on natural particles can result in significant spectral shifts (DellaGuardia and Thomas, 1983; De Mayo et al., 1985; Gäb et al., 1977) along with changes in quantum efficiencies and photoproduct distributions (Nakamura and Thomas, 1985; Miller and Zepp, 1979b). Kinetic and product studies of photoreactions of extremely hydrophobic sorbed nonionic organic chemicals indicated that the chemicals were in a microenvironment that was less polar than water and that was a considerably better hydrogen donor (Miller and Zepp, 1979b). This result is consistent with sorption studies (Westall, this volume, Chapter 1; Karickhoff, 1984) which indicate that such hydrophobic chemicals preferentially sorb in organic components of sediments.

4.3. Indirect Heterogeneous Photoreactions

Indirect photoreactions involving particles have also received considerable attention during the past few years. Most of this research can be subdivided into two broad areas: reactions involving reactive intermediates produced by surface photolysis, such as semiconductor-type mechanisms, and reactions involving electronically excited surface species.

Heterogeneous organic photocatalysis involving particles with semiconducting properties has received considerable attention recently. This research area has been reviewed by Fox (1983). Semiconductors such as titanium dioxide and zinc oxide have been shown to effectively catalyze both the reductions of halogenated chemicals and the oxidation of various other pollutants (Carey and Oliver, 1980; Pruden and Ollis, 1983; Hidaka et al., 1985; Draper and Crosby, 1985).

All semiconductors have a characteristic optical band gap, which corresponds to the light energy required to promote an electron from the valence band to the conduction band. Several metal oxides that occur in the environment can be activated by absorption of ground-level solar radiation (Table 16.5). When the electron is so promoted, it leaves behind a positively charged "hole" or vacancy in the conduction band. Oxidation of a reactant can occur through electron transfer from reactant to hole on the surface. Likewise, reduction can occur on the surface through transfer of the promoted electron to the reactant. Oxidation can also occur via reactions of the holes with water and the electrons with protons or oxygen to produce hydroperoxyl and hydroxyl radicals, which in turn can rapidly react with solution-phase reactant. The hydroperoxyl radicals also can react to form hydrogen peroxide. Studies of the photoproduction of H_2O_2 at zinc oxide surfaces date back to the early 1950s (Calvert et al., 1954).

Semiconductors are versatile reagents that show great promise for waste treatment. Their role in photochemistry in the aquatic environment appears to be very limited, however. For example, studies by Oliver and coworkers (1979) failed to find any photoproduction of hydroxyl radicals or chloroorganic

Table 16.5. Band Gap Energies for Selected Semiconducting Oxides That Absorb Sunlight

Oxide	Band Gap, eV	Wavelength of Light Equivalent to Band Gap, nm
ZnO	3.35	370
PbO	2.76	499
TiO_2 (rutile)	3.0–3.3	376–413
β-MnO_2	0.26	4770
β-PbO_2	1.7	729
α-Fe_2O_3 (hematite)	2.34	530
CdS	2.4	517
β-HgS	0.54	2300

Source: Strehlow and Cook (1973); Faust (1985).

photoreduction on various natural sediments and clay minerals. Leighton (1961) long ago reached much the same conclusion with regard to the role of semiconducting metal oxides in the formation of photochemical smog. Nonetheless, it appears possible that natural semiconducting particles may contribute to the photochemical generation of the oxidant hydrogen peroxide in the hydrosphere.

Surface photoreactions involving excited sensitizer molecules provide another indirect photolysis pathway. Humic substances, common constituents of natural particles, are known to photosensitize a variety of organic photoreactions (Zepp et al., 1985; Haag et al., 1984), including those mediated by singlet oxygen, an excited form of dioxygen. Recent studies by Gohre and Miller (1983, 1985) have confirmed that singlet oxygen photoproduction occurs on soil and other natural surfaces that are exposed to sunlight. Based on results of recent studies of sensitized reactions involving humic substances (Zepp et al., 1985), it is likely that future research will demonstrate that other types of surface photosensitized reactions involving excited triplet states occur as well.

The indirect photolysis of anilines and organophosphorothioates sorbed on algae provides another example that does not involve the intermediacy of singlet oxygen (Zepp and Schlotzhauer, 1983). The rate constants for these photoreactions were virtually unaffected by heat killing the cells and by addition of the metabolic inhibitor DCMU. The aniline reaction is described by the hyperbolic rate expression discussed in Section 2.2 (Fig. 16.2). The reaction also exhibited a pronounced hydrophobic effect; rate constants for reaction of aniline, toluidine, and butylaniline were proportional to octanol–water partition coefficients for these compounds (Fig. 16.12). This hydrophobic effect is exactly the opposite of that observed in the nitroaromatic reduction in anaerobic sediments (Fig. 16.7), underscoring the interesting diversity of results obtained in kinetic studies of heterogeneous reactions.

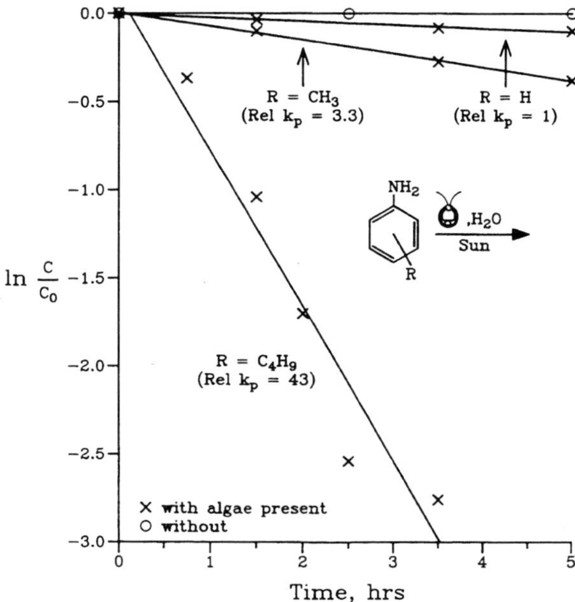

Figure 16.12. Kinetic plots for heterogeneous oxidation of aniline and alkyl-substituted anilines in suspensions of *Chlamydomonas* sp. (Zepp, 1985).

5. CONCLUSIONS

The models and equations presented in this chapter provide a starting point for kinetic analysis of thermal and photochemical reactions in heterogeneous systems of environmental interest. Particularly under the short time domains used in laboratory studies, equilibrium concepts can sometimes be applied to describe the partitioning of the reactant between phases in sediment suspensions, for example, in the case of the neutral hydrolysis of nonionic organic compounds.

Heterogeneous reactions, especially those of chemicals that sorb to particles by site-specific or chemisorption mechanisms, are often affected by site saturation. Under steady-state conditions such reactions are described by hyperbolic rate expressions; they can be first-order or zero-order with respect to reactant concentration (or some intermediate order) depending on reactant and particle concentration. Reductive photodissolutions of metal oxides and indirect photoreactions on algae surfaces are two types of reactions that involve site saturation.

Reactions in heterogeneous systems also can be affected by intraparticle or interparticle transport. Such transport effects are particularly likely to occur in the case of tightly sorbed organic cations such as methylviologen (paraquat) or in the case of extremely hydrophobic nonionic organic reactants that have been equilibrated for long time periods prior to initiation of reactions such as photolysis or alkaline hydrolysis. Semilog plots of concentration versus time for such reactions are usually characterized by a decrease in slope with increasing conversion. Two-component or dispersed kinetics models are used to treat the kinetic results in these systems.

Through application of these concepts, it was shown that the neutral hydrolysis rates of various organic chemicals are approximately unaffected by sorption on certain sediments but that the alkaline hydrolysis rates of esters are strongly retarded by sediment sorption. In anaerobic sediment–water systems, the reduction rates of nitroaromatic compounds exhibit sorptive retardation, an effect that is possibly attributable to competition between sorption to unreactive sediment components and reaction with some reductant in the aqueous phase.

Photoreaction rates in particle–water systems are influenced by effects of the particles on the irradiance, by the fraction of the reactant that is sorbed on photoreactive sites, and by the quantum efficiency of the photoreaction on the particle surface. Two types of surface photoreactions occur: those involving direct light adsorption by the sorbed reactant and those involving initiation of photolysis through light absorption by photoreactive components of the particle. Photoreactions of surface inner-sphere complexes between metal oxides and organic ligands provide examples of important direct reactions. Reactions involving semiconductor mechanisms, such as TiO_2-mediated oxidations, and photosensitized reactions on surfaces, such as surface photooxidations involving singlet oxygen, are two types of indirect photoprocesses that are relevant to wastewater treatment or environmental organic oxidation.

Acknowledgments

We appreciate the many helpful discussions with S. W. Karickhoff concerning sorption kinetics and kinetics in heterogeneous systems. We also thank A. T. Stone, T. D. Waite, and P. Gschwend for prepublication copies of manuscripts that were discussed here, and B. C. Faust for a copy of his dissertation and other reference sources.

SYMBOLS

C_R Concentration of reactant sorbed on reactive phase of particles (mol kg^{-1}).

C_W Reactant concentration dissolved in aqueous phase of particle suspension (mol L^{-1} or M).

$C_{A,i}$ Concentration of acid i in a system with multiple acids (M).

$C_{B,i}$ Concentration of base i in a system with multiple bases (M).

F_j^e Fraction of reactant in a multiphase system that is in phase j at equilibrium (unitless).

I_a Rate of light absorption by a photoreactive chemical (einstein or mol L^{-1} s^{-1}).

k_W Rate constant for reaction in aqueous phase of particle suspension (s^{-1}).

k_S Rate constant for reaction in reactive phase of sorbed state (s^{-1}).

k_1 Rate constant for sorption of reactant to reactive sites on particle [L (mol of reactive sites)$^{-1}$ s^{-1}].

k_{-1} Rate constant for desorption of reactant from reactive sites on particle (s^{-1}).

k_U Rate constant for intraparticle diffusion of sorbed reactant into unreactive sites (s^{-1}).

k_2 Rate constant for release of chemical sorbed in nonlabile phase of sediment (hr^{-1}).

k_N Rate constant for neutral hydrolysis (s^{-1}).

$k_{B,i}$ Rate constant for base-catalyzed hydrolysis by base i (M^{-1} s^{-1}).

$k_{A,i}$ Rate constant for acid-catalyzed hydrolysis by acid i (M^{-1} s^{-1}).

k_{hy} Overall rate constant for hydrolysis (s^{-1}).

K_j Partition coefficient expressed as ratio of concentration in phase j of a multiphase system to concentration in reference phase at equilibrium.

K Equilibrium constant of sorption defined as k_1/k_{-1} [L (mol of reactive sites)$^{-1}$].

K_p Equilibrium partition coefficient for sorption (L kg^{-1}).

v Rate of abiotic transformation, expressed as change in total concentration per unit time (M s^{-1}).

v_{hy} Rate of hydrolysis (M s^{-1}).

ϕ Quantum yield; fraction of absorbed light that results in a certain photoprocess (unitless).

ρ_j Relative amounts of phase j and some reference phase in a multiphase system. Units selected to be compatible with units of K_j.

ρ Particle (e.g., sediment) concentration (kg L^{-1}).

R_t Concentration of reactive sites on particles (mol of reactives sites kg^{-1}).

θ Fraction of particle's reactive sites that are occupied (unitless).

REFERENCES

Albery, W. J., Bartlett, P. N., Wilde, C. P., and Darwent, J. R. (1985); A general model for dispersed kinetics in heterogeneous systems, *J. Amer. Chem. Soc.* **107,** 1854–1858.

Baker, K. S., and Smith, R. C. (1982), Bio-optical classification and model of natural waters 2, *Limnol. Oceanogr.* **27,** 500–509.

Brown, G. T., Darwent, J. R., and Fletcher, P. D. I. (1985), Interfacial electron transfer in TiO_2 colloids, *J. Amer. Chem. Soc.* **107,** 6446–6451.

Calvert, J. G., Theurer, K., Rankin, G. T., and MacNevin, W. M. (1954), A study of the mechanism of the photochemical synthesis of hydrogen peroxide at zinc oxide surfaces, *J. Amer. Chem. Soc.* **76,** 2575–2578.

Carey, J. H., and Oliver, B. G. (1980), The photochemical treatment of wastewater by ultraviolet irradiation of semiconductors, *Water Poll. Res. J. Canada* **15,** 157–185.

Clark, A. (1970), *The Theory of Adsorption and Catalysis,* Academic Press, New York, pp. 233–401.

Cunningham, K. M., Goldberg, M. C., and Weiner, E. R. (1985), The aqueous photolysis of ethylene glycol adsorbed on goethite, *Photochem. Photobiol.* **41,** 409–416.

DellaGuardia, R. A., and Thomas, J. K. (1983), Photoprocesses on colloidal clay systems. Tris(2,2′-bipyridine)ruthenium(II) bound to colloidal kaolin and montmorillonite, *J. Phys. Chem.* **87,** 990–998.

De Mayo, P., Natarajau, L. V., and Ware, W. R. (1985), "Surface Photochemistry: Temperature Effects on the Emission of Aromatic Hydrocarbons Adsorbed on Silica Gel," in M. A. Fox, Ed., *Organic Phototransformations in Nonhomogeneous Media,* ACS Symp. Ser., Washington, DC, pp. 1–19.

Draper, R. B., Jr., and Crosby, D. G. (1985), "Enhanced Photodegradation of Thiocarbamate Herbicides," presented in part at the 189th Natl. Meeting American Chem. Soc., Miami, FL, April 1985.

Faust, B. C. (1985), Photo-induced reductive dissolution of hematite (α-Fe_2O_3) by S(IV) oxyanions, Ph.D. Thesis, California Institute of Technology, Pasadena, CA.

Faust, B. C., and Hoffmann, M. R. (1986), Photo-induced dissolution of α-Fe_2O_3 by bisulfite, *Environ. Sci. Technol.,* **20,** 943–948.

Fendler, J. H., and Fendler, E. J. (1975), *Catalysis in Micellar and Macromolecular Systems,* Academic Press, New York.

Fox, M. A. (1983), Organic heterogeneous photocatalysis: Chemical conversions sensitized by irradiated semiconductors, *Acc. Chem. Res.* **16,** 314–321.

Gäb, S., Schmitzer, J., Thamm, H. W., Parlar, H., and Korte, F. (1977), Photomineralisation rate of organic compounds adsorbed on particulate matter, *Nature* **270,** 331–333.

Gaines, G. L., Jr., and Wise, G. (1983), "Insiders, Outsiders, and Surfaces: Irving Langmuir's Contribution to Catalysis," in B. M. Davis and W. P. Hettinger, Jr., Eds., *Heterogeneous Catalysis: Selected American Histories,* ACS Symp. Ser. 222, Washington, DC, pp. 13–22.

Gohre, K., and Miller, G. C. (1983), Singlet oxygen generation on soil surfaces, *J. Agr. Food Chem* **31,** 1104–1108.

Gohre, K., and Miller, G. C. (1985), Photochemical generation of singlet oxygen on nontransition-metal oxide surfaces, *J. Chem. Soc., Faraday Trans. 1* **81,** 793–800.

Haag, W. R., Hoigné, J., Gassman, E., and Braun, A. M. (1984), Singlet oxygen in surface waters. Part II: Quantum yields of its production by some natural humic materials as a function of wavelength, *Chemosphere* **13,** 641–650.

Hidaka, H., Kubota, H., Graetzel, M., Serpone, N., and Pelizetti, E. (1985), Photodegradation of the sodium dodecylbenzenesulfonate surfactant in aqueous semiconductor dispersions, *Nouv. J. Chim.* **9,** 67–69.

Imboden, D. M., and Schwarzenbach, R. P. (1985), "Spatial and Temporal Distribution of Chemical Substances in Lakes: Modeling Concepts," in W. Stumm, Ed., *Chemical Processes in Lakes,* Wiley, New York, pp. 1–30.

Jerlov, N. (1976), *Marine Optics,* Elsevier, Amsterdam.

Karickhoff, S. W. (1980), "Sorption Kinetics of Hydrophobic Pollutants in Natural Sediments," in R. A. Baker, Ed., *Contaminants and Sediments,* Vol. II, Ann Arbor Science, Ann Arbor, MI, pp. 193–205.

Karickhoff, S. W. (1984), Organic pollutant sorption in aquatic systems, *J. Hydraul. Eng.* **110,** 707–735.

Karickhoff, S. W., and Brown, D. S. (1978), Paraquat sorption as a function of particle size in natural sediments, *J. Environ. Qual.* **7,** 246–252.

Karickhoff, S. W., and Morris, K. R. (1985), Sorption dynamics of hydrophobic pollutants in sediment suspensions, *Environ. Toxicol. Chem.* **4,** 469–479.

Laidler, K. J., and Bunting, P. S. (1973), *The Chemical Kinetics of Enzyme Action,* Clarendon Press, London, pp. 382–412.

Leighton, P. (1961), *Photochemistry of Air Pollution,* Academic Press, New York, pp. 96–98.

Mabey, W., and Mill, T. (1978), Critical review of hydrolysis of organic compounds in water under environmental conditions, *J. Phys. Chem. Ref. Data* **7,** 383.

Macalady, D. L., Tratynek, P. G., and Grundl, T. J. (1986), Abiotic reduction reactions of anthropogenic organic chemicals in anaerobic systems: A critical review, *J. Hydrol. Contam.,* **1,** 1–28.

Macalady, D. L., and Wolfe, N. L. (1984), "Abiotic Hydrolysis of Sorbed Pesticides," in R. F. Krueger and J. N. Seiber, Eds., *Treatment and Disposal of Pesticide Wastes,* ACS Symp. Ser. 259, Washington, DC, pp. 221–244.

Macalady, D. L., and Wolfe, N. L. (1985), Effects of sediment sorption on abiotic hydrolyses. I. Organophosphorothioate esters, *J. Agr. Food Chem.* **33,** 167–173.

Miller, G. C., and Zepp, R. G. (1979a), Effects of suspended sediments on photolysis rates of dissolved pollutants, *Water Res.* **13,** 453–459.

Miller, G. C., and Zepp, R. G. (1979b), Photoreactivity of aquatic pollutants sorbed on suspended sediments, *Environ. Sci. Technol.* **13,** 860–863.

Moore, J. W., and Pearson, R. G. (1981), *Kinetics and Mechanism,* Wiley, New York, pp. 284–333.

Morel, A., and Smith, R. C. (1982), Terminology and units in optical oceanography, *Marine Geodesy* **5,** 335–349.

Nakamura, T., and Thomas, J. K. (1985), Photochemistry of materials adsorbed on clay systems. Effect of the nature of the adsorption on the kinetic description of the reactions, *Langmuir* **1,** 568–573.

Oliver, B. G., Cosgrove, E. G., and Carey, J. H. (1979), Effect of suspended sediments on the photolysis of organics in water, *Environ. Sci. Technol.* **13,** 1075–1077.

Pasciak, W. J., and Gavis, J. (1975), Transport limited nutrient uptake rates in *Ditylum brightwellii, Limnol. Oceanogr.* **20,** 604–617.

Perdue, E. M., and Wolfe, N. L. (1983), Prediction of buffer catalysis in field and laboratory studies of pollutant hydrolysis reactions, *Environ. Sci. Technol.* **17,** 635–642.

Pruden, A. L., and Ollis, D. F. (1983), Degradation of chloroform by photoassisted heterogeneous catalysis in dilute aqueous suspensions of titanium dioxide, *Environ. Sci. Technol.* **17,** 628–631.

Sanders, P., and Wolfe, N. L. (1985), "Reduction of Nitroaromatic Compounds in Anaerobic, Sterile Sediments," presented in part at the 190th Natl. Meeting Amer. Chem. Soc., Chicago, IL, September 1985.

Satterfield, C. N., and Sherwood, T. K. (1963), *The Role of Diffusion in Catalysis,* Addison-Wesley, Reading, MA.

Smith, R. C., Baker, K. S., and Fahy, J. B. (1983), Effects of suspended sediments on penetration of solar radiation into natural waters, U.S. Environmental Protection Agency Report EPA-600/S3-83-060, Sept. 1983, NTIS Order No. PB 83-238 188.

Stone, A. T. (1986), "Adsorption of Organic Reductants and Subsequent Electron Transfer on Metal Oxide Surfaces," in J. A. Davis and K. F. Hayes, Eds., *Geochemical Processes at Mineral Surfaces,* ACS Symp. Ser. 323, Washington, DC, 446–461.

Stone, A. T., and Morgan, J. J. (1987), "Reductive Dissolution of Metal Oxides," this volume, Chapter 9.

Strehlow, W. H., and Cook, E. L. (1973), Compilation of energy band gaps in elemental and binary compound semiconductors and insulators, *J. Phys. Chem. Ref. Data* **2,** 163–199.

Sunda, W. G., Huntsman, S. A., and Harvey, G. R. (1983), Photoreduction of manganese oxides in seawater and its geochemical and biological implications, *Nature* **301,** 234–236.

Theng, B. K. G. (1979), *Formation and Properties of Clay–Polymer Complexes,* Elsevier, New York.

Veith, G. (1985), U.S. Environmental Protection Agency, Duluth, MN, personal communication.

Waite, T. D. (1986), "Photoredox Chemistry of Colloidal Metal Oxides," in J. A. Davis and K. F. Hayes, Eds., *Geochemical Processes at Mineral Surfaces,* ACS Symp. Ser. 323, Washington, DC, 426–445.

Waite, T. D., and Morel, F. M. M. (1984), Photoreductive dissolution of colloidal iron oxide: Effect of citrate, *J. Colloid Interface Sci.* **102,** 121.

Westall, J. C. (1987), "Adsorption Mechanisms in Aquatic Surface Chemistry," this volume, Chapter 1.

Wolfe, N. L. (1980), "Determining the Role of Hydrolysis in the Fate of Organics in Natural Waters," in R. Haque, Ed., *Dynamics, Exposure, and Hazard Assessment of Toxic Chemicals,* Ann Arbor Science, Ann Arbor, MI, pp. 163–177.

Wolfe, N. L., Kitchens, B. E., Macalady, D. L., and Grundl, T. J. (1987), Physical and chemical factors that influence the anaerobic degradation of methyl parathion in sediment systems, *Environ. Toxicol. Chem.* in press.

Wu, S., and Gschwend, P. M. (1986), Sorption kinetics of hydrophobic organic compounds to natural sediments and soils, *Environ. Sci. Technol.* **20,** 717–725.

Zepp, R. G. (1985), "Photobiological Transformations of Xenobiotics in Natural

Waters: Role of Microalgae," presented in part at the 189th Natl. Meeting Amer. Chem. Soc., Miami, FL.

Zepp, R. G., and Cline, D. M. (1977), Rates of direct photolysis in aquatic environments, *Environ. Sci. Technol.* **11,** 359–366.

Zepp, R. G., and Schlotzhauer, P. F. (1981), Effects of equilibration time on photoreactivity of the pollutant DDE sorbed on natural sediments, *Chemosphere* **10,** 453–460.

Zepp, R. G., and Schlotzhauer, P. F. (1983), Influence of algae on photolysis rates of chemicals in water, *Environ. Sci. Technol.* **17,** 462–468.

Zepp, R. G., Schlotzhauer, P. F., and Sink, R. M. (1985), Photosensitized transformations involving electronic energy transfer in natural waters: Role of humic substances, *Environ. Sci. Technol* **19,** 74–81.

17

THE ROLE OF PARTICLES IN REGULATING THE COMPOSITION OF SEAWATER

Michael Whitfield and David R. Turner

Marine Biological Association of the United Kingdom, The Laboratory, Citadel Hill, Plymouth, United Kingdom

Abstract

The formation and dissolution of particles in natural waters influence the dissolved concentrations not only of the major constituents of the particle matrix but also of any trace components that might have been incorporated incidentally. Particles also provide active surfaces for adsorption–desorption reactions and can act as sites for biologically and photochemically induced redox reactions. The oceans present a suitable paradigm for studying such processes in natural waters since the ocean reservoir has a long holding time and the main removal process is via sedimentary deposition and burial.

The oceanic particle cycle is driven by the biological production of particles in the surface layers which subsequently fall into the deep ocean where they are partially redissolved. This gives rise to a significant standing population of fine particles, which are occasionally removed from the water column by interactions with large, rapidly sinking particles and which provide sites for the scavenging of reactive elements from deep water. The processes of particle production and particle dissolution are well separated in space and time so that much can be learned about particle–water interactions from the interpretation of vertical profiles of the dissolved and particulate constituents in the deep ocean. The mean oceanic residence time (τ_Y) can be used to divide the elements into three categories that exhibit characteristically different deep-sea profiles resulting from progressively stronger particle–water interactions.

1. INTRODUCTION

The composition of natural waters is governed by the balance struck between the rate of addition of dissolved components and their rate of removal. Such exchanges occur via uptake or release by mineral phases or by the biota or via the transport of volatiles across the air–sea interface. The initial interactions between natural waters and mineral phases occur during the weathering of rocks, but such processes are inherently slow and further reactions proceed as the rock fragments undergo modification in streams, rivers, lakes, estuaries, and eventually in the sea. During this process the composition of natural waters can be influenced directly by the formation or dissolution of particles or indirectly by adsorption–desorption reactions at particle surfaces. The inorganic precipitation of minerals in natural waters (e.g., the formation of evaporite salt deposits) is, however, an unusual and localized event, and most particulate matter produced in situ is of biological origin. Indeed it is difficult to separate living from nonliving particles, and such distinction might be misleading since biological processes often control the course of inorganic reactions in natural systems. Bacteria actively colonize detrital material and are largely responsible for its degradation, while, conversely, encapsulated bacteria can act as sites for mineral deposition and for the accumulation of metal oxyhydroxide aggregates (Cowen and Bruland, 1985).

Biological processes are also responsible for the precipitation of more than two dozen mineral phases either as distinct deposits within organisms or as exoskeletons (Lowenstam, 1981). All the particles produced biologically are capable of interacting with dissolved components via adsorption–desorption reactions, and their numbers will be augmented by detrital mineral phases introduced by wash-off or sediment resuspension and by aeolian dust, including man-made aerosol particles. Because of the complex nature of the particle populations and the intimate interplay between the biological and inorganic cycles of the elements, it is often difficult to unravel the contributions made by the various processes to the composition of natural waters. To help elucidate some of the ground rules governing the exchange of elements between the dissolved and particulate phases, it is useful to consider the oceans as a paradigm.

The ocean reservoir has a long holding time so that the individual behavior of the various elements can be viewed against a time scale of several hundred million years (the age of the oldest deep-sea sediments). For most elements the main removal process is via sedimentary deposition and burial, although in some instances hydrothermal interactions with deep-sea basalts are also important. The major input of particulate material, either detrital or biologically generated, is in the surface layers of the ocean, and the particulate material is able to undergo further interactions as it sinks through the water column before settling in the sediment. Further significant exchanges, driven by particle–water interactions, can be observed across the sediment–water interface for many elements. The processes of particle production and dissolution are consequently well-separated in space and time, and much can be learned about particle–water interactions from the interpretation of vertical profiles of the dissolved and particulate constituents in the deep ocean. Furthermore, advances in clean sampling and handling techniques and in the development of sensitive analytical methods (some working in the 10^{-14} *M* range) have enabled oceanographically consistent profiles to be obtained for the majority of the elements in the periodic table over the last decade. Such wide coverage enables the involvement of the elements in particle–water interactions to be considered against the well-established background of relationships between the chemical periodicity of the elements and their biological utilization and chemical speciation in solution.

The contemporary alignment of the ocean basins also provides interesting contrasts, since the age of the deep water increases progressively as it passes from the Atlantic via the Indian Ocean to the Pacific. Consequently, elements that are released into deep water by the dissolution of particles will be enriched in the deep Pacific relative to the Atlantic, whereas elements that are scavenged from deep water will be depleted. Deep-sea profiles from the Atlantic Ocean are usually more difficult to interpret simply in terms of vertical processes than those obtained in the Pacific, since the production of large volumes of cold, dense water at the northern and southern extremities of the Atlantic basin ensures active deep-water circulation. The influence of riverine and aeolian inputs is also exaggerated in the Atlantic Ocean, since the ratio of the input fluxes relative to the surface area is greater than in the Pacific Ocean. Furthermore, anthropogenic effects, which accentuate the input of elements such as Pb, Sn, and Te via aerosols, are at their highest in the north Atlantic.

We intend to take advantage of these factors to show how the particle–water interactions that control the composition of seawater reflect the chemical periodicity of the elements. Before considering the individual elements in detail, we must discuss briefly the characteristics of the oceanic particulate cycle.

2. PARTICLES IN THE OCEAN

2.1. The Biological Particle Cycle

The overwhelming bulk of the particulate matter falling from the surface layers into the deep ocean is produced by biological processes (Fig. 17.1; Broecker and Peng, 1982). The particles are generated primarily by photosynthesis in the sunlit zone, and in addition to the major organic components (C, N, and P), a wide range of trace elements are incorporated either intentionally or incidentally. In the process, elements with variable oxidation states that are present in the oxidized form are often reduced to the lower oxidation state. As the organic matrices are broken down by respiration, these ele-

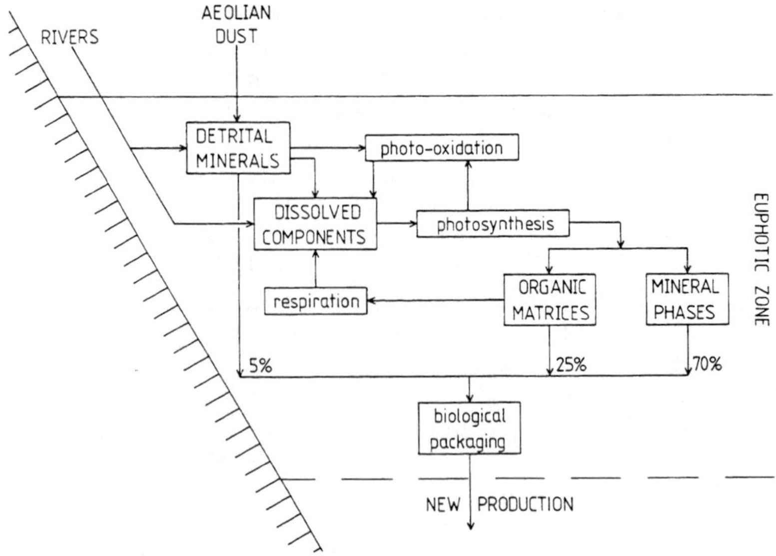

Figure 17.1. Schematic diagram of the biologically driven particle cycle in the surface layers of the ocean (see Section 2.1). The percentages on the arrows are the weight percentage contributions of the various phases to the particle flux.

ments are released in the reduced form so that significant concentrations of Mn(II), I(−I), Cr(III), As(III), and Se(IV) are observed. Since the inorganic reactions leading to the reoxidation of these reduced forms are usually very slow, they may persist sufficiently to establish their own characteristic deep-sea profiles. The cycling of Se provides a particularly clear example of this process. The thermodynamically stable Se(VI) is reduced to Se(IV) and eventually to Se(−II) for incorporation into seleno-amino acids. The reduced forms are released on degradation and are reoxidized only slowly (Cutter and Bruland, 1984). During respiration, significant quantities of dissolved and fine particulate organic matter are also released. These may in turn be oxidized by photochemical processes in which the oxidized forms of inorganic redox couples may be used as electron acceptors, thus releasing the reduced forms. Significant quantities of Fe(II), Mn(II), and I(−I) may be introduced into the surface waters by this mechanism. In relatively shallow productive regions along the ocean margins, the particulate matter produced by photosynthesis may reach the sediment with little alteration en route. The interstitial waters of such organic-rich sediments rapidly become anoxic, and the oxidized forms of the predominant redox couples are once again used as electron acceptors, this time by bacteria involved in the fermentative oxidation of organic matter. These sediments are consequently important sites for manganese recycling and for the release of elements associated with the organic phase (e.g., Cu; see Boyle et al., 1981; Bruland, 1980) or for the removal of elements which are insoluble in the reduced form (e.g., Cr, U). The horizontal transport of surface waters ensures that the influence of these marginal processes can be observed in the open ocean. Care must be taken to make due allowance for these advective processes when interpreting vertical profiles of trace element concentration in the deep ocean.

The phytoplankton are actively grazed by zooplankton which package their own waste products into fecal pellets. Although most of the material processed in this way is effectively recycled in the surface layers, a small proportion (18%; Collier and Edmond, 1984) of the material (known as *new production*) sinks into the deep ocean, mainly in the form of relatively large (>100 μm), rapidly sinking (100–1000 m day^{-1}) fecal pellets. On average, this particulate material contains (by weight) 70% biogenic mineral phases (60% $CaCO_3$, 10% silica), 25% organic material, and 5% detrital mineral phases. Despite the preponderance of mineral phases, the bulk of the trace elements removed from the surface layers appears to be incorporated into the organic fraction (Morel and Hudson, 1985; Morel and Morel-Laurens, 1983; Collier and Edmond, 1984; Sigg, Chapter 12, this volume). As the particles fall through the water column they provide a food source for successive populations of filter feeders so that the material in them is repackaged many times en route to the sediment. The fecal material also carries with it its own microbial flora and fauna so that the process of degradation continues in the deep water and the elements so effectively sequestered in the surface layers are once again released into solution. Although natural organic particles appear to fall more rapidly than might be expected from Stokes' law (Chase, 1979) the degradation process is so efficient that less than 1% of the particulate organic matter leaving the surface actually reaches the sea bed (Table 17.1). Under productive zones, the bulk of the organic material is broken down within or just below the thermocline at depths of 500–1000 m. The intense respiration which results produces an oxygen–minimum zone and on occasion provides

Table 17.1. Characteristics of Biogenic Minerals

Phase	Formula	Source	Saturation Index[a] (1 atm, 25°C)	Final Burial (Wt %)
Calcite	$CaCO_3$	Foraminifera Coccolithophorids	5.5	12
Aragonite	$CaCO_3$	Pteropods	3.7	
Celestite	$SrSO_4$	Acantharids Radiolaria	0.09	<1
Barite	$BaSO_4$	Radiolaria(?)	0.5	12
Silica	SiO_2	Diatoms Radiolaria	1.7 (quartz) 0.09 (silica gel)	2

[a] Saturation index = (activity product in seawater)/(solubility product). For values less than unity, the water column is undersaturated with respect to the mineral phase.
Source: Whitfield and Watson (1983).

a further site for the production of the reduced forms of redox-active elements. At times, however, sufficient numbers of large particles leave the surface layers to impart a seasonal pattern to the particulate flux even at depths of 3000 m or more (Bacon et al., 1985). Organic-rich sediments are, however, rare in the deep ocean. This vigorous, biologically driven cycle has a profound influence on the distribution of most of the trace elements in seawater and hence on the composition of any individual water sample. However, the *re*cycling is so efficient in many instances that it is not clear to what extent the biologically produced flux controls the *mean* concentration of individual elements in seawater (i.e., the ratio of total number of moles in the ocean to the total ocean volume).

The biogenic mineral phases associated with the particle flux (Table 17.1) are also dissolved with increasing efficiency as the particles encounter progressively higher pressures on sinking to greater depths. Since the organic phases tend to be recycled rapidly by biological processes just below the sunlit zone whereas the mineral phases are dissolved by chemical processes in deep water, there is a progressive shift in the mean composition of the particulate matter with depth (Simpson, 1982; Broecker and Peng, 1982). The nature of the underlying sediments is dictated by the relative survival of the different components of the particulate flux. This in turn depends on the productivity of the overlying water, the magnitude of any external fluxes, and the depth and composition of the water column. Organic-rich sediments underlie productive zones and are usually found in shallow waters in coastal or upwelling regions. Siliceous and carbonaceous sediments are also found under productive zones and are restricted to depths below which the rate of dissolution exceeds the rate of supply. Where such processes do not predominate, shales and red clays are found which largely reflect the composition of the aeolian input.

The dissolution of the settling particles releases the trace components and the major particulate components (C, N, P, Si) into the water column in characteristic ratios known as *Redfield ratios*. As a consequence, the deep-water concentrations of many trace elements are strongly correlated with the concentrations of elements characteristic of the carrier phases (notably P for the organic phase and Si for the mineral phase, Table 17.2; see also Morel and Hudson, 1985). Although it is tempting to interpret such correlations as indicating transport by organic material or by opaline silica, some care must be exercised. Si and P are often themselves correlated in the deep ocean (Measures et al., 1983), and advective processes in the deep ocean can obscure the signatures of individual sources of dissolved components (Edmond, 1974).

2.2. Deep-Sea Particles

The breakdown of many of the larger particles falling from the sunlit zone by dissolution processes releases a cloud of finer, more refractory particles

Table 17.2. Redfield Correlations between Trace Elements and Structural Components of Marine Particulate Matter ([X] = $a + b$[P] $+ c$[Si])[a]

X	a	b	c	Location	Reference
C_{ic} (μM)[b]	1161	—	0.555	All oceans	Watson & Whitfield (1985)
Cd (nM)	−0.068	0.347	—	N. Pacific	Bruland (1980, 1983)
Cr(VI) (nM)	3.6	—	0.066	N. Atlantic	Campbell and Yeats (1981)
Ge (pM)	3.8	—	0.69	All oceans	Froelich et al. (1985)
Gd (pM)	1.33	1.29	—	N. Pacific[c]	de Baar et al. (1985b)
La (pM)	3.57	8.87	—	N. Pacific[c]	de Baar et al. (1985b)
Lu (pM)	0.435	—	0.0306	N. Atlantic	de Baar et al. (1983)
	−1.82	—	0.027	Pacific	de Baar et al. (1985a)
	0.163	0.28	—	N. Pacific[c]	de Baar et al. (1985b)
NO_3^-	−1.25	15.0	—	All oceans	Watson & Whitfield (1985)
Ra[d]	7.3		0.1	Atlantic[e]	Broecker et al. (1976)
	13	—	0.072	Antarctic	Ku & Lin (1976)
Se VI (nM)	63	4.21	50	All oceans	Measures et al. (1983)
Se IV (nM)	448	3.32	181	All oceans	Measures et al. (1983)
Yb (pM)	0.95	1.72	—	N. Pacific[c]	de Baar et al. (1985b)
Zn (nM)	0.02		0.0535	N. Pacific	Bruland (1980)

[a] Concentrations of P and Si in μM.
[b] Total carbonate released from $CaCO_3$ dissolution.
[c] Above 900 m.
[d] Concentration of Ra in d.p.m. per 100 kg.
[e] Above 200 m.

with much slower settling times (1–100 m day^{-1}). This continuous supply of particles at the smaller end of the size spectrum is matched by a continuous process of Brownian (<1 μm) or shear-controlled coagulation between particles of similar size and by the capture of smaller particles by those that are sufficiently large (2–4 mm, "marine snow") to fall through the water column without significant degradation (Simpson, 1982; McCave, 1984). There are also biological repackaging mechanisms at work at intermediate depths below the sunlit zone, although the nature and distribution of the organisms responsible are still unclear. The continuous cycle of aggregation and disaggregation produces a particle-size distribution in the oceans that is flattened, with no peaks that can be identified with specific particle populations. The size distribution may be described by the equation

$$N = kd^{-\beta} \tag{17.1}$$

where N is the cumulative number of particles with diameters greater than d (Hunt, 1980; McCave, 1984). For deep-sea particulates in areas well away from the sea bed, $\beta = 3$ with d in the range 1–100 μm, so that approximately equal volumes of material are found in logarithmically increasing size grades.

For particles with $d > 100$ μm, β appears to become larger (up to 4), possibly because the particle-counting techniques do not adequately account for the relatively rare large particles (Simpson, 1982). Little attention has been paid to particles for which $d < 1$ μm, although Harris (1977) showed that $\beta = 1.62$ in this size range, indicating proportionately higher particle numbers.

The small, slowly settling particles provide sites for adsorption–desorption reactions and a possible pathway for the removal of the more particle-reactive trace elements. Although the data are rather sparse, it is possible to give a rough "thumbnail sketch" of the characteristics of particles most likely to be involved in such reactions. They will tend to be <5 μm in diameter with settling times significantly less than 1 m day^{-1}. The overall concentration of such particles in the ocean will be in the range of 20–50 μg L^{-1} (Simpson, 1982) with an effective surface area of 10 m^2 g^{-1} and a total concentration of surface sites in the range 1–2 m*M* g^{-1} (Balistrieri and Murray, 1984; Nyffeler et al., 1984). Despite their slow individual settling rates, the average lifetime of these particles in the water column is only 7.5 yr (Bacon and Anderson, 1982), indicating how effectively they are removed via interaction with the marine snow particles.

2.3. Summary

It would appear, then, that the biological cycle of photosynthesis and respiration is responsible for the transport of a wide range of trace elements out of the surface layers into the deep ocean. Here the particles are largely redissolved with rapid loss of the organic phases and a much slower, depth-dependent loss of the biogenic mineral phases. As a consequence of the particle breakdown, the deep waters support a significant population of small, slowly settling particles that can act as sites for adsorption–desorption reactions and that are removed to the sediment from time to time following encounters with large, rapidly falling organic particles.

A number of studies (Turner et al., 1980; Turner and Whitfield, 1983; Li, 1981, 1982a) have indicated that the gross particle–water partition coefficients for a wide range of elements can be related within an order of magnitude or so to the chemical periodicity of the elements by considering a variety of chemical bonding parameters. These correlations address the problem of the overall control of the concentrations of the elements over millions of years. Here we are specifically interested in the processes driven by the biota which effect a redistribution of trace element concentrations within the oceans over much shorter time scales. Consequently, rather than probe these correlations further, we would like to relate the information that has been obtained over the past five years on the concentration profiles of more than 60 elements in the deep ocean to the chemical periodicity of the elements.

3. CLASSIFICATION OF DEEP-OCEAN PROFILES

3.1. The Mean Oceanic Residence Time

Since incorporation in particulate matter is the predominant route for the removal of elements from seawater, we can obtain a rough measure of the intensity of the particle–water interactions experienced by the different elements by considering their mean oceanic residence times (Whitfield, 1979; Tables 17.3 and 17.4):

$$\tau_Y = \frac{\text{total no. of moles of Y in ocean}}{\text{rate of addition or removal}} \qquad (17.2)$$

There are often considerable uncertainties in both terms in this definition, and so we are simply going to use τ to subdivide the elements into three categories that can be represented by three different profile types in the deep ocean (Table 17.3).

Elements that interact weakly with the solid phase (*accumulated* elements) have long residence times ($\tau_Y > 10^5$ yr) and exhibit high concentrations in seawater relative to their crustal abundance. Since on average they remain in the oceans for periods well in excess of the mean oceanic stirring time ($\tau_R = 10^3$ yr), their concentrations maintain a constant ratio to one another and are influenced only by the gain or loss of water from the oceans. These elements are known as the "salinity elements," and they show uniform depth profiles in the ocean when normalized to seawater conductivity (Tables 17.3 and 17.4).

Elements with intermediate residence times ($\tau_Y = 10^3$–10^5 yr, *recycled* elements) tend to be incorporated into the biological particle cycle. They consequently exhibit vertical concentration profiles which show significant surface depletion (Tables 17.3 and 17.4) and deep regeneration. Despite their active involvement in particle–water interactions, they remain in the oceans for considerable periods of time because the particles in which they are initially incorporated are redissolved in deep water. The vertical profiles of these elements usually show coherent behavior from one location to another because their residence times are generally greater than τ_R.

The elements that show the strongest interactions with the particulate phase have very short residence times ($\tau_Y < 10^3$ yr, *scavenged* elements). In fact, the mean oceanic residence time concept itself is not applicable to these elements because they are removed on time scales far shorter than a single stirring cycle of the ocean reservoir. The scavenged elements are present in seawater at very low concentrations relative to their crustal abundances, and they frequently exhibit profiles that show a surface maximum related to their primary input into the oceans via aeolian dust or from the ocean margins, with concentrations decreasing to very low levels in the deep ocean (Tables

Table 17.3. Classification of the Elements according to Their Oceanic Profiles

Element Type	Mean Oceanic Residence Time	Concentration[a] Range (M)	$\frac{[X]_{\text{deep Atlantic}}}{[X]_{\text{deep Pacific}}}$	Profile Type
Accumulated	$>10^5$	10^{-8}–10^{-1}	1	—[X]→ ←depth—
Recycled	10^3–10^5	10^{-11}–10^{-5}	<1	—[X]→ ←depth—
Scavenged	$<10^3$	10^{-14}–10^{-11}	>1	—[X]→ ←depth—

[a] These ranges show significant overlap, since the concentrations of the elements also depend on crustal abundance.

17.3 and 17.4). The deep-ocean profiles of the scavenged elements show considerable variability from site to site, since they retain the signatures of localized input or removal processes because of their very short residence times.

τ_Y gives us a pragmatic view of the relative particle reactivities of the elements based on the ease with which they are removed from the ocean reservoir. To provide a more specific link between particulate removal and the position of the elements in the periodic table, we need to select some appropriate chemical bonding parameters.

Table 17.4. Characteristics of the Elements

Element	Oxidation State	Concentration: Atlantic Surf	Atlantic Deep	Pacific Surf	Pacific Deep	τ_Y (yr)	$-\log R'$[a]	Biol. Util.[b]	Reference
				Accumulated Elements[c]					
B	III	.42 m*M*				1E7	4.7	B	Bruland (1983)
Br	−I	.84 m*M*				1E8		C	Whitfield (1981)
Cl	−I	.53 *M*				4E8		B	Whitfield (1981)
Cs	I	2.3 n*M*				6E5	<0	D	Bruland (1983)
F	−I	68 μ*M*				4E5		C	Whitfield (1981)
K	I	10 m*M*				5E6	−0.5	A	Bruland (1983)
Li	I	2.6 μ*M*				2E6	0.4	D	Stoffyn-Egli & Mackenzie (1984)
Mg	II	53 m*M*				1E7	2.5	A	Bruland (1983)
Mo	VI	107 n*M*				6E5		B	Collier (1985)
Na	I	.47 *M*				1E8	−0.2	B	Bruland (1983)
Rb	I	1.4 μ*M*				8E5	<0	D	Bruland (1983)
S	VI	28 m*M*				8E6		A	Whitfield (1981)
Tl	I	69 p*M*				1E4	0.5	D	Flegal & Patterson (1983)
U	VI	13.5 n*M*				3E5	1.4	D	Anderson (1982)
				Recycled Elements					
Ag	I			1 p*M*	23 p*M*	5E3	−3.3	D	Martin et al. (1983)
As	V	20 n*M*	21 n*M*	20 n*M*	24 n*M*	9E4		D	Andreae (1979)
									Burton et al. (1983)
Ba	II	35 n*M*	70 n*M*	35 n*M*	150 n*M*	1E4	0.4	D	Collier & Edmond (1984)
Be	II	10 p*M*	20 p*M*	4 p*M*	25 p*M*	4E3	5.9	D	Measures & Edmond (1982)
C	IV	2.0 m*M*	2.2 m*M*	2.0 m*M*	2.4 m*M*	8E5		A	Collier & Edmond (1984)
Ca	II	10 m*M*	11 m*M*	10 m*M*	11.3 m*M*	1E6	1.3	B	Bruland (1983)

Table 17.4. (*Continued*)

Element	Oxidation State	Concentration: Atlantic Surf	Atlantic Deep	Pacific Surf	Pacific Deep	τ_Y (yr)	$-\log R'$[a]	Biol. Util.[b]	Reference
Cd	II	10 p*M*	.35 n*M*	10 p*M*	1 n*M*	3E4	3.3	D	Collier & Edmond (1984)
Cr	VI	3.5 n*M*	4.5 n*M*	3 n*M*	5 n*M*	1E4		C	Campbell & Yeats (1981)
									Murray et al. (1983)
Cu	II?	1.3 n*M*	2 n*M*	1.3 n*M*	4.5 n*M*	3E3	5.0	A	Collier & Edmond (1984)
Dy	III	5 p*M*	6.1 p*M*			300	5.1	D	Elderfield & Greaves (1982)
Er	III	3.6 p*M*	5.3 p*M*			400	5.0	D	Elderfield & Greaves (1982)
Eu	III	0.6 p*M*	1 p*M*	0.7 p*M*	1.8 p*M*	500	5.5	D	Elderfield & Greaves (1982)
									de Baar et al. (1985a)
Fe	III	2 n*M*	7 n*M*	0.2 n*M*	2 n*M*	98	5.8	A	Symes & Kester (1985)
									Collier & Edmond (1984)
Gd	III	3.4 p*M*	6.1 p*M*	4 p*M*	10 p*M*	300	5.0	D	Elderfield & Greaves (1982)
									de Baar et al. (1985a)
Gè	IV	1 p*M*	20 p*M*	5 p*M*	100 p*M*	2E4		D	Froelich et al. (1985)
Ho	III	1.5 p*M*	1.8 p*M*	1 p*M*	3.6 p*M*		5.0	D	de Baar et al. (1983, 1985a)
I	V	0.2 μ*M*	0.45 μ*M*	0.35 μ*M*	0.47 μ*M*	3E5		C	Elderfield & Truesdale (1980)
									Wong & Brewer (1974)
La	III	13 p*M*	28 p*M*	19 p*M*	51 p*M*	200	5.1	D	Elderfield & Greaves (1982)
									de Baar et al. (1985a)
Lu	III	0.8 p*M*	1.2 p*M*	0.35 p*M*	2.4 p*M*	4E3	5.1	D	de Baar et al. (1983, 1985a)
N	V	5 n*M*	20 μ*M*	5 n*M*	40 μ*M*	6E3		A	Takahashi et al. (1981)
Nd	III	13 p*M*	23 p*M*	13 p*M*	34 p*M*	500	5.3	D	Elderfield & Greaves (1982)
									de Baar et al. (1985a)
Ni	II	2 n*M*	7 n*M*	2 n*M*	10 n*M*	8E4	4.2	C	Collier & Edmond (1984)
P	V	50 n*M*	1.4 μ*M*	50 n*M*	2.8 μ*M*	1E5		A	Collier & Edmond (1984)
Pd	II			0.18 p*M*	0.66 p*M*	5E4	5.8	D	Lee (1983)

Pr	III	3 pM	5 pM	3.2 pM	7.3 pM		5.3	D	de Baar et al. (1983, 1985a)
Pt	II			0.6 pM	1.4 pM		−1.0	D	Hodge et al. (1985)
Ra[d]	II	8	20	10	35		0	D	Broecker et al. (1976) Chung & Craig (1980)
Sc	III	14 pM	20 pM	8 pM	18 pM	5E3	6.0	D	Bruland (1983)
Se	IV	0.1 nM	0.9 nM	0.07 nM	0.9 nM	3E4		C	Cutter & Bruland (1984) Measures et al. (1984)
Se	VI	0.5 nM	1.5 nM	0.13 nM	1.25 nM			C	Cutter & Bruland (1984) Measures et al. (1984)
Si	IV	1 μM	30 μM	1 μM	150 μM	3E4		C	Collier & Edmond (1984)
Sm	III	2.7 pM	4.4 pM	2.7 pM	6.8 pM	200	5.3	D	Elderfield & Greaves (1982) de Baar et al. (1985a)
Sr	II	89 μM	90 μM	89 μM	90 μM	4E6	0.6	D	Bruland (1983)
Tb	III	0.7 pM	1 pM	0.5 pM	1.6 pM		5.3	D	de Baar et al. (1983, 1985a)
Tm	III	0.8 pM	1 pM	0.4 pM	2 pM		5.2	D	de Baar et al. (1983, 1985a)
V	V	23 nM		32 nM	36 nM	5E4	2.1	C	Collier (1984) Morris (1975)
Yb	III	3 pM	4.5 pM	2.2 pM	13 pM	400	5.0	D	Elderfield & Greaves (1982) de Baar et al. (1985a)
Zn	II	0.8 nM	1.6 nM	0.8 nM	8.2 nM	5E3	4.7	A	Collier & Edmond (1984)
				Scavenged Elements					
Al	III	37 nM	20 nM	5 nM	0.5 nM	150	5.6	D	Orians & Bruland (1985)
As[e]	III			0.3 nM	70 pM		4.6	D	Andreae (1979)
Bi	III	0.25 pM		0.2 pM	0.02 pM		6.1	D	Lee (1982) Measures et al. (1984)

Table 17.4. (*Continued*)

Element	Oxidation State	Concentration				τ_Y (yr)	$-\log R'$[a]	Biol. Util.[b]	Reference
		Atlantic		Pacific					
		Surf	Deep	Surf	Deep				
Ce	III	66 p*M*	19 p*M*	11 p*M*	4 p*M*	100	5.0	D	Elderfield & Greaves (1982)
									de Baar et al. (1985a)
Co	II			0.12 n*M*	0.02 n*M*	40	4.1	B	Knauer et al. (1982)
Cr[e]	III			0.2 n*M*	0.05 n*M*		6.5	C	Cranston & Murray (1978)
									Murray et al. (1983)
Hg?	II	2.5 p*M*	2.5 p*M*	1.7 p*M*	1.7 p*M*		−3.8	D	Dalziel & Yeats (1985)
									Gill & Fitzgerald (1985)
I[e]	−I	0.2 μ*M*	10 p*M*	90 p*M*	60 p*M*			C	Bruland (1983)
									Elderfield & Truesdale (1980)
Mn	II	1.9 n*M*	1.8 n*M*	1.9 n*M*	0.8 n*M*	50	3.2	A	Collier & Edmond (1984)
Pb	II	0.15 n*M*	20 p*M*	50 p*M*	5 p*M*	50	4.8	D	Bruland (1983)
Sn	IV	20 p*M*	5 p*M*				8.4	D	Byrd & Andreae (1982)
Te	VI	0.9 p*M*	0.4 p*M*	1 p*M*	0.4 p*M*		0.2	D	Lee & Edmond (1985)
Te	IV	0.4 p*M*	0.2 p*M*	0.5 p*M*	1 p*M*			D	Lee & Edmond (1985)
^{232}Th[f]	IV	87	3		20	50	6.0	D	Huh & Bacon (1985)

[a] Scavenging index, see Eq. 17.11 and Figure 17.6.

[b] Biological utilization. A, essential in all species tested; B, widely but not universally required; C, required in a limited group of species; D, not required (McClendon, 1976).

[c] Concentrations normalized to a salinity of 35.

[d] Concentration units d.p.m. per 100 kg.

[e] These profiles may result from homogeneous oxidation (see Section 4.3.1).

[f] Concentration units d.p.m. per 10^6 kg.

3.2. Chemical Bonding Parameters

Most particulate phases encountered in the oceans, whether oxides, carbonates, silicates, oxyhydroxides, or organic matrices, will have significant numbers of surface hydroxyl sites. The stability constants for the interaction of cations with these sites have been shown to be proportional (on a log/log basis) to the corresponding stability constants for hydrolysis (Dugger et al., 1964; Balistrieri et al., 1981). The Born electrostatic theory provides a particularly simple picture of cation hydrolysis which has proved useful in discussing the chemical speciation of the elements in seawater (Turner et al., 1981). According to this theory, the intensity of the ion–water interactions can be related to the *electrostatic energy* Z_i^2/r_i associated with the encounter. Z_i is the charge on the central cation, and r_i is its ionic radius. A plot of the degree of hydrolysis (α_{OH}) of the cations versus the electrostatic energy (Fig. 17.2) provides a clear separation of the elements into weakly, strongly, and totally hydrolyzed categories that will prove useful in our subsequent discussions. The degree of hydrolysis (α_{OH}) is defined by (Turner et al., 1981);

$$\alpha_{OH} = 1 + \sum_j {}_j\beta_{M,OH}[OH]^j \tag{17.3}$$

where ${}_j\beta_{M,OH}$ is the stability constant for the formation of the jth hydroxy complex. To complement the electrostatic viewpoint it is also important to consider a parameter that reflects the covalent contribution to chemical bonding. Although a variety of parameters have been considered (see the review by Turner and Whitfield, 1983), the simplest index for seawater is provided by (Turner et al., 1981)

$$\Delta\beta = \log \beta^0_{MF} - \log \beta^0_{MCl} \tag{17.4}$$

where the β values represent the stability constants of the monofluoro and monochloro complexes, respectively. The (a)-type cations ($\Delta\beta > 2$) bond most strongly to hard anions (e.g., F^-, OH^-, SO_4^{2-}) via electrostatic interactions, and (b)-type cations ($\Delta\beta < 2$) bond most strongly to soft anions (e.g., Cl^-, Br^-, HS^-) via covalent interactions. A plot of log (Z_i^2/r_i) versus $\Delta\beta$ yields the *complexation field diagram* (CF diagram, Turner et al., 1981; Fig. 17.3), which enables the elements to be grouped according to their chemical speciation patterns in seawater.

We will now use these bonding parameters to look at the elements associated with each of the three categories identified in Tables 17.3 and 17.4 in an attempt to rationalize their behavior. As the overall pattern becomes clearer, we will turn our attention to those elements whose deep-sea profiles have not yet been studied and try to predict their behavior.

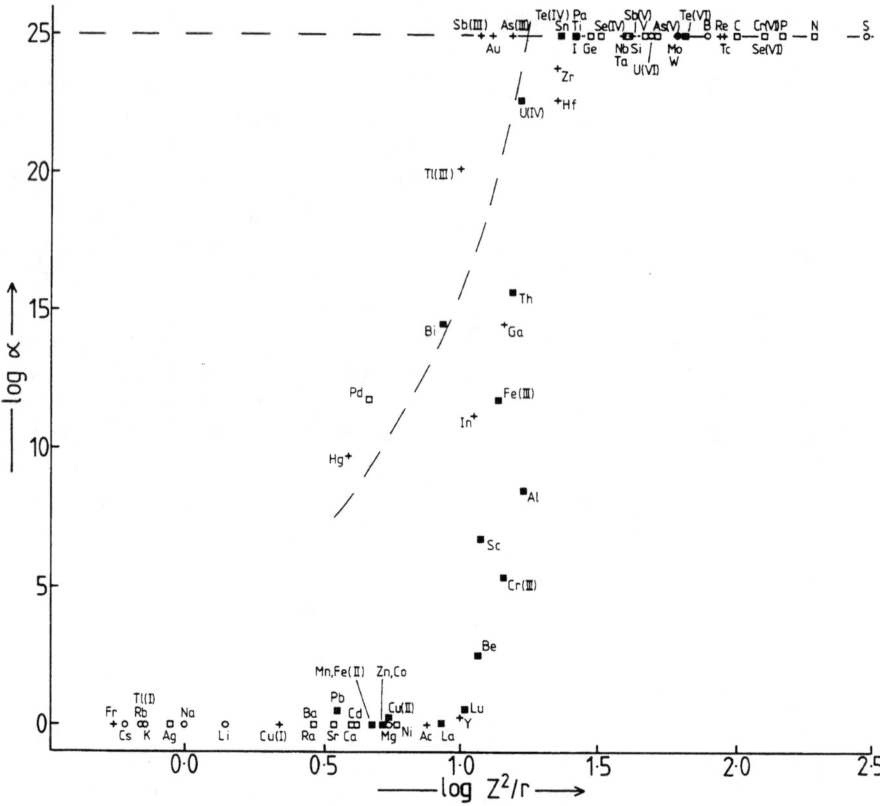

Figure 17.2. Plot of the hydrolysis side-reaction coefficient (α, Eq. 17.3) at pH 8.2 versus the electrostatic energy Z_i^2/r_i (the Born diagram). The horizontal dashed line at log α = 25 indicates the effective limit of α, and all fully hydrolyzed elements are plotted on this line. The curved dashed line separates the more polarizing (b)-type cations, which are hydrolyzed exceptionally strongly, from the main sequence (Turner et al., 1981). (○) Accumulated elements (Section 4.1); (□) recycled elements not subject to scavenging (Sections 4.2.1 and 4.2.2); (■) scavenged elements (Sections 4.2.3 and 4.3); (+) behavior in the oceans unknown (Section 5).

4. DEEP-OCEAN PROFILES OF THE ELEMENTS

4.1. Accumulated Elements

The majority of the elements that are effectively accumulated in the oceans are elemental ions with a small electrostatic energy (Fig. 17.2) and an inert gas structure (e.g., Li^+, Na^+, K^+, Rb^+, Cs^+, Mg^{2+}, F^-, Cl^-, Br^-, Fig. 17.4). These ions interact only weakly with mineral phases, usually via ion exchange, and are present predominantly as free ions in solution. They tend to be removed from the oceans either via entrapment in the interstitial waters of consolidating sediments or by interactions with hot basalts. For example, Li

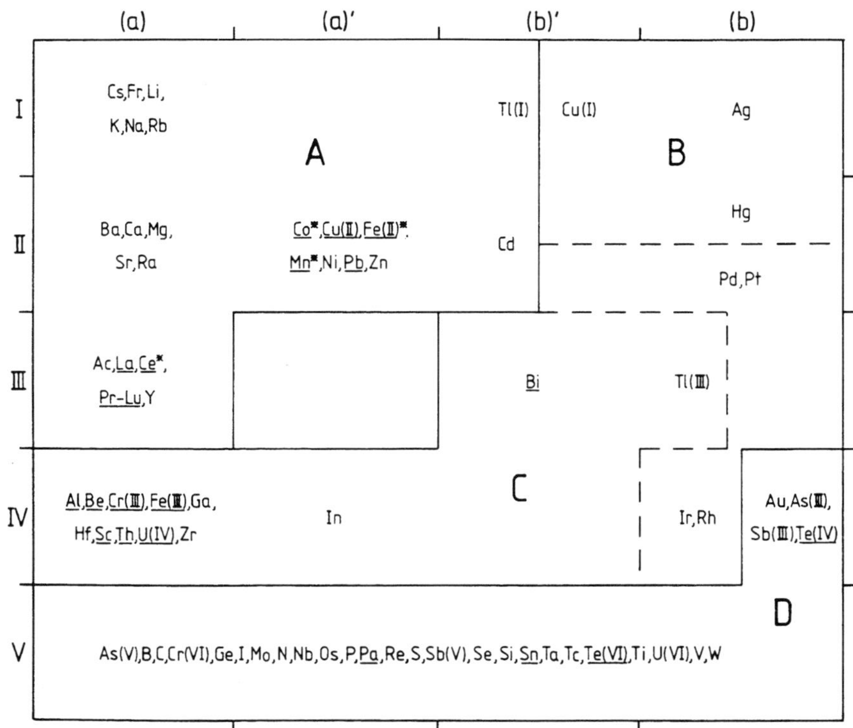

Figure 17.3. Complexation field (CF) diagram for the elements in seawater. The horizontal groupings I, II, III, IV, and V correspond to increasing values of Z^2/r (Fig. 17.2), and the vertical groupings (a), (a)′, (b), and (b)′ correspond to increasing covalent character (decreasing $\Delta\beta$, Eq. 17.4). The quantitative basis for these divisions is discussed by Turner et al. (1981). The diagram is divided into four sectors: A, weakly complexed elements with >1% present as free cation; B, elements dominated by chloride complexes; C, strongly hydrolyzed elements (transition zone of Fig. 17.2); D, fully hydrolyzed elements. The dashed boundaries between sectors B and C reflect a lack of data on the hydrolysis of the platinum group metals. Elements that are scavenged in the oceans are underlined, and those that are subject to oxidative scavenging are marked with an asterisk.

is removed by low-temperature (150°C) basalt–water interactions away from the ridge crests (Stoffyn-Egli and Mackenzie, 1984), and both K and Mg show significant removal in deep-sediment profiles (Gieskes, 1983). In addition to the inert gas anions, there are two oxyanions (SO_4^{2-}, $B(OH)_4^-$) which also interact weakly in solution and with the solid phase. Although a number of these elements (notably Na, K, Mg, B, S, Cl, Fig. 17.4) are biologically important, their seawater concentrations are too high for their utilization to have any impact on their distribution in the oceans.

In addition to the major sea salt components, there are three elements (Tl, U, Mo) that are also accumulated in the oceans (Table 17.4), each for

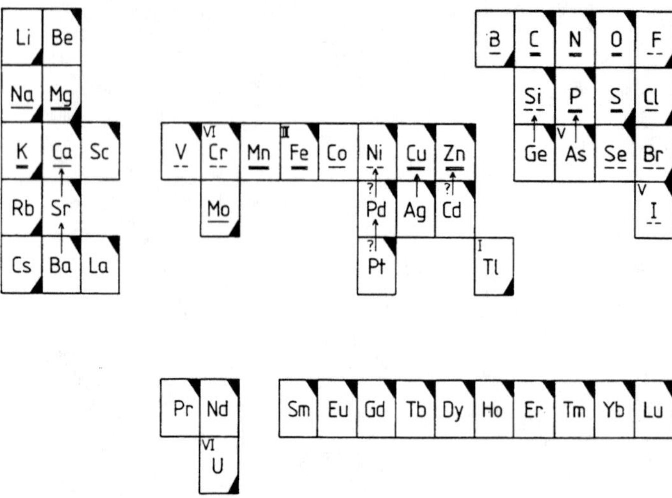

Figure 17.4. Periodic table showing the accumulated and recycled elements. Oxidation states are identified, where necessary, in the top left-hand corner of each box. Vertical arrows indicate analogies which lead to the inclusion of nonessential elements in the biological particle cycle (Section 4.2.2). Corner shadings: lower right corner (e.g., Li), accumulated elements; upper right corner (e.g., Be), recycled elements. Biological requirements (McClendon, 1976) are indicated by underlining: double underline, essential in all species tested; single underline, widely but not universally required; dashed underline, required in a limited group of species.

a different reason. Although earlier equilibrium calculations suggested that Tl might be present in seawater as Tl(III) (Turner et al., 1981), the observed distribution indicates that it is actually present as Tl(I) (Flegal and Patterson, 1985). For all the elements from Hg to Po along this row of the periodic table (Fig. 17.5), the lower oxidation state is stabilized by the presence of a filled 6*s* shell. Tl^+ therefore has properties similar to those of the alkali metals. Although Tl^+ has an ionic radius similar to that of Rb^+, its mean oceanic residence time is two orders of magnitude lower (Table 17.4). Since Tl is enriched in ferromanganese nodules, it is likely that it is removed by slow oxidative scavenging (see Section 4.3.1). More detailed studies might reveal small but consistent deficits of Tl in deep water (cf. lithium, Stoffyn-Egli and Mackenzie, 1984). U is unique among the natural elements in seawater in being present as an oxocation which interacts strongly with carbonate ions to yield a large, inert, negatively charged complex. Although U is removed from the surface layers in the organic phase of detrital particles, less than 0.1% of the total U is entrained in this way, and this value falls to one part in 10^5–10^6 in the deep ocean (Anderson, 1982). The associated flux of U is too small to influence the dissolved U profile. It is likely that U is removed as U(IV) in anoxic sediments along the ocean margins. Mo can also be removed in anoxic sediments as the insoluble Mo(VI) sulfide (Bertine, 1972). It is present in seawater as the molybdate anion (MoO_4^{2-}), which is analogous to sulfate

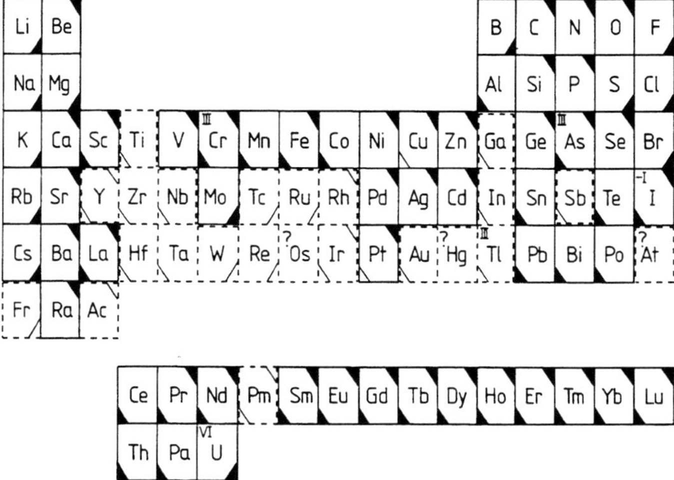

Figure 17.5. Periodic table showing the scavenged elements and the elements whose oceanic cycles have not been investigated. The accumulated and recycled elements are also shown. The oxidation states given refer to scavenging processes and therefore differ from those given in Figure 17.4. Corner shadings: lower right, accumulated element; upper right, recycled element; lower left, scavenged element. Elements whose oceanic cycles have not yet been investigated are enclosed by dashed lines, and their predicted behavior is indicated by open corner hatching. As(III) and I(− I) are shown as being scavenged (see also Table 17.4), but their characteristic "scavenging" profiles may well be caused by homogeneous oxidation.

and interacts only weakly with cations in solution. Analyses of deep-sea profiles in the north Pacific and north Atlantic reveal identical Mo concentrations throughout the water column when corrected for salinity (Morris, 1975; Collier, 1985). It is possible that both U and Mo will show fine structure associated with removal at ocean margins when more detailed profiles become available.

4.2. Recycled Elements

4.2.1. Active Biological Uptake. Since the driving force behind the recycling of the elements in the water column is the formation and dissolution of biogenic particulate matter, one might expect the recycled elements to coincide with those for which there is a specific biological requirement (Egami, 1974; McClendon, 1976; Williams, 1981). The biota have a preference for the lighter, more abundant elements in the periodic table, and only 4 of the 24 biologically important elements (Se, Mo, Br, I) have atomic numbers greater than that of Zn (N = 30). All the essential cationic elements are found grouped together in sector A of the CF diagram (Fig. 17.3). We have already seen that a number of biologically important elements (Na, K, Mg, B, S, F, Cl, Br, Mo) are accumulated in the oceans, and their concentrations are consequently too high to be effectively influenced by biological cycling. Of the remaining elements specifically required by the biota, most [C, N, Si, P, Ca, V, Cr(VI), Fe, Ni, Cu, Zn, Se, I(V)] show deep-sea profiles characteristic of the recycled elements (Table 17.3 and Fig. 17.4). These elements tend to be enriched in the deep Pacific relative to the deep Atlantic (Table 17.4), with the degree of enrichment being related to the intensity with which the elements are recycled (Broecker and Peng, 1982). Oxygen acts as an "antinutrient," since it is released during photosynthesis and consumed during respiration. Its deep-sea profiles are therefore a mirror image of those of N and P. The biologically important elements will be actively taken up by the phytoplankton so that only in the broadest sense can this aspect of their biogeochemical cycling be attributed to particle–water interactions. Their active removal from the surface layers is usually very effective, so that their availability can actually control new phytoplankton production. N and P are normally cast in the role of limiting elements in this sense, although the production of diatoms can also be restricted by the availability of Si for shell formation. This nutrient limitation strictly refers to the extent of new (or exportable) production, since productivity per se can be maintained in the presence of extremely low nutrient concentrations by efficient recycling. With the exception of C, Si, and Ca, the elements in this group are all transported out of the surface layers predominantly in the organic phase (Collier and Edmond, 1984; Morel and Morel-Laurens, 1983; Morel and Hudson, 1985). It has been suggested that a number of these elements (notably Fe, Ni, Cu, Zn) might act in concert with N and P (and occasionally Si) as colimiting nutrients and that the biological systems in the surface layers have evolved to come to some sort of accommodation with the natural availability of this nutrient suite (Morel and Hudson, 1985). This also implies that the trace metal nutrients, in common with N, C, P, and Si, should maintain a reasonably constant ratio to one another in the resulting particulate matter. This interesting hypothesis remains to be tested, since we do not yet have adequate data either on the degree to which phytoplankton growth is limited by the availability of trace metals or on the trace metal composition of detrital particulate matter. Within this group one might expect the closest correlations between trace elements and structural components of the major carrier phases (e.g., P, Si). Although a number of such correlations are observed (Table 17.2), some of the essential trace metals (notably Fe, Ni, Cu) show more complex behavior (see Section 4.2.3).

4.2.2. Uptake by Analogy. In addition to the biologically important elements, there are a number of elements that have no apparent biological function but show deep-sea profiles characteristic of the recycled elements [Fig. 17.4; Be, Sc, Ge, As, Sr, Ba, Pd, Pt, Ag, Cd, and the rare earth elements (REE)]. A significant number of elements in this category may be considered as being assimilated by analogy with essential elements because of a lack of discrimination in the biological uptake mechanism. A classic example of this behavior is provided by Ge. Although more than four-fifths of the Ge in seawater is present as methyl-Ge (Lewis et al., 1985), the inorganic Ge present shows a constant ratio to Si (0.7×10^{-6}) throughout the oceans and in biogenic Si (Froelich et al., 1985). Ge mimics the behavior of Si so closely that it is used as a "superheavy isotope" of Si to trace its biological utilization. Similar cases of mistaken identity on the part of organisms may occur with $HAsO_4^{2-}$, which is assimilated as HPO_4^{2-} (Andreae, 1979), and Ag, which follows Cu (Martin et al. 1983). Arsenic is not very effectively cycled by the biota since there is little depletion in the surface layers and the concentration in the deep Pacific is only slightly cnhanced over that in the deep Atlantic (Table 17.4). The main influence of the biota on the cycling of As is via the formation of As(III), which might be scavenged in the deep ocean (Section 4.3.1). Although few data are available, the distribution of Ag appears to follow that of Cu (Martin et al., 1983). This suggests that Cu, like Ag, is assimilated in the monovalent form (sector B, Fig. 17.3) and is held in the organic matrix predominantly by the formation of strong metal–sulfur bonds. The Ag is certainly tightly held in the organic phase, being released only after acid digestion in Teflon bombs, and a constant Ag:C ratio of 1.7×10^{-7} is observed. Cd, another (b)-type cation (Fig. 17.3) is also held strongly via sulfur bonds, and its distribution in the oceans is so closely tied to the organic phase that an approximately constant relationship is found between Cd and P concentrations throughout the world oceans (Table 17.2; Bruland, 1980; Bruland et al., 1978; Danielsson, 1980; Burton et al., 1983; Collier and Edmond, 1984). The correlation is so close that the Cd contents of foraminifera have been used to estimate the P content of paleo-oceans (Boyle and Kegwin, 1982). It is possible that Cd finds its way into the organisms via the pathway used for Zn, and once inside it is effectively sequestered in the sulfur "trap." A similar suggestion has been made for the assimilation, via the pathway used for Ni, of Pd (Lee 1983) and, by inference, Pt (Hodge et al., 1985). However, this link must be rather tenuous since, unlike Ni, the solution speciation of Pd(II) is dominated by the formation of strong hydroxide and chloride complexes ($\log \alpha = 11.6$) with only a negligible proportion of free metal ion remaining in solution (Fig. 17.3), and the limited data available also suggest similar behavior for Pt(II).

The uptake of Sr and Ba is analogous to that of Ca in that all three elements are involved in the formation of biogenic minerals (Whitfield and Watson, 1983; Table 17.1). These (a)-type cations interact strongly with hard anions such as sulfate and carbonate to produce insoluble salts. Of the minerals produced, only $CaCO_3$ (as calcite or aragonite) is supersaturated in the surface layers (Table 17.1). A negligible fraction of the $SrSO_4$ (celestite) produced in the surface layers actually reaches the sediments, and most of the Sr removed from the oceans is probably incorporated into $CaCO_3$ or possibly $BaSO_4$ lattices. The cycling of Ba has received considerable attention since it provides a useful analog for ^{226}Ra. A careful study of the various contributions to the vertical flux of Ba in the oceans (Collier and Edmond, 1984) has revealed that the known carriers associated with large (>40 μm) particles are insufficient, by an order of magnitude, to account for the observed deep-water enrichment of Ba. However, small discrete crystals of barite ($BaSO_4$) have been observed in the suspended matter in the deep ocean (Dehairs et al., 1980) associated with the smaller size fraction. In addition, measurements made on particulate matter in the deep sea indicate that there is successive enrichment of Ba with respect to Al at depths down to 500 m, with increases in the range 5–30-fold (Collier and Edmond, 1984). The identity of the solid phases involved and the nature of the processes producing them both remain a mystery.

4.2.3. Uptake with Scavenging. Cu and Fe have the shortest residence times of the biologically important elements in the "recycled" group. Although they show characteristic deep-sea profiles with surface depletion and deep-sea enrichment, the increase in concentration with depth is only gradual, suggesting the action of an additional process which removes the elements from the water column after they have been released by the dissolution of the primary carrier phase. This is clearly demonstrated by the concave plots obtained when the concentration of Cu in deep water is plotted against the salinity or the depth-corrected temperature in situations where the horizontal transport of Cu is negligible (Craig, 1974; Bruland, 1980). Cu and Fe are apparently scavenged from the water column by adsorption onto the surfaces of the resident population of fine particles (Section 2.2) and are thence transported to the sediment via interactions with the larger, rapidly settling particles. The Cu scavenging rate is well correlated with the primary productivity in the overlying water column (Collier and Edmond, 1984). Within the sediment some remobilization of Cu can occur, giving rise to localized near-bottom enrichments (Klinkhammer, 1980). Cu forms exceptionally strong complexes with organic ligands in comparison with other elements in the first long row of the periodic table because of Jahn-Teller stabilization effects (Irving-Williams rule; Williams, 1979). The distribution of Fe in the oceans has only been sketchily considered because of the problems raised by the very low concentration levels and the high risk of contamination. Fe is thermodynamically stable as Fe(III), which forms insoluble oxyhydroxides in the seawater medium, and its behavior is unusual in that it is present predominantly in the particulate form in the deep sea (Symes and Kester, 1985). It is likely that the fraction of Fe presently considered as dissolved (i.e., <0.4 μm) is actually colloidal. The inclusion of Fe in the "recycled elements" on the basis of its deep-sea profiles is therefore fortuitous, and it properly belongs with the "scavenged elements" (Section 4.3).

The final group of elements displaying nutrient-type profiles (Be, Sc, REE) have no apparent biological function nor can they be considered as presenting any compelling analogies with biologically important elements. However, they do exhibit relatively large electrostatic energies and, together with Cr(III), form a coherent group on the Born diagram (Fig. 17.2). It is likely that these elements are incidentally associated with the particulates leaving the surface layers by virtue of their ability to form strong electrostatic bonds with surface groups. Like Cu and Fe, they have short residence times, and once released from their primary carrier phase, they are scavenged in the deep water. The behavior of Be closely matches that of Cu, showing both a concave concentration versus salinity plot and near-bottom enrichment associated with sedimentary recycling (Measures and Edmond, 1982, 1983). Sc and the REE (with the exception of Ce) do not show such direct evidence of scavenging, and in fact good Redfield correlations are observed on occasion with the heavy REE tracking Si and the light REE tracking P (Table 17.2; Elderfield and Greaves, 1982; de Baar et al., 1985a). However, the correlation with Si shows rapidly increasing curvature with decreasing atomic number, indicating progressively more effective scavenging. There are also a number of other clues which indicate a systematic deep-sea removal. The REE have relatively short residence times, and their concentrations in seawater increase systematically with increasing atomic number in a manner (Table 17.4) that closely parallels the shale normalized pattern of REE distribution. This increase also matches the increasing degree of complexation of the REE in seawater with atomic number (Turner et al., 1981; de Baar et al., 1985c)—including an anomalous blip at Gd where the $4f$ shell is half-filled. For elements in this region of the CF diagram (Fig. 17.3), one might expect the scavenging process to proceed via the uptake of the free metal ion, and hence its efficiency should decrease as the metal ions become more effectively complexed in solution with increasing atomic number. This effect is revealed not only in the increase in the seawater concentrations with increasing atomic number but also in the progressive enrichment observed in the deep Pacific relative to the deep Atlantic (de Baar et al., 1985a; Table 17.4).

The similarity in behavior between Cu and the REE is also reflected in the effective recycling of the REE in deep-ocean sediments (Elderfield and Greaves, 1982) and in anoxic environments (German et al., 1985). We will now consider the chemistry of the elements whose distribution is *dominated* by scavenging processes (Fig. 17.5).

4.3. Scavenged Elements

4.3.1. Oxidative Scavenging. The deep-sea concentration profiles of the elements with the lowest concentrations and shortest residence times in the ocean are characterized by surface maxima and by a rapid falloff to low concentration levels in the deep sea (Tables 17.3 and 17.4, Fig. 17.5). Such profiles are generated by the active addition of the elements to the surface layers coupled with their vigorous removal throughout the water column. Although scavenging onto particulate matter is usually the dominant removal process, such profiles can also be produced by chemical species that are formed in the surface layers by biological or photochemical reduction and are removed by gradual oxidation to dissolved forms in deep water [e.g., I(−I), As(III)]. Here we will focus on those elements that are removed by scavenging onto particulate matter.

The most striking example of removal via particulate matter among the "recycled elements" was provided by iron, which was soluble in the reduced form Fe(II) but was rapidly oxidized in the water column to produce insoluble Fe(III). Similar behavior is observed for Mn, which is soluble as the thermodynamically unstable Mn(II) but oxidizes to form insoluble Mn(III) and Mn(IV) oxyhydroxides. Whereas the lifetime of Fe(II) in the water column is only a matter of hours, the lifetime of Mn(II) is of the order of 70 years in open ocean water (Yeats and Bewers, 1985). Consequently, any Mn(II) introduced into the surface layers as a result of photooxidation (Sunda et al., 1983) or diagenesis in anoxic or suboxic sediments along the ocean margins (Martin et al., 1985; also see Section 2.1) can accumulate and contribute to a surface maximum (Landing and Bruland, 1980; Klinkhammer and Bender, 1980; Yeats and Bewers, 1985). Secondary features may be introduced into the Mn profiles by the release of Mn from particulate matter in the oxygen-minimum zone underlying productive waters (Klinkhammer and Bender, 1980) and by the release of Mn from active centers along the midoceanic ridge. The removal of Mn by oxidative scavenging becomes effective in deep waters, yielding a relatively short scavenging residence time (Table 17.5). Consequently a large proportion (up to 30%, Yeats and Bewers, 1985) of the total Mn is present in the particulate phase. The scavenging of Fe and Mn from the water column by virtue of the insolubility of their thermodynamically stable higher oxidation states explains why they are present in such low concentrations in seawater even though they are the most abundant transition metals in the earth's crust. Where processes such as hydrothermal release or sediment diagenesis produce a significant flux of Fe(II) or Mn(II), large concretions, known as ferromanganese nodules, can be produced (Glasby and Read, 1976). In the sediment they represent the fourth most abundant marine deposit after aluminosilicate clays, calcium carbonate, and biogenic silica (Glasby, 1984). As the nodules form they also incorporate a range of trace metals (Li, 1982b). In some instances [e.g., Co(II), Ce(III), Pb(II)], the Mn(IV) is able to oxidize trace metals to higher oxidation states which are incorporated into the ferromanganese nodules (Hem, 1980). Thus Co(II) can be oxidized to Co^{3+}, a d^6 cation able to exhibit octahedral coordination which has an ionic radius almost identical with that of Fe^{3+} and Mn^{4+} (Glasby and Thijssen, 1982). It will therefore substitute within the lattice of the ferromanganese nodule rather than in the interlayer spacings as is the case for Ni, Cu, and Zn. The oxidative incorporation of Co into ferromanganese nodules has been demonstrated experimentally (Hem, 1980; Crowther et al., 1983), and Co(III) has been directly identified in ferromanganese nodules by X-ray photoelectron spectroscopy (Dillard et al., 1982). The Co:Mn ratio in seawater (2.5×10^{-2}) is closely similar to that observed in crustal rock (2.6×10^{-2}), suggesting that these elements are removed from the oceans by a similar process (Knauer et al., 1982). Although the data on the oceanic distribution of Co are sparse and are subject to advective influences (Knauer et al., 1982), scavenging removal with a short residence time is indicated (Table 17.5).

The shale normalized patterns of the concentrations of the REE in seawater indicate a strong negative anomaly for Ce, which is present in seawater as Ce(III), although Ce(IV) is the thermodynamically stable oxidation state (de Baar et al., 1985a). Cerium is unique among the REE in having such a stable +4 oxidation state, and ferromanganese nodules frequently exhibit Ce enrichment. It is probable that Ce, like Co, is incorporated into ferromanganese nodules as the highly insoluble higher oxide which is produced by the oxidation of Ce(III) at the nodule surface. The deep-sea concentration profiles of Ce show a surface maximum with deep-water depletion, and the deep-water values are considerably lower in the Pacific than in the Atlantic, indicating an active scavenging process. Like manganese, Ce shows enhanced levels in the oxygen–minimum zone and positive anomalies where advective processes might transport water from the turnover of anoxic sediments (de Baar et al., 1983, 1985a).

Pb also shows a marked surface maximum (Bruland, 1983; Flegal and Patterson, 1983) which is associated, to a large degree, with the input of anthropogenic lead via aeolian dust. Lead has a short scavenging residence time (Table 17.5) and exhibits lower concentrations in the deep Pacific than in the deep Atlantic. Studies of the oxidation state of trace elements in

Table 17.5. Scavenging Residence Times of the Elements with Respect to Particulate Removal in the Deep Sea

Element	Scavenging Residence Time (yr)	Reference
Sn	10	Balistrieri et al. (1981)
Th	33	Bacon and Anderson (1982)
Fe	40	Morel and Hudson (1985)
Co	40	Knauer et al. (1982)
Po	40	Bacon et al. (1985)
Ce	50	de Baar et al. (1985a)
Mn	51	Weiss (1977)
Pb	54	Craig et al. (1973)
Pa	67	Bacon and Anderson (1982)
Sm	200	Elderfield and Greaves (1983)
Cu	650	Boyle et al. (1977)
Sc	2500	Craig (1974)
Be	3700	Measures and Edmond (1982)
Lu	4000	Elderfield and Greaves (1983)

ferromanganese nodules (Dillard et al., 1982) indicate the presence of Pb(IV), and it is therefore possible that Pb also follows an oxidative removal path (see Hem, 1980). The elements that have been shown to be susceptible to oxidative scavenging all fall in sector A of the CF diagram (Fig. 17.3).

The enrichment of Pt relative to Pd in ferromanganese nodules and the much greater stability of Pt(IV) relative to Pd(IV) have prompted Hodge et al. (1985) to suggest that Pt might also be subject to oxidative removal. There is as yet insufficient data on the distribution of the platinum group metals to assess this hypothesis. The slow removal of Tl via the oxidative scavenging of Tl(III) is possible (Section 4.1; Flegal and Patterson, 1985) although there is no direct evidence for the presence of this oxidation state in ferromanganese nodules.

4.3.2. Hydrolytic Scavenging. The other elements which show characteristic scavenging profiles [Al, Bi, Cr(III), ^{232}Th, Sn, Te] have no obvious oxidative removal pathway. Of these elements, Al, Bi, Sn, and Te have a significant aeolian input with a dominant anthropogenic component for Sn and Te. Chromium as Cr(III) is introduced into the surface layers by the biological or photochemical reduction of Cr(VI), and ^{232}Th is contributed by river input. In addition we must consider ^{210}Po (Cochran et al., 1983), ^{231}Pa (Anderson et al., 1983), and the shorter-lived isotopes of Th (^{228}Th, ^{230}Th, and ^{234}Th, Bacon and Anderson, 1982), which show marked scavenging behavior although they do not have typical scavenging profiles since their oceanic distribution is dominated by that of their parent elements. The efficiency with which these elements are removed by particulate matter will depend on the balance struck between the opposing effects of solution complexation and surface adsorption. The strongly hydrolyzed elements of the group [Al, Bi, Cr(III), ^{232}Th] all fall in the transition zone of the Born diagram (Fig. 17.2) and in sector C of the CF diagram (Fig. 17.3) in common with Be and Sc—recycled elements that exhibit deep-water scavenging. The elements whose deep-sea concentration profiles are *dominated* by scavenging are invariably more strongly hydrolyzed than the recycled elements in this zone. This link between solution hydrolysis and surface adsorption has been given a quantitative expression by the surface complexation model (Schindler, 1975; Balistrieri et al., 1981; Hunter, 1983), which is based on the simple relationship

$$\frac{\tau_M}{\tau_P} = \frac{[M]_s}{[M]_p} \qquad (17.5)$$

where τ_M and τ_P are the scavenging residence time of M and the particle residence time, respectively, and the subscripts s and p represent the concentrations of the elements in the solution and particulate phase, respectively. This relationship appears to hold good for a range of scavenged elements (Cu, Pb, Mn, Pa, Th) with τ_P = 7.5 yr and a total particle concentration of

10 μg L^{-1}, suggesting removal by a single population of particles (Bacon and Anderson, 1982). Balistrieri et al. (1981) expressed Eq. 17.5 in the form

$$\frac{\tau_M}{\tau_P} - 1 = \frac{\alpha_M}{aK_M\{S\text{—}OH\}} \tag{17.6}$$

where a is the concentration of particles, {S—OH} is the concentration of available surface sites, α_M is the overall side-reaction coefficient (Turner et al., 1981), and K_M is the equilibrium constant for adsorption. Attempts have been made to relate the observed values of τ_M to complexation via organic surface coatings on the particles (Balistrieri et al., 1981; Hunter, 1983). The agreement between theory and experiment deteriorates if corrections are made for the competitive complexing of Mg and Ca (Morel and Hudson, 1985) but improves if allowance is made for the low concentrations of particulate matter assumed in the original models (2 μg L^{-1}). Nonetheless, the case for complexation via organic coatings remains equivocal. We will take a simpler approach and use Eq. 17.6 to define an index of scavenging R given by

$$R = \{S\text{—}OH\}\, a \left(\frac{\tau_M}{\tau_P} - 1\right) = \frac{\alpha_M}{K_M} \tag{17.7}$$

or

$$\log R = \log \alpha_M - \log K_M \tag{17.8}$$

The index R clearly represents the balance between solution complexation and adsorption. It is also proportional to τ_M if $\tau_M/\tau_P \gg 1$ and if {S—OH}, a, and τ_P are constant as suggested by Bacon and Anderson (1982). The intrinsic constants for cation adsorption (K_{int}) can be related to the first hydrolysis constant of the cation (K_{OH}) by (Dugger et al., 1964; Balistrieri et al., 1981):

$$\log K_{int} = a_0 + a_1 \log K_{OH} \tag{17.9}$$

Now,

$$\log K_M = \log K_{int} - \frac{nF\psi}{2.303RT} \tag{17.10}$$

where ψ is the potential in the adsorption plane. For simplicity we will assume that ψ is small and write

$$\log R' = \log \alpha_M - \log K_{OH} \tag{17.11}$$

where log R' is linearly related to log R and thus still provides an index of scavenging.

This analysis, while useful, covers only those elements whose equilibrium speciation in seawater includes 1% or more of free cation (sector A, Fig. 17.3). We can include the strongly and fully hydrolyzed elements that are most effectively scavenged by recalling that the interpretation of the linear relationship between log K_{int} and log K_{OH} is based on the similarity of the reactions

$$S{-}O^- + M^{n+} \rightleftharpoons S{-}O^-M^{n+} \quad \text{(adsorption)} \qquad (17.12)$$

$$H{-}O^- + M^{n+} \rightleftharpoons H{-}O^-M^{n+} \quad \text{(hydrolysis)} \qquad (17.13)$$

The surface complexation concept can be extended to include the adsorption of hydrolysis products as well as cations (Davies and Leckie, 1978), for example,

$$S{-}O^- + MOH^{(n-1)+} \rightleftharpoons S{-}O^-MOH^{(n-1)+} \qquad (17.14)$$

which is analogous to the second hydrolysis reaction of M^{n+}. We can therefore extend Eq. 17.11 to include all elemental cations and neutral or positively charged hydroxides which can add a further hydroxyl group and thus provide an analogous log K_{OH} for the calculation. α_M will then be the ratio of total metal concentration to that of the cation or hydroxide being considered.

This approach still excludes (1) strong acid anions (e.g., NO_3^-, SO_4^{2-}, Cl^-, SeO_4^{2-}, Br^-, RuO_4^-, I^-, ReO_4^-), since they have no neutral or cationic species which can interact with a negatively charged surface (Eqs. 17.12 and 17.14), and (2) those acids which ionize by loss of H^+ rather than by addition of OH^-. Following the equations given by Baes and Mesmer (1976), the latter group comprises H_2CO_3, $Si(OH)_4$, H_3PO_4, H_3VO_4, H_2CrO_4, $Ge(OH)_4$, H_3AsO_4, H_2SeO_4, H_2MoO_4, $Te(OH)_4$, $Te(OH)_6$, HIO_3, and H_2WO_4. However, V and Te(IV) are considered in terms of adsorption of VO_2^+ and $Te(OH)_3^+$, respectively. These groups include Si, Ge, and Te and all the condensed oxyacids [i.e., those formed by loss of water from the hypothetical $M(OH)_n$]. Inclusion of the chloro-dominated elements (sector B, Fig. 17.3) is also problematical, since these elements have very low concentrations of the free cation but also form cationic and neutral complexes such as MCl_n, $MCl_{(n-1)}^+$. We have not attempted to deal with the interactions of these species with the particle surfaces and have only considered the interactions of the free cations.

We have calculated values of log R' for all cations and positively charged and neutral hydroxides of the natural elements (Fig. 17.6). Only the lowest value of log R' (i.e., the strongest scavenging) is plotted in cases where several possible adsorbed species can exist. The correlation between the values of log R' and the known scavenging behavior of the elements is striking. There appears to be a transition zone in the region of log $R' = -5.0$. At lower log

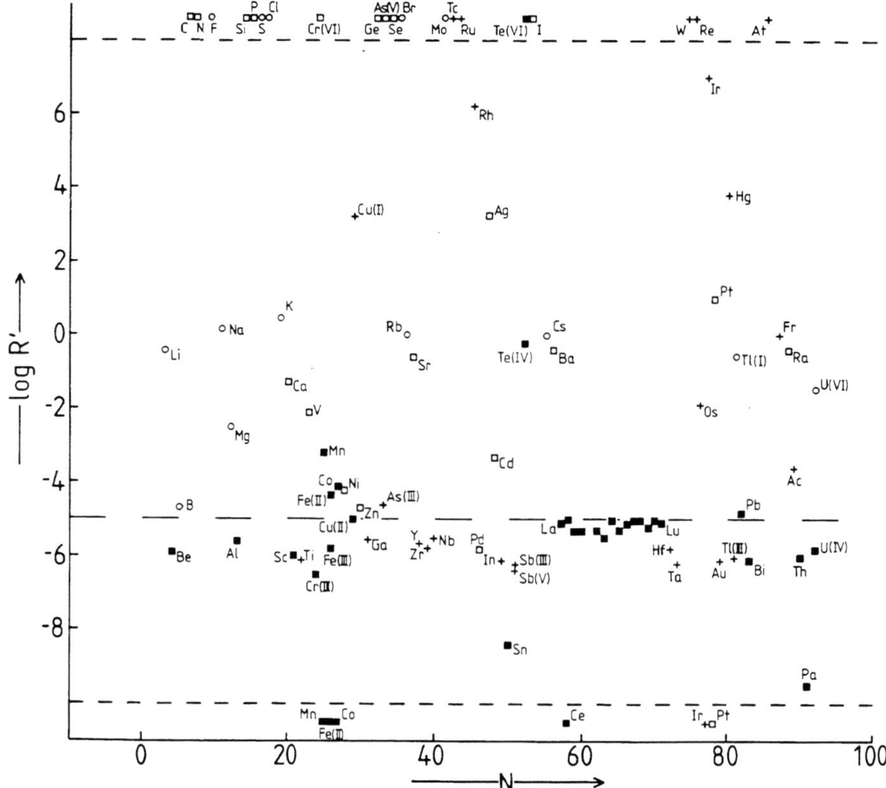

Figure 17.6. Plot of the scavenging index (log R', Eq. 17.11) versus atomic number N. Elements considered to have no adsorbable species are collected above the upper dashed line, and elements for which oxidative scavenging may occur (Section 4.3.1) are collected below the lower dashed line. The horizontal line at log $R' = 5$ represents an approximate boundary between scavenged and nonscavenged elements. (○) Accumulated elements; (□) recycled elements not subject to scavenging; (■) scavenged elements; (+) behavior in the oceans unknown.

R' values almost all elements whose behavior is known are scavenged, while at higher log R' values no scavenging behavior has been detected except in cases where oxidative precipitation may occur (Section 4.3.1). The REE form an intriguing group in this context. They lie in the borderline area, and their removal by particulate scavenging appears to be secondary to their recycling by biological processes. Furthermore, their deep-sea concentrations increase with atomic number in a sequence which parallels the increasing strength of their solution complexes—suggesting that it is the free cation which is adsorbed (de Baar et al., 1985a,c).

The very high log R' values calculated for the chloro-dominated elements are unrealistic, since we have neglected the adsorption of cationic and neutral chloro complexes, and the hydrolysis constants are probably not reliable for Pd and are nonexistent for Pt, Rh, and Ir (Baes and Mesmer, 1976). Nevertheless, the limited evidence that is available suggests an absence of scavenging for Ag and Pd. The anomalous position of Pd in Figure 17.6 probably reflects uncertainties in the Pd hydrolysis constants. It is possible that Pt may be removed by oxidative scavenging, while the unique chemistry of Hg is no doubt responsible for the rather featureless profiles observed (Table 17.4; Dalziel and Yeats, 1985; Gill and Fitzgerald, 1985). As well as being methylated biologically, Hg might undergo reduction to Hg(0), which is soluble in water to 10^{-5} M and is also volatile (Fitzgerald et al., 1983).

Another intriguing element (Fig. 17.6) is Te, which, along with Ge and Si, was excluded on the grounds that its uncondensed hydroxides ionize by the loss of H^+ rather than by the addition of OH^- (Baes and Mesmer, 1976) so that the anion does not provide an analogy to Eq. 17.14. However, if it is assumed that the adsorbed species are $S{-}O{-}Ge(OH)_4$, $S{-}O{-}Si(OH)_4$, $S{-}O{-}Te(OH)_4$, and $S{-}O{-}Te(OH)_6$, then the log R' values would be -4.7 (Ge), -4.3 (Si), -5.3 [Te(IV)], and -5.2 [Te(VI)], which would be consistent with the observed scavenging of Te (Lee and Edmond, 1985) and also with the "recycled" profiles of Si and Ge (Section 4.2). It is possible, therefore, that Te(IV) and Te(VI) are able to bind five and seven oxygen atoms, respectively, and that perhaps $Te(OH)_5^-$ and $Te(OH)_7^-$ should be considered as the structures of the aqueous anions.

5. PREDICTIONS

So far we have tried to provide a rationalization of the oceanic distribution of those elements for which reliable deep-sea concentration profiles are available. The fact that there remain a number of natural elements whose oceanic distribution patterns are not known (elements marked + in Fig. 17.6) allows us to test the predictive power of the scheme of rationalization that we have adopted (Fig. 17.5). Our predictions will be based on the elemental relationships exhibited in the CF diagram (Fig. 17.3) and the log R' diagram (Fig. 17.6). The uncertainty surrounding the mode of incorporation of trace elements into the organic matrix will make it difficult to predict the exact nature of the profiles to be expected, since the chemical tendencies of the elements might be overridden by incidental biological cycling. For example, our present scheme would predict that Sr should be accumulated and La should be scavenged. In fact, they both show "recycled" profiles (Table 17.4), albeit with very different surface/deep concentration ratios. Since none of the "unknown" elements is biologically essential, we will divide our predictions between the "accumulated" and "scavenged" categories (Tables 17.3 and 17.4). If the elements are incorporated incidentally into biogenic particles, this subdivision should give some indication of the relative intensity of vertical fractionation.

Additions to the "accumulated" elements might be provided by the strong acid anions (Tc, Ru, W, Re, Fig. 17.6) and Fr, which will show only weak particle interactions. At the other end of the scale, thermodynamic calculations predict oxidative scavenging (in the +4 state) for Pt and Ir, and log R' values (Fig. 17.6) predict hydrolytic scavenging for Ga, Y, Ti, Zr, Hf, Nb, Ta, Au, and Sb. The chemistry of Au is complicated by the possibility of forming colloidal native gold. The chemistry of Sb suggests that it should be subject to hydrolytic scavenging (Fig. 17.6), although measurements made in artificial mesocosms (Santschi et al., 1983) and on trace metal exchange with natural particles (Li et al., 1984) indicate that this element has a weak affinity for particles. The time scales of these measurements may be too short to model processes in the deep ocean adequately, and it will be interesting to see what is revealed by oceanographically consistent profiles when they become available.

Rh is not predicted to be scavenged (Fig. 17.6) and is likely to show a "recycled" profile like other strongly chloro-complexed cations such as Pd and Ag (Section 4.2.2), although the influence of hydrolysis is unclear. Ac is similar in chemistry to the lanthanides, but since it has a significantly larger ionic radius it is much more weakly hydrolyzed and less likely to be scavenged. In common with Y, it is likely to show a "recycled" profile. Very little information is available about the chemistry of At, but it might behave in a similar manner to IO_3^- and show a "recycled" profile.

Os is unique in that it is predicted by thermodynamic calculations to form only the +8 oxidation state in seawater. OsO_4 is a covalent tetrahedral molecule which hydrolyzes only weakly (Baes and Mesmer, 1976) and thus has a high log R' value consistent with nonscavenged behavior (Fig. 17.6). However, the likely volatility of OsO_4 and its affinity for biological tissue (it is used as a staining agent) could give rise to a very interesting oceanic osmium cycle (cf. mercury, Section 4.3.2).

The various predictions are summarized in Fig. 17.5. We have not attempted to extend our predictions to the transuranic elements at this stage. The behavior of the lighter transuranic elements (Np, Pu, Am), which is complicated by the occurrence of several oxidation states, is of considerable importance, and extension of our approach to these elements would certainly be of interest.

NOTE ADDED IN PROOF

Since preparing this paper, a note on the oceanic distribution of ^{227}Ac has been drawn to our attention (Nozaki, 1984). This indicates that ^{227}Ac, unlike its parent ^{231}Pa, shows significant enrichment in deep water within 1–2 km of the bottom, suggesting that it is recycled in the sediment and that it has a relatively long scavenging residence time (3000 y, cf. 4000 y for Lu, Table 17.5). This is in agreement with our predictions.

REFERENCES

Anderson, R. F. (1982), Concentration, vertical flux and mineralisation of particulate uranium in seawater, *Geochim. Cosmochim. Acta* **46,** 1293–1299.

Anderson, R. F., Bacon, M. P., and Brewer, P. G. (1983), Removal of ^{230}Th and ^{231}Pa at ocean margins, *Earth Planet. Sci. Lett.* **66,** 73–90.

Andreae, M. O. (1979), Arsenic speciation in seawater and in interstitial waters: the influence of biological-chemical interactions on the chemistry of a trace element, *Limnol. Oceanogr.* **24,** 440–452.

Bacon, M. P., and Anderson, R. F. (1982), Distribution of thorium isotopes between dissolved and particulate forms in the deep sea, *J. Geophys. Res.* **87,** 2045–2056.

Bacon, M. P., Huh, C.-A., Fleer, A. P., and Deuser, W. G. (1985), Seasonality in the flux of natural radionuclides and plutonium in the deep Sargasso Sea, *Deep-Sea Res.* **32A,** 273–286.

Baes, C. F., and Mesmer, R. E. (1976), *The Hydrolysis of Cations,* Wiley, New York, 489pp.

Balistrieri, L. S., and Murray, J. W. (1984), Marine scavenging: Trace metal adsorption by interfacial sediment from MANOP site H, *Geochim. Cosmochim. Acta* **48,** 921–929.

Balistrieri, L. S., Brewer, P. G., and Murray, J. W. (1981), Scavenging residence times of trace metals and surface chemistry of sinking particles in the deep ocean, *Deep-Sea Res.* **28A,** 101–121.

Bertine, K. K. (1972), The deposition of molybdenum in anoxic waters. *Marine Chem.* **1,** 43–53.

Boyle, E. A., and Kegwin, L. D. (1982), Deep circulation of the North Atlantic over the last 2000 years: Geochemical evidence, *Science* **218,** 784–786.

Boyle, E. A., Sclater, F. R., and Edmond, J. M. (1977), The distribution of dissolved copper in the Pacific, *Earth Planet. Sci. Lett.* **37,** 38–54.

Boyle, E. A., Huested, S. S., and Jones, S. P. (1981), On the distribution of copper, nickel and cadmium in the surface waters of the North Atlantic and North Pacific Oceans, *J. Geophys. Res.* **86,** 8048–8066.

Broecker, W. S., and Peng, T-H. (1982), *Tracers in the Sea,* Lamont-Doherty Geological Observatory, New York, 690pp.

Broecker, W. S., Goddard, J., and Sarmiento, J. L. (1976), The distribution of ^{226}Ra in the Atlantic Ocean, *Earth Planet. Sci. Lett.* **32,** 220–235.

Bruland, K. W. (1980), Oceanographic distributions of cadmium, zinc, nickel and copper in the North Pacific, *Earth Planet. Sci. Lett.* **47,** 176–198.

Bruland, K. W. (1983), "Trace Elements in Seawater," in J. P. Riley and R. Chester, Eds., *Chemical Oceanography,* Vol. 8, Academic Press, London, pp. 157–220.

Bruland, K. W., Knauer, G. A., and Martin, J. H. (1978), Cadmium in Northeast Pacific waters, *Limnol. Oceanogr.* **23,** 618–625.

Burton, J. D., Maher, W. A., and Statham, P. J. (1983), "Some Recent Measurements of Trace Metals in Atlantic Ocean Waters," in C. S. Wong, E. A. Boyle, K. W. Bruland, J. D. Burton, and E. D. Goldberg, Eds., *Trace Metals in Seawater,* Plenum, New York, pp. 415–426.

Byrd, J. T., and Andreae, M. O. (1982), Tin and methyltin species in seawater: Concentrations and fluxes, *Science* **218,** 565–569.

Campbell, J. A., and Yeats, P. A. (1981), Dissolved chromium in the Northwest Atlantic Ocean, *Earth Planet. Sci. Lett.* **53,** 427–433.

Chase, R. R. P. (1979), Settling behaviour of natural aquatic particles, *Limnol. Oceanogr.* **24,** 417–426.

Chung, Y., and Craig, H. (1980), ^{226}Ra in the Pacific Ocean, *Earth Planet. Sci. Lett.* **49,** 267–292.

Cochran, J. K., Bacon, M. P., Krishnaswami, S., and Turekian, K. K. (1983), ^{210}Po and ^{210}Pb distributions in the central and eastern Indian Ocean, *Earth Planet. Sci. Lett.* **65,** 433–452.

Collier, R. W. (1984), Particulate and dissolved vanadium in the North Pacific Ocean, *Nature* **309,** 441–444.

Collier, R. W. (1985), Molybdenum in the Northeast Pacific Ocean, *Limnol. Oceanogr.* **30,** 1351–1354.

Collier, R. W., and Edmond, J. M. (1984), The trace element geochemistry of marine biogenic particulate matter, *Progr. Oceanogr.* **13,** 113–199.

Cowen, J. P., and Bruland, K. W. (1985), Metal deposits associated with bacteria: Implications for Fe and Mn marine biogeochemistry, *Deep-Sea Res.* **32,** 253–272.

Craig, H. (1974), A scavenging model for trace elements in the deep sea, *Earth Planet. Sci. Lett.* **23,** 149–159.

Craig, H., Krishnaswami, S., and Somayajulu, B. L. K. (1973), ^{210}Pb–^{226}Ra: Radioactive disequilibrium in the deep sea, *Earth Planet. Sci. Lett.* **17,** 295–305.

Cranston, R. E., and Murray, J. W. (1978), The determination of chromium species in natural waters, *Anal. Chim. Acta* **99,** 275–282.

Crowther, D. L., Dillard, J. G., and Murray, J. W. (1983), The mechanism of Co(II) oxidation on synthetic birnessite, *Geochim. Cosmochim. Acta* **47,** 1399–1403.

Cutter, G. A., and Bruland, K. W. (1984), The marine biogeochemistry of selenium: A re-evaluation, *Limnol. Oceanogr.* **29,** 1179–1192.

Dalziel, J. A., and Yeats, P. A. (1985), Reactive mercury in the central North Atlantic Ocean, *Marine Chem.* **15,** 357–361.

Danielsson, L.-G. (1980), Cadmium, cobalt, copper, iron, lead, nickel and zinc in Indian Ocean water, *Marine Chem.* **8,** 199–215.

Davis, J. A., and Leckie, J. O. (1978), Surface ionisation and complexation at the oxide/water interface. II. Surface properties of amorphous iron oxyhydroxide and adsorption of metal ions, *J. Colloid Interface Sci.* **67,** 90–107.

de Baar, H. J. W., Brewer, P. G., and Bacon, M. P. (1983), Rare-earth distributions with a positive Ce anomaly in the Western North Atlantic Ocean, *Nature* **301,** 324–327.

de Baar, H. J. W., Bacon, M. P., Brewer, P. G., and Bruland, K. W. (1985a). Rare earth elements in the Pacific and Atlantic Oceans, *Geochim. Cosmochim. Acta* **49,** 1943–1959.

de Baar, H. J. W., Elderfield, H., and Bruland, K. W. (1985b), Rare earth elements at VERTEX IV in the central North Pacific Gyre, *Trans. Am. Geophys. Union* (*EOS*) **66,** 1291.

de Baar, H. J. W., Brewer, P. G., and Bacon, M. P. (1985c), Anomalies in rare earth distributions in seawater: Gd and Tb, *Geochim. Cosmochim. Acta* **49,** 1961–1969.

Dehairs, F., Chesselet, R., and Jedwab, J. (1980), Discrete suspended particles of barite and the barium cycle in the open ocean, *Earth Planet. Sci. Lett.* **49,** 528–550.

Dillard, J. G., Crowther, D. L., and Murray, J. W. (1982), The oxidation state of cobalt and selected metals in Pacific ferromanganese nodules, *Geochim. Cosmochim. Acta* **46,** 755–759.

Dugger, D. L., Stanton, J. H., Irby, B. N., McConnell, B. L., Cummings, W. W., and Maatman, R. W. (1964), The exchange of twenty metal ions with the weakly acidic silanol group of silica gel, *J. Phys. Chem.* **68,** 757–760.

Edmond, J. M. (1974), On the dissolution of carbonate and silicate in the deep ocean, *Deep-Sea Res.* **21,** 455–480.

Egami, F. (1974), Minor elements and evolution, *J. Mol. Evol.* **4,** 113–120.

Elderfield, H., and Truesdale, V. W. (1980), On the biophilic nature of iodine in seawater, *Earth Planet. Sci. Lett.* **50,** 105–114.

Elderfield, H., and Greaves, M. J. (1982), The rare earth elements in seawater, *Nature* **296,** 214–219.

Fitzgerald, W. F., Gill, G. A., and Hewitt, A. D. (1983), "Air–Sea Exchange of Mercury", in C. S. Wong, E. A. Boyle, K. W. Bruland, J. D. Burton, and E. D. Goldberg, Eds., *Trace Metals in Seawater,* Plenum, New York, pp. 297–315.

Flegal, A. R., and Patterson, C. C. (1983), Vertical concentration profiles of lead in the central Pacific at 15°N and 20°S, *Earth Planet Sci. Lett.* **64,** 19–32.

Flegal, A. R., and Patterson, C. C. (1985), Thallium concentrations in seawater, *Marine Chem.* **15,** 327–331.

Froelich, P. N., Hambrick, G. A., Andreae, M. O., Mortlock, R. A., and Edmond, J. M. (1985). The geochemistry of inorganic germanium in natural waters, *J. Geophys. Res.* **90,** 1133–1141.

German, C., de Baar, H. J. W., Elderfield, H., and Bacon, M. P. (1985), Concentration gradients for rare earth elements in suspended particles and in solution in an anoxic basin, the Cariaco Trench, *Trans. Am. Geophys. Union* (*EOS*) **66,** 1326.

Gieskes, J. M. (1983), "The Chemistry of Interstitial Waters of Deep-Sea Sediments: Interpretation of Deep-sea Drilling Data," in J. P. Riley and R. Chester, Eds., *Chemical Oceanography,* Vol. 8, Academic Press, London, pp. 221–269.

Gill, G. A., and Fitzgerald, W. F. (1985), Mercury sampling of open ocean waters at the picomolar level, *Deep-Sea Res.* **32A,** 287–297.

Glasby, G. P. (1984), Manganese in the marine environment, *Oceanogr. Marine Biol. Ann. Rev.* **22,** 169–194.

Glasby, G. P., and Read, A. J. (1976), "Deep-sea Manganese Nodules," in K. H. Wolf, Ed., *Handbook of Stratabound and Stratiform Ore Deposits,* Elsevier, Amsterdam, pp. 295–340.

Glasby, G. P., and Thijssen, T. (1982), Control of the mineralogy and composition of marine manganese nodules by the supply of divalent transition metal ions, *N. Jb. Miner. Abh.* **145,** 291–307.

Harris, J. E. (1977), Characteristics of suspended matter in the Gulf of Mexico. II. Particle size analysis of suspended matter from deep water, *Deep-Sea Res.* **24,** 1055–1061.

Hem, J. D. (1980), "Redox Coprecipitation Mechanisms of Manganese Oxides," in M. C. Kavanaugh and J. O. Leckie, Eds., *Particulates in Water: Characterisation, Fate, Effects and Removal,* American Chemical Society, Washington, DC, pp. 45–72.

Hodge, V. F., Stallard, M., Koide, M., and Goldberg, E. D. (1985), Platinum and the platinum anomaly in the marine environment, *Earth Planet. Sci. Lett.* **72,** 158–162.

Huh, C.-A., and Bacon, M. P. (1985), Thorium-232 in the Eastern Caribbean Sea, *Nature* **316,** 718–721.

Hunt, J. R. (1980), "Prediction of Oceanic Particulate Size Distribution from Coagulation and Sedimentation Mechanisms," in M. C. Kavanaugh and J. O. Leckie, Eds., *Particulates in Water: Characterisation, Fate, Effects and Removal,* American Chemical Society, Washington, DC, pp. 243–257.

Hunter, K. A. (1983), The adsorptive properties of sinking particles in the deep ocean, *Deep-Sea Res.* **30A,** 669–675.

Klinkhammer, G. P. (1980), Early diagenesis in sediments from the Eastern Equatorial Pacific. II. Porewater metal results, *Earth Planet. Sci. Lett.* **46,** 361–384.

Klinkhammer, G. P., and Bender, M. L. (1980), The distribution of Mn in the Pacific Ocean, *Earth Planet. Sci. Lett.* **46,** 361–384.

Knauer, G. A., Martin, J. H., and Gordon, R. M. (1982), Cobalt in Northeast Pacific Waters, *Nature* **297,** 49–51.

Ku, T.-L., and Lin, M.-C. (1976), ^{226}Ra distribution in the Antarctic Ocean, *Earth Planet. Sci. Lett.* **32,** 236–248.

Landing, W. M., and Bruland, K. W. (1980), Manganese in the North Pacific, *Earth Planet. Sci. Lett.* **49,** 45–56.

Lee, D. S. (1982), Determination of bismuth in environmental samples by flameless atomic absorption spectroscopy with hydride generation, *Anal. Chem.* **54,** 1682–1686.

Lee, D. S. (1983), Palladium and nickel in Northeast Pacific waters, *Nature* **305,** 47–48.

Lee, D. S., and Edmond, J. M. (1985), Tellurium species in seawater, *Nature* **313,** 782–785.

Lewis, B. L., Froelich, P. N., and Andreae, M. O. (1985), Methylgermanium in natural waters, *Nature* **313,** 303–305.

Li, Y.-H. (1981), Ultimate removal mechanisms of elements from the ocean, *Geochim. Cosmochim. Acta* **45,** 1659–1664.

Li, Y.-H. (1982a), Ultimate removal mechanisms of elements from the ocean (reply to a comment by M. Whitfield and D. R. Turner), *Geochim. Cosmochim. Acta* **46,** 1993–1995.

Li, Y.-H. (1982b), Interelement relationships in abyssal Pacific ferromanganese nodules and associated pelagic sediments, *Geochim. Cosmochim. Acta* **46,** 1053–1060.

Li, Y.-H., Burkhardt, L., Buchholtz, M., O'Hara, P., and Santschi, P. H. (1984), Partition of radiotracers between suspended particles and seawater, *Geochim. Cosmochim. Acta* **48,** 2011–2019.

Lowenstam, H. A. (1981), Minerals formed by organisms, *Science* **211,** 1126–1131.

McCave, I. N. (1984), Size spectra and aggregation of suspended particles in the deep ocean, *Deep-Sea Res.* **31A,** 329–352.

McClendon, J. H. (1976), Elemental abundance as a factor in the origins of mineral nutrient requirements, *J. Mol. Evol.* **8,** 175–195.

Martin, J. H., Knauer, G. A., and Broenkow, W. W. (1985), VERTEX: the lateral transport of manganese in the Northeast Pacific, *Deep-Sea Res.* **32,** 1405–1427.

Martin, J. H., Knauer, G. A., and Gordon, R. M. (1983), Silver distributions and fluxes in Northeast Pacific waters, *Nature* **305,** 306–309.

Measures, C. I., and Edmond, J. M. (1982), Beryllium in the water column of the central North Pacific, *Nature* **297,** 51–53.

Measures, C. I., and Edmond, J. M. (1983), The geochemical cycle of ^{9}Be: A reconnaissance, *Earth Planet. Sci. Lett.* **66,** 101–110.

Measures, C. I., Grant, B. C., Khadem, M., Lee, D. S., and Edmond, J. M. (1984), Distribution of Be, Al, Se and Bi in the surface waters of the Western North Atlantic and Caribbean, *Earth Planet. Sci. Lett.* **71,** 1–12.

Measures, C. I., Grant, B. C., Mangum, B. J., and Edmond, J. M. (1983), "The Relationship of the Distribution of Dissolved Selenium(IV) and (VI) in Three Oceans to Physical and Biological Processes," in C. S. Wong, E. A. Boyle, K. W. Bruland, J. D. Burton, and E. D. Goldberg, Eds., *Trace Metals in Seawater,* Plenum, New York, pp. 73–83.

Morel, F. M. M., and Morel-Laurens, N. M. L. (1983), "Trace Metals and Plankton in the Oceans: Facts and Speculations," in C. S. Wong, E. A. Boyle, K. W. Bruland, J. D. Burton, and E. D. Goldberg, Eds., *Trace Metals in Seawater,* Plenum, New York, pp. 841–869.

Morel, F. M. M., and Hudson, R. J. M. (1985), "The Geobiological Cycle of Trace Elements in Aquatic Systems: Redfield Revisited", in W. Stumm, Ed., *Chemical Processes in Lakes,* Wiley, New York, pp. 251–281.

Morris, A. W. (1975), Dissolved molybdenum and vanadium in the Northeast Atlantic Ocean, *Deep-Sea Res.* **22,** 49–54.

Murray, J. W., Spell, B., and Paul, B. (1983), "The Contrasting Geochemistry of Manganese and Chromium in the Eastern Tropical Pacific Ocean," in C. S. Wong, E. A. Boyle, K. W. Bruland, J. D. Burton and E. D. Goldberg, Eds., *Trace Metals in Seawater,* Plenum, New York, pp. 643–669.

Nozaki, Y. (1984), Excess ^{227}Ac in deep ocean water, *Nature* **310,** 486.

Nyffeler, U. P., Li, Y-H., and Santschi, P. H. (1984), A kinetic approach to describe trace-element partitioning between particles and solution in natural aquatic systems, *Geochim. Cosmochim. Acta* **48,** 1513–1522.

Orians, K. J., and Bruland, K. W. (1985), Dissolved aluminium in the central North Pacific, *Nature* **316,** 427–429.

Santschi, P. H., Adler, D. M., and Amdurer, M. (1983), "The Fate of Particles and Particle-Reactive Trace Metals in Coastal Waters: Radioisotope Studies in Microcosms," in C. S. Wong, E. A. Boyle, K. W. Bruland, J. D. Burton, and E. D. Goldberg, Eds., *Trace Metals in Seawater,* Plenum, New York, pp. 331–349.

Schindler, P. W. (1975), Removal of trace metals from the oceans: A zero order model, *Thalassia Jugosl.* **11,** 101–111.

Simpson, W. R. (1982), Particulate matter in the oceans—Sampling methods, concentration, size distribution and particle dynamics, *Oceanogr. Marine Biol. Ann. Rev.* **20,** 119–172.

Stoffyn-Egli, R., and Mackenzie, F. T. (1984), Mass balance of lithium in the oceans, *Geochim. Cosmochim. Acta* **48,** 859–872.

Sunda, W. G., Huntsman, S. A., and Harvey, G. R. (1983), Photoreduction of manganese oxides in seawater and its geochemical and biochemical implications, *Nature* **301,** 234–239.

Symes, J. L., and Kester, D. R. (1985), The distribution of iron in the Northwest Pacific, *Marine Chem.* **17,** 57–74.

Takahashi, T., Broecker, W. S., and Bainbridge, A. (1981), "Supplement to the Alkalinity and Total Carbon Dioxide Concentration in the World Oceans," in B. Bolin, Ed., *SCOPE 16: Carbon Cycle Modelling,* Wiley, New York, pp. 159–200.

Turner, D. R., and Whitfield, M. (1983), Inorganic controls on the biogeochemical cycling of the elements in the ocean, *Ecol. Bull.* (*Stockholm*) **35,** 9–37.

Turner, D. R., Dickson, A. G., and Whitfield, M. (1980), Water–rock partition coefficients and the composition of natural waters—A reassessment. *Marine Chem.* **9,** 211–218.

Turner, D. R., Whitfield, M., and Dickson, A. G. (1981), The equilibrium speciation of dissolved components in freshwater and seawater at 25°C and 1 atmosphere pressure, *Geochim. Cosmochim. Acta* **45,** 855–881.

Watson, A. J., and Whitfield, M. (1985), Composition of particles in the global ocean, *Deep-Sea Res.* **32,** 1023–1039.

Weiss, R. F. (1977), Hydrothermal manganese in the deep sea: Scavenging residence times and manganese/helium-3 relationships, *Earth Planet. Sci. Lett.* **37,** 257–262.

Whitfield, M. (1979), The mean oceanic residence time (MORT) concept—A rationalisation, *Marine Chem.* **8,** 101–123.

Whitfield, M. (1981), The world ocean—Mechanism or machination?, *Interdisc. Sci. Rev.* **6,** 12–35.

Whitfield, M., and Watson, A. J. (1983), "The Influence of Biomineralisation on the Composition of Seawater," in P. Westbroek and E. W. de Jong, Eds., *Biomineralisation and Biological Metal Accumulation,* D. Reidel, Amsterdam, pp. 57–72.

Williams, A. F. (1979), *A Theoretical Approach to Inorganic Chemistry,* Springer, Berlin, 316pp.

Williams, R. J. P. (1981), Natural selection of the elements, *Proc. Roy. Soc. Ser. B* **213,** 361–397.

Wong, G. T. F., and Brewer, P. G. (1974), The determination and distribution of iodate in South Atlantic waters, *J. Marine Res.* **32,** 25–36.

Yeats, P. A., and Bewers, J. M. (1985), Manganese in the Western North Atlantic, *Marine Chem.* **17,** 255–263.

18

FROM MOLECULES TO PLANETARY ENVIRONMENTS: UNDERSTANDING GLOBAL CHANGE

William S. Fyfe

Department of Geology, The University of Western Ontario, London, Ontario, Canada

Abstract

One of the greatest future tasks of earth scientists is to understand the processes which control and moderate change in the environment of the biosphere. With satellites we can observe change on the planetary scale with increasing precision. We are beginning to quantify the natural fluxes in the surface related to energy transfer from the interior. Ocean chemistry and probably some continental waters are significantly influenced by these processes. We are increasingly concerned with the anthropogenic changes which must accelerate for decades. But the understanding of the buffer systems and the response of aquatic systems involves the understanding of the interactions at the solid–liquid–gas interfaces at the micro level. Very fine particulates formed by both inorganic and biological processes must have the most sensitive response to change.

1. INTRODUCTION

I recently read a manuscript from Wallace Broeker of Columbia University in which he posed the very basic question as to why our environment had remained so remarkably stable over the 3.8 billion years for which we have some record. To the best of our knowledge the earth has never totally frozen, never had a runaway greenhouse, and while there have been substantial oscillations, the system appears to eventually return to some limit tolerable to life. When one considers that there have been substantial changes in the solar spectrum and intensity, vast changes in the distribution of continents and continental elevations, and presumably equally great changes in ocean current patterns, and changes in climate related to all the above, the overall stability of the environmental system is quite remarkable.

As it has become clear that the energy input into our planet is strongly influenced by solar orbits and atmospheric chemistry, it is clear that there must be finely tuned buffer systems which appear to follow Le Chatelier's law (1888): "That if a system is disturbed, it will readjust in such a way as to tend to neutralize the disturbance and restore equilibrium." On a global scale it is clear that to maintain appropriate atmospheric chemistry and hydrospheric chemistry, very efficient interface processes must occur. The materials, both inorganic and biological, that cover our surface (sediments, soils, the biomass) tend often to be of very small particle size, and perhaps the ultimate systems which prevent environmental catastrophe involve such small particles. Only in the last decade or so have the techniques become fully available to study such materials and their surfaces at an adequate atomic–molecular level.

2. CHANGE

Students who studied the earth sciences 50 years ago were taught that things happen slowly on planet earth. Sure there were ice ages and volcanoes, occasional extinction of species, and even the fantastic notion that continents might move, but it was all rather slow and unlikely to have much influence on my life or that of my grandchildren. But all this has changed. The recent data from the study of ice cores and their contained record of atmospheric chemistry has shown that important gases like CO_2–CH_4 have changed dramatically in recent time while carbon dioxide has gone through rapid oscillations in periods of 100 years or less over the past 100,000 years. And these changes can be correlated with solar orbits and major climate changes. The recent El Niño phenomenon has shown how rapidly ocean temperatures can swing. Recent work on the great Cretaceous–Tertiary boundary extinctions have recorded how rapidly major biological change can occur, and the great debate continues as to whether this is caused by internal (volcanic) or external (meteoric) events (Officer et al., 1986).

As we have become increasingly aware of rapid pre-human variations, we are now increasingly concerned with changes occurring because of human activity. With human population increasing by 85 million per year, and with our energy-dependent technology, we have slowly become aware that we are now perhaps the greatest agent of change in the outer geospheres (e.g., freshwater chemistry, atmospheric chemistry, soil erosion, sea level rise, species extinction and redistribution, etc.). It is concern with the scale of such change that is leading to proposals for an increased international effort on the study of global change focused on the theme: "To describe and understand the interactive physical, chemical, and biological processes that regulate the Earth's unique environment for life, the changes that are occurring in this system, and the manner in which they are influenced by human actions" (Eddy, 1986; Malone and Roederer, 1985).

It is now clear that change on planet earth involves a complex interaction between the inorganic physical processes and the biological processes that occur at the micro level and influence the macro systems. The understanding of change implies that we have a precise record of the past and present processes. Given adequate models, the hope is that it will be possible to project with some confidence for decades or centuries at least. It is also clear that such a study requires communication among a very wide range of experts, from solar physicists to biochemists.

Part of the present concern with change involves concerns over irreversibility of some forms of change. There have been spectacular recent examples [e.g., the predicted extinction of 13–25% of all species in the next two decades, the introduction of monocultures into forests and agriculture, the removal of tropical forests (Enger et al., 1986)]. The question of reversibility of change is rarely considered when new development is planned, and given present increases in human population such changes can be on a grand scale.

3. OBSERVATION

The advances in the power of modern planetary observation are remarkable and now set the stage for the methodology of the study of change. We can now see the earth. The first records of continental movement, ice cover, cloud cover, global surface temperature, ocean winds and currents, global albedo, even ocean bottom topography, are now becoming available. Satellites can measure integrated marine and land bioproductivity or photosynthetic activity. There are still serious gaps (e.g., global erosion rates), but these are under consideration. As the record is extended and the quality of data is improved by ground-based correlation, there is little doubt that the large-scale record of change will provide the background data for the selection of local regions requiring intensive ground study to reveal the processes leading to change.

And while the large scale tools for the study and monitoring of change improve (the meter to kilometer scale), the same is equally true for study of the interface processes at the micro scale. With the host of new spectroscopic tools to examine surfaces (see Motschi, Chapter 5 of this volume), techniques for the analysis of liquids and gases at the micro level, and modern electron microscopy, we can virtually see what and how fast interface processes occur at the molecular level. The combination of the large- and small-scale observations is necessary to describe the systems which regulate the planetary environment.

4. EARTH CONVECTION: MEGAPROCESSES: GEOCHEMICAL CYCLES

The stage on which anthropogenic change occurs is set by the basic cooling processes of the earth. At present about half the total heat loss occurs by conduction through the lithosphere and about half by convective heat-mass transfer, the processes that at times produce volcanic phenomena, change surface topography, and cause the surface motions of the lithospheric plates. Observations over the past two decades have shown that during these fundamental processes very-large-scale chemical exchange processes occur across the hydrosphere–crust system, processes related to water-cooling processes in the upper 10 km or so of the crust, where perhaps almost half the energy is transferred to fluids. As surface fluids also exchange mass and energy with the atmosphere, and provide the environment and nutrient sources for the biosphere, we can begin to appreciate that such interactions influence the physical and chemical state of all the outer geospheres which form the environment.

New crust and lithosphere are created at ridges from the rise of hot asthenosphere. Part of the energy associated with this process ($\approx 5 \times 10^{19}$ cal yr^{-1} is transferred to circulating seawater (5×10^{17} g if discharge is at 100°C, 1.4×10^{17} g if discharge is at 350°C). The hot-water discharge flux > 100 km^3 can be compared with the river flux of 36,000 km^3 yr^{-1}. As the solubility of many components is much greater in hot, reduced saline water than in cold oxygenated water, the ridge flux has a significant influence on ocean chemistry. Even for common species like SiO_2, Ca, and K, the hydrothermal flux makes up a significant fraction of the river flux (Thompson, 1983). A simple example of the influence of the two great fluxes into the oceans is provided by the $^{87}Sr/^{86}Sr$ variation with time (present oceans, 0.709; river input, 0.711; hydrothermal flux, 0.702). This great example shows us that deep fluids may be important also in some continental aquatic systems, particularly in regions where regional heat flow is large due to volcanic or other tectonic processes. The ocean ridge systems have been the center of intensive study over the past decades, and gradually the chemistry of the fluxes is becoming well known (see Rona et al., 1983; Emiliani, 1981).

Another interesting example of possible deep fluid influences on environmental chemistry comes from the great debate over extinctions, in particular the Cretaceous–Tertiary boundary. While attention has focused on the influence of objects from space, others have shown that periods of intense volcanism must also be considered, periods where the mass of volcanism may be 10× "normal" (Officer et al., 1986).

While the ridge processes are rather well understood, the off-ridge or conveyer belt processes are not. Davis and Lister (1977) considered the possible style of slow convection in permeable basaltic ocean crust beneath a less permeable sediment blanket. Anderson et al. (1979) showed from heat flow studies that this process continues in crust older than 50 million years. Fyfe

(1974) emphasized that the exothermic hydration of periodotite to form serpentine could provide an energy source to drive slow convection. Study of fluid inclusions from deep levels of the Cyprus ophiolite which are highly saline provides evidence that seawater penetration may well go almost to the oceanic Moho, and water removal from seawater during serpentinization produces very saline discharge fluids. The flow mechanisms have been discussed by MacDonald and Fyfe (1985). Gas fluxes due to ridge processes (CO_2, CO, CH_4) may also be significant to some parts of global cycles (see Owen and Rae, 1985).

In the general plate tectonic cycle, lithosphere creation at ridges is similar in mass to the return flow into the mantle at subduction zones. The exchange processes at ridges and on the conveyer belt modify the original mantle product which is later subducted, and in this way continental components fed to the oceans (and H_2O, CO_2, C, SO_4, S^{2-}) are in part returned to the mantle. It is now quite certain that during subduction substantial volumes of sediments ($\simeq$ km^3 yr^{-1}) trapped in the roughness of bending lithosphere are also recycled. In particular, it is the recycling of volatiles that leads to the magmatic processes that occur above subduction zones and in continents, processes that recycle volatiles and many other species back to the surface. Volcanism and magma production above subduction zones comprise about 10% of the global phenomena. This perhaps implies that hydrothermal fluxes into surface aquatic systems will be significant, but as fluids will generally be less saline than in the marine case they will transport lower concentrations of most metals to the surface (there are exceptions, as for gold). In general, the global flux from these processes (examples are provided by Wairakei, New Zealand, and Yellowstone, U.S.A.; see Ellis and Mahon, 1977) are not well quantified.

A process associated with subduction, and one which profoundly influences global hydrogeology, is that of continental collision, the process which forms the great mountain ranges of the Alpine or Himalayan type. The thrust processes which greatly thicken continental crust lead to another set of major fluid flux processes in the crust and in part into surface systems. The processes included in a thrusting process are shown schematically in Figure 18.1, a phenomenon which Oliver (1986) terms the "squeegee" effect. During thrusting, pore fluids are pushed out ahead of the thrusts and lubricate or float the thrusts (Hubbert and Rubey, 1959). During thermal equilibration of overthickened crust, fluids in the basal crust are expelled by metamorphic processes. Fyfe (1986) has discussed such processes for an orogen on the scale of the Himalayas, a process where the metamorphic flux involves a mass of about 1% of the ocean mass or a mass similar to the ice caps. Impressive quantities of carbon dioxide may also be evolved. Such processes will be periodic, and flow out along thrust planes and faults could have periodic influence on surface water chemistry. Such fluids involved in seismic pumping phenomena have been directly observed (Fyfe et al., 1978; Chi-Yu King, 1985). Given the mass of salt (evaporites) in continental crust, it is also

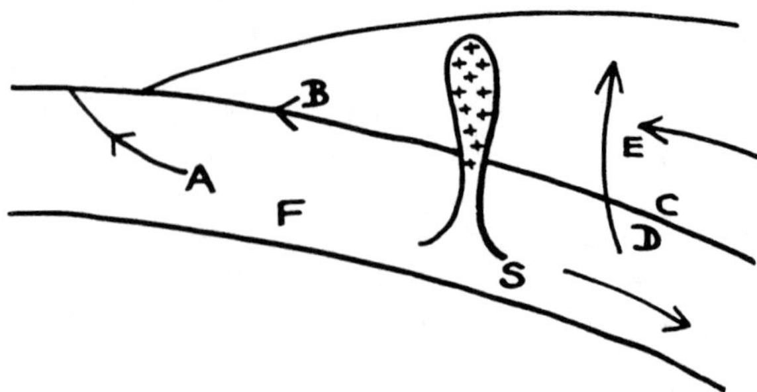

Figure 18.1. Continental crust is doubled in thickness during a continent–continent collision of the Himalayan type. Fluids are expelled ahead of the thrust (A) by compaction. Other fluids (B) move along and lubricate the thrusts while some penetrate the overthrust (E) forming vein swarms or shear zones. If the overthickened crust is thick, melting may occur at the base (S), forming granitic magmas. The entire underthrust block tends to lose most of its volatiles due to heating and compression.

possible that collisions may influence local or even global salinity on a significant scale.

5. EROSION

One of the greatest processes on earth which is driven by the creation of topography involves erosion by solution and particle transport. It is this process which provides most of the nutrient supply to the biosphere and, along with biota, forms the great array of small particles. The total suspended sediment load transported to the oceans at the present time is about 1.35×10^{16} g yr^{-1} (Milliman and Meade, 1983), while the dissolved load is about 3.6×10^{15} g yr^{-1}. The mass of material eroded off the continents annually is thus of an order of magnitude similar to that of the rate of crust formation and subduction. There is growing concern that human activity is greatly increasing erosion, and soil erosion is considered by some to be the world's most serious environmental problem (Toy, 1982).

The action of water on minerals is one of the most important processes which produce extremely high surface area and reactive materials in the surface environment, and the characterization of such materials is becoming one of our great challenges in understanding the chemistry of aquatic systems. Here I wish to mention a few recent examples studied in our laboratories.

We have recently been concerned with the chemistry and mineralogy of tropical soils in environments such as the Amazon Basin. In this situation the residual soils show extreme leaching and approach the end member Fe_2O_3–Al_2O_3–SiO_2–H_2O system (Kronberg et al., 1979). Because of a gibbsite-kaolin–dominated mineralogy, soil water is acid. As the interface with primary

Figure 18.2. Silicate fibers, clay precursors, formed on feldspar surfaces during weathering. Their composition is dominated by Si–Al–Fe–O. TEM photo by K. Tazaki; see Kronberg et al. (1985) and Tazaki and Fyfe (1986).

minerals is approached, extensive etching is observed (see Lasaga, Chapter 10 of this volume), but by careful study of feldspar surfaces quite strange materials are observed (Kronberg et al., 1985; Tazaki, 1986). Extremely thin fibers ~20 Å thick appear to coat the surfaces, and the fibers eventually roll up to form what could be termed clay precursors on a scale of a few hundred angstroms (Fig. 18.2). The materials are X-ray amorphous. High-resolution EDAX (Energy Dispersive Analysis of X-rays) indicates that the major species are Si, Al, Fe. The surface area of this material is quite extraordinary. We do not know how such materials form, but it appears possible that Al–Si–O strands may be produced from the original feldspar structure, possibly associated with the formation of etch pits.

A second example of such remarkable clay particle size has been observed in deeper layers of marine sediments north of Hawaii. Here (Fig. 18.3), wind-blown smectite clays in the pelagic sediments appear to slowly react with Mg in the pore waters to produce the magnesium clay palygorskite. The clay needles (or rolled-up tubes) have thickness dimensions in the 100-Å range. If this process is common, then it provides another major sink for Mg in marine sediments (Tazaki et al., 1986).

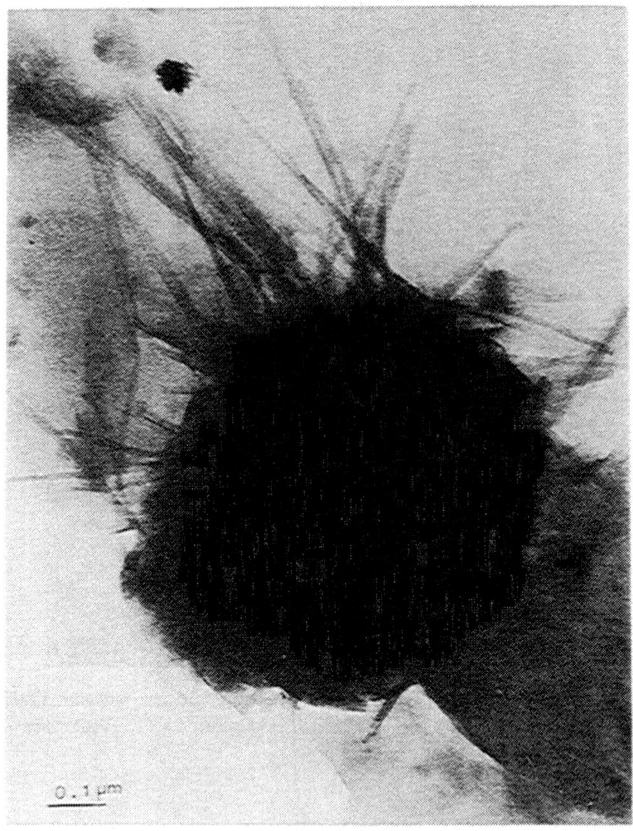

Figure 18.3. Palygorskite fibers, Mg-rich clay, formed on montmorillonite in a deep-sea sediment. TEM photograph by K. Tazaki (see Tazaki et al., 1986).

In the soil-forming process, remarkable chemical separations are also observed. We have recently described uranium-rich laterites formed on siltstones containing fossils. In what we term inorganic chromatography, uranium is enriched in the microenvironment of fossils, but even more remarkably, almost pure cerium–neodymium oxides occur in micrometer-sized particles (Leonardos et al., 1985). Gilles and Bancroft (1985) have shown the type of micromechanism that can lead to gold enrichment in the same environment. Gold can be solubilized in low-temperature oxygenated, acid chloride waters or waters containing species like thiosulfate. When such fluids pass reduced zones, particularly those with sulfide minerals, native gold is instantaneously precipitated. They have observed the formation of gold spheres on ZnS and FeS surfaces in times of minutes by ESCA and SEM. Typical particle sizes are in the micrometer range when observed. Such variable surfaces and redox conditions in soils, and the particulates formed, have been little studied.

6. BIOPRECIPITATION

Goldschmidt (1937) was one of the first geochemists to appreciate the great importance of the biomass in element concentration processes. Given that about 5×10^{17} g of carbon is present in living biomass and given that the biomass contains inorganic materials in the percent range, it is clear that such biomass must be of significance, particularly as it is now known that at least 30 chemical elements are bioessential (Mertz, 1981). Lowenstam (1981) has studied the wide range of normally rather simple mineral phases precipitated in organisms, a list which grows steadily. While knowledge of mineral precipitation in macroorganisms is extensive, the situation with microorganisms is less well known.

Beveridge (1974) studied the uptake of simple ions on bacteria and the cell wall structure of bacteria. His work showed both the high exchange capacity and specificity of such materials (Fig. 18.4). It is clear that bacteria and algae live in almost every possible system where liquid water is present below 110°C or so, from hot springs (Brock, 1978) to the subzero saline lakes of Antarctica (Young, 1981).

Beveridge and Murray (1976) showed the nucleation of gold crystals in the 0.01 μm range on bacterial cell walls in dilute gold solutions. Mann (1984)

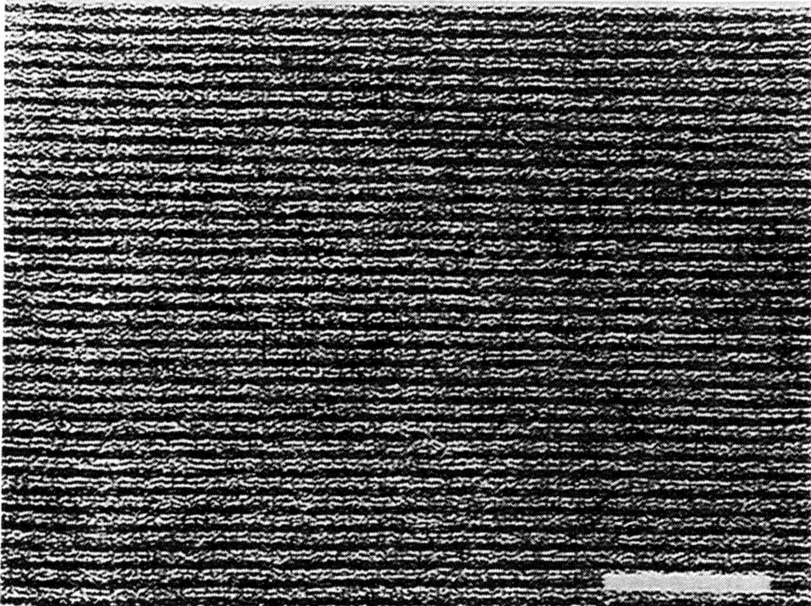

Figure 18.4. A remarkable TEM photograph of the "crystallinity" of a bacterial cell wall. The scale bar is 0.1 μm. From Beveridge (1974), with permission. Such materials form excellent but selective fixation agents for metal ions.

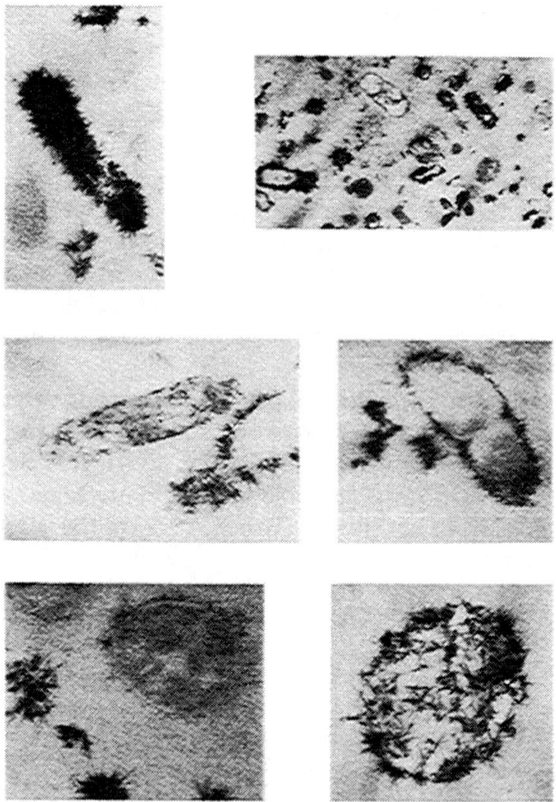

Figure 18.5. Remarkable examples of iron mineralization associated with bacteria in muds in acid mine drainage. TEM photographs by H. Mann.

and Mann and Fyfe (1986) showed the remarkable (and specific) ability of some algae to complex with metals, and again, in dilute uranium solutions, uranium oxides were precipitated in the cell walls with typical crystal sizes in the micrometer range. Mann et al. (1986) have described the massive precipitation of lepidocrocite (FeOOH) in algae in acid mine tailings waters. Typical crystal sizes are around 100 Å. The iron content of the acid waters is ~100 ppb, and algae contain up to 40% iron by dry weight. They also contain remarkable concentrations of some other metals (Al, 14,000 ppm; Ti, 400–900 ppm; Pb, up to 2000 ppm). While algae are prolific in the acid waters, bacteria dominate in the sediments. A few centimeters beneath the water interface, bacteria are heavily mineralized with an as yet unidentified iron phase (Fig. 18.5).

A recent finding (Ferris et al., 1986) is that thermophilic bacteria living near 60°C in the hot springs of Yellowstone National Park appear to provide the nucleation sites for iron–silica phases. In Figure 18.6, spherical crystallites,

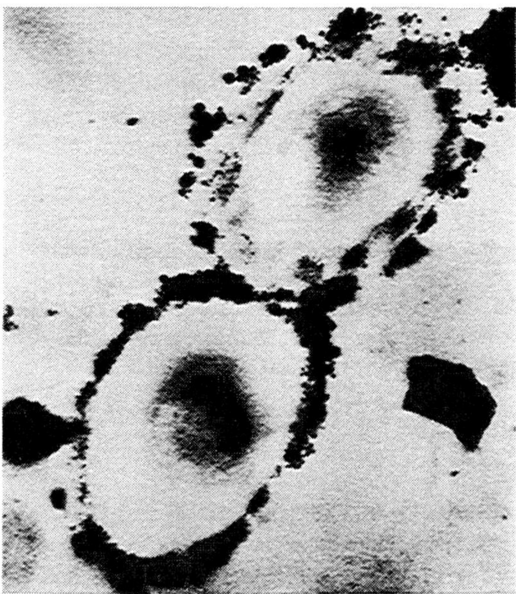

Figure 18.6. Nucleation of Si–Fe oxide particles on bacteria in warm waters at Yellowstone National Park; magnification 36,000× (See Ferris et al., 1986). TEM photo by G. Ferris.

dominated by silica, are shown decorating bacterial cell walls. Such crystallites are again in the submicrometer range. It appears that nucleation occurs on the cell wall, and once grown to a certain size, the spherical particles detach and grow. Such observations clearly show that microorganisms must play a major role in the formation of very fine particulates in aquatic systems, where frequently their walls serve as major complexing agents for the collection of metal ions. The microparticulates are frequently in a size range that would pass conventional filters.

In the examples illustrated above, materials are formed with sizes ~100 Å, a few unit cells. When one considers the general relations between chemical potential, particle size, and solubility (Stumm and Morgan, 1981), it seems that such particles should have very high solubilities, and while they form in supersaturated environments the degree of supersaturation and the apparent stability of the particles appear anomalous. It also seems that the biological microenvironment (or the environment of biological debris) is optimal for formation of such particles. One is almost forced to the conclusion that very strong surface complexing by organic molecules may eventually provide the explanation to the enigma in that the surfaces are complexed very strongly and the fine particles are stabilized. The pathways, crystal + solvent → simple ions → clusters → colloids → macrocrystals, all involve interactions with the host solute, and whether the reaction steps are fast or delayed will depend on the interface chemistry. It is obvious that as a first step, description of such fine particles, through the use of high resolution electron microscopy, is essential to our understanding of the behavior of many elements in aquatic systems and is essential in bridging the gaps from simple ions and complex ions to the particles separated on conventional microfilters.

Acknowledgments

I gratefully acknowledge the assistance of various colleagues and in particular H. Mann, F. G. Ferris, and K. Tazaki of this department. R. G. Murray of our Medical School, and T. J. Beveridge, Department of Microbiology at Guelph University. Their skills and electron microscopes have provided the illustrations for this paper. I thank J. Vibetti for providing recent data on fluids in the Cyprus Ophiolite.

REFERENCES

Anderson, R. N., Hobart, M. A., and Langseth, M. C. (1979), Geothermal convection through oceanic crust and sediments in the Indian Ocean, *Science* **204,** 828.

Beveridge, T. J. (1974), "Structure of the Macromolecular Superficial Wall Layers of *Spirillium* Species," Ph.D. dissertation, The University of Western Ontario, 432 pp.

Beveridge, T. J., and Murray, R. G. E. (1976), Uptake and retention of metals by cell walls of *Bacillus subtilis, J. Bacteriol.* **127,** 1502.

Brock, T. D. (1978), *Thermophilic Microorganisms and Life at High Temperatures,* Springer-Verlag, Berlin, 468 pp.

Davis, E. E., and Lister, C. R. B. (1977), Heat flow measured over the Juan de Fuca ridge: Evidence for widespread hydrothermal circulation in a highly heat transportive crust, *J. Geophys. Res.* **82,** 4845.

Enger, E. D., Kormelink, J. R., Smith, B. F., and Smith, R. J. (1986), *Environmental Science,* Wm. C. Brown, Dubuque, Iowa.

Eddy, J. A. (1986), *Global Change in the Geosphere-Biosphere. Initial Priorities for an IGBP,* National Academy Press, Washington, DC.

Ellis, A. J., and Mahon, W. A. J. (1977), *Chemistry and Geothermal Systems,* Academic, New York.

Emiliani, C. (1981), *The Oceanic Lithosphere, Vol. 7, The Sea,* Wiley, New York.

Ferris, F. G., Beveridge, T. J., and Fyfe, W. S. (1986), Iron-silica crystallite nucleation by bacteria in a geothermal sediment, *Nature,* **320,** 609–611.

Fyfe, W. S. (1974), Heats of chemical reactions and submarine heat production, *Geophys. J. Roy. Astron. Soc.* **37,** 213.

Fyfe, W. S. (1986), Fluids in deep continental crust, *Am. Geophys. Union, Geodynamics Ser.* **14,** 33.

Fyfe, W. S., Price, N. J., and Thompson, A. B. (1978), *Fluids in the Earth's Crust,* Elsevier, New York.

Gilles, E. J., and Bancroft, G. M. (1985), An XPS and SEM study of gold deposition at low temperatures on sulphide mineral surfaces: Concentration of gold by adsorption/reduction, *Geochim. Cosmochim. Acta* **49,** 979.

Goldschmidt, V. M. (1937), The principles of the distribution of chemical elements in minerals and rocks, *J. Chem. Soc.* 655.

Hubbert, M. K., and Rubey, W. W. (1959), Role of fluid pressure in mechanics of over-thrust faulting, *Geol. Soc. Am. Bull.* **70,** 115.

King, C.-Y. (1985), *Earthquake Hydrology and Chemistry,* Birkhauser, Basel.

Kronberg, B. I., Fyfe, W. S., Leonardos, O. H., and Santos, A. M. (1979), The chemistry of some Brazilian soils: Element mobility during intense weathering, *Chem. Geol.* **24,** 211.

Kronberg, B. I., Tazaki, K., and Melfi, A. J. (1985), "Detailed Geochemical Studies of the Initial Stages of Weathering of Alkaline Rocks—Ilha de Sao Sebastino, Brazil," in Y. Ogura, Ed., *International Seminar on Laterite,* Mining and Metallurgical Institute of Japan, Tokyo, p. 223.

Leonardos, O. H., Fernandes, S. M., Fyfe, W. S., and Powell, M. (1985), "The Micro Chemistry of Uraniferous Laterites from Brazil: A Natural Example of Inorganic Chromatography," in Y. Ogura, Ed., *International Seminar on Laterite,* Mining and Metallurgical Institute of Japan, Tokyo, p. 465.

Lowenstam, H. A. (1981), Minerals formed by organisms, *Science* **211,** 1126.

MacDonald, A. H., and Fyfe, W. S. (1985), Rates of serpentinization in seafloor environments, *Tectonophysics* **116,** 123.

Malone, H. T. F., and Roederer, J. G., Eds. (1985), *Global Change,* Cambridge University Press, Cambridge, England.

Mann, H. (1984), "Algal Uptake of U, Ba, Co, Ni and V: Studies of Natural and Experimental Systems," Ph.D. dissertation, The University of Western Ontario, London, Ontario, Canada, 318 pp.

Mann, H., and Fyfe, W. S. (1986), Uranium uptake by algae: experimental and natural environments, *Can. J. Earth Sci.* **22,** 1899.

Mann, H., Tazaki, K., Fyfe, W. S., Beveridge, T. J., and Humphrey, R. (1986), Cellular lepidocrocite precipitation and heavy metal sorption in *Euglena* sp.: Implications for biomineralization, *Chem. Geol.,* in press.

Mertz, W. (1981), The essential trace elements, *Science* **213,** 1332.

Milliman, J. D., and Meade, R. H. (1983), World-wide delivery of river sediments to the oceans, *J. Geol.* **91,** 1.

Officer, C. B., Drake, C. L., and Devine, J. D. (1986), Volcanism and Cretaceous/Tertiary extinctions, *Paleoceanography* (in press).

Oliver, J. (1986), Fluids expelled tectonically from orogenic belts: Their role in hydrocarbon migration and other geologic phenomena, *Geology* **14,** 99.

Owen, R. M., and Rae, D. K. (1985), Sea-floor hydrothermal activity links climate to tectonics: The Eocene carbon dioxide greenhouse, *Science* **227,** 166.

Rona, P. A., Boström, K., Laubier, L., and Smith, K. L., Eds. (1983), *Hydrothermal Processes at Seafloor Spreading Centres,* Plenum, New York.

Stumm, W., and Morgan, J. J. (1981), *Aquatic Chemistry,* Wiley, New York.

Tazaki, K. (1986), Observations of primitive clay precursors during microdine weathering, *Contrib. Mineral. Petrol.* **92,** 86.

Tazaki, K., Fyfe, W. S., and Heath, G. R. (1986), Palygorskite formed on montmorillonite in North Pacific deep-sea sediments, *Clay Sci.* (in press).

Thompson, G. (1983), "Basalt–Seawater Interaction," in Rona et al. (1983), 225–278.

Toy, T. J. (1982), Accelerated erosion: Processes, problems and prognosis, *Geology* **10,** pp. 524–529.

Young, P. (1981), Thick layers of life blanket lake bottoms in Antarctica valleys, *Smithsonian J.* **12,** 52.

INDEX

ENVIRONMENTAL SCIENCE AND TECHNOLOGY

A Wiley-Interscience Series of Texts and Monographs

Edited by ROBERT L. METCALF, *University of Illinois*
WERNER STUMM, *Eidgenössische Technische Hochschule, Zurich*

ENVIRONMENT, POWER, AND SOCIETY
Howard T. Odum

PHYSIOCHEMICAL PROCESSES FOR WATER QUALITY CONTROL
Walter J. Weber, Jr., Editor

pH AND pION CONTROL IN PROCESS AND WASTE STREAMS
F. G. Shinskey

CHEMICAL CONTROL OF INSECT BEHAVIOR: Theory and Application
H. H. Shorey and John J. McKelvey, Jr., Editors

AIR POLLUTION CONTROL, Part III: Measuring and Monitoring Air Pollutants
Werner Strauss, Editor

POLLUTANTS AND HIGH RISK GROUPS: The Biological Basis of Human Hypersusceptibility to Environmental and Occupational Pollutants
Edward J. Calabrese

SULFUR IN THE ENVIRONMENT
Part I: The Atmospheric Cycle
Part II: Ecological Impacts
Jerome O. Nriagu, Editor

METHODOLOGICAL APPROACHES TO DERIVING ENVIRONMENTAL AND OCCUPATIONAL HEALTH STANDARDS
Edward J. Calabrese

CHEMICAL CONCEPTS IN POLLUTANT BEHAVIOR
Ian J. Tinsley

RESOURCE RECOVERY AND RECYCLING
A. F. M. Barton

PLANT DISEASE CONTROL: Resistance and Susceptibility
Richard C. Staples and Gary H. Toenniessen, Editors

STRAUSS' AIR POLLUTION CONTROL, Part IV
Gordon M. Bragg and Jennifer Strauss, Editors

MODELING WASTEWATER RENOVATION: Land Treatment
I. K. Iskandar, Editor

ENVIRONMENTAL ENGINEERING AND SANITATION, Third Edition
Joseph A. Salvato, Jr.

AIR AND WATER POLLUTION CONTROL: A Benefit Cost Assessment
A. Myrick Freeman, III

INTRODUCTION TO INSECT PEST MANAGEMENT, Second Edition
Robert L. Metcalf and William H. Luckman, Editors

INDOOR AIR POLLUTION: Characterization, Prediction, and Control
Richard A. Wadden and Peter A. Scheff

PRINCIPLES OF ANIMAL EXTRAPOLATION
Edward J. Calabrese

SYSTEMS ECOLOGY: An Introduction
Howard T. Odum

NONCONVENTIONAL ENERGY RESOURCES
Philip R. Pryde

WASTES IN THE OCEAN

Volume 1: Industrial and Sewage Wastes in the Ocean
Iver W. Duedall, Bostwick H. Ketchum, P. Kilho Park, and Dana R. Kester, Editors

Volume 2: Dredged Material Disposal in the Ocean
Dana R. Kester, Bostwick H. Ketchum, Iver W. Duedall, and P. Kilho Park, Editors

Volume 3: Radioactive Wastes and the Ocean
P. Kilho Park, Dana R. Kester, Iver W. Duedall, and Bostwick H. Ketchum, Editors

Volume 4: Energy Wastes in the Ocean
Iver W. Duedall, Bostwick H. Ketchum, P. Kilho Park, and Dana R. Kester, Editors

Volume 5: Deep Sea Waste Disposal
Dana R. Kester, Bostwick H. Ketchum, Iver W. Duedall, and P. Kilho Park, Editors

Volume 6: Near-Shore Waste Disposal
Bostwick H. Ketchum, P. Kilho Park, Dana R. Kester, and Iver W. Duedall, Editors

LEAD AND LEAD POISONING IN ANTIQUITY
Jerome O. Nriagu

INTEGRATED MANAGEMENT OF INSECT PESTS OF POME AND STONE FRUITS
B. A. Croft and S. C. Hoyt, Editors

VIBRIOS IN THE ENVIRONMENT
Rita R. Colwell, Editor

WATER RESOURCES: Distribution, Use and Management
John R. Mather

COGENETICS: Genetic Variation in Susceptibility to Environmental Agents
Edward J. Calabrese

GROUNDWATER POLLUTION MICROBIOLOGY
Gabriel Bitton and Charles P. Gerba, Editors

CHEMISTRY AND ECOTOXICOLOGY OF POLLUTION
Des W. Connell and Gregory J. Miller

SALINITY TOLERANCE IN PLANTS: Strategies for Crop Improvement
Richard C. Staples and Gary H. Toenniessen, Editors

ECOLOGY, IMPACT ASSESSMENT, AND ENVIRONMENTAL PLANNING
Walter E. Westman

CHEMICAL PROCESSES IN LAKES
Werner Stumm, Editor

INTEGRATED PEST MANAGEMENT IN PINE-BARK BEETLE ECOSYSTEMS
William E. Waters, Ronald W. Stark, and David L. Wood, Editors

PALEOCLIMATE ANALYSIS AND MODELING
Alan D. Hecht, Editor

BLACK CARBON IN THE ENVIRONMENT: Properties and Distribution
E. D. Goldberg

GROUND WATER QUALITY
C. H. Ward, W. Giger, and P. L. McCarty, Editors

TOXIC SUSCEPTIBILITY: Male/Female Differences
Edward J. Calabrese

ENERGY AND RESOURCE QUALITY: The Ecology of the Economic Process
Charles A. S. Hall, Cutler J. Cleveland, and Robert Kaufmann

AGE AND SUSCEPTIBILITY TO TOXIC SUBSTANCES
Edward J. Calabrese

(continued on back)

ECOLOGICAL THEORY AND INTEGRATED PEST MANAGEMENT PRACTICE
Marcos Kogan, Editor

AQUATIC SURFACE CHEMISTRY: Chemical Processes at the Particle Water Interface
Werner Stumm, Editor